T0212715

Lecture Notes in Computer Science 8894

Commenced Publication in 1973
Founding and Former Series Editors:
Gerhard Goos, Juris Hartmanis, and Jan van Leeuwen

More information about this series at http://www.springer.com/series/7407

Marek Cygan · Pinar Heggernes (Eds.)

Parameterized and Exact Computation

9th International Symposium, IPEC 2014
Wroclaw, Poland, September 10–12, 2014
Revised Selected Papers

 Springer

Editors

Marek Cygan
Faculty of Mathematics, Informatics
    and Mechanics
University of Warsaw
Warsaw
Poland

Pinar Heggernes
Department of Informatics
University of Bergen
Bergen
Norway

ISSN 0302-9743   ISSN 1611-3349 (electronic)
Lecture Notes in Computer Science
ISBN 978-3-319-13523-6   ISBN 978-3-319-13524-3   (eBook)
DOI 10.1007/978-3-319-13524-3

Library of Congress Control Number: 2014956520

LNCS Sublibrary: SL1 – Theoretical Computer Science and General Issues

Springer Cham Heidelberg New York Dordrecht London

Printed on acid-free paper

Springer International Publishing AG Switzerland is part of Springer Science+Business Media
(www.springer.com)

Preface

The International Symposium on Parameterized and Exact Computation (IPEC, formerly IWPEC) is an international symposium series that covers research in all aspects of parameterized and exact algorithms and complexity. Started in 2004 as a biennial workshop, it became an annual event in 2008. This volume contains the papers presented at IPEC 2014: the 9th International Symposium on Parameterized and Exact Computation held during September 10–12, 2014, in Wrocław, Poland. The symposium was part of ALGO 2014, which also hosted six other workshops and symposia, including the Annual European Symposium on Algorithms (ESA 2014). The seven previous meetings of the IPEC/IWPEC series were held in Bergen, Norway (2004), Zürich, Switzerland (2006), Victoria, Canada (2008), Copenhagen, Denmark (2009), Chennai, India (2010), Saarbrücken, Germany (2011), Ljubljana, Slovenia (2012), and Sophia Antipolis, France (2013).

The invited plenary talk was given by Hans Bodlaender (Utrecht University) on "Lowerbounds for Kernelization." The keynote speaker, together with coauthors, Rodney G. Downey, Michael R. Fellows, Danny Hermelin, Lance Fortnow, and Rahul Santhanam, were awarded the EATCS-IPEC Nerode Prize 2014 for outstanding papers in the area of multivariate algorithmics. These proceedings contain an extended abstract of the invited talk. Additionally, a tutorial on "Backdoors, Satisfiability, and Problems Beyond NP" was given by Stefan Szeider (Vienna University of Technology). We thank the speakers for accepting our invitation.

In response to the call for papers, 42 papers were submitted. Each submission was reviewed by at least three reviewers. The reviewers were either Program Committee members or invited external reviewers. The Program Committee held electronic meetings using the EasyChair system, went through extensive discussions, and selected 27 of the submissions for presentation at the symposium and inclusion in this LNCS volume.

We would like to thank the Program Committee, together with the external reviewers, for their commitment in the difficult paper selection process. We also thank all the authors who submitted their work for our consideration. Finally, we are grateful to the local organizers of ALGO, in particular to Marcin Bieńkowski and Jarek Byrka, for the effort they put to make chairing IPEC an enjoyable experience.

September 2014

Marek Cygan
Pinar Heggernes

Organization

Program Committee

Hans L. Bodlaender	Utrecht University, The Netherlands
Marek Cygan	University of Warsaw, Poland
Holger Dell	Université Paris Diderot, France
Martin Grohe	RWTH Aachen University, Germany
Pinar Heggernes	University of Bergen, Norway
Marcin Kaminski	University of Warsaw, Poland
Petteri Kaski	Aalto University, Finland
Ken-Ichi Kawarabayashi	National Institute of Informatics, Japan
Michael A. Langston	University of Tennessee, USA
Jesper Nederlof	Maastricht University, The Netherlands
Marcin Pilipczuk	University of Bergen, Norway
Saket Saurabh	Institute of Mathematical Sciences, India
Ildikó Schlotter	Budapest University of Technology and Economics, Hungary
Jan Arne Telle	University of Bergen, Norway

Additional Reviewers

Belmonte, Rémy	Otachi, Yota
Bonsma, Paul	Phillips, Charles
Dawar, Anuj	Pilipczuk, Michal
Drucker, Andrew	Raymond, Jean-Florent
Fernau, Henning	Saeidinvar, Reza
Fertin, Guillaume	Santhanam, Rahul
Ganian, Robert	Schmid, Markus L.
Golovach, Petr	Schweitzer, Pascal
Hagan, Ron	Suchy, Ondrej
Halldorsson, Magnus M.	Trotignon, Nicolas
Iwata, Yoichi	van Leeuwen, Erik Jan
Kakimura, Naonori	Wahlström, Magnus
Lokshtanov, Daniel	Wang, Kai
Mouawad, Amer	Xiao, Mingyu

Contents

Lower Bounds for Kernelization

Hans L. Bodlaender[(✉)]

Department of Information and Computing Science,
Utrecht University, P.O. Box 80.089, 3508 TB Utrecht, The Netherlands
h.l.bodlaender@uu.nl

Abstract. Kernelization is the process of transforming the input of a combinatorial decision problem to an equivalent instance, with a guarantee on the size of the resulting instances as a function of a parameter. Recent techniques from the field of fixed parameter complexity and tractability allow to give lower bounds for such kernels. In particular, it is discussed how one can show for parameterized problems that these do not have polynomial kernels, under the assumption that $coNP \nsubseteq NP/poly$.

1 Introduction

In this paper, a number of recent techniques for lower bounds for kernelization are surveyed. The study of kernelization is motivated in two ways: first, it allows a precise mathematical analysis what can be achieved with polynomial time preprocessing of combinatorial problems. Second, a kernelization algorithm for a (decidable) problem also gives that the problem is fixed parameter tractable.

An important driving force behind much algorithm research is the intractability of many (combinatorial) problems, coming from practical applications and from theoretical investigations. One of the approaches when we ask for exact solutions is to first preprocess the instances before applying an exact solver: the former is typically fast, and the latter is typically slow (e.g., using integer linear programming, branch and bound, a SAT-solver). In kernelization, we make the assumption that the preprocessing takes polynomial time, is *safe* (in the sense that the answer for the problem instance is the same as, or can be derived from, the answer for the reduced instance), and we ask if there is a guaranteed upper bound of the size of the reduced instance. This upper bound is expressed as a function of some parameter of the input, possibly the target value, or some structural parameter of the input.

Parameterization is very useful for the analysis of preprocessing. The following lemma illustrates the limitations of a setting without parameterization.

Lemma 1 (Folklore). *Let Q be an NP-hard decision problem. If we have a polynomial time procedure, that given an input s, either decides if $s \in Q$, or produces an input s' with $s \in Q \Leftrightarrow s' \in Q$, then $P = NP$.*

© Springer International Publishing Switzerland 2014
M. Cygan and P. Heggernes (Eds.): IPEC 2014, LNCS 8894, pp. 1–14, 2014.
DOI: 10.1007/978-3-319-13524-3_1

Proof. Suppose we have such Q. Given an input, repeatedly apply Q on its own output till we decide. This gives a polynomial time algorithm for an NP-hard problem.   □

The study of fixed parameter tractability is motivated from the observation that often, when we have a problem that is intractable, actual instances may be much easier due to the fact that some parameter of these instances is small. Again, this parameter may be the target value, or some structural parameter of the input. E.g., a combinatorial problem arising from facility location problem may be NP-hard, but may be still polynomial time solvable when we know that the number of facilities to be placed is at most three (e.g., by exhaustive search). Another example is that many NP-hard problems become linear time solvable on graphs of bounded treewidth (see e.g., [6].) A decidable problem has a kernel, if and only if it is fixed parameter tractable (see Lemma 2). This strong relationship between the notions of kernelization and FPT is an important motivation behind the research on kernelization. Kernels of smaller size lead to faster FPT algorithms, and thus an important question is: what is the smallest size that we can obtain for a kernel for some given parameterized problem?

For several parameterized problems, kernels of small size are known: e.g.: VERTEX COVER has a kernel with at most $2k$ vertices (and $O(k^2)$ bits) (see e.g. [1,34]) FEEDBACK VERTEX SET has a kernel with $O(k^2)$ vertices (and $O(k^2)$ bits) [36]. There nowadays are many problems for which kernels of polynomial size are known. But also, for many problems, no such kernels are known. Current lower bound techniques explain why: it is shown that the problem has no polynomial kernel (or no kernel at all) unless a currently widely believed complexity theoretic assumption does not hold. Such lower bounds are useful for the algorithm designer: like an NP-hardness proofs guides us away from trying to design a polynomial time algorithm for a problem, here lower bounds can guide us away from trying to design (polynomial) kernels.

In this survey, we discuss a number of techniques to show such conditional lower bounds for kernelization.

2   Preliminaries

Throughout the paper, we assume that Σ is some finite alphabet. For a string $s \in \Sigma^*$, we denote with $|s|$ its length. A parameterized problem is a subset of $\Sigma^* \times \mathbf{N}$.

To a parameterized problem $Q \subseteq \Sigma^* \times \mathbf{N}$ we associate its *classic variant* $Q^c \subseteq (\Sigma \cup \{(,1,)\})^*$, which is obtained from Q by writing the second parameter (k) in unary.

In the literature, small variations on the definition of FPT are used. We use here the notion of strongly uniform FPT (see [18, Section 2.1]).

Definition 1. *FPT (fixed parameter tractable) is the class of parameterized problems Q such that there is an algorithm A, that decides for a given instance*

$(s, k) \in \Sigma^* \times \mathbf{N}$ *if* $(s, k) \in Q$ *in* $O(f(k) \cdot |s|^c)$ *time, for a computable function* f *and constant* $c \in \mathbf{N}$.

Many parameterized problems are known to be fixed parameter tractable (in FPT); the design of efficient parameterized algorithms is a very active field of study. E.g., see [18, 19, 21, 35].

Definition 2. *A* kernelization algorithm *or in short, a* kernel *for a parameterized problem* Q *is an algorithm* A, *that, given an instance* $(s, k) \in \Sigma^* \times \mathbf{N}$, *outputs an instance* $(s', k') \in \Sigma^* \times \mathbf{N}$, *such that there are computable functions* f *and* g, *and a constant* c, *with*

1. *A uses* $O(f(k)|s|^c)$ *time;*
2. $(s, k) \in A$, *if and only if* $(s', k') \in A$;
3. $|s'| \leq g(k)$, $k' \leq g(k)$.

Thus, a kernelization algorithm is a polynomial time algorithm, that transforms an input for parameterized problem Q to an equivalent input, but with the size of the latter bounded by a (computable) function in the parameter. The function g is said to be the *size* of the kernel.

There are minor variations on the definition of kernelization, and also more general notions have been studied. Some of these are reviewed in Sect. 6. The notion of kernelization is tightly bound to the notion of fixed parameter tractability, as the following well known result shows.

Lemma 2 (Folklore). *Let* $Q \subseteq \Sigma^* \times \mathbf{N}$ *be a decidable problem. Then* $Q \in FPT$, *if and only if* Q *has a kernel.*

Proof. Suppose $Q \in FPT$. Let A be an algorithm that decides on Q in $f(k)n^c$ time, for some computable function f and constant c. Suppose we have an instance (x, k) of size n. Run algorithm A for n^{c+1} steps. If A terminates within this time, then output a trivial $O(1)$ size yes- or no-instance. Otherwise, $n \geq f(k)$: output (x, k). This fulfils the definition of a kernel.

Suppose we have a kernelization algorithm A. First run A on the input, and then, as Q is decidable, run any decision algorithm on the remaining reduced instance. The time of the latter step is bounded by a function of the parameter k, and the time of the former step is polynomial. The combination of the steps is an FPT algorithm.   □

Lemma 2 is important for two reasons. First, it shows us that we can turn a kernelization algorithm directly in an FPT algorithm, and the method (first build a small equivalent instance, and then solve that instance) follows the approach discussed in the introduction for hard problems. Kernels of smaller size give faster FPT algorithms, and this thus motivates the search for kernels of small size. Second, it shows that if we have reason to believe that no FPT algorithm exists for a problem Q, then we also have reason to believe that there is no kernelization algorithm for Q. The latter is precisely the case for parameterized problems that are $W[1]$-hard. If a $W[1]$-hard problem belongs to the class FPT, then we have that $FPT = W[1]$, and from that it follows that the Exponential Time Hypothesis (ETH) does not hold [10]. Thus a corollary of Lemma 2 is that no $W[1]$-hard problem has a kernel, assuming the ETH.

As discussed above, we are interested in kernels of small size. An important class of kernels are the *polynomial kernels*: a kernel is polynomial, if the function g in Definition 2 (i.e., the upper bound for the resulting instances (s', k') on $|s'|$ and of k') is polynomial, i.e., there is a constant c' with $g(n) = O(n^{c'})$.

In this survey, we look at a number of techniques to give conditional proofs that such polynomial kernels do not exist. The results usually depend on the assumption that $NP \nsubseteq coNP/poly$, or, equivalently, that $coNP \nsubseteq NP/poly$. If this assumption would not hold, the polynomial time hierarchy would collapse to its third level [37]. For many parameterized problem, unconditional proofs that no polynomial kernel exists cannot reasonably be expected. E.g., if $P = NP$, then the parameterized variants of NP-complete problems (like LONG PATH, TREEWIDTH) have kernels of size $O(1)$.

3   Compositions

In this section, we discuss the first technique to show that problems do not have polynomial kernels: *composition*. Actually, composition comes in two flavours: OR-composition, and AND-composition. In several cases, compositionality gives simple and sometimes even trivial proofs for parameterized problems that they do not have polynomial kernels, assuming $NP \nsubseteq coNP/poly$. In several other cases, such proofs can be hard and lengthy. We start this section with describing the intuition behind the ideas. Then we introduce the main notions, stating the main theorems, and proving the main theorem for the case of OR-composition. We end the section with showing for some parameterized problems that they are compositional, and conclude that they have no polynomial kernel, again under the assumption that $NP \nsubseteq coNP/poly$.

3.1   Intuition

To get the intuition behind the approach, we consider the LONG PATH problem. A k-path is a simple path with at least k edges. In the LONG PATH problem, we are given an undirected graph G, integer k, and must decide if G has a k-path; k is the parameter.

Let us look at the situation that we would have a kernel of polynomial size for the LONG PATH problem, say a polynomial time algorithm A that reduces an instance of LONG PATH to one with k^c bits to describe it. Now, suppose we have a graph with k^{c+1} connected components. G contains a simple path with k edges, if and only if at least one of its connected components has a simple path with k edges. But these different connected components can be regarded as separate instances of LONG PATH, and thus, we would have a manner of reducing k^{c+1} different instances of LONG PATH to k^c bits in total: significantly fewer bits than the number of components. Such reduction seems only possible when we are able to *solve* the LONG PATH problem on some of the components; that is unlikely in polynomial time as the problem in its classic variant is NP-complete.

With this intuition in mind, we now look at the formal notion of compositions, and see how this can be used to prove conditional lower bounds for kernelization in Sect. 3.2.

3.2 Compositionality and Lower Bounds

The techniques and results in this section are mostly due to Bodlaender et al. [3] and Fortnow and Santhanam [22].

The notion of composition comes in two flavours: or-composition and and-composition. We give the definition of or-composition in full, and explain the difference with the definition of and-composition.

Definition 3. *An* or-composition *for a parameterized problem $Q \subseteq \Sigma^* \times \mathbf{N}$ is an algorithm A that gets as input a sequence of instances $(s_1, k), \ldots (s_r, k)$, and outputs one instance (s', k'), such that*

1. *A uses time that is bounded by a polynomial in $\sum_{i=1}^{r} |s_i| + k$;*
2. *$(s', k') \in Q$, if and only if there exists an i, $1 \leq i \leq k$ with $(s_i, k) \in A$;*
3. *k' is bounded by a polynomial in k.*

I.e., the or-composition algorithm transforms a sequence of instances with the same parameter to one instance, the latter being a yes-instance for Q if and only if at least one of the former instances is a yes-instance; it uses time that is polynomial in the total size of the instances in the sequence; and the resulting parameter must be polynomially bounded in the parameter in the original instances. We see examples of compositions in Sect. 3.3.

And-compositions are defined in the same way; we change the second condition in Definition 3 to

– *$(s', k') \in Q$, if and only if for all i, $1 \leq i \leq k$ with $(s_i, k) \in A$.*

If a problem has an or-composition, we say it is *or-compositional*; similarly for *and-compositional*. Combining the results of three different papers, we obtain the following central result, which provides us with a powerful tool to show that problems are likely not to have a polynomial kernel. In Sect. 3.4 we sketch the proof for part (a).

Theorem 1. *(a)* [Bodlaender et al. [3], Fortnow and Santhanam [22]] *Let Q be a parameterized problem that is or-compositional and whose classic variant is NP-hard. Then Q does not have a polynomial kernel unless $coNP \subseteq NP/poly$.*
(b) [Bodlaender et al. [3], Drucker [20]] *Let Q be a parameterized problem that is and-compositional and whose classic variant is NP-hard. Then Q does not have a polynomial kernel unless $coNP \subseteq NP/poly$.*

3.3 Compositional Problems

As a graph has a k-path, if and only if at least one of its connected components has a k-path, the LONG PATH problem is trivially or-compositional: just take the disjoint union of the graphs of the instances, and do not change parameter k. Similarly, if we have a graph parameter like TREEWIDTH which is for each graph the maximum of the parameters value over all connected components, disjoint union gives a trivial and-composition. A direct corollary of Theorem 1 and the corresponding NP-hardness results is that LONG PATH and TREEWIDTH have no polynomial kernel unless $coNP \subseteq NP/poly$.

In many other cases, compositions are far from trivial. See for example [13, 17, 28, 32].

Compositionality of DISJOINT FACTORS. An example of a non-trivial or-composition is the following, from Bodlaender et al. [8]. In the DISJOINT FACTORS problem, we are given a string $s \in \{1, \ldots, k\}$, and ask for a collection of k substrings $s_1, \ldots, s_k$ of s, that do not overlap, and for each i, s_i starts and ends with an i. The size of the alphabet k is the parameter of this problem.

For example, 1231331212 is a positive instance, with substrings 1231, 212 and 33; and 1221 is a negative instance.

An or-composition for DISJOINT FACTORS can be obtained as follows. First, we notice that the problem is solvable in $O(kn \cdot 2^k)$ time with standard dynamic programming techniques for strings of length n. Suppose we have instances $s^1, \ldots, s^r \in \{1, 2, \ldots, k\}$. If $r > 2^k$, we can solve all these instances in polynomial time, so assume $r \leq 2^k$. By possibly adding dummy instances, we can assume that $r = 2^k$. Now, add $k+1, \ldots, 2k$ to our alphabet. Instead of formally defining the composition, the following examples will make the scheme hopefully clear: if $k = 2$, take $34s_1 4s_2 43s_4 3$; if $k = 3$, take

$$456s + 16s_2 656s_3 6s_4 65456s_5 6s_6 656s_7 6s_8 654.$$

The resulting string has a solution, if and only if at least one s_i has a solution: the factor with $k+1$ 'disables' either the first or second half of the strings s_i; the next factor disables half of the remaining ones, and once we found factors for $k+1$ till $2k$, only one s_i remains to find the factors for $1, \ldots, k$. For a more detailed explanation, see [8] or [33]. Now, as DISJOINT FACTORS is NP-complete [8], it follows from Theorem 1, that it has no polynomial kernel unless $coNP \subseteq NP/poly$.

3.4 Proof Sketch For Lower Bounds With Or-Composition

Below, we give a proof for Theorem 1(a). The original proof was via a notion of distillation; we give here a direct proof without this intermediate step; it follows the proof method from [22, Theorem 3.1]. Druckers proof [20] for the case of and-composition is much more involved; a new and possibly simpler proof was very recently given by Dell [14].

Proof of Theorem 1(a). Suppose we have a parameterized problem Q that is compositional, whose classic variant is NP-hard, and that has a kernel with inputs with parameter k mapped to an equivalent input with at most k^c bits for some constant c.

Throughout the following proof, we will switch without notification between the parameterized and classic variants of the problem, and ignore some simple but technical details on how instances of the two versions are mapped. We always assume that the parameter is given in unary.

Denote the complement of the classic variant of Q by not-Q. (E.g., if Q is the problem, given a graph G and integer parameter k, to decide if G has a simple cycle of length at least k, then not-Q is the problem, given a pair (G, k) to decide if all simple cycles in G have length at most $k - 1$.) As (the classic variant of) Q is NP-hard, we have that not-Q is coNP-hard. Thus, if we show that not-Q belongs to $NP/poly$, we have that $coNP \subseteq NP/poly$ and proved the result.

Hence, what remains to be done is to give a nondeterministic algorithm for not-Q that can access polynomial advice. I.e., for each input size n, the algorithm can consult a string advice(n), whose size is bounded by a polynomial in n. In our case, the advice will consist of $O(n^2)$ instances, each of size at most $n^{cc'}$, thus the advice has size $O(n^{cc'+2})$ bits. Each element of the advice will belong to not-Q.

The algorithm will have the following form:

- Suppose an instance (x, k) of size n is given. (We have that $k \leq n$.)
- Set $r = n^{cc'}$.
- Non-deterministically guess a sequence of r instances, each of size n, and with parameter k.
- If (x, k) is not in the sequence, then reject.
- Compute the composition of the sequence, say (y, k'). By assumption on the composition, we have that $k' \leq k^{c'} \leq n^{c'}$.
- Compute the kernel (z, k'') of (y, k'). By assumption on the kernelization algorithm, we have that the size of this instance is at most $k'^c \leq n^{cc'} = r$.
- Check if (z, k'') is in advice(n). If so, accept; otherwise, reject.

Each element in the advice will be a negative instance of Q (or, equivalently, a positive instance of not-Q.) If $(x, k) \in Q$, then the properties of or-composition and kernelization imply that $(y, k') \in Q$ and thus that $(z, k'') \in Q$. As the advice only contains elements from not-Q, $(z, k'') \notin$ advice(n), and we correctly reject. What remains now is to show that there is a sufficiently small advice set such that for each $(x, k) \in$ not-Q of size n, there is a guess that gives a kernel in the advice. Lemma 3 shows that such a set indeed exists.

We say that an instance $(x, k) \in$ not-Q of size n is *covered* by an instance (y, k), if there is a sequence $\mathbf{x} = (x_1, k), (x_2, k), \ldots, (x_{k^c}, k)$ such that

1. There is an i, $1 \leq i \leq k^c$ with $x = x_i$. (I.e., (x, k) is part of the sequence.)
2. The kernelization algorithm, applied to the result of the or-composition algorithm, applied to the sequence $\mathbf{x}$ belongs to advice(n).

Lemma 3. *For sufficiently large n, there exists a set advice(n) of $O(n^2)$ instances of size at most $n^{cc'}$, such that*

- *Each instance $(y, k') \in$ advice(n) belongs to not-Q.*
- *Each instance (x, k) of size n is covered by an element of advice(n).*

Proof. We build the advice incrementally, starting with an empty set. Repeat the following step: add to the advice an instance of size at most $n^{cc'}$ in not-Q that covers the largest number of instances from not-Q of size n that are not yet covered by the advice.

Claim. There is an instance in not-Q of size at most $n^{cc'}$ that covers a constant fraction of all uncovered instances of size n in not-S.

Proof. Recall that $r = n^{cc'}$. Let A be the set of uncovered instances of size n in not-S. These form $|A|^r$ r-tuples. Each tuple is mapped by composition and kernelization to an instance of size at most r, of which there less than $2 \cdot 2^r$. By pigeon-hole principle, one of the latter is the image of $\frac{|A|^r}{2 \cdot 2^r}$ tuples, and thus covers at least $(\frac{|A|^r}{2 \cdot 2^r})^{1/r} \geq |A|/4$ instances from A.   □

As there are at most 2^n instances of size n in not-S, the claim shows that $O(\log 2^n) = O(n)$ elements are sufficient for the advice.   □

Lemma 3 shows that the advice is polynomial, and thus completes the proof of Theorem 1(a).

4   Transformations

A second technique to show conditional lower bounds for kernels is based upon using transformations. The technique is quite similar to usual NP-completeness proofs, with the specific twist here that the transformation should map an instance with parameter k to a new instance whose parameter is polynomially bounded in k. The technique was independently observed by several groups of authors [2, 8, 17]; the formalization is taken from [8], while the terminology was proposed by Lokshtanov.

Definition 4. *A* polynomial parameter transformation (ppt) *from a parameterized problem Q to a parameterized problem R is an algorithm A, that given an instance of Q, outputs an instance of R, such that*

1. *For all instances (x, k), $(x, k) \in Q \Leftrightarrow A(x, k) \in R$.*
2. *Given an instance (x, k), A uses time, polynomial in $|x| + k$.*
3. *There is a constant c, such that for all instances (x, k), if $A(x, k) = (x', k')$, then $k' \leq k^c$.*

The following theorem can be easily proven, and gives a direct method to lift lower bounds for kernels for some problem to similar lower bounds for other problems. See e.g. [8] or [5] for the proof.

Theorem 2. *If there is a ppt from Q to R and a polynomial time reduction from the classic variant of R to the classic variant of Q, and R has a polynomial kernel, then Q has a polynomial kernel.*

Corollary 1. *Suppose we have parameterized problems Q and R, with the classic variant of Q NP-hard, and the classic variant of R in NP. If Q has no polynomial kernel, then R has no polynomial kernel.*

As an example, we consider the DISJOINT CYCLES problem: determine, given an undirected graph G and integer k (the parameter), whether G has at least k vertex disjoint cycles. This problem is well known to be NP-complete. A ppt from DISJOINT CYCLES to DISJOINT FACTORS is as follows. Given a string $s_1 s_2 \cdots s_n \in \{1, 2, \ldots, k\}^n$, we build a graph with $n + k$ vertices. We first take a path with n vertices, each vertex representing a character from the string. For each symbol in the alphabet $i \in \{1, \ldots, k\}$, we add a vertex v_i, and make v_i incident to all path vertices that represent a character with this symbol. See Fig. 1. It is not hard to observe that the resulting graph has k disjoint cycles, if and only if the string has the required set of factors. (Each cycle needs to use one of the vertices not on the path, the remainder of the cycle corresponds to a factor, and as the cycles must be disjoint, the factors may not overlap.) From the earlier observed lower bound for kernels for DISJOINT FACTORS, it thus follows that DISJOINT CYCLES has no polynomial kernel assuming $coNP \not\subseteq NP/poly$,

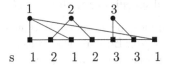

Fig. 1. Example of transformation: DISJOINT FACTORS to DISJOINT CYCLES

5   Cross Composition

Suppose we want to show that parameterized problem Q has no polynomial kernel under the usual assumption that $NP \not\subseteq coNP/poly$. An (and- or or-) composition as discussed in the previous chapter takes a collection of instances of Q and 'merges' these to one instance of the problem. The notion of cross composition, introduced by Bodlaender et al. [5], allows to start with a collection of instances of some (other) problem Q' (which should be NP-hard), and transforms this collection to one instance of Q. In this way, the notion of cross composition gives a more powerful tool to proof the conditional lower bounds for kernels.

The first ingredient for the notion of cross composition is a polynomial equivalence relation.

Definition 5. *A polynomial equivalence relation is an equivalence relation $\sim$ on Σ^*, such that*

– *Given two strings s_1 and s_2, we can decide if $s_1 \sim s_2$ in time, polynomial in $|s_1| + |s_2|$.*
– *There is a polynomial p, such that The number of equivalence classes of $\sim$ that contain strings of length at most r is bounded by $p(r)$.*

An example is the following. We consider instances of a graph problem, and two instances are equivalent if and only if they have the same number of vertices and the same number of edges.

We now come to the definition of OR cross composition, and then briefly give the difference with the definition of AND cross composition.

Definition 6. *Suppose we have a parameterized problem $Q \subset \Sigma^* \mathbf{N}$, a language $L \subseteq \Sigma^*$ and a polynomial equivalence relation $\sim$ on Σ^*. An OR cross composition of L to Q with respect to $\sim$ is an algorithm A, such that*

– *The input of A is a sequence instances $s_1, \ldots, s_r$ of L that belong to the same equivalence class of $\sim$.*
– *A uses time, polynomial in $\sum_{i=1}^{r} |s_i|$.*
– *A outputs one instance (s', k') of Q, with k' bounded by a polynomial in $\max |s_i| + \log k$.*
– *$(s', k') \in Q$ if and only if there is an i with (s_i) in L.*

The notion of *AND cross composition* is defined in exactly the same way, except that the last condition in Definition 6 is replaced as follows.

$(s', k') \in Q$ if and only if for all an i with (s_i) in L.

Building upon the techniques and results of Bodlaender et al. [3], Fortnow and Santhanam [22] and Drucker [20], Bodlaender, Jansen, and Kratsch [5] obtained the following result, which provides us with a powerful mechanism to show conditional kernel lower bounds.

Theorem 3 (Bodlaender et al. [5]). *Let $L \subseteq \Sigma^*$ be an NP-hard language, let $\sim$ be a polynomial equivalence relation, and $Q \subseteq \Sigma^* \times \mathbf{N}$ be a parameterized language. Suppose there exists an OR cross composition or an AND cross composition from L to Q with respect to $\sim$. Then Q does not have a polynomial kernel, unless $NP \subseteq coNP/poly$.*

For three reasons, this result gives more possibilities to show conditional kernel lower bounds:

– We can start with a collection of instances of any NP-hard problem, instead of having to use instances of the problem we want to prove a bound for itself.
– The polynomial equivalence relation allows us to make several additional assumptions on this collection of instances.
– The bound on k' helps to bound the number of instances of L we have to compose.

Cross compositions were used for kernel lower bounds for e.g., TREEWIDTH and PATHWIDTH [4], CLIQUE COVER [12], VERTEX COVER [28], and TEST COVER [23].

6   Other Models and Extensions

This survey so far only discussed results that some problems do not have polynomial kernels under the assumption of $coNP \nsubseteq NP/poly$. Several extensions and variants have been studied, that will be briefly mentioned below.

Many if not all of the lower bounds we discussed hold for more general settings: when we compress instances of a compositional language Q into instances of some other language R, compression in a setting of protocols, etc.

Many of the lower bounds also hold when we compress to a different target language. Also, lower bounds can be stated for the amount of information sent in certain types of protocols, for details see e.g. [16]. Drucker also proved his lower bounds for more general settings, e.g., probabilistic and quantum [20].

Earlier, lower bounds for a different model of compression with applications in e.g., cryptography were investigated by Harnik and Naor [24].

Non-increasing parameters. In interesting technique for lower bounds for kernels that do not increase the parameter was introduced by Chen et al. [11]. Combining composition and branching, one can show for several problems (including LONG PATH) that they do not have a polynomial kernel which does not increase the parameter, assuming that $P \neq NP$. Chen et al. [9] also augment the framework discussed in Sect. 3.2 to obtain stronger lower bounds.

Co-nondeterminism. Kratsch [29] and Kratsch et al. [31] showed that one can use *co-nondeterministic composition* to prove lower bounds, thus extending the power of the framework to a more general notion of composition. E.g., [29] gives a lower bound for kernels for the problem to decide whether a given graph contains a vertex set of size k that is independent or a clique; the problem is in FPT as direct consequence of Ramsey theory.

6.1   Polynomial Lower Bounds

An important result was obtained by Dell and van Melkebeek [16], who extended the techniques to obtain sharp polynomial lower bounds for problems with a polynomial kernel for several well known parameterized problems.

For instance, they showed that if there is a polynomial time algorithm that gives a kernel for VERTEX COVER or FEEDBACK VERTEX SET with $O(k^{2-\epsilon})$ bits for some $\epsilon > 0$, then $coNP \subseteq NP/poly$. Thus we have existing kernels for these problems (see e.g. [1,36]) are to be expected to be sharp with respect to the number of edges. The technique has been used by Dell and Marx [15] to obtain lower bounds for kernels for a number of packing problems. See also [26].

An interesting application of the technique was found by Kratsch et al. [30], who show that a simple quadratic kernel for the problem to cover a point set in the plane with k straight lines is essentially tight.

Parametric duality. Chen et al. [9] introduce the technique of parametric duality, which allows to give linear lower bounds for several problems, e.g., a bound of $2k$ vertices for VERTEX COVER, assuming that $P \neq NP$.

6.2   Turing Kernelization

A different model for kernelization is Turing kernelization. A Turing kernel is an algorithm that solves a parameterized problem in polynomial time, but the algorithm in addition has access to an oracle that decides instances of size $f(k)$ in one time step. With a proof similar to that of Lemma 2, one sees that a decidable problem has a Turing kernel iff it is in FPT.

There are several problems that have a polynomial Turing kernel, but (assuming $coNP \not\subseteq NP/poly$) polynomial kernel. See e.g., [2,27]. A lower bound theory for Turing kernelization has been set up by Hermelin et al. [25].

7   Conclusions

This paper gives a compact and incomplete survey on a number of techniques to show lower bounds (under a complexity theoretic assumption) for kernels were discussed. An excellent and more extensive survey was made by Misra et al. [33].

The framework helps in the classification of the complexity of parameterized problems: is the problem in P (regardless of parameter), has it a polynomial kernel, does it belong to FPT, is it in XP, or is it already NP-hard for some fixed value of the parameter?

I would like to end the survey with a practical warning: even when we know that for parameter, the problem at hand has no polynomial kernel, it still can be very useful to preprocess the problem. An illustrative example is for the problem of TREEWIDTH. Treewidth parameterized by the target value is and-compositional, and thus not likely to have a polynomial kernel. However, pre-processing rules were seen to be very effective for many instances from real-world applications [7]. In [4,26], (part of) a theoretical explanation is provided: with different parameterizations, treewidth has kernels of small (polynomial) size.

Acknowledgments. This survey would not have been possible without the discussions and cooperation with several collegues: Rod Downey, Mike Fellows, Bart Jansen, Danny Hermelin, Stefan Kratsch, Stéphan Thomassé and Andres Yeo. Thank you very much! I apologize to all whose work was inadvertingly or due to space was not or insufficiently discussed here.

References

1. Abu-Khzam, F.N., Collins, R.L., Fellows, M.R., Langston, M.A., Suters, W.H., Symons, C.T.: Kernelization algorithms for the vertex cover problem: theory and experiments. In: Proceedings of the 6th Workshop on Algorithm Engineering and Experimentation and the 1st Workshop on Analytic Algorithmics and Combinatorics, ALENEX/ANALCO 2004, pp. 62–69. ACM-SIAM (2004)
2. Binkele-Raible, D., Fernau, H., Fomin, F.V., Lokshtanov, D., Saurabh, S., Villanger, Y.: Kernel(s) for problems with no kernel: on out-trees with many leaves. ACM Trans. Algorithms **8**(5), 38 (2012)
3. Bodlaender, H.L., Downey, R.G., Fellows, M.R., Hermelin, D.: On problems without polynomial kernels. J. Comput. Syst. Sci. **75**, 423–434 (2009)

4. Bodlaender, H.L., Jansen, B.M.P., Kratsch, S.: Preprocessing for treewidth: a combinatorial analysis through kernelization. SIAM J. Discrete Math. **27**, 2108–2142 (2013)
5. Bodlaender, H.L., Jansen, B.M.P., Kratsch, S.: Kernelization lower bounds by cross-composition. SIAM J. Discrete Math. **28**, 277–305 (2014)
6. Bodlaender, H.L., Koster, A.M.C.A.: Combinatorial optimization on graphs of bounded treewidth. Comput. J. **51**(3), 255–269 (2008)
7. Bodlaender, H.L., Koster, A.M.C.A., Eijkhof, F.: Pre-processing rules for triangulation of probabilistic networks. Comput. Intell. **21**(3), 286–305 (2005)
8. Bodlaender, H.L., Thomassé, S., Yeo, A.: Kernel bounds for disjoint cycles and disjoint paths. Theoret. Comput. Sci. **412**, 4570–4578 (2011)
9. Chen, J., Fernau, H., Kanj, I.A., Xia, G.: Parametric duality and kernelization: lower bounds and upper bounds on kernel size. SIAM J. Comput. **37**, 1077–1106 (2007)
10. Chen, J., Huang, X., Kanj, I.A., Xia, G.: Strong computational lower bounds via parameterized complexity. J. Comput. Syst. Sci. **72**, 1346–1367 (2006)
11. Chen, Y., Flum, J., Müller, M.: Lower bounds for kernelizations and other preprocessing procedures. Theory Comput. Syst. **48**(4), 803–839 (2011)
12. Cygan, M., Kratsch, S., Pilipczuk, M., Pilipczuk, M., Wahlström, M.: Clique cover and graph separation: new incompressibility results. In: Czumaj, A., Mehlhorn, K., Pitts, A., Wattenhofer, R. (eds.) ICALP 2012, Part I. LNCS, vol. 7391, pp. 254–265. Springer, Heidelberg (2012)
13. Cygan, M., Pilipczuk, M., Pilipczuk, M., Wojtaszczyk, J.O.: Kernelization hardness of connectivity problems in d-degenerate graphs. Discrete Appl. Math. **160**, 2131–2141 (2012)
14. Dell, H.: A simple proof that AND-compression of NP-complete problems is hard. In: Proceedings IPEC 2014 (2014)
15. Dell, H., Marx, D.: Kernelization of packing problems. In: Proceedings of the 22nd Annual ACM-SIAM Symposium on Discrete Algorithms, SODA 2012, pp. 68–81 (2012)
16. Dell, H., van Melkebeek, D.: Satisfiability allows no nontrivial sparsification unless the polynomial-time hierarchy collapses. In: Schulman, L.J. (ed.) Proceedings of the 42nd Annual Symposium on Theory of Computing, STOC 2010, pp. 251–260 (2010)
17. Dom, M., Lokshtanov, D., Saurabh, S.: Incompressibility through colors and IDs. In: Albers, S., Marchetti-Spaccamela, A., Matias, Y., Nikoletseas, S., Thomas, W. (eds.) ICALP 2009, Part I. LNCS, vol. 5555, pp. 378–389. Springer, Heidelberg (2009)
18. Downey, R.G., Fellows, M.R.: Parameterized Complexity. Springer, Berlin (1999)
19. Downey, R.G., Fellows, M.R.: Fundamentals of Parameterized Complexity. Texts in Computer Science. Springer, Berlin (2013)
20. Drucker, A.: New limits to classical and quantum instance compression. In: Proceedings of the 53rd Annual Symposium on Foundations of Computer Science, FOCS 2012, pp. 609–618 (2012)
21. Flum, J., Grohe, M.: Parameterized Complexity Theory. Springer, Berlin (2006)
22. Fortnow, L., Santhanam, R.: Infeasibility of instance compression and succinct PCPs for NP. J. Comput. Syst. Sci. **77**, 91–106 (2011)
23. Gutin, G., Muciaccia, G., Yeo, A.: Non-existence of polynomial kernels for the test cover problem. Inf. Process. Lett. **113**, 123–126 (2013)
24. Harnik, D., Naor, M.: On the compressibility of $\mathcal{NP}$ instances and cryptographic applications. SIAM J. Comput. **39**, 1667–1713 (2010)

25. Hermelin, D., Kratsch, S., Sołtys, K., Wahlström, M., Wu, X.: A completeness theory for polynomial (Turing) kernelization. In: Gutin, G., Szeider, S. (eds.) IPEC 2013. LNCS, vol. 8246, pp. 202–215. Springer, Heidelberg (2013)
26. Jansen, B.M.P.: On sparsification for computing treewidth. In: Gutin, G., Szeider, S. (eds.) IPEC 2013. LNCS, vol. 8246, pp. 216–229. Springer, Heidelberg (2013)
27. Jansen, B.M.P.: Turing kernelization for finding long paths and cycles in restricted graph classes. In: Schulz, A.S., Wagner, D. (eds.) ESA 2014. LNCS, vol. 8737, pp. 579–591. Springer, Heidelberg (2014)
28. Jansen, B.M.P., Bodlaender, H.L.: Vertex cover kernelization revisited - upper and lower bounds for a refined parameter. Theory Comput. Syst. **53**, 263–299 (2013)
29. Kratsch, S.: Co-nondeterminism in compositions: a kernelization lower bound for a Ramsey-type problem. ACM Trans. Algorithms **10**(4), 19 (2014)
30. Kratsch, S., Philip, G., Ray, S.: Point line cover: the easy kernel is essentially tight. In: Proceedings of the 24th Annual ACM-SIAM Symposium on Discrete Algorithms, SODA 2014, pp. 1596–1606 (2014)
31. Kratsch, S., Pilipczuk, M., Rai, A., Raman, V.: Kernel lower bounds using co-nondeterminism: finding induced hereditary subgraphs. In: Fomin, F.V., Kaski, P. (eds.) SWAT 2012. LNCS, vol. 7357, pp. 364–375. Springer, Heidelberg (2012)
32. Kratsch, S., Wahlström, M.: Two edge modification problems without polynomial kernels. Discrete Optim. **10**, 193–199 (2013)
33. Misra, N., Raman, V., Saurabh, S.: Lower bounds on kernelization. Discrete Optim. **8**, 110–128 (2011)
34. Nemhauser, G.L., Trotter, L.E.: Vertex packing: structural properties and algorithms. Math. Program. **8**, 232–248 (1975)
35. Niedermeier, R.: Invitation to Fixed-Parameter Algorithms. Oxford Lecture Series in Mathematics and Its Applications. Oxford University Press, Oxford (2006)
36. Thomassé, S.: A $4k^2$ kernel for feedback vertex set. ACM Trans. Algorithms **6**(2), 1–8 (2010)
37. Yap, H.P.: Some Topics in Graph Theory. London Mathematical Society Lecture Note Series. Cambridge University Press, Cambridge (1986)

On the Parameterized Complexity of Associative and Commutative Unification

Tatsuya Akutsu[1]([✉]), Jesper Jansson[1,2], Atsuhiro Takasu[3],
and Takeyuki Tamura[1]

[1] Bioinformatics Center, Institute for Chemical Research, Kyoto University,
Gokasho, Uji, Kyoto 611-0011, Japan
{takutsu,jj,tamura}@kuicr.kyoto-u.ac.jp
[2] The Hakubi Project, Kyoto University, Sakyo-ku, Kyoto 606-8501, Japan
[3] National Institute of Informatics, Tokyo 101-8430, Japan
takasu@nii.ac.jp

Abstract. This paper studies the unification problem with associative, commutative, and associative-commutative functions. The parameterized complexity is analyzed with respect to the parameter "number of variables". It is shown that both the associative and associative-commutative unification problems are $W[1]$-hard. For commutative unification, a polynomial-time algorithm is presented in which the number of variables is assumed to be a constant. Some related results for the string and tree edit distance problems with variables are also presented.

1 Introduction

Unification is an important concept in many areas of computer science such as automated theorem proving, program verification, natural language processing, logic programming, and database query systems [14,17,18]. The unification problem is, in its fundamental form, to find a substitution for all variables in two given terms that make the terms identical, where terms are built up from function symbols, variables, and constants [18]. As an example, the two terms $f(x,y)$ and $f(g(a), f(b,x))$ with variables x and y and constants a and b become identical by substituting x by $g(a)$ and y by $f(b, g(a))$. When one of the two input terms contains no variables, the unification problem is called *matching*.

Unification has a long history beginning with the seminal work of Herbrand in 1930 (see, e.g., [18]). It is becoming an active research area again because of *math search*, an information retrieval (IR) task where the objective is to find all documents containing a specified mathematical formula and/or all formulas similar to a query formula [16,19,20]. Also, math search systems such as Wolfram Formula Search and Wikipedia Formula Search have been developed. Since mathematical formulas are typically represented by rooted trees, it seems natural to measure the similarity between formulas by measuring the structural

---

This work was partially supported by the Collaborative Research Programs of National Institute of Informatics.

M. Cygan and P. Heggernes (Eds.): IPEC 2014, LNCS 8894, pp. 15–27, 2014.
DOI: 10.1007/978-3-319-13524-3_2

similarity of their trees. However, methods based on approximate tree matching like the *tree edit distance* (see, e.g., the survey in [7]) alone are not sufficient since every label is treated as a constant. For example, the query $x^2 + x$ has the same tree edit distance to each of the formulas $y^2 + z$ and $y^2 + y$ although $y^2 + y$ is mathematically the same as $x^2 + x$, but $y^2 + z$ is not.

An exponential-time algorithm for the unification problem was given in [22] and a faster, linear-time algorithm [9,21] appeared a few years later. Various extensions of unification have also been considered in the literature [5,17,18]. Three of them, unification with commutative, associative, and associative-commutative functions (where a function f is called *commutative* if $f(x, y) = f(y, x)$ always holds, *associative* if $f(x, f(y, z)) = f(f(x, y), z)$ always holds, and *associative-commutative* if it is both associative and commutative), are especially relevant for math search since many functions encountered in practice have one of these properties. However, when allowing such functions, there are more ways to match nodes in the two corresponding trees, and as a result, the computational complexity of unification may increase. Indeed, each of the associative, commutative, and associative-commutative unification (and matching) problems is NP-hard [5,10,17], and polynomial-time algorithms are known only for very restricted cases [2,5,17]; e.g., associative-commutative matching can be done in polynomial time if every variable occurs exactly once [5]. Due to the practical importance of these (and other) extensions of unification, heuristic algorithms have been proposed, sometimes incorporating approximate tree matching techniques [13,14].

This paper studies the parameterized complexity of associative, commutative, and associative-commutative unification with respect to the parameter "number of variables appearing in the input", denoted from here on by k. (We choose this parameter because the number of variables is often much smaller than the size of the terms.) In addition, we introduce and study the string and tree edit distance problems with variables. The following table summarizes our new results:

	Matching	Unification	DO-matching	DO-unification
SEDV	$W[2]$-hard (Theorem 1)	$O(|\Sigma|^k poly)$ (Proposition 1)	–	P (Theorem 4)
OTEDV	$W[1]$-hard (Theorem 3)	–	–	P (Theorem 4)
Associative	$W[1]$-hard (Theorem 5) (NP-complete [5])	–	P [5]	P (Theorem 6)
Commutative	NP-hard [5] FPT^a (Theorem 7)	XP (Theorem 8)	P [5]	P (Proposition 3)
Associative and commutative	$W[1]$-hard (Theorem 9) (NP-hard [5])	–	P [5]	P (Proposition 4)

^{a}Under the assumption that Conjecture 1 holds

Here, SEDV = the string edit distance problem with variables, OTEDV = the ordered tree edit distance problem with variables, and DO = distinct occurrences of all variables. $W[1]$-hard and FPT mean with respect to the parameter k. For simplicity, the algorithms described in this paper only determine if two terms are unifiable, but they may be modified to output the corresponding substitutions (when unifiable) by using standard traceback techniques. We remark that associative unification is in PSPACE and both commutative unification and associative-commutative unification are in NP [6]; although it means that all problems can be solved in single exponential time of the size of the input, it does not necessarily mean single exponential-time algorithms with respect to the number of variables.

2 Unification of Strings

Let Σ be an alphabet and Γ a set of variables. A *substitution* is a mapping from Γ to Σ. For any string s over $\Sigma \cup \Gamma$ and substitution θ, let $s\theta$ denote the string over Σ obtained by replacing every occurrence of a variable $x \in \Gamma$ in s by the symbol $\theta(x)$. (We write x/a to express that x is substituted by a.) Two strings s_1 and s_2 are called *unifiable* if there exists a substitution θ such that $s_1\theta = s_2\theta$.

Example 1. Suppose $\Sigma = \{a, b, c\}$ and $\Gamma = \{x, y, z\}$. Let $s_1 = abxbx$, $s_2 = ayczc$, and $s_3 = ayczb$. Then s_1 and s_2 are unifiable since $s_1\theta = s_2\theta = abcbc$ holds for $\theta = \{x/c, y/b, z/b\}$. On the other hand, s_1 and s_3 are not unifiable since there does not exist any θ with $s_1\theta = s_3\theta$. □

We shall use the following notation. For any string s, $|s|$ is the length of s. For any two strings s and t, the string obtained by concatenating s and t is written as $s\,t$. Furthermore, for any positive integers i, j with $1 \leq i \leq j \leq |s|$, $s[i]$ is the ith character of s and $s[i \ldots j]$ is the substring $s[i]\,s[i+1] \cdots s[j]$. (Thus, $s = s[1..|s|]$.) The *string edit distance* (see, e.g., [15]) between two strings s_1, s_2 over Σ, denoted by $d_S(s_1, s_2)$, is the length of a shortest sequence of edit operations that transforms s_1 into s_2, where an *edit operation* on a string is one of the following three operations: a *deletion* of the character at some specified position, an *insertion* of a character at some specified position, or a *replacement* of the character at some specified position by a specified character.[1] For example, $d_S(bcdfe, abgde) = 3$ because $abgde$ can be obtained from $bcdfe$ by the deletion of f, the replacement of c by g, and the insertion of an a, and no shorter sequence can accomplish this. By definition, $d_S(s_1, s_2) = \min_{ed\,:\,ed(s_1)=s_2} |ed| = \min_{ed\,:\,ed(s_2)=s_1} |ed|$ holds, where ed is a sequence of edit operations.

We generalize the string edit distance to two strings s_1, s_2 over $\Sigma \cup \Gamma$ by defining $\hat{d}_S(s_1, s_2) = \min_{ed\,:\,(\exists\theta)\,(ed(s_1)\theta = s_2\theta)} |ed|$. The *string edit distance problem with variables* takes as input two strings s_1, s_2 over $\Sigma \cup \Gamma$, and asks for the value of $\hat{d}_S(s_1, s_2)$. (To the authors' knowledge, this problem has not been

---

[1] In the literature, "replacement" is usually referred to as "substitution". Here, we use "replacement" to distinguish it from the "substitution" of variables defined above.

studied before. Note that it differs from the *pattern matching with variables problem* [11], in which one of the two input strings contains no variables and each variable may be substituted by any string over Σ, but no insertions or deletions are allowed.) Let k be the number of variables appearing in at least one of s_1 and s_2. Although $d_S(s_1, s_2)$ is easy to compute in polynomial time (see [15]), computing $\hat{d}_S(s_1, s_2)$ is $W[2]$-hard with respect to the parameter k:

Theorem 1. *The string edit distance problem with variables is $W[2]$-hard with respect to k when the number of occurrences of every variable is unrestricted.*

Proof. We present an FPT-reduction [12] from the longest common subsequence problem (LCS) to a decision problem version of the edit distance problem with variables. LCS is, given a set of strings $R = \{r_1, r_2, \ldots, r_q\}$ over an alphabet Σ_0 and an integer l, to determine whether there exists a string r of length l such that r is a subsequence of r_i for every $r_i \in R$, where r is called a *subsequence* of r' if r can be obtained by performing deletion operations on r'. It is known that LCS is $W[2]$-hard with respect to the parameter l (problem "LCS-2" in [8]).

Given any instance of LCS, we construct an instance of the string edit distance problem with variables as follows. Let $\Sigma = \Sigma_0 \cup \{\#\}$, where $\#$ is a symbol not appearing in $r_1, r_2, \ldots, r_q$, and $\Gamma = \{x_1, x_2, \ldots, x_l\}$. Clearly, R has a common subsequence of length l if and only if there exists a θ such that $x_1 x_2 \cdots x_l \theta$ is a common subsequence of R. Now, construct s_1 and s_2 by setting:

$$s_1 = x_1 x_2 \cdots x_l \# x_1 x_2 \cdots x_l \# \cdots \# x_1 x_2 \cdots x_l$$
$$s_2 = r_1 \# r_2 \# \cdots \# r_q$$

where the substring $x_1 x_2 \cdots x_l$ occurs q times in s_1. By the construction, there exists a θ such that $x_1 x_2 \cdots x_l \theta$ is a common subsequence of R if and only if there exists a θ such that $s_1 \theta$ is a subsequence of s_2. The latter statement holds if and only if $\hat{d}_S(s_1, s_2) = (\sum_{i=1}^{q} |r_i|) - ql$. Since $k = l$, this is an FPT-reduction. □

The above proof can be extended to prove the $W[1]$-hardness of a restricted case with a bounded number of occurrences of each variable (omitted in the conference proceedings version).

Theorem 2. *The string edit distance problem with variables is $W[1]$-hard with respect to k, even if the total number of occurrences of every variable is 2.*

Note that in the special case where every variable in the input occurs exactly once, the problem is equivalent to approximate string matching with don't-care symbols, which can be solved in polynomial time [3].

On the positive side, the number of possible θ is bounded by $|\Sigma|^k$. This immediately yields a fixed-parameter algorithm w.r.t. k when Σ is fixed:

Proposition 1. *The string edit distance problem with variables can be solved in $O(|\Sigma|^k poly(m, n))$ time, where m and n are the lengths of the two input strings.*

3   Unification of Terms

We now consider the concept of unification for structures known as *terms* that are more general than strings [18]. From here on, Σ is a set of function symbols, where each function symbol has an associated *arity*, which is an integer describing how many arguments the function takes. A function symbol with arity 0 is called a *constant*. Γ is a set of variables. A *term* over $\Sigma \cup \Gamma$ is defined recursively as: (i) A constant is a term; (ii) A variable is a term; (iii) If $t_1, \ldots, t_d$ are terms and f is a function symbol with arity $d > 0$ then $f(t_1, \ldots, t_d)$ is a term.

Every term is identified with a rooted, ordered, node-labeled tree in which every internal node corresponds to a function symbol and every leaf corresponds to a constant or a variable. The tree identified with a term t is also denoted by t. For any term t, $N(t)$ is the set of all nodes in its tree t, $r(t)$ is the root of t, and $\gamma(t)$ is the function symbol of $r(t)$. The *size* of t is defined as $|N(t)|$. For any $u \in N(t)$, t_u denotes the subtree of t rooted at u and hence corresponds to a subterm of t. Any variable that occurs only once in a term is called a *DO-variable*, where "DO" stands for "distinct occurrences", and a term in which all variables are DO-variables is called a *DO-term* [5]. A term that consists entirely of elements from Σ is called *variable-free*.

Let $\mathcal{T}$ be a set of terms over $\Sigma \cup \Gamma$. A *substitution* θ is defined as any partial mapping from Γ to $\mathcal{T}$ (where we let x/t indicate that the variable x is mapped to the term t), under the constraint that if $x/t \in \theta$ then t is not allowed to contain the variable x. For any term $t \in \mathcal{T}$ and substitution θ, $t\theta$ is the term obtained by simultaneously replacing its variables in accordance with θ. For example, $\theta = \{x/y, y/x\}$ is a valid substitution, and in this case, $f(x, y)\theta = f(y, x)$.

Two terms $t_1, t_2 \in \mathcal{T}$ are said to be *unifiable* if there exists a θ such that $t_1\theta = t_2\theta$, and such a θ is called a *unifier*. In this paper, the *unification problem* is to determine whether two input terms t_1 and t_2 are unifiable. (Other versions of the unification problem have also been studied in the literature, but will not be considered here.) Unless otherwise stated, m and n denote the sizes of the two input terms t_1 and t_2. The unification problem can be solved in linear time [9,21]. The important special case of the unification problem where one of the two input terms is variable-free is called the *matching problem*.

Example 2. Let $\Sigma = \{a, b, f, g\}$, where a and b are constants, f has arity 2, and g has arity 3, and let $\Gamma = \{w, x, y, z\}$. Define the terms $t_1 = f(g(a, b, a), f(x, x))$, $t_2 = f(g(y, b, y), z)$, and $t_3 = f(g(a, b, a), f(w, f(w, w)))$. Then t_1 and t_2 are unifiable since $t_1\theta_1 = t_2\theta_1 = f(g(a, b, a), f(x, x))$ holds for $\theta_1 = \{y/a, z/f(x, x)\}$. Similarly, t_2 and t_3 are unifiable since $t_2\theta_2 = t_3\theta_2 = f(g(a, b, a), f(w, f(w, w)))$ with $\theta_2 = \{y/a, z/f(w, f(w, w))\}$. However, t_1 and t_3 are not unifiable because it is impossible to simultaneously satisfy $x = w$ and $x = f(w, w)$.   □

Similar to what was done in Sect. 2, we can combine the *tree edit distance* with unification to get what we call the *tree edit distance problem with variables*. Let $d_T(t_1, t_2)$ be the tree edit distance between two node-labeled (ordered or unordered) trees t_1 and t_2 (see [7] for the definition). We generalize

$d_T(t_1, t_2)$ to two trees, i.e., two terms, over $\Sigma \cup \Gamma$ by defining $\hat{d}_T(t_1, t_2) = \min_{ed\,:\,(\exists\theta)\,(ed(t_1)\theta = t_2\theta)} |ed|$. The *tree edit distance problem with variables* takes as input two (ordered or unordered) trees t_1, t_2 over $\Sigma \cup \Gamma$, and asks for the value of $\hat{d}_T(t_1, t_2)$.

As before, let k be the number of variables appearing in at least one of t_1 and t_2. By combining the proofs of Theorems 2 and 5 below, we obtain:

Theorem 3. *The tree edit distance problem with variables is $W[1]$-hard with respect to k, both for ordered and unordered trees, even if the number of occurrences of every variable is bounded by 2.*

As demonstrated in [5], certain matching problems are easy to solve for DO-terms. The next theorem, whose proof is omitted in this version, states that the ordered tree edit distance problem with variables also becomes polynomial-time solvable for DO-terms. (In contrast, the classic *unordered* tree edit distance problem is already NP-hard for variable-free terms; see, e.g., [7].)

Theorem 4. *The ordered tree edit distance problem with variables can be solved in polynomial time when t_1 and t_2 are DO-terms.*

4   Associative Unification

A function f with arity 2 is called *associative* if $f(x, f(y, z)) = f(f(x, y), z)$ always holds. Associative unification is a variant of unification in which functions may be associative. This section assumes that all functions are associative although all results are valid by appropriately modifying the details even if usual (non-associative) functions are included.

Associative matching was shown to be NP-hard in [5] by a simple reduction from 3SAT. However, the proof in [5] does not show the parameterized hardness.

Theorem 5. *Associative matching is $W[1]$-hard with respect to the number of variables even for a fixed Σ.*

Proof. As in the proof of Theorem 1, we reduce from LCS.

First consider the case of an unrestricted Σ. Let $(\{r_1, \ldots, r_q\}, l)$ be any given instance of LCS. For each $i = 1, \ldots, q$, create a term u^i as follows: $u^i = f(y_{i,1}, f(x_1, f(y_{i,2}, f(x_2, \cdots f(y_{i,l}, f(x_l, f(y_{i,l+1}, g(\#, \#))) \cdots))))$, where $\#$ is a character not appearing in $r_1, \ldots, r_q$. Create a term t_1 by replacing the last occurrence of $\#$ in each u^i by u^{i+1} for $i = 1, \ldots, q-1$, thus concatenating $u^1, \ldots, u^q$, as shown in Fig. 1. Next, transform each r_i into a string r_i' of length $1 + 2 \cdot |r_i|$ by inserting a special character $\&$ in front of each character in r_i, and appending $\&$ to the end of r_i, where each $\&$ is considered to be a distinct constant (i.e., $\&$ does not match any symbol but can match any variable). Represent each r_i' by a term t^i defined by: $t^i = f(r_i'[1], f(r_i'[2], f(r_i'[3], f(\cdots, f(r_i'[1 + 2 \cdot |r_i'|], g(\#, \#)) \cdots))))$. Finally, create a term t_2 by concatenating $t^1, \ldots, t^q$. (Again, see Fig. 1.) Now, t_1 and t_2 are unifiable if and only if there exists a

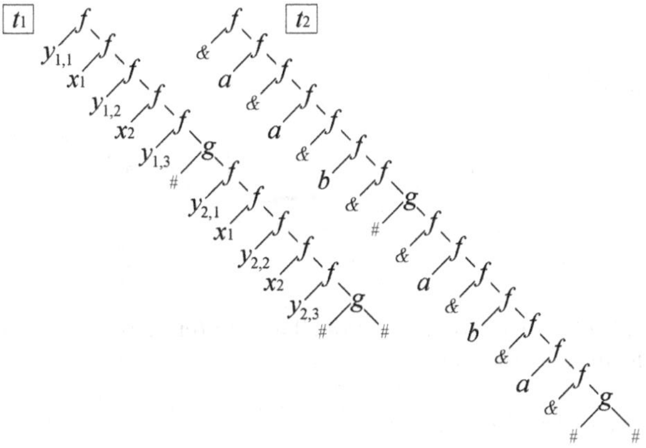

Fig. 1. Illustrating the reduction in Theorem 5. Here, $s_1 = aab$, $s_2 = aba$, and $l = 2$.

common subsequence of $\{r_1, \ldots, r_q\}$ of length l. Since the number of variables in t_1 is $(l+1)q + l = lq + l + q$, it is an FPT-reduction and thus the problem is $W[1]$-hard.

For the case of a fixed Σ, represent each constant by a distinct term using a special function symbol h and binary encoding (e.g., the 10th symbol among 16 symbols can be represented as $h(1, h(0, h(1, 0))))$. □

We next consider associative unification for DO-terms, which has some similarities with DO-associative-commutative matching [5]. For any term t, define the *canonical form* of t (called the "flattened form" in [5]) as the term obtained by contracting all edges in t whose two endpoints are labeled by the same function symbol. For example, both $f(f(a, b), f(g(c, f(d, f(e, h)), e))$ and $f(a, f(b, f(g(c, f(f(d, e), h)), e)))$ are transformed into $f(a, b, g(c, f(d, e, h)), e)$. As another example, the canonical form of $f(g(a, b), f(c, d))$ is $f(g(a, b), c, d)$. It is known [5] that the canonical form of t can be computed in linear time.

We begin with the simplest case in which every term is variable-free.

Proposition 2. *Associative unification for variable-free terms can be done in linear time.*

Proof. Transform the two terms into their canonical forms in linear time as above. Then it suffices to test if the canonical forms are isomorphic. The rooted ordered labeled tree isomorphism problem is trivially solvable in linear time. □

To handle the more general case of two DO-terms t_1 and t_2, we transform them into their canonical forms t^1 and t^2 and apply the following procedure, which returns 'true' if and only if t^1 and t^2 are unifiable. See Fig. 2 for an illustration. The procedure considers all $u \in N(t^1)$, $v \in N(t^2)$ in bottom-up order, and assigns $D[u, v] = 1$ if and only if $(t^1)_u$ and $(t^2)_v$ are unifiable.

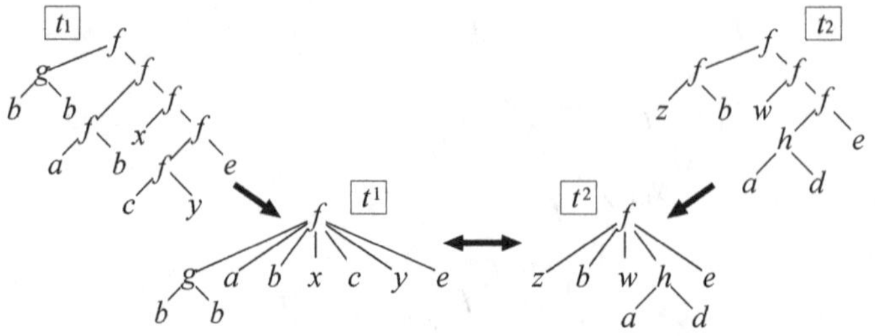

Fig. 2. An example of associative unification. The DO-terms t_1, t_2 are transformed into their canonical forms and then unified by $\theta = \{y/h(a,d),\ z/f(g(b,b),a),\ w/f(x,c)\}$.

Procedure $AssocMatchDO(t^1, t^2)$
    for all $u \in N(t^1)$ **do**   /* in post-order */
      for all $v \in N(t^2)$ **do**   /* in post-order */
        if $(t^1)_u$ or $(t^2)_v$ is a constant **then**
          if $(t^1)_u$ and $(t^2)_v$ are unifiable   (#)
          then $D[u,v] \leftarrow 1$ **else** $D[u,v] \leftarrow 0$;
        else if $(t^1)_u$ or $(t^2)_v$ is a variable **then**
          $D[u,v] \leftarrow 1$;
        else   /* $(t^1)_u = f_1((t^1)_{u_1}, \ldots, (t^1)_{u_p})$, $(t^2)_v = f_2((t^2)_{v_1}, \ldots, (t^2)_{v_q})$ */
          if $f_1 = f_2$ and $\langle (t^1)_{u_1}, \ldots, (t^1)_{u_p} \rangle$ can match $\langle (t^2)_{v_1}, \ldots, (t^2)_{v_q} \rangle$
          then $D[u,v] \leftarrow 1$ **else** $D[u,v] \leftarrow 0$;
    if $D[r(t_1), r(t_2)] = 1$ **then return** true **else return** false.

Step (#) takes $O(1)$ time because here $(t^1)_u$ and $(t^2)_v$ are unifiable if and only if they are the same constant or one of $(t^1)_u$ and $(t^2)_v$ is a variable.

When both u and v are internal nodes, we need to check if $\langle (t^1)_{u_1}, \ldots, (t^1)_{u_p} \rangle$ and $\langle (t^2)_{v_1}, \ldots, (t^2)_{v_q} \rangle$ can be matched. This may be done efficiently by regarding the two sequences as strings and applying string matching with variable-length don't-care symbols [4], while setting the difference to 0 and allowing don't-care symbols in both strings. Here, $(t^1)_{u_i}$ (resp., $(t^2)_{v_j}$) is regarded as a don't-care symbol that can match any substring of length at least 1 if it is a variable, otherwise $(t^1)_{u_i}$ can match $(t^2)_{v_j}$ if and only if $D[u_i, v_j] = 1$ (details are omitted in this version). It is to be noted that a variable in a term may partially match two variables in the other term. For example, consider the two terms $f(t_1, x, t_2)$ and $f(y, z)$. Here, $\theta = \{x/f(t_3, t_4),\ y/f(t_1, t_3),\ z/f(t_4, t_2)\}$ is a unifier. However, in this case, a simpler unifier is $\theta' = \{x/t_3,\ y/f(t_1, t_3),\ z/t_2\}$ because each variable occurs only once. Therefore, we can use approximate string matching with variable-length don't-care symbols, which also shows the correctness of the algorithm.

The **for**-loops are iterated $O(mn)$ times and string matching with variable-length don't-care symbols takes polynomial time, so we obtain:

Theorem 6. *Associative unification for DO-terms takes polynomial time.*

5   Commutative Unification

A function f with arity 2 is called *commutative* if $f(x, y) = f(y, x)$ always holds. Commutative unification is a variant of unification in which functions are allowed to be commutative. Commutative matching was shown to be NP-hard in [5] (by another reduction from 3SAT than the one referred to above).

First note that commutative unification is easy to solve when both t_1 and t_2 are variable-free because in this case, it reduces to the rooted unordered labeled tree isomorphism problem which is solvable in linear time (see, e.g., p. 86 in [1]):

Proposition 3. *Commutative unification for variable-free terms can be done in linear time.*

Next, we consider commutative matching. We will show how to construct a 0-1 table $D[u, v]$ for all node pairs $(u, v) \in N(t_1) \times N(t_2)$, such that $D[u, v] = 1$ if and only if $(t_1)_u$ and $(t_2)_v$ are unifiable, by applying bottom-up dynamic programming. It is enough to compute these table entries for pairs of nodes with the same depth only. We also construct a table $\Theta[u, v]$, where each entry holds a set of possible substitutions θ such that $(t_1)_u\theta = (t_2)_v$.

Let $\theta_1 = \{x_{i_1}/t_{i_1}, \ldots, x_{i_p}/t_{i_p}\}$ and $\theta_2 = \{x_{j_1}/t_{j_1}, \ldots, x_{j_p}/t_{j_q}\}$ be substitutions. θ_1 is said to be *compatible* with θ_2 if there exists no variable x such that $x = x_{i_a} = x_{j_b}$ but $t_{i_a} \neq t_{j_b}$. Let Θ_1 and Θ_2 be sets of substitutions. We define $\Theta_1 \bowtie \Theta_2 = \{\theta_i \cup \theta_j : \theta_i \in \Theta_1 \text{ is compatible with } \theta_j \in \Theta_2\}$. For any node u, u_L and u_R denote the left and right child of u. The algorithm is as follows:

Procedure $CommutMatch(t_1, t_2)$
    for all pairs $(u, v) \in N(t_1) \times N(t_2)$ with the same depth
    do   /* in bottom-up order */
        if $(t_1)_u$ is a variable **then**
           $\Theta[u, v] \leftarrow \{\{(t_1)_u/(t_2)_v\}\}$; $D[u, v] \leftarrow 1$
        else if $(t_1)_u$ does not contain any variables **then**
           $\Theta[u, v] \leftarrow \emptyset$;
           if $(t_1)_u = (t_2)_v$ **then** $D[u, v] \leftarrow 1$ **else** $D[u, v] \leftarrow 0$
        else if $\gamma((t_1)_u) \neq \gamma((t_2)_v)$ **then**
           $\Theta[u, v] \leftarrow \emptyset$; $D[u, v] \leftarrow 0$ /* recall: $\gamma(t)$ is a function symbol of $r(t)$ */
        else
           $\Theta[u, v] \leftarrow \emptyset$; $D[u, v] \leftarrow 0$;
           for all $(u_1, u_2, v_1, v_2) \in \{(u_L, u_R, v_L, v_R), (u_R, u_L, v_L, v_R)\}$ **do**   (#)
              if $D[u_1, v_1] = 1$ and $D[u_2, v_2] = 1$ and $\Theta_1[u_1, v_1] \bowtie \Theta_2[u_2, v_2] \neq \emptyset$
              then $\Theta[u, v] \leftarrow \Theta[u, v] \cup (\Theta_1[u_1, v_1] \bowtie \Theta_2[u_2, v_2])$; $D[u, v] \leftarrow 1$;
    if $D[r(t_1), r(t_2)] = 1$ **then return** true **else return** false.

Let B_i denote the maximum size of $\Theta[u, v]$ when the number of (distinct) variables in $(t_1)_u$ is i. Then, we have the following conjecture.

Conjecture 1. $B_1 = 1$ and $B_{i+j} = 2B_i B_j$ hold, from which $B_i = 2^{i-1}$ follows.

Theorem 7. *If Conjecture 1 holds, commutative matching can be done using* $O(2^k poly(m, n))$ *time, where* k *is the number of variables in* t_1.

Proof. The correctness follows from the observation that each variable is substituted by a term without variables and the property $f(x, y) = f(y, x)$ is taken into account at step (#). As for the time complexity, first consider the number of elements in $\Theta[u, v]$. A crucial observation is that if $(t_1)_{u_L}$ does not contain a variable then $|\Theta[u, v]| \leq \max(|\Theta[u_R, v_L]|, |\Theta[u_R, v_R]|)$ holds (and analogously for $(t_1)_{u_R}$). Assuming that Conjecture 1 is true, $\Theta_1[u_1, v_1] \bowtie \Theta_2[u_2, v_2]$ can be computed in $O(2^k poly(m, n))$ time by using 'sorting' as in usual 'join' operations. Thus, the total running time is also $O(2^k poly(m, n))$. □

Finally, we consider the case where both t_1 and t_2 contain variables. As in [21], we represent two variable-free terms t_1 and t_2 by a directed acyclic graph (DAG) $G(V, E)$, where t_1 and t_2 respectively correspond to r_1 and r_2 of indegree 0 ($r_1, r_2 \in V$). Then, testing whether r_1 and r_2 represent the same term takes polynomial time (in the size of G) by using the following procedure, where t_u denotes the term corresponding to a node u in G:

```
Procedure TestCommutIdent(r₁, r₂, G(V, E))
    for all u ∈ V do                              /* in post-order */
        for all v ∈ V do                          /* in post-order */
            if u = v then D[u, v] ← 1; continue;
            if tu or tv is a constant then
                if tu = tv then D[u, v] ← 1 else D[u, v] ← 0;
            else
                Let u = f₁(uL, uR) and v = f₂(vL, vR);
                if f₁ = f₂ then
                    if (D[uL, vL] = 1 and D[uR, vR] = 1) or
                       (D[uL, vR] = 1 and D[uR, vL] = 1)
                    then D[u, v] ← 1 else D[u, v] ← 0
                else D[u, v] ← 0;
    if D[r₁, r₂] = 1 then return true else return false.
```

To cope with terms involving variables, we need to consider all possible mappings from the set of variables to $N(t_1) \cup N(t_2)$. For each such mapping, we replace all appearances of the variables by the corresponding nodes, resulting in a DAG to which we apply $TestCommutIdent(r_1, r_2, G(V, E))$. The following pseudocode describes the procedure for terms with variables:

```
Procedure CommutUnify(t₁, t₂)
    for all mappings M from a set of variables to nodes in t₁ and t₂ do
        if there exists a directed cycle (excluding a self-loop) then continue;
        Replace each variable having a self-loop with a distinct constant symbol;
        Replace each occurrence of a variable node u with node M(u);
            /* if M(u) = v and M(v) = w then u is replaced by w */
        Let G(V, E) be the resulting DAG;
```

Let r_1 and r_2 be the nodes of G corresponding to t_1 and t_2;
    if $TestCommutIdent(r_1, r_2, G(V, E)) = $ true **then return** true;
return false.

In summary, we have the following theorem, which implies that commutative unification belongs to the class XP [12].

Theorem 8. *Commutative unification can be done in $O((m + n)^{k+2})$ time.*

Proof. The correctness of $TestCommutIdent(r_1, r_2, G(V, E))$ follows from the fact that $f_1(t_1, t_2)$ matches $f_2(t'_1, t'_2)$ if and only if f_1 and f_2 are identical function symbols and either (t_1, t_2) matches (t'_1, t'_2) or (t_1, t_2) matches (t'_2, t'_1). It is clear that this procedure runs in $O(mn)$ time. Therefore, commutative matching of two variable-free terms can be done in polynomial time.

Next, we consider $CommutUnify(t_1, t_2)$. For an illustration of how it works, see Fig. 3. To prove the correctness, it is straightforward to see that if there exists some mapping M which returns 'true', then t_1 and t_2 are commutatively unifiable and such a mapping gives a substitution θ satisfying $t_1\theta = t_2\theta$. Conversely, suppose that t_1 and t_2 are commutatively unifiable. Then there exist unifiable non-commutative terms t'_1 and t'_2 that are obtained by exchanging the left and right arguments in some terms in t_1 and t_2. Let θ be the substitution satisfying $t'_1\theta = t'_2\theta$. Then, $t_1\theta = t_2\theta$ holds. We assign distinct constants to variables appearing in $t_1\theta$. We also construct a mapping from the remaining variables to $N(t_1) \cup N(t_2)$ by regarding $x/t \in \theta$ as a mapping of x to t. We construct $G(V, E)$ according to this mapping. Then, it is obvious that $TestCommutIdent(r_1, r_2, G(V, E)) = $ true holds.

Since the number of possible mappings is bounded by $(m+n)^k$, where k is the number of variables in t_1 and t_2, $CommutUnify(t_1, t_2)$ runs in $O((m + n)^{k+2})$ time. □

6 Associative-Commutative Unification

Associative-commutative unification is the variant of unification in which some functions can be both associative and commutative. The next theorem, whose proof is omitted in this version, shows that associative-commutative matching is $W[1]$-hard even if every function is associative and commutative.

Theorem 9. *Matching is $W[1]$-hard with respect to the number of variables even if every function symbol is associative and commutative.*

On the other hand, associative-commutative matching can be done in polynomial time if t_1 is a DO-term [5]. We can extend this algorithm to the special case of unification where both terms are DO-terms by adding a condition in the algorithm that $f((t_1)_{u_1}, \ldots, (t_1)_{u_p})$ and $f((t_2)_{v_1}, \ldots, (t_2)_{v_q})$ can be unified if $(t_1)_{u_i}$ and $(t_2)_{v_j}$ are variables for some i, j. This yields:

Proposition 4. *Associative-commutative unification can be done in polynomial time if both t_1 and t_2 are DO-terms.*

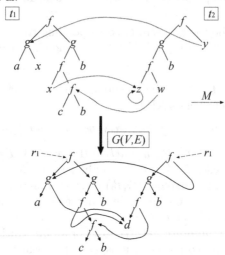

Fig. 3. Example of a DAG $G(V, E)$ for *CommutIdent* and for the proof of Theorem 8.

7   Concluding Remarks

This paper has studied the parameterized complexity of unification with associative and/or commutative functions with respect to the number of variables. Determining whether each of commutative unification and the matching version of Theorem 2 (i.e., where all variables occur in one of the strings and the number of occurrences of each variable is at most 2), is $W[1]$-hard or FPT and whether associative unification is in XP, as well as any nontrivial improvements of the presented results, are left as open problems.

References

1. Aho, A.V., Hopcroft, J.E., Ullman, J.D.: The Design and Analysis of Computer Algorithms. Addison-Wesley, Reading (1974)
2. Aikou, K., Suzuki, Y., Shoudai, T., Uchida, T., Miyahara, T.: A polynomial time matching algorithm of ordered tree patterns having height-constrained variables. In: Apostolico, A., Crochemore, M., Park, K. (eds.) CPM 2005. LNCS, vol. 3537, pp. 346–357. Springer, Heidelberg (2005)
3. Akutsu, T.: Approximate string matching with don't care characters. Inf. Process. Lett. **55**, 235–239 (1995)
4. Akutsu, T.: Approximate string matching with variable length don't care characters. IEICE Trans. Inf. Syst. **E79–D**, 1353–1354 (1996)
5. Benanav, D., Kapur, D., Narendran, P.: Complexity of matching problems. J. Symbolic Comput. **3**, 203–216 (1987)
6. Baader, F., Snyder, W.: Unification theory. In: Robinson, J.A., Voronkov, A. (eds.) Handbook of Automated Reasoning, pp. 447–533. Elsevier, Amsterdam (2001)
7. Bille, P.: A survey on tree edit distance and related problem. Theoret. Comput. Sci. **337**, 217–239 (2005)

8. Bodlaender, H.L., Downey, R.G., Fellows, M.R., Wareham, H.T.: The parameterized complexity of sequence alignment and consensus. Theoret. Comput. Sci. **147**, 31–54 (1995)
9. de Champeaux, D.: About the paterson-wegman linear unification algorithm. J. Comput. Syst. Sci. **32**, 79–90 (1986)
10. Eker, S.: Single elementary associative-commutative matching. J. Autom. Reasoning **28**, 35–51 (2002)
11. Fernau, H., Schmid, M.L.: Pattern matching with variables: A multivariate complexity analysis. In: Fischer, J., Sanders, P. (eds.) CPM 2013. LNCS, vol. 7922, pp. 83–94. Springer, Heidelberg (2013)
12. Flum, J., Grohe, M.: Parameterized Complexity Theory. Springer, Berlin (2006)
13. Gilbert, D., Schroeder, M.: FURY: fuzzy unification and resolution based on edit distance. In: Proceedings of 1st IEEE International Symposium on Bioinformatics and Biomedical Engineering, pp. 330–336 (2000)
14. Iranzo, P.J., Rubio-Manzano, C.: An efficient fuzzy unification method and its implementation into the Bousi∼Prolog system. In: Proceedings of 2010 IEEE International Conference on Fuzzy Systems, pp. 1–8 (2010)
15. Jones, N.C., Pevzner, P.A.: An Introduction to Bioinformatics Algorithms. MIT Press, Cambridge (2004)
16. Kamali, S., Tompa, F.W.: A new mathematics retrieval system. In: Proceedings of ACM Conference on Information and Knowledge Management, pp. 1413–1416 (2010)
17. Kapur, D., Narendran, P.: Complexity of unification problems with associative-commutative operators. J. Autom. Reasoning **28**, 35–51 (2002)
18. Knight, K.: Unification: a multidisciplinary survey. ACM Comput. Surv. **21**, 93–124 (1989)
19. Nguyen, T.T., Chang, K., Hui, S.C.: A math-aware search engine for math question answering system. In: Proceedings of ACM Conference on Information and Knowledge Management, pp. 724–733 (2012)
20. NTCIR: http://research.nii.ac.jp/ntcir/ntcir-10/conference.html (2013)
21. Paterson, M.S., Wegman, M.N.: Linear unification. J. Comput. Syst. Sci. **16**, 158–167 (1978)
22. Robinson, J.A.: A machine-oriented logic based on the resolution principle. J. ACM **12**, 23–41 (1965)

On Polynomial Kernelization
of $\mathcal{H}$-FREE EDGE DELETION

N.R. Aravind, R.B. Sandeep$^{(\boxtimes)}$, and Naveen Sivadasan

Department of Computer Science and Engineering,
Indian Institute of Technology Hyderabad, Hyderabad, India
{aravind,cs12p0001,nsivadasan}@iith.ac.in

Abstract. For a set of graphs $\mathcal{H}$, the $\mathcal{H}$-FREE EDGE DELETION problem asks to find whether there exist at most k edges in the input graph whose deletion results in a graph without any induced copy of $H \in \mathcal{H}$. In [3], it is shown that the problem is fixed-parameter tractable if $\mathcal{H}$ is of finite cardinality. However, it is proved in [4] that if $\mathcal{H}$ is a singleton set containing H, for a large class of H, there exists no polynomial kernel unless $coNP \subseteq NP/poly$. In this paper, we present a polynomial kernel for this problem for any fixed finite set $\mathcal{H}$ of connected graphs and when the input graphs are of bounded degree. We note that there are $\mathcal{H}$-FREE EDGE DELETION problems which remain NP-complete even for the bounded degree input graphs, for example TRIANGLE-FREE EDGE DELETION [2] and CUSTER EDGE DELETION(P_3-FREE EDGE DELETION) [15]. When $\mathcal{H}$ contains $K_{1,s}$, we obtain a stronger result - a polynomial kernel for K_t-free input graphs (for any fixed $t > 2$). We note that for $s > 9$, there is an incompressibility result for $K_{1,s}$-FREE EDGE DELETION for general graphs [5]. Our result provides first polynomial kernels for CLAW-FREE EDGE DELETION and LINE EDGE DELETION for K_t-free input graphs which are NP-complete even for K_4-free graphs [23] and were raised as open problems in [4,19].

1 Introduction

For a graph property Π, the Π EDGE DELETION problem asks whether there exist at most k edges such that deleting them from the input graph results in a graph with property Π. Numerous studies have been done on edge deletion problems from 1970s onwards dealing with various aspects such as hardness [1,2,7–9,14,20–23], polynomial-time algorithms [13,21,22], approximability [1,21,22], fixed-parameter tractability [3,10], polynomial problem kernels [2,10–12] and incompressibility [4,5,16].

There are not many generalized results on the NP-completeness of edge deletion problems. This is in contrast with the classical result by Lewis and Yannakakis [18] on the vertex counterparts which says that Π VERTEX DELETION problems are NP-complete if Π is non-trivial and hereditary on induced

---

R.B. Sandeep—supported by TCS Research Scholarship.

M. Cygan and P. Heggernes (Eds.): IPEC 2014, LNCS 8894, pp. 28–38, 2014.
DOI: 10.1007/978-3-319-13524-3_3

subgraphs. By a result of Cai [3], the Π EDGE DELETION problem is fixed-parameter tractable for any hereditary property Π that is characterized by a finite set of forbidden induced subgraphs. We observe that polynomial problem kernels have been found only for a few parameterized Π EDGE DELETION problems.

In this paper, we study a subset of Π EDGE DELETION problems known as $\mathcal{H}$-FREE EDGE DELETION problems where $\mathcal{H}$ is a set of graphs. The objective is to find whether there exist at most k edges in the input graph such that deleting them results in a graph with no induced copy of $H \in \mathcal{H}$. In the natural parameterization of this problem, the parameter is k. In this paper, we give a polynomial problem kernel for parameterized version of $\mathcal{H}$-FREE EDGE DELETION where $\mathcal{H}$ is any fixed finite set of connected graphs and when the input graphs are of bounded degree. In this context, we note that TRIANGLE-FREE EDGE DELETION [2] and CUSTER EDGE DELETION(P_3-FREE EDGE DELETION) [15] are NP-complete even for bounded degree input graphs. We also note that, under the complexity theoretic assumption $coNP \nsubseteq NP/poly$, there exist no polynomial problem kernels for the H-FREE EDGE DELETION problems when H is 3-connected but not complete, or when H is a path or cycle of at least 4 edges [4]. When the input graph has maximum degree at most Δ and if the maximum diameter of graphs in $\mathcal{H}$ is D, then the number of vertices in the kernel we obtain is at most $2\Delta^{2D+1} \cdot k^{pD+1}$ where $p = \log_{\frac{2\Delta}{2\Delta-1}} \Delta$. Our kernelization consists of a single rule which removes vertices of the input graph that are 'far enough' from all induced $H \in \mathcal{H}$ in G.

When $\mathcal{H}$ contains $K_{1,s}$, we obtain a stronger result - a polynomial kernel for K_t-free input graphs (for any fixed $t > 2$). Let $s > 1$ be the least integer such that $K_{1,s} \in \mathcal{H}$. Then the number of vertices in the kernel we obtain is at most $8d^{3D+1} \cdot k^{pD+1}$ where $d = R(s, t-1) - 1$, $R(s, t-1)$ is the Ramsey number and $p = \log_{\frac{2d}{2d-1}} d$. We note that CLAW-FREE EDGE DELETION and LINE EDGE DELETION are NP-complete even for K_4-free input graphs [23]. As a corollary of our result, we obtain the first polynomial kernels for these problems when the input graphs are K_t-free for any fixed $t > 2$. The existence of a polynomial kernel for CLAW-FREE EDGE DELETION and LINE EDGE DELETION were raised as open problems in [4,19]. We note that for $s > 9$, there is an incompressibility result for $K_{1,s}$-FREE EDGE DELETION for general graphs [5].

1.1 Related Work

Here, we give an overview of various results on edge deletion problems.

NP-completeness: It has been proved that Π EDGE DELETION problems are NP-complete if Π is one of the following properties: without cycle of any fixed length $l \geq 3$, without any cycle of length at most l for any fixed $l \geq 4$, connected with maximum degree r for every fixed $r \geq 2$, outerplanar, line graph, bipartite, comparability [23], claw-free (implicit in the proof of NP-completeness of the LINE EDGE DELETION problem in [23]), P_l-free for any fixed $l \geq 3$ [7], circular-arc, chordal, chain, perfect, split, AT-free [21], interval [9], threshold [20] and complete [14].

Fixed-parameter Tractability and Kernelization: Cai proved in [3] that parameterized Π EDGE DELETION problem is fixed-parameter tractable if Π is a hereditary property characterized by a finite set of forbidden induced subgraphs. Hence $\mathcal{H}$-FREE EDGE DELETION is fixed-parameter tractable for any finite set of graphs $\mathcal{H}$. Polynomial problem kernels are known for chain, split, threshold [12], triangle-free [2], cograph [11] and cluster [10] edge deletions. It is proved in [4] that for 3-connected H, H-FREE EDGE DELETION admits no polynomial kernel if and only if H is not a complete graph, under the assumption $coNP \not\subseteq NP/poly$. Under the same assumption, it is proved in [4] that for H being a path or cycle, H-FREE EDGE DELETION admits no polynomial kernel if and only if H has at least 4 edges. Unless $NP \subseteq coNP/poly$, H-FREE EDGE DELETION admits no polynomial kernel if H is $K_1 \times (2K_1 \cup 2K_2)$ [16].

2   Preliminaries and Basic Results

We consider only simple graphs. For a set of graphs $\mathcal{H}$, a graph G is $\mathcal{H}$-free if there is no induced copy of $H \in \mathcal{H}$ in G. For $V' \subseteq V(G)$, $G \setminus V'$ denotes the graph $(V(G) \setminus V', E(G) \setminus E')$ where $E' \subseteq E(G)$ is the set of edges incident to vertices in V'. Similarly, for $E' \subseteq E(G)$, $G \setminus E'$ denotes the graph $(V(G), E(G) \setminus E')$. For any edge set $E' \subseteq E(G)$, $V_{E'}$ denotes the set of vertices incident to the edges in E'. For any $V' \subseteq V(G)$, the closed neighbourhood of V', $N_G[V'] = \{v : v \in V'$ or $(u, v) \in E(G)$ for some $u \in V'\}$. In a graph G, distance from a vertex v to a set of vertices V' is the shortest among the distances from v to the vertices in V'.

A parameterized problem is *fixed-parameter tractable*(FPT) if there exists an algorithm to solve it which runs in time $O(f(k)n^c)$ where f is a computable function, n is the input size, c is a constant and k is the parameter. The idea is to solve the problem efficiently for small parameter values. A related notion is *polynomial kernelization* where the parameterized problem instance is reduced in polynomial (in $n + k$) time to a polynomial (in k) sized instance of the same problem called *problem kernel* such that the original instance is a yes-instance if and only if the problem kernel is a yes-instance. We refer to [6] for an exhaustive treatment on these topics. A kernelization rule is *safe* if the answer to the problem instance does not change after the application of the rule.

In this paper, we consider $\mathcal{H}$-FREE EDGE DELETION[1] which is defined as given below.

$\mathcal{H}$-FREE EDGE DELETION
Instance: A graph G and a positive integer k.
Problem: Does there exist $E' \subseteq E(G)$ with $|E'| \leq k$ such that $G \setminus E'$ does not contain $H \in \mathcal{H}$ as an induced subgraph.
Parameter: k

---

[1] We leave the prefix 'parameterized' henceforth as it is evident from the context.

We define an $\mathcal{H}$ *deletion set (HDS)* of a graph G as a set $M \subseteq E(G)$ such that $G \setminus M$ is $\mathcal{H}$-free. The *minimum $\mathcal{H}$ deletion set (MHDS)* is an HDS with smallest cardinality. We define a partition of an MHDS M of G as follows.

$M_1 = \{e : e \in M$ and e is part of an induced $H \in \mathcal{H}$ in $G\}$.

$M_j = \{e : e \in M \setminus \bigcup_{i=1}^{i=j-1} M_i$ and e is part of an induced $H \in \mathcal{H}$ in $G \setminus \bigcup_{i=1}^{i=j-1} M_i\}$, for $j > 1$.

We define the *depth* of an MHDS M of G, denoted by l_M, as the least integer such that $|M_i| > 0$ for all $1 \le i \le l_M$ and $|M_i| = 0$ for all $i > l_M$. Proposition 1 shows that this notion is well defined.

Proposition 1.   *1. $\{M_j\}$ forms a partition of M.*
   2. There exists $l_M \ge 0$ such that $|M_i| > 0$ for $1 \le i \le l_M$ and $|M_i| = 0$ for $i > l_M$.

Proof. If $i \ne j$ and M_i and M_j are nonempty, then $M_i \cap M_j = \emptyset$. For $i \ge 1$, $M_i \subseteq M$. Assume there is an edge $e \in M$ and $e \notin \bigcup M_j$. Delete all edges in $\bigcup M_j$ from G. What remains is an $\mathcal{H}$-free graph. As M is an MHDS, there can not exist such an edge e. Now let j be the smallest integer such that M_j is empty. Then from definition, for all $i > j$, $|M_i| = 0$. Therefore $l_M = j - 1$. $\square$

We observe that for an $\mathcal{H}$-free graph, the only MHDS M is $\emptyset$ and hence $l_M = 0$. For an MHDS M of G with a depth l_M, we define the following terms.
$S_j = \bigcup_{i=j}^{i=l_M} M_j$ for $1 \le j \le l_M + 1$.
$T_j = M \setminus S_{j+1}$ for $0 \le j \le l_M$.
$V_{\mathcal{H}}(G)$ is the set of all vertices part of some induced $H \in \mathcal{H}$ in G.
We observe that $S_1 = T_{l_M} = M$, $S_{l_M} = M_{l_M}$, $T_1 = M_1$ and $S_{l_M+1} = T_0 = \emptyset$.

Proposition 2. *For a graph G, let $E' \subseteq E(G)$ such that at least one edge in every induced $H \in \mathcal{H}$ in G is in E'. Then, at least one vertex in every induced $H \in \mathcal{H}$ in $G \setminus E'$ is in $V_{E'}$.*

Proof. Assume that there exists an induced $H \in \mathcal{H}$ in $G \setminus E'$ with the vertex set V'. For a contradiction, assume that $|V' \cap V_{E'}| = 0$. Then, V' induces a copy of H in G. Hence, E' must contain some of its edges. $\square$

Lemma 1. *Let G be the input graph of an $\mathcal{H}$-FREE EDGE DELETION problem instance where $\mathcal{H}$ is a set of connected graphs with diameter at most D. Let M be an MHDS of G. Then, every vertex in V_M is at a distance at most $(l_M - 1)D$ from $V_{\mathcal{H}}(G)$ in G.*

Proof. For $2 \le j \le l_M$, from definition, at least one edge in every induced $H \in \mathcal{H}$ in $G \setminus T_{j-2}$ is in M_{j-1}. Hence by Proposition 2, at least one vertex in every induced $H \in \mathcal{H}$ in $G \setminus T_{j-1}$ is in $V_{M_{j-1}}$. By definition, every vertex in V_{M_j} is part of some induced $H \in \mathcal{H}$ in $G \setminus T_{j-1}$. This implies every vertex in V_{M_j} is at a distance at most D from $V_{M_{j-1}}$. Hence every vertex in $V_{M_{l_M}}$ is at a distance at most $(l_M - 1)D$ from V_{M_1}. By definition, $V_{M_1} \subseteq V_{\mathcal{H}}(G)$. Hence the proof. $\square$

Lemma 2. *Let G be a graph with maximum degree at most Δ and M be an MHDS of G. Then, for $1 \le j \le l_M$, $(2\Delta - 1) \cdot |M_j| \ge |S_{j+1}|$.*

Proof. For $1 \le j \le l_M$, from definition, M_j has at least one edge from every induced $H \in \mathcal{H}$ in $G \backslash T_{j-1}$. Let M'_j be the set of edges incident to vertices in V_{M_j} in $G \setminus T_{j-1}$. We observe that $(G \setminus T_{j-1}) \setminus M'_j$ is $\mathcal{H}$-free and hence $|T_{j-1} \cup M'_j|$ is an HDS of G. Clearly, $|M'_j| \le \Delta |V_{M_j}| \le 2\Delta |M_j|$. Since M is an MHDS, $|T_{j-1} \cup M'_j| = |T_{j-1}| + |M'_j| \ge |M| = |T_{j-1}| + |S_j|$. Therefore $|M'_j| \ge |S_j|$. Hence, $2\Delta |M_j| \ge |S_j| = |M_j| + |S_{j+1}|$. $\qquad\square$

Now we give an upper bound for the depth of an MHDS in terms of its size and maximum degree of the graph.

Lemma 3. *Let M be an MHDS of G. If the maximum degree of G is at most $\Delta > 0$, then $l_M \le 1 + \log_{\frac{2\Delta}{2\Delta-1}} |M|$.*

Proof. The statement is clearly true when $l_M \le 1$. Hence assume that $l_M \ge 2$. The result follows from repeated application of Lemma 2.

$$|M| = |S_1| = |M_1| + |S_2| \ge \frac{|S_2|}{2\Delta - 1} + |S_2|$$

$$\ge |S_{l_M}| \left(\frac{2\Delta}{2\Delta - 1} \right)^{l_M - 1}$$

$$\ge \left(\frac{2\Delta}{2\Delta - 1} \right)^{l_M - 1} \quad [\because |S_{l_M}| \ge 1] \qquad\qquad\square$$

Corollary 1. *Let (G, k) be a yes-instance of $\mathcal{H}$-FREE EDGE DELETION where G has maximum degree at most $\Delta > 0$. For any MHDS M of G, $l_M \le 1 + \log_{\frac{2\Delta}{2\Delta-1}} k$.* $\qquad\square$

Lemma 4. *Let $\mathcal{H}$ be a set of connected graphs with diameter at most D. Let $V' \supseteq V_{\mathcal{H}}(G)$ and let $c \ge 0$. Let G' be obtained by removing vertices of G at a distance more than $c + D$ from V'. Furthermore, assume that if G' is a yes-instance then there exists an MHDS M' of G' such that every vertex in $V_{M'}$ is at a distance at most c from V' in G'. Then (G, k) is a yes-instance if and only if (G', k) is a yes-instance of $\mathcal{H}$-FREE EDGE DELETION.*

Proof. Let G be a yes-instance with an MHDS M. Then $M' = M \cap E(G')$ is an HDS of G' such that $|M'| \le k$. Conversely, let G' be a yes-instance. By the assumption, there exists an MHDS M' of G' such that every vertex in $V_{M'}$ is at a distance at most c from V' in G'. We claim that M' is an MHDS of G. For contradiction, assume $G \setminus M'$ has an induced $H \in \mathcal{H}$ with a vertex set V''. As G and G' has same set of induced copies of graphs in $\mathcal{H}$, at least one edge in every induced copy of graphs in $\mathcal{H}$ in G is in M'. Then, by Proposition 2, at least one vertex in V'' is in $V_{M'}$. We observe that for every vertex in G' the distance from V' is same in G and G'. Hence every vertex in V'' is at a distance at most $c + D$ from V' in G. Then, V'' induces a copy of H in $G' \setminus M'$ which is a contradiction. $\qquad\square$

Lemma 5. *Let G be a graph and let $d > 1$ be a constant. Let $V' \subseteq V(G)$ such that all vertices in G with degree more than d is in V'. Partition V' into V_1 and V_2 such that V_1 contains all the vertices in V' with degree at most d and V_2 contains all the vertices with degree more than d. If every vertex in G is at a distance at most $c > 0$ from V', then $|V(G)| \leq |V_1| \cdot d^{c+1} + |N_G(V_2)| \cdot d^c$.*

Proof. To enumerate the number of vertices in G, consider the d-ary breadth first trees rooted at vertices in V_1 and in $N_G[V_2]$.

$$|V(G')| \leq |V_1| \left(\frac{d^{c+1} - 1}{d - 1} \right) + |N_G[V_2]| \left(\frac{d^c - 1}{d - 1} \right)$$
$$\leq |V_1| d^{c+1} + |N_G[V_2]| d^c \qquad \square$$

3   Polynomial Kernels

In this section, we assume that $\mathcal{H}$ is a fixed finite set of connected graphs with diameter at most D. First we devise an algorithm to obtain polynomial kernel for $\mathcal{H}$-FREE EDGE DELETION for bounded degree input graphs. Then we prove a stronger result - a polynomial kernel for K_t-free input graphs (for some fixed $t > 2$) when $\mathcal{H}$ contains $K_{1,s}$ for some $s > 1$.

We assume that the input graph G has maximum degree at most $\Delta > 1$ and G has at least one induced copy of H. We observe that if these conditions are not met, obtaining polynomial kernel is trivial.

Now we state the kernelization rule which is the single rule in the kernelization.

Rule 0: Delete all vertices in G at a distance more than $(1 + \log_{\frac{2\Delta}{2\Delta-1}} k)D$ from $V_{\mathcal{H}}(G)$.

We note that the rule can be applied efficiently with the help of breadth first search from vertices in $V_{\mathcal{H}}(G)$. Now we prove the safety of the rule.

Lemma 6. *Rule 0 is safe.*

Proof. Let G' be obtained from G by applying Rule 0. Let M' be an MHDS of G'. If G' is a yes-instance, then by Lemma 1 and Corollary 1, every vertex in $V_{M'}$ is at a distance at most $D \log_{\frac{2\Delta}{2\Delta-1}} k$ from $V_{\mathcal{H}}(G')$. Hence, we can apply Lemma 4 with $V' = V_{\mathcal{H}}(G)$ and $c = D \log_{\frac{2\Delta}{2\Delta-1}} k$. $\qquad \square$

Lemma 7. *Let (G, k) be a yes-instance of $\mathcal{H}$-FREE EDGE DELETION. Let G' be obtained by one application of Rule 0 on G. Then, $|V(G')| \leq (2\Delta^{2D+1} \cdot k^{pD+1})$ where $p = \log_{\frac{2\Delta}{2\Delta-1}} \Delta$.*

Proof. Let M be an MHDS of G such that $|M| \leq k$. We observe that every vertex in $V_{\mathcal{H}}(G)$ is at a distance at most D from V_{M_1} in G. Hence, by construction, every vertex in G' is at a distance at most $(2 + \log_{\frac{2\Delta}{2\Delta-1}} k)D$ from V_{M_1} in G and in G'. We note that $|V_{M_1}| \leq 2k$. To enumerate the number of vertices in G', we apply Lemma 5 with $V' = V_{M_1}$, $c = (2 + \log_{\frac{2\Delta}{2\Delta-1}} k)D$ and $d = \Delta$.

$$|V(G')| \leq 2k\Delta^{(2+\log_{\frac{2\Delta}{2\Delta-1}} k)D+1}$$
$$\leq 2\Delta^{2D+1} \cdot k^{pD+1} \qquad \square$$

Now we present the algorithm to obtain a polynomial kernel. The algorithm applies Rule 0 on the input graph and according to the number of vertices in the resultant graph it returns the resultant graph or a trivial no-instance.

Kernelization for $\mathcal{H}$-FREE EDGE DELETION
($\mathcal{H}$ is a finite set of connected graphs with maximum diameter D)
Input:(G, k) where G has maximum degree at most Δ.

1. Apply Rule 0 on G to obtain G'.
2. If the number of vertices in G' is more than $2\Delta^{2D+1} \cdot k^{pD+1}$ where $p = \log_{\frac{2\Delta}{2\Delta-1}} \Delta$, then return a trivial no-instance $(H, 0)$ where H is the graph with minimum number of vertices in $\mathcal{H}$. Else return (G', k).

Theorem 1. *The kernelization for $\mathcal{H}$-FREE EDGE DELETION returns a kernel with the number of vertices at most $2\Delta^{2D+1} \cdot k^{pD+1}$ where $p = \log_{\frac{2\Delta}{2\Delta-1}} \Delta$.*

Proof. Follows from Lemmas 6 and 7 and the observation that the number of vertices in the trivial no-instance is at most $2\Delta^{2D+1} \cdot k^{pD+1}$. $\qquad \square$

3.1 A Stronger Result for a Restricted Case

Here we give a polynomial kernel for $\mathcal{H}$-FREE EDGE DELETION when $\mathcal{H}$ is a fixed finite set of connected graphs and contains a $K_{1,s}$ for some $s > 1$ and when the input graphs are K_t-free, for any fixed $t > 2$.

It is proved in [17] that the maximum degree of a {claw, K_4}-free graph is at most 5. We give a straight forward generalization of this result for {$K_{1,s}, K_t$}-free graphs. Let $R(s, t)$ denote the Ramsey number. Remember that the Ramsey number $R(s, t)$ is the least integer such that every graph on $R(s, t)$ vertices has either an independent set of order s or a complete subgraph of order t.

Lemma 8. *For integers $s > 1, t > 1$, any {$K_{1,s}, K_t$}-free graph has maximum degree at most $R(s, t - 1) - 1$.*

Proof. Assume G is $\{K_{1,s}, K_t\}$-free. For contradiction, assume G has a vertex v of degree at least $R(s, t-1)$. By the definition of the Ramsey number there exist at least s mutually non-adjacent vertices or $t-1$ mutually adjacent vertices in the neighborhood of v. Hence there exist either an induced $K_{1,s}$ or an induced K_t in G. $\qquad\square$

We modify the proof technique used for devising polynomial kernelization for $\mathcal{H}$-FREE EDGE DELETION for bounded degree graphs to obtain polynomial kernelization for K_t-free input graphs for the case when $\mathcal{H}$ contains $K_{1,s}$ for some $s > 1$.

Let $s > 1$ be the least integer such that $\mathcal{H}$ contains $K_{1,s}$. Let $t > 2$, G be K_t-free and M be an MHDS of G. Let $d = R(s, t-1) - 1$. Let D be the maximum diameter of graphs in $\mathcal{H}$. We define the following.

$M_0 = \{e : e \in M$ and e is incident to a vertex with degree at least $d + 1\}$.
$V_R(G) = \{v : v \in V(G)$ and v has degree at least $d + 1$ in $G\}$.

Lemma 9. $G \setminus M_0$ has degree at most d and every vertex in G with degree at least $d + 1$ is incident to at least one edge in M_0.

Proof. As $G \setminus M$ is $\{K_{1,s}, K_t\}$-free and every edge in M which is incident to at least one vertex of degree at least $d + 1$ is in M_0, the result follows from Lemma 8. $\qquad\square$

Lemma 10. Let M be an MHDS of G. Let $M' = M \setminus M_0$ and $G' = G \setminus M_0$. Then, M' is an MHDS of G' and every vertex in V_M is at a distance at most $Dl_{M'}$ from $V_{\mathcal{H}}(G) \cup V_R(G)$ in G.

Proof. It is straight forward to verify that M' is an MHDS of G'. By Lemma 1, every vertex in $V_{M'}$ is at a distance at most $(l_{M'} - 1)D$ from $V_{\mathcal{H}}(G')$ in G'. Every induced $H \in \mathcal{H}$ in G' is either an induced H in G or formed by deleting M_0 from G. Therefore, every vertex in $V_{\mathcal{H}}(G')$ is at a distance at most D from $V_{\mathcal{H}}(G) \cup V_R(G)$ in G'. Hence, every vertex in $V_{M'}$ is at a distance at most $Dl_{M'}$ from $V_{\mathcal{H}}(G) \cup V_R(G)$ in G'. The result follows from the fact $M = M' \cup M_0$. $\qquad\square$

The single rule in the kernelization is:

Rule 1: Delete all vertices in G at a distance more than $(2 + \log_{\frac{2d}{2d-1}} k)D$ from $V_{\mathcal{H}}(G) \cup V_R(G)$ where $d = R(s, t-1) - 1$.

Lemma 11. *Rule 1 is safe.*

Proof. Let G' be obtained from G by applying Rule 1. Let M' be an MHDS of G'. If G' is a yes-instance, then by Lemma 10 and Corollary 1, every vertex in $V_{M'}$ is at a distance at most $D(1 + \log_{\frac{2d}{2d-1}} k)$ from $V_{\mathcal{H}}(G') \cup V_R(G')$ in G'. We note that $V_{\mathcal{H}}(G) = V_{\mathcal{H}}(G')$ and $V_R(G) = V_R(G')$. Hence, we can apply Lemma 4 with $V' = V_{\mathcal{H}}(G) \cup V_R(G)$, $c = D(1 + \log_{\frac{2d}{2d-1}} k)$ and $d = R(s, t-1) - 1$. $\qquad\square$

Lemma 12. *Let (G,k) be a yes-instance of $\mathcal{H}$-FREE EDGE DELETION where G is K_t-free. Let G' be obtained by one application of Rule 1 on G. Then, $|V(G')| \leq 8d^{3D+1} \cdot k^{pD+1}$ where $p = \log_{\frac{2d}{2d-1}} d$.*

Proof. Let M be an MHDS of G such that $|M| \leq k$. We observe that every vertex in $V_{\mathcal{H}}(G)$ is at a distance at most D from V_{M_1} in G. Hence, by construction, every vertex in G' is at a distance at most $D(3 + \log_{\frac{2d}{2d-1}} k)$ from $V_{M_1} \cup V_R(G)$. Clearly $|V_{M_1}| \leq 2k$. Using Lemma 9 we obtain $|N[V_R(G)]| \leq 2k(d+2)$. To enumerate the number of vertices in G', we apply Lemma 5 with $V' = V_{M_1} \cup V_R(G)$, $c = D(3 + \log_{\frac{2d}{2d-1}} k)$ and $d = R(s, t-1) - 1$.

$$|V(G')| \leq 2kd^{D(3+\log_{\frac{2d}{2d-1}} k)+1} + 2k(d+2)d^{D(3+\log_{\frac{2d}{2d-1}} k)}$$
$$\leq 8d^{3D+1} \cdot k^{pD+1}$$

$\square$

Now we present the algorithm.

Kernelization for $\mathcal{H}$-FREE EDGE DELETION
($\mathcal{H}$ contains $K_{1,s}$ for some $s > 1$)
Input:(G,k) where G is K_t-free for some fixed $t > 2$.
Let $s > 1$ be the least integer such that $\mathcal{H}$ contains $K_{1,s}$.

1. Apply Rule 1 on G to obtain G'.
2. If the number of vertices in G' is more than $8d^{3D+1} \cdot k^{pD+1}$ where $d = R(s, t-1) - 1$ and $p = \log_{\frac{2d}{2d-1}} d$, then return a trivial no-instance $(K_{1,s}, 0)$. Else return (G', k).

For practical implementation, we can use any specific known upper bound for $R(s, t-1)$ or the general upper bound $\binom{s+t-3}{s-1}$.

Theorem 2. *The kernelization for $\mathcal{H}$-FREE EDGE DELETION when $K_{1,s} \in \mathcal{H}$ and the input graph is K_t-free returns a kernel with the number of vertices at most $8d^{1+3D} \cdot k^{1+pD}$ where $d = R(s, t-1) - 1$ and $p = \log_{\frac{2d}{2d-1}} d$.*

Proof. Follows from Lemmas 11 and 12. $\square$

It is known that line graphs are characterized by a finite set of connected forbidden induced subgraphs including a claw ($K_{1,3}$). Both CLAW-FREE EDGE DELETION and LINE EDGE DELETION are NP-complete even for K_4-free graphs [23].

Corollary 2. CLAW-FREE EDGE DELETION *and* LINE EDGE DELETION *admit polynomial kernels for K_t-free input graphs for any fixed $t > 3$.* $\square$

We observe that the kernelization for $\mathcal{H}$-FREE EDGE DELETION when $K_{1,s} \in \mathcal{H}$ and the input graph is K_t-free works for the case when $K_t \in \mathcal{H}$ and the input graph is $K_{1,s}$-free.

Theorem 3. $\mathcal{H}$-FREE EDGE DELETION *admits polynomial kernelization when $\mathcal{H}$ is a finite set of connected graphs, $K_t \in \mathcal{H}$ for some $t > 2$ and the input graph is $K_{1,s}$-free for some fixed $s > 1$.*

4 Concluding Remarks

Our results may give some insight towards a dichotomy theorem on incompressibility of $\mathcal{H}$-FREE EDGE DELETION raised as an open problem in [4]. We conclude with an open problem: does $\mathcal{H}$-FREE EDGE DELETION admit polynomial kernel for planar input graphs?

References

1. Alon, N., Shapira, A., Sudakov, B.: Additive approximation for edge-deletion problems. Ann. Math. **170**(1), 371–411 (2009)
2. Brügmann, D., Komusiewicz, C., Moser, H.: On generating triangle-free graphs. Electron. Notes Discrete Math. **32**, 51–58 (2009)
3. Cai, L.: Fixed-parameter tractability of graph modification problems for hereditary properties. Inf. Process. Lett. **58**(4), 171–176 (1996)
4. Cai, L., Cai, Y.: Incompressibility of H-free edge modification. In: Gutin, G., Szeider, S. (eds.) IPEC 2013. LNCS, vol. 8246, pp. 84–96. Springer, Heidelberg (2013)
5. Cai, Y.: Polynomial kernelisation of H-free edge modification problems. Master's thesis, Department of Computer Science and Engineering, The Chinese University of Hong Kong, Hong Kong SAR, China (2012)
6. Downey, R.G., Fellows, M.R.: Fundamentals of Parameterized Complexity. Springer, London (2013)
7. El-Mallah, E.S., Colbourn, C.J.: The complexity of some edge deletion problems. IEEE Trans. Circ. Syst. **35**(3), 354–362 (1988)
8. Garey, M.R., Johnson, D.S., Stockmeyer, L.: Some simplified NP-complete problems. In: Proceedings of the Sixth Annual ACM Symposium on Theory of Computing, pp. 47–63. ACM (1974)
9. Goldberg, P.W., Golumbic, M.C., Kaplan, H., Shamir, R.: Four strikes against physical mapping of DNA. J. Comput. Biol. **2**(1), 139–152 (1995)
10. Gramm, J., Guo, J., Hüffner, F., Niedermeier, R.: Graph-modeled data clustering: fixed-parameter algorithms for clique generation. In: Petreschi, R., Persiano, G., Silvestri, R. (eds.) CIAC 2013. LNCS, pp. 108–119. Springer, Heidelberg (2003)
11. Guillemot, S., Paul, C., Perez, A.: On the (non-)existence of polynomial kernels for P_l-free edge modification problems. Algorithmica **65**(4), 900–926 (2012)
12. Guo, J.: Problem kernels for NP-complete edge deletion problems: split and related graphs. In: Tokuyama, T. (ed.) ISAAC 2007. LNCS, vol. 4835, pp. 915–926. Springer, Heidelberg (2007)
13. Hadlock, F.: Finding a maximum cut of a planar graph in polynomial time. SIAM J. Comput. **4**(3), 221–225 (1975)

14. Karp, R.M.: Reducibility among combinatorial problems. In: Miller, R.E., Thatcher, J.W., Bohlinger, J.D. (eds.) Complexity of Computer Computations, pp. 85–103. Plenum Press, New York (1972)
15. Komusiewicz, C., Uhlmann, J.: Alternative parameterizations for cluster editing. In: Černá, I., Gyimóthy, T., Hromkovič, J., Jefferey, K., Královič, R., Vukolić, M., Wolf, S. (eds.) SOFSEM 2011. LNCS, vol. 6543, pp. 344–355. Springer, Heidelberg (2011)
16. Kratsch, S., Wahlström, M.: Two edge modification problems without polynomial kernels. In: Chen, J., Fomin, F.V. (eds.) IWPEC 2009. LNCS, vol. 5917, pp. 264–275. Springer, Heidelberg (2009)
17. Le, V.B., Mosca, R., Müller, H.: On stable cutsets in claw-free graphs and planar graphs. J. Discrete Algorithms 6(2), 256–276 (2008)
18. Lewis, J.M., Yannakakis, M.: The node-deletion problem for hereditary properties is NP-complete. J. Comput. Syst. Sci. 20(2), 219–230 (1980)
19. Kowalik, L., Cygan, M., Pilipczuk, M.: Open problems from workshop on kernels. Worker 2013 (2013)
20. Margot, F.: Some complexity results about threshold graphs. Discrete Appl. Math. 49(1), 299–308 (1994)
21. Natanzon, A., Shamir, R., Sharan, R.: Complexity classification of some edge modification problems. Discrete Appl. Math. 113(1), 109–128 (2001)
22. Shamir, R., Sharan, R., Tsur, D.: Cluster graph modification problems. Discrete Appl. Math. 144(1), 173–182 (2004)
23. Yannakakis, M.: Edge-deletion problems. SIAM J. Comput. 10(2), 297–309 (1981)

Solving Linear Equations Parameterized by Hamming Weight

Vikraman Arvind[1], Johannes Köbler[2](✉), Sebastian Kuhnert[2],
and Jacobo Torán[3]

[1] Institute of Mathematical Sciences, Chennai, India
arvind@imsc.res.in
[2] Institut für Informatik, Humboldt-Universität zu Berlin, Berlin, Germany
{koebler,kuhnert}@informatik.hu-berlin.de
[3] Institut für Theoretische Informatik, Universität Ulm, Ulm, Germany
jacobo.toran@uni-ulm.de

Abstract. Given a system of linear equations $Ax = b$ over the binary field $\mathbb{F}_2$ and an integer $t \geq 1$, we study the following three algorithmic problems:

1. Does $Ax = b$ have a solution of weight at most t?
2. Does $Ax = b$ have a solution of weight exactly t?
3. Does $Ax = b$ have a solution of weight at least t?

We investigate the parameterized complexity of these problems with t as parameter. A special aspect of our study is to show how the maximum multiplicity k of variable occurrences in $Ax = b$ influences the complexity of the problem. We show a sharp dichotomy: for each $k \geq 3$ the first two problems are W[1]-hard (which strengthens and simplifies a result of Downey et al. [SIAM J. Comput. 29, 1999]). For $k = 2$, the problems turn out to be intimately connected to well-studied matching problems and can be efficiently solved using matching algorithms.

1 Introduction

There are well known efficient methods, like Gaussian elimination, to solve systems of linear equations $Ax = b$ over $\mathbb{F}_2$. The problem becomes harder when we are seeking for a solution u with certain constraints placed on its Hamming weight $wt(u)$. This problem has been extensively studied in the context of error correcting codes as it is closely related to the minimum weight codeword problem: given a linear code defined by $Ax = 0$, what is the minimum weight of a non-zero codeword in it? This problem is known to be NP-hard [12], and even hard to approximate within any constant factor, assuming NP $\neq$ RP [5]. There are three related decision problems of interest for systems of linear equations $Ax = b$ over $\mathbb{F}_2$:

This work was supported by the Alexander von Humboldt Foundation in its research group linkage program. The third author was supported by DFG grant KO 1053/7-2.

© Springer International Publishing Switzerland 2014
M. Cygan and P. Heggernes (Eds.): IPEC 2014, LNCS 8894, pp. 39–50, 2014.
DOI: 10.1007/978-3-319-13524-3_4

1. $(A, b, t) \in \text{LINEQ}_\leq$ if $Ax = b$ admits a solution u with $1 \leq wt(u) \leq t$.
2. $(A, b, t) \in \text{LINEQ}_=$ if $Ax = b$ admits a solution u with $wt(u) = t$.
3. $(A, b, t) \in \text{LINEQ}_\geq$ if $Ax = b$ admits a solution u with $wt(u) \geq t$.

Berlekamp et al. [2] show that both $\text{LINEQ}_\leq$ and $\text{LINEQ}_=$ are NP-complete. When b is the all zeros vector, $\text{LINEQ}_\leq$ is the minimum weight codeword problem which is NP-hard [12], as already mentioned. Ntafos et al. [8] show that $\text{LINEQ}_\geq$ is NP-complete (also see [13]). See [7] for a nice discussion of these hardness results.

When the weight threshold t is considered as parameter, we denote the resulting parameterized versions of these problems by $\text{LINEQ}_{\leq,t}$, $\text{LINEQ}_{=,t}$ and $\text{LINEQ}_{\leq,t}$, respectively. Downey et al. [4] studied special cases of $\text{LINEQ}_{\leq,t}$ and $\text{LINEQ}_{=,t}$: when the vector b is either the all zeros vector or b is the all ones vector. These two special cases are called EVEN and ODD, respectively, for the weight at most t version. As argued in Remark 2.1 below, all other cases for vector b are in fact equivalent to either one of them. Observe that in the EVEN case, setting all variables to zero is always a solution; this is why $\text{LINEQ}_{\leq,t}$ and EVEN ask for solutions of weight at least 1. For the weight exactly t version, the problems are called EXACT EVEN and EXACT ODD. It turns out via a complicated proof in [4], that ODD, EXACT ODD and EXACT EVEN are W[1]-hard. Whether EVEN is also W[1]-hard remains open. The problem $\text{LINEQ}_{\geq,t}$, to our knowledge, has not been studied in the parameterized setting before. We show in Sect. 6 that in contrast to the other two, this problem is in FPT.

Our main contribution is the study of $\text{LINEQ}_{\leq,t}$ and $\text{LINEQ}_{=,t}$ in the light of some additional parameters: the maximum number k of occurrences of a variable in the system and the maximum size s of an equation. When k and s are restricted or used as an additional parameter, we denote this by an additional subscript to the respective problem. For example, k is treated as an additional parameter (besides t) in $\text{LINEQ}_{\leq,t,k}$, and bounded by k_{max} in $\text{LINEQ}_{\leq,t,k \leq k_{max}}$.

Concerning parameter k, we show a sharp dichotomy in the complexity of the problem. We prove that $\text{LINEQ}_{\leq,t,k \leq k_{max}}$ and $\text{LINEQ}_{=,t,k \leq k_{max}}$ are fpt tractable for $k_{max} \leq 2$, whereas for each $k_{max} \geq 3$, both problems are W[1]-hard. For the weight exactly t version, the hardness also holds for $b = 0$, while this case remains open for the weight at most t version.

Our hardness proof is a direct reduction from the parameterized clique problem. It strengthens and is much simpler than the proofs in [3,4] that (for unbounded occurrence multiplicity of the variables) go over a series of reductions running into nearly 10 pages. Furthermore, it gives alternative proofs of hardness for their results for EXACT EVEN, ODD and EXACT ODD.

For $k_{max} = 2$, we establish a connection between the equation systems and graph matching problems. We show that $\text{LINEQ}_{\leq,k \leq 2}$ and $\text{LINEQ}_{\geq,k \leq 2}$ are solvable in polynomial time, while $\text{LINEQ}_{=,k \leq 2}$ is solvable in randomized NC (RNC). The latter result follows from an interesting connection between $\text{LINEQ}_{=,k \leq 2}$ and RED-BLUE PERFECT MATCHING [10] (also known as EXACT MATCHING), which is known to be solvable in RNC [9] but not known to be in P. We show in Sect. 4 that both problems are equivalent under logarithmic space reductions.

Table 1. Summary of results.

Problem	Parameter/restriction list α					
	t	$t, k \leq 3$	$t, k \leq 2$	$k \leq 2$	s, t	$s \leq 2$
LinEq$_{\leq, \alpha}$	W[1]-hard	W[1]-hard	FPT	P	FPT	L-complete
	[4]	Theorem 3.1	Theorem 4.1	Theorem 4.1	Theorem 5.1	Theorem 5.5
LinEq$_{=, \alpha}$	W[1]-hard[a]	W[1]-hard[a]	FPT	RNC	FPT	L-complete
	[4]	Theorem 3.1	Theorem 4.5	Corollary 4.4	Theorem 5.4	Theorem 5.5
LinEq$_{\geq, \alpha}$	FPT	FPT	FPT	P	FPT	L-complete
	Theorem 6.1	Theorem 6.1	Theorem 6.1	Theorem 4.2	Theorem 6.1	Theorem 5.5

[a]Remains W[1]-hard for $b = 0$.

Hence, proving that LinEq$_{=, k=2}$ is in P would imply that RED-BLUE PERFECT MATCHING is also in P, solving a long standing open question. Further we show that LinEq$_{=, t, k \leq 2}$ is fixed parameter tractable.

If the maximum equation size s is an additional parameter then, as we show in Sect. 5, all three problems are fixed parameter tractable. In particular, if $s \leq 2$ then even the parameter-free versions of all three problems are solvable in logarithmic space. A summary of the results is given in Table 1.

Our fpt algorithms involve standard techniques like color coding (Theorems 4.5 and 5.4), depth-bounded search trees (Theorem 5.1), and reduction to problem kernels (Theorem 6.1).

For space reasons some proofs are omitted; the full version can be found at http://eccc.hpi-web.de/report/2014/096/.

2 Basic Transformations

In this section we describe some basic transformations between various linear equation system problems. First, note that LinEq$_{\leq}$ is polynomial-time reducible to LinEq$_{=}$ via $(A, b, t) \mapsto \{(A, b, t') : 1 \leq t' \leq t\}$. We next show that (EXACT) EVEN fpt reduces to (EXACT) ODD, taking the focus away from the "mixed" case (when b is neither the all zeros nor the all ones vector).

Remark 2.1. A system of linear equations $Ax = b$ over $\mathbb{F}_2$ can be easily transformed into an equivalent system $A'x' = 1$: Add a new variable x_0 and equation $x_0 = 1$. Convert each 0-equation into an equivalent 1-equation by adding x_0 to it. Then $Ax = b$ has a weight t solution if and only if $A'x' = 1'$ has a weight $t + 1$ solution.

In the non-parameterized setting, the "mixed" case is also reducible to EVEN.

Remark 2.2. A system $Ax = b$ over $\mathbb{F}_2$ with n variables can be transformed into a system $A'x' = 0$ such that $Ax = b$ has a weight t solution if and only if $A'x' = 0$ has a weight $t+n+1$ solution: add a new variable x_0 and a new equation $x_0 = 1$. Convert each 1-equation into an equivalent 0-equation by adding x_0 to it. Introduce n new variables $y_1, \ldots, y_n$ and replace the $x_0 = 1$ equation by the equations $x_0 + y_i = 0$ for $i = 1, \ldots, n$.

Lemma 2.3. *Let $Ax = b$ be a system of linear equations and let k be the maximum number of occurrences of any variable in it. Then an equivalent system $A'y = b'$ with at most three occurrences of each variable can be constructed in polynomial time, where equivalent means that a weight t solution for $Ax = b$ induces a weight kt solution for $A'y = b'$ and any weight t' solution for $A'y = b'$ induces a weight t'/k solution for $Ax = b$.*

The proof idea for the above lemma is to introduce k copies of each variable, to replace each occurrence with a different copy, and to force the copies to take equal values using additional equations.

As a consequence of Lemma 2.3 we can reduce all linear equation problems to the case $k \leq 3$. For example, it follows that $\text{LINEQ}_{\leq,t,k}$ is fpt reducible to $\text{LINEQ}_{\leq,t,k\leq3}$ and that $\text{LINEQ}_=$ is polynomial-time reducible to $\text{LINEQ}_{=,k\leq3}$.

To facilitate the presentation of some of our proofs, it is convenient to consider a more general problem in which each variable x_i occurring in $Ax = b$ has a positive integer weight w_i (encoded in unary). The weight t of a solution is the sum of the weights of the variables assigned value 1. The next lemma shows that the weighted case is polynomial-time reducible to the unweighted case (where all variables have weight 1).

Lemma 2.4. *Let $Ax = b$ be a system of linear equations with variable weights given in unary. Then an equivalent unweighted system $A'y = b'$ can be constructed in polynomial time, where equivalent means that a weight t solution for $Ax = b$ induces a weight t solution for $A'y = b'$ and vice versa. Moreover,*

(i) *if all variables of $Ax = b$ occur in exactly 2 equations then all variables of $A'y = b'$ occur in exactly 2 equations.*

(ii) *if all variables of $Ax = b$ occur in exactly 3 equations and have odd weight, then all variables of $A'y = b'$ occur in exactly 3 equations.*

We close this section by giving a useful graph theoretical interpretation of the linear equation problems.

Remark 2.5. We will consider systems $Ax = b$ with m variables and n equations, that is, A is an $n \times m$ matrix over $\mathbb{F}_2$. It will be convenient to interpret A as the incidence matrix of a hypergraph. With this interpretation each equation becomes a vertex and each variable becomes a hyperedge that consists of all vertices (equations) in which it occurs. Note that this might give a multi-hypergraph since different variables might occur in exactly the same equations.

A vertex v_j will be called *even* if $b_j = 0$, and *odd* if $b_j = 1$. A solution of weight t is a selection of t hyperedges that covers each even vertex with an even number of hyperedges and each odd vertex with an odd number of hyperedges. Observe that in the case that every variable appears exactly twice in the equation system we get a standard multi-graph in which each edge connects two vertices.

3   At Most Three Occurrences of Each Variable

This section is devoted to our main result showing that $\text{LINEQ}_{\leq,t,k\leq k_{\max}}$ and $\text{LINEQ}_{=,t,k\leq k_{\max}}$ are W[1]-hard for each $k_{\max} \geq 3$.

Theorem 3.1. $\text{LINEQ}_{\leq,t,k\leq 3}$ and $\text{LINEQ}_{=,t,k\leq 3}$ are W[1]-hard. The hardness even holds for the case that each variable occurs exactly three times.

To prove Theorem 3.1 we make use of the hypergraph interpretation of a linear system of equations as explained in Remark 2.5. The key step is the design of a selector gadget, which can be used to select a specified number of vertices from a given vertex set $V = \{v_1, \ldots, v_n\}$. Besides the vertices in V, the gadget contains a special *start vertex* a and a set U of internal vertices, i.e., the vertex set is $V \cup U \cup \{a\}$. We say that a set $\mathcal{S}$ of hyperedges *activates* a vertex if $\mathcal{S}$ covers it an odd number of times. Further, we call $\mathcal{S}$ *admissible* if it activates the start vertex a but no internal vertex in U. Using this notation we will construct the hyperedge set $\mathcal{E}$ of the gadget $\text{Sel}_{k,V}^{a}$ in such a way that the minimal admissible subsets $\mathcal{S}$ of $\mathcal{E}$ activate besides a exactly the k-element subsets of V.

The construction of $\text{Sel}_{k,V}^{a}$ is illustrated in Fig. 1. The set of internal vertices is $U = \{u_{\ell,i} : 1 < \ell < k \wedge \ell \leq i \leq n - k + \ell\}$. The intended semantics is that if a minimal admissible subset $\mathcal{S}$ covers the vertex $u_{\ell,i}$, then v_i is the ℓth smallest of the activated vertices from V. The hyperedge set of $\text{Sel}_{k,V}^{a}$ is $\mathcal{E} = \bigcup_{\ell=1}^{k-1} \mathcal{E}_k$, where the *level 1 hyperedges* are $\mathcal{E}_1 = \{\{a, v_i, u_{2,i'}\} : 1 < i < i' \leq n - k + 2\}$, the *level ℓ hyperedges* are $\mathcal{E}_\ell = \{\{u_{\ell,i}, v_i, u_{\ell+1,i'}\} : \ell \leq i < i' \leq n - k + \ell + 1\}$ for $\ell = 2, \ldots, k - 2$, and the *level $k - 1$ hyperedges* are $\mathcal{E}_{k-1} = \{\{u_{k-1,i}, v_i, v_{i'}\} : n - k \leq i < i' \leq n - 1\}$. In the weighted version $\text{Sel}_{k,V}^{a,w}$ of the gadget, all its hyperedges have weight w. The following lemma summarizes its properties.

Lemma 3.2. Let $V = \{v_1, \ldots, v_n\}$ and let k and w be positive integers. For any subset $W \subseteq V$ of size k, there is an admissible set $\mathcal{S} \subseteq \mathcal{E}$ of weight $(k-1)w$ for the selector gadget $\text{Sel}_{k,V}^{a,w}$ that activates exactly a and the vertices in W. Moreover, any admissible set $\mathcal{S} \subseteq \mathcal{E}$ of weight less than $(k+1)w$ for $\text{Sel}_{k,V}^{a,w}$ has weight exactly $(k-1)w$ and activates exactly k of the vertices in V.

Proof of Theorem 3.1. We reduce from the W[1]-complete clique problem which asks whether a given graph has a clique of size k, where k is treated as parameter.

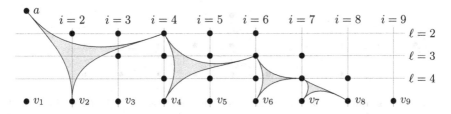

Fig. 1. The vertices of the selector gadget $\text{Sel}_{5,\{v_1,\ldots,v_9\}}^{a}$ and the minimal admissible subset of hyperedges that leads to the activation of $\{v_2, v_4, v_6, v_7, v_8\}$.

Let $G = (V, E)$ and k be the given instance. We will construct an equation system $Ax = b$ with exactly one 1-equation where each variable occurs exactly three times. Continuing with the hypergraph view, we will use several instances of the selector gadget; each uses its own internal vertices. Besides the internal vertices, the hypergraph contains one special start vertex a (which is the only odd vertex), one vertex for each graph vertex in V, and one vertex for each graph edge in E. Let $w = k^2$ if k is odd, and $w = k^2 + 1$ otherwise. Add the selector gadget $\mathrm{Sel}_{k,V}^{a,w}$ to the constructed hypergraph. For each graph vertex $v \in V$, let $E(v)$ denote the set of edges incident to it, and add the selector gadget $\mathrm{Sel}_{k-1,E(v)}^{v,1}$. Its role is to ensure that if v is selected by $\mathrm{Sel}_{k,V}^{a,w}$, then v must be adjacent to all other selected vertices. See Fig. 2 for an illustration of this construction. As the selector gadget has only hyperedges of size 3 and as w is odd, Lemma 2.4 implies that the weights can be removed while maintaining 3-uniformity.

We show that for $t = (k-1)w + k(k-2)$, the graph G has a clique of size k iff the equation system described by the constructed hypergraph has a solution of weight at most t. If G contains a k-clique C, choose the admissible hyperedge subset of $\mathrm{Sel}_{k,V}^{a,w}$ that activates exactly the vertices in C. Then, for each clique vertex $v \in C$, add the admissible hyperedge subset for $\mathrm{Sel}_{k-1,E(v)}^{v,1}$ that activates $\{e \in E(v) : e \subseteq C\}$. Combining these hyperedge sets yields a solution of weight t.

Now consider any solution to the equation system of weight at most t. As $(k+1)w \geq k^3 + k^2 > k^3 - k - 1 \geq t$, Lemma 3.2 implies that this solution contains exactly $(k-1)$ hyperedges from $\mathrm{Sel}_{k,V}^{a,w}$, which activate a set C of exactly k vertices in V. As these have to be covered an even number of times, each has to be covered an odd number of times from within its selector gadget. So for each $v \in C$, the solution must include at least $k-2$ hyperedges of the selector gadget $\mathrm{Sel}_{k-1,E(v)}^{v,1}$. As this accounts for the remaining weight permitted by t, the solution cannot include further hyperedges. In particular, all vertices from E that are covered at all are graph edges that are incident to a vertex in C. As all vertices in E are even, these edges have to be covered twice by the solution, implying that every vertex $v \in C$ has $k-1$ neighbors in C, thus C is a k-clique.

Finally, note that the constructed equation system admits a solution of weight at most t if and only if it admits one of weight exactly t.   □

Using the construction of Remark 2.1, we obtain alternative proofs that ODD and EXACT ODD are W[1]-hard. To generalize it to EXACT EVEN, we can multiply

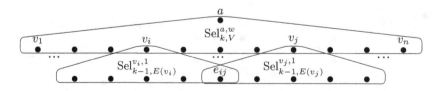

Fig. 2. Hypergraph view of the equation system that has a weight t solution if and only if the underlying graph G has a k-clique. The gadgets $\mathrm{Sel}_{k-1,E(v_i)}^{v_i,1}$ and $\mathrm{Sel}_{k-1,E(v_j)}^{v_j,1}$ share the vertex e_{ij} iff the graph edge e_{ij} connects the graph vertices v_i and v_j.

all weights by 2, add a new hyperedge $\{a\}$ of weight one, and ask for a solution of weight $2t + 1$.

4 At Most Two Occurrences of Each Variable

We show that the three problems are easier when every variable appears at most twice in $Ax = b$. It turns out that these problems can be solved using standard matching algorithms. $\text{LINEQ}_{\leq, k \leq 2}$ and $\text{LINEQ}_{\geq, k \leq 2}$ have deterministic polynomial-time algorithms, whereas $\text{LINEQ}_{=, k \leq 2}$ has a randomized polynomial time algorithm (in fact a randomized NC algorithm).

Firstly, we note that we can easily transform the instance to the case when every variable in the system $Ax = b$ appears in *exactly* two equations without any change in the parameter t. We include the new equation $\sum_{i=1}^{n} \sum_{j=1}^{n} A_{ij} x_j = \sum_{i=1}^{n} b_i$ (obtained by adding up all equations in $Ax = b$) to obtain a new system $A'x = b'$. Note that $Ax = b$ and $A'x = b'$ have identical solutions. Furthermore, the new equation $\sum_{i=1}^{n} \sum_{j=1}^{n} A_{ij} x_j = \sum_{i=1}^{n} b_i$ has on its left-hand side precisely the sum of all single occurrence variables of the system $Ax = b$. Hence every variable in $A'x = b'$ occurs exactly twice.

We will use the multi-graph interpretation of Remark 2.5 to design the algorithms in this subsection.

Theorem 4.1. $\text{LINEQ}_{\leq, k \leq 2} \in \mathsf{P}$.

Proof. Given an instance (A, b, t) of $\text{LINEQ}_{\leq, k \leq 2}$, we construct the graph G associated with $Ax = b$. The set of edges with value 1 in a solution to the system consists of an edge disjoint set of paths connecting the odd vertices by pairs and possibly some edge disjoint cycles.

If there are odd vertices we do not need to consider the cycles, since we are searching for a solution of minimum weight. Such a solution corresponds to a set of edge disjoint paths of minimum total length pairing the odd vertices. This can be obtained by computing the minimum distance between all pairs of odd vertices in the graph. With this we can construct a weighted clique in the following way: each vertex in the clique represents an odd vertex in the graph. The edge between two clique vertices is weighted with the minimum distance between the corresponding odd vertices in the original graph. We claim that a perfect matching with minimum weight in the clique defines a solution of minimum weight in the system. To see this, observe that if two edges $\{a_1, a_2\}$ and $\{b_1, b_2\}$ in the perfect matching of minimum weight would correspond to paths that share at least one edge in G, then the total length of the shortest paths between a_1 and one of the b-vertices and a_2 and the other b-vertex would be smaller than $d(a_1, a_2) + d(b_1, b_2)$ since the new paths would not contain the common edge. This implies that a perfect matching of minimum weight corresponds to a minimum weight solution of the system. Since minimum weight perfect matching can be solved in polynomial time, the result follows.

When all vertices are even, we need to ensure that at least one variable is set to 1. In this case, a minimum weight non-trivial solution is just a cycle in G with a minimum number of edges. This also can be computed in polynomial time. $\quad\square$

Using similar ideas we can show that the weight at least t version of the problem can also be solved in polynomial time.

Theorem 4.2. $\text{LINEQ}_{\geq,k\leq 2} \in \mathsf{P}$.

We show next that $\text{LINEQ}_{=,k=2}$ is equivalent to RED-BLUE PERFECT MATCH-ING (RBPM), a problem introduced by Papadimitriou et al. [10]. This problem is defined in the following way: Given a graph G with blue and red edges and a number t, is there a perfect matching in G with exactly t red edges? RBPM can be solved in randomized NC [9], but until now, no deterministic polynomial time algorithm for it is known. In fact, not even the parameterized version of this problem (with t as parameter) is known to lie in FPT.

Theorem 4.3. $\text{LINEQ}_{=,k\leq 2}$ *and* RBPM *are many-one equivalent under loga-rithmic space reductions.*

Proof (forward direction). Let $Ax = b$ be a system of equations in which every variable appears exactly twice and let $G = (V, E)$ be its interpretation as a graph.

We first assume that $b = 0$. Let u be a solution of weight t. Since u selects for each vertex v an even number of all edges incident to v, u corresponds to a union of edge disjoint cycles in G with exactly t edges. Now consider the graph $G' = (V', E')$ that is obtained from G by expanding each vertex $v \in V$ with degree d_v into d_v new vertices $v_1, \ldots, v_{d_v}$ and connecting all pairs of these vertices by red edges. The original edges incident with v in G are each connected to one of the new vertices and are all colored blue (see Fig. 3). Notice that in G', u corresponds to a union of vertex disjoint cycles with exactly $2t$ edges, where each cycle consists of alternating red and blue edges. Hence, the t red edges on these cycles form a matching that can be extended to a perfect matching of G' by adding all blue edges that are not lying on any cycle of u. Conversely, any perfect matching of G' with t red edges yields a solution of weight t by taking its symmetric difference with the set of all blue edges. This shows that $Ax = b$ has a solution of weight t if and only if G' has a perfect matching with t red edges.

If G has $r > 0$ odd vertices, each solution u corresponds to a union of edge disjoint cycles and paths with exactly t edges, where exactly the endpoints of the paths are odd vertices. We construct G' as before but expand each odd vertex into a red clique of size $d_v + 1$ by adding a special clique vertex v_0 that is connected via red edges to the other d_v clique vertices $v_1, \ldots, v_{d_v}$. In this

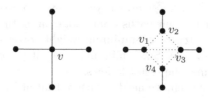

Fig. 3. Expansion of a vertex v of degree 4. The dotted edges are red edges.

graph, each solution u corresponds to a union of edge disjoint cycles and paths with exactly t blue and $t + r/2$ red edges, where exactly the endpoints of the paths are special clique vertices. Similarly to the even case, the $t + r/2$ red edges of u form a matching that can be extended to a perfect matching of G' by adding all blue edges that are not lying on any cycle or path. Conversely, any perfect matching M in G' has to match each special clique vertex via a red edge, implying that an odd number of the blue edges connected to its clique does not belong to M. Hence, if M has $t + r/2$ red edges, taking the symmetric difference of M with the set of all blue edges yields again a solution of weight t.   □

Interestingly, whereas the forward reduction also works in the parameterized setting (with t as parameter), this is not true for the converse reduction. Further, observe that the variants of RED-BLUE PERFECT MATCHING in which we ask for a matching with at most or at least t red edges are known to be in P. This provides alternative proofs for Theorems 4.1 and 4.2.

As RBPM is in randomized NC [9] we obtain the following corollary.

Corollary 4.4. LINEQ$_{=,k\leq 2}$ ∈ RNC.

We close this section by showing that in the parameterized setting, a solution of weight t can be found in fpt time when each variable occurs at most twice.

Theorem 4.5. LINEQ$_{=,t,k\leq 2}$ ∈ FPT.

Proof. Let (A, b, t) be the input instance and let $G = (V, E)$ be the corresponding graph. If $b = 0$, let u_0 be the empty solution. Otherwise, using the algorithm of Theorem 4.1, we compute a solution u_0 of minimum weight for $Ax = b$. If $|u_0| \geq t$, we are done. Otherwise, observe that every solution of $Ax = b$ can be written as a sum (modulo 2) of u_0 and some edge-disjoint cycles of G (that might overlap with u_0). To find a suitable set of cycles, we use the color coding method introduced in [1]. Each edge in u_0 receives its own unique color (recall that $|u_0| < t$). Let C_{u_0} be the set of colors of the edges in u_0. The remaining edges are colored uniformly at random using t new different colors. In case that there is a solution of weight exactly t, the probability that all the edges in the solution have different colors depends only on t and it is at least $\frac{t!}{t^t}$. A *color pattern* for a cycle is a sequence of colors to be encountered on the cycle. Now consider each possible set C of disjoint color patterns (their number only depends on t). For any set of disjoint cycles that realizes C, the corresponding solution has weight equal to the number of colors that appear in C or in C_{u_0} but not in both. If C leads to solutions of weight t, it remains to check if each color pattern $c_1, \ldots, c_k$ in C can be realized in G. The latter can be checked dynamically by computing sets $S_i(v)$, with $v \in V$ and $0 \leq i \leq k$, such that $u \in S_i(v)$ if and only if there is a path from u to v that realizes $c_1, \ldots, c_i$. Initially, $S_0(v) = \{v\}$ for each $v \in V$. For $i \in \{1, \ldots, k\}$, the set $S_i(v)$ is the union of all $S_{i-1}(u)$ for which $\{u, v\}$ is an edge of color c_i. There is a cycle realizing $c_1, \ldots, c_k$ if and only if there is a vertex v with $v \in S_k(v)$.

The probabilistic part in the previous algorithm can be derandomized using a perfect hash family as explained in [1].   □

5 Using the Equation Size as an Additional Parameter

In this section we show that the weight at most t and the weight exactly t versions of the problem become fixed parameter tractable when we treat the maximum equation size s as an additional parameter. For the weight at least t version we show in Sect. 6 that even $\text{LINEQ}_{\geq,t}$ is in FPT.

We call a solution $u \neq 0$ of a system $Ax = b$ *minimal* if for any solution $u' \neq 0$ with $u'_i \leq u_i$ for all i it holds that $u' = u$.

Theorem 5.1. $\text{LINEQ}_{\leq,t,s} \in$ FPT. *Moreover, for each instance, all minimal solutions of weight at most t can be found in fpt time.*

Proof. The algorithm traverses the following search tree to find all minimal solutions of weight at most t. If $b = 0$, the first branch is to select a variable, set it to 1 and continue with the resulting system over the remaining $m-1$ variables. This m-way branching is only needed once to avoid the trivial all zeroes solution. If $b \neq 0$ and the number of variables set to 1 so far is smaller than t, we pick the first equation with $b_j = 1$ and branch over all variables that occur in this equation. In each branch, we set the chosen variable to 1 and continue with the system over the remaining variables. As soon as all equations are satisfied by setting the remaining variables to 0 (i.e., $b = 0$), we reach at a successful leaf providing a solution of weight at most t. If already t variables have been set to 1 and $b \neq 0$, the current node is declared to be an unsuccessful leaf.

Since for every minimal solution u of weight at most t there is a path that selects at each node one more variable from u, the tree enumerates any such solution. Further, the tree can be traversed in fpt time as its depth is bounded by t and the number of its leaves is bounded by $s^{t-1}m$. □

To solve the weight exactly t case, we will again design a color coding algorithm similar to that in Theorem 4.5, where minimal solutions take the role of cycles. The following lemma shows that any solution $u \neq 0$ of a system $Ax = 0$ is the sum of disjoint minimal solutions.

Lemma 5.2. *Any solution $u \neq 0$ of a homogeneous system $Ax = 0$ over $\mathbb{F}_2$ is the sum of disjoint minimal solutions.*

Next, we observe the following colored variant of Theorem 5.1. When the variables are colored, we say that a solution u *respects* a set C of colors if u contains exactly one variable of each color in C, and no other variables.

Lemma 5.3. *Given a system $Ax = b$ over $\mathbb{F}_2$, a coloring of its variables and a set C of colors, all minimal solutions that respect C can be found in fpt time when $|C|$ and the maximum size s of the equations are treated as parameters.*

Now, the following theorem can be proved along the same lines as Theorem 4.5.

Theorem 5.4. $\text{LINEQ}_{=,t,s} \in$ FPT.

We close this section by considering restrictions on the parameter s. In the case $s \leq 2$ we can assume that all equations contain exactly 2 variables. Let G be the graph that has one vertex for each literal and an edge between each pair of literals that are forced to be equivalent by some equation. If the system is satisfiable, the connected components of G can be grouped into pairs of complementary equivalence classes. Define the size of an equivalence class as the number of positive literals in it. By choosing the one of smaller/larger size from each pair of complementary equivalence classes gives a minimum/maximum weight solution. Furthermore, the weight exactly t version reduces to UNARY SUBSET SUM: It suffices to check whether a subset of the size differences of all pairs sums up to t minus the size of a minimum solution. As both UNDIRECTED CONNECTIVITY and UNARY SUBSET SUM can be solved in logarithmic space [6,11], we have the upper bounds of the following theorem.

Theorem 5.5. LINEQ$_{\leq,s\leq2}$, LINEQ$_{=,s\leq2}$, and LINEQ$_{\geq,s\leq2}$ are all L-complete.

Complementing this result, the following lemma shows that the general case can be reduced to the case $s \leq 3$, implying that all three problems remain NP-hard under this restriction.

Lemma 5.6. Given a system of linear equations $Ax = b$ and a number t we can construct a new system $A'y = b'$ with equations of size at most 3 and a number t' so that there is a solution of weight t for the first system if and only if there is a solution of weight t' for the second one.

6 The Weight at Least t Version

In this section we give an fpt algorithm finding solutions of weight at least t.

Theorem 6.1. LINEQ$_{\geq,t} \in$ FPT.

Proof. Let $Ax = b$ be the given equation system and let t be the given weight threshold. Using Gaussian elimination, we can decide whether $Ax = b$ is feasible and compute the dimension d of the solution space of $Ax = 0$ in polynomial time. If d exceeds $t \log m$ then there must be a solution of weight at least t, since there can be only

$$\sum_{i=0}^{t-1} \binom{m}{i} < 2^{t \log m}$$

solutions of weight less than t. Otherwise, we compute a linearly independent spanning set $\{v_1, v_2, \ldots, v_d\}$ of at most $t \log m$ solutions for $Ax = 0$, along with a particular solution u of $Ax = b$. Any solution to $Ax = b$ is of the form $u + \sum_{i=1}^{d} \alpha_i v_i$, where $\alpha_i \in \mathbb{F}_2$. If one of the solution vectors u and $u + v_i$ for $i = 1, \ldots, d$ has support at least t, it witnesses that the input is a positive instance. Otherwise all but $t^2 \log m$ coordinates are always zero in any solution vector. Hence, the number of relevant variables (that can take value 1 in any solution) is bounded by $t^2 \log m$. Discarding the other variables, we now have a system of linear equations with at most $t^2 \log m$ variables. We can now brute-force search for all solutions of Hamming weight at most $t - 1$. Note that the number S of such solutions is bounded by

$$\sum_{i=0}^{t-1} \binom{t^2 \log m}{i} < (t^2 \log m)^t.$$

Keeping in mind the easily checked fact that $(\log m)^{O(t)} = t^{O(t)} \mathrm{poly}(m)$, this search takes fpt time. Finally, comparing S with the dimension d, if $S = 2^d$, then all solutions have weight less than t, otherwise there must be a solution of weight at least t. □

The previous algorithm may not always construct a solution of weight at least t if it exists. However, using self-reduction in a standard way, we can also solve the search problem in fpt time.

References

1. Alon, N., Yuster, R., Zwick, U.: Color coding. J. ACM **42**(4), 844–856 (1995)
2. Berlekamp, E.R., McEliece, R.J., van Tilborg, H.C.A.: On the inherent intractability of certain coding problems. IEEE Trans. Inform. Theory **24**, 384–386 (1978)
3. Downey, R.G., Fellows, M.R.: Parameterized Complexity. Springer, Berlin (1999)
4. Downey, R.G., Fellows, M.R., Vardy, A., Whittle, G.: The parametrized complexity of some fundamental problems in coding theory. SIAM J. Comput. **29**(2), 545–570 (1999)
5. Dumer, I., Micciancio, D., Sudan, M.: Hardness of approximating the minimum distance of a linear code. IEEE Trans. Inform. Theory **49**(1), 22–37 (2003)
6. Elberfeld, M., Jakoby, A., Tantau, T.: Logspace versions of the theorems of Bodlaender and Courcelle. In: FOCS Conference, pp. 143–152 (2010)
7. Johnson, D.S.: The NP-completeness column. ACM Trans. Algorithms **1**(1), 160–176 (2005)
8. Ntafos, S.C., Hakimi, S.L.: On the complexity of some coding problems. IEEE Trans. Inform. Theory **27**(6), 794–796 (1981)
9. Mulmuley, K., Vazirani, U., Vazirani, V.: Matching is as easy as matrix inversion. Combinatorica **7**, 105–113 (1987)
10. Papadimitriou, C., Yannakakis, M.: The complexity of restricted spanning tree problems. J. ACM **29**, 285–309 (1982)
11. Reingold, O.: Undirected connectivity in log-space. J. ACM **55**(4), 1–24 (2008)
12. Vardy, A.: The intractability of computing the minimum distance of a code. IEEE Trans. Inform. Theory **43**, 1757–1766 (1997)
13. Vardy, A.: Algorithmic complexity in coding theory and the minimum distance problem. In: Proceeding of 29th ACM symposium on theory of computing, pp. 92–109 (1997)

The Parameterized Complexity of Geometric Graph Isomorphism

Vikraman Arvind and Gaurav Rattan[✉]

Institute of Mathematical Sciences, Chennai, India
{arvind,grattan}@imsc.res.in

Abstract. We study the parameterized complexity of *Geometric Graph Isomorphism* (It is known as *Point Set Congruence* problem in computational geometry): given two sets of n points $A, B \subset \mathbb{Q}^k$ in k-dimensional euclidean space, with k as the fixed parameter, the problem is to decide if there is a bijection $\pi : A \rightarrow B$ such that for all $x, y \in A$, $\|x - y\| = \|\pi(x) - \pi(y)\|$, where $\| \cdot \|$ is the euclidean norm. Our main results are the following:

- We give a $O^*(k^{O(k)})$ time (The $O^*(\cdot)$ notation here, as usual, suppresses polynomial factors) FPT algorithm for Geometric Isomorphism. In fact, we show the stronger result that canonical forms for finite point sets in $\mathbb{Q}^k$ can also be computed in $O^*(k^{O(k)})$ time. This is substantially faster than the previous best time bound of $O^*(2^{O(k^4)})$ for the problem [1].
- We also briefly discuss the isomorphism problem for other l_p metrics. We describe a deterministic polynomial-time algorithm for finite point sets in $\mathbb{Q}^2$.

1 Introduction

Given two finite n-point sets A and B in a metric space (X, d), we say A and B are *isomorphic* if there is a *distance-preserving* bijection between A and B. The *Geometric Graph Isomorphism* problem for this metric space, denoted GGI, is to decide if A and B are isomorphic. The most well-studied version of this general problem, which is also the main focus in this paper, is the standard k-dimensional *euclidean space* $(\mathbb{R}^k, l_2)$ equipped with the l_2 metric. This problem is also known as the *Point Set Congruence* problem in the computational geometry literature [2–4]. It is called "Geometric Graph Isomorphism" by Evdokimov and Ponomarenko in [1], which we find more suitable as the problem is closely related to Graph Isomorphism.

When k is constant, there is an easy polynomial-time algorithm for the problem [4]. When $k = n$, Papadimitriou and Safra [5] note that the problem is polynomial-time equivalent to the standard Graph Isomorphism problem. The interesting case is when the dimension k is much smaller than n. In the computational geometry literature, a randomized algorithm running in time $O(n^{\frac{k-1}{2}} \cdot \log n)$ was given in [2]. This was improved to an $O(n^{\lceil \frac{k}{3} \rceil} \cdot \log n)$ algorithm

© Springer International Publishing Switzerland 2014
M. Cygan and P. Heggernes (Eds.): IPEC 2014, LNCS 8894, pp. 51–62, 2014.
DOI: 10.1007/978-3-319-13524-3_5

in [3]. We note that both these results are for a random access model of computation, used in computational geometry, which allows for arbitrary precision real arithmetic.

For point sets $A, B \subset \mathbb{Q}^k$, when the dimension k treated as a fixed parameter, an FPT algorithm for the GGI problem, running in time $(2^{k^4} nM)^{O(1)}$, was given by Evdokimov and Ponomarenko [1], where M upper bounds the binary encodings of the rational numbers in the input point sets. Their algorithm uses concepts from cellular algebras and is technically nontrivial.[1]

Our Results

As our first result we give a $O^*(k^{O(k)})$ time FPT algorithm for Geometric Graph Isomorphism. (Here, the $O^*(\cdot)$ notation hides polynomial factors in the input size.) Indeed, we actually give a $O^*(k^{O(k)})$ time algorithm that computes *canonical forms* for point sets in $\mathbb{Q}^k$. Our main contribution here is an intuitive geometric approach based on integer lattices — surprisingly not used in earlier work on the problem — to the Geometric Graph Isomorphism problem. Once we formulate the approach, it turns out that well-known algorithmic results for integer lattices can be applied. Specifically, using a suitable algorithm for computing all shortest vectors in an integer lattice [6] easily yields a $2^{O(k^2)} \text{poly}(nM)$ time FPT algorithm for GGI, where M upper bounds the binary encodings of the numbers in the input. This already improves the $O^*(2^{O(k^4)})$ time algorithm of [1]. Then, using key ideas from the recent lattice isomorphism algorithm of Haviv and Regev [7], we improve the FPT algorithm to $k^{O(k)} \text{poly}(nM)$ running time.

At this point we recall the definition of *canonical forms*. Computing canonical forms for structures is a fundamental algorithmic problem. *Graph Canonization*, which is the problem computing canonical forms for graphs, is closely connected to Graph Isomorphism. For a graph class $\mathcal{K}$, a mapping $f : \mathcal{K} \to \mathcal{K}$ is a *canonizing function* if $f(X)$ is isomorphic to X for each graph X in $\mathcal{K}$, and for any other graph X' in the class, $f(X) = f(X')$ if and only if X and X' are isomorphic. We say that $f(X)$ is the *canonical form* assigned by f to the isomorphism class containing X. For example, $f(X)$ can be defined as the lex-first graph in $\mathcal{K}$ isomorphic to X. This canonizing function is known to be NP-hard to compute. Whether there is *some* polynomial-time computable canonizing function for graphs is open. It is also open whether graph canonization is polynomial-time equivalent to graph isomorphism. Graph classes with efficient isomorphism tests are often known to have canonization algorithms [8] of comparable complexity.

Analogously, corresponding to geometric isomorphism, we can define canonical forms and the canonization problem for an n-point sets $A \subset \mathbb{Q}^k$. Given $A \subset \mathbb{Q}^k$ as input, a *canonizing function* $f : A \mapsto f(A)$ outputs an isomorphic point set $f(A)$ such that $f(A) = f(B)$ if and only if A and B are isomorphic point sets.

---

[1] To the best of our knowledge, this paper appears to be unknown in the Computational Geometry literature.

We actually show that canonical forms for point sets in $\mathbb{Q}^k$ can also be computed in $O^*(k^{O(k)})$ time, which yields the $O^*(k^{O(k)})$ time algorithm for GGI. This result is presented in Sect. 3.

Other metrics. In Sect. 4, we briefly examine GGI for other l_p metrics. For the 2-dimensional case $\mathbb{Q}^2$ we show that the problem is in deterministic polynomial time. This is by a reduction to the problem of isomorphism of colored graphs with color classes of size 2 (BCGI$_2$), which is known to be solvable in polynomial time [9]. For higher dimensions we do not have any nontrivial upper bounds better than general Graph Isomorphism.

2   Preliminaries

Let $[k]$ denote the set $\{1, \ldots, k\}$. As we consider points with rational coordinates in the euclidean space $\mathbb{R}^k$, we are effectively working with $\mathbb{Q}^k$. The projection of a vector $v \in \mathbb{Q}^k$ on a subspace $S \subset \mathbb{Q}^k$ is denoted v_S. The *inner product* of vectors $u = (u_1, \ldots, u_k)$ and $v = (v_1, \ldots, v_k)$ is $\langle u, v \rangle = \sum_{i \in [k]} u_i v_i$. The *euclidean norm*, $\|u\|$, of a vector u, is $\sqrt{\langle u, u \rangle}$, and the *distance* between two points u and v in $\mathbb{R}^k$ is $\|u - v\|$. Vectors u, v are *orthogonal* if $\langle u, v \rangle = 0$. In general, for $p \geq 1$, the p-norm of a vector $x = (x_1, \ldots, x_k)$ is $\|x\|_p = (\|x_1\|^p + \cdots + \|x_k\|^p)^{1/p}$, and the ∞-norm $\|x\|_\infty$ is $max\{|x_1|, \ldots, |x_k|\}$. The euclidean norm is the 2-norm.

For a vector set $S = \{u_1, \ldots, u_n\}$, the $n \times n$ *Gram matrix* of S is $G(S)_{i,j} = \langle u_i, u_j \rangle$. Two sets S and T have the same Gram matrix if and only if there is an orthogonal matrix O such that $T = OS$. Furthermore, a Gram matrix G is known to be Cholesky decomposable as LL^T for a unique lower triangular matrix L. The Cholesky decomposition is polynomial-time computable.

Given two point sets A and B in $\mathbb{Q}^k$, a bijection $\pi : A \to B$ is a *geometric isomorphism* if for every $x, y \in A$, $\|x - y\| = \|\pi(x) - \pi(y)\|$. Given two subspaces U and V of $\mathbb{Q}^k$, a bijection $\tau : U \to V$ is an *isometry* if for every $x, y \in U$, $\|x - y\| = \|\pi(x) - \pi(y)\|$. If τ is also a linear map then it is a *linear isometry*. It is natural to ask whether an isomorphism between point sets can also be extended to an isometry between the vector spaces which are spanned by these sets. In Sect. 3, we note that this holds for the euclidean metric in Lemmas 3.1 and 3.2 and use it to obtain our algorithmic results.

We now recall some definitions and properties of integer lattices [10]. A lattice $\mathcal{L}_B$ in $\mathbb{R}^k$ is the set of all integer linear combinations of a finite *basis* set of vectors $B = \{b_1, \ldots, b_m\} \subset \mathbb{Q}^k$. The number k is the *dimension* of the lattice. We will assume that the vectors b_i have rational entries with standard binary encodings (bounded by M). Then, we can compute a linearly independent basis of $r \leq k$ vectors for the lattice in time polynomial in k, M and m [10]. The number r is the *rank* of the lattice $\mathcal{L}_B$.

A fundamental quantity for a lattice $\mathcal{L}$ is the length $\lambda_1(\mathcal{L})$ of a shortest vector in it. There are several algorithms for exactly computing shortest vectors and for approximating them in the literature [10]. We recall an important result due to Micciancio and Voulgaris [6] for enumerating all the shortest vectors in a given lattice.

Theorem 2.1. ([6], **Corollary 5.8).** *There is a deterministic algorithm that takes as input a basis of some lattice $\Lambda \subset \mathbb{Q}^k$, and a target vector $\boldsymbol{t} \in \mathbb{Q}^k$, and an integer $p \geq 2$, and in time $\tilde{O}((4p)^k) \cdot poly(M, n)$ it outputs all vectors in Λ within distance $p\lambda_1(\Lambda)$ from $\boldsymbol{t}$. (The $\tilde{O}(\cdot)$ notation suppresses polylogarithmic factors).*

We also recall a well-known bound on the number of short vectors in a lattice (see [6]).

Lemma 2.2. *In a lattice $\mathcal{L}$ of rank k, the number of vectors of length at most $p\lambda_1(\mathcal{L})$ is bounded by $(2p+1)^k$.*

Haviv and Regev, in [7], study the lattice isomorphism problem under orthogonal transformations. In the process, they develop a general *isolation lemma* which they apply to lattice isomorphism and give a $O^*(k^{O(k)})$ time algorithm for checking if two rank-k lattices are isomorphic under orthogonal transformations. They introduce the notion of a *linearly independent chain* in a given set of vectors. We recall the definition as we will require it to describe our canonical forms algorithm for point sets in $\mathbb{Q}^k$. For a finite set $A \subseteq \mathbb{Q}^k$ and a vector $v \in \mathbb{Q}^k$, we say that v *uniquely defines a linearly independent chain of length m* in A if there are m vectors $x_1, \ldots, x_m \in A$ such that for every $1 \leq j \leq m$, the minimum inner product of v with vectors in $A \backslash Span(x_1, \ldots, x_{j-1})$ is uniquely achieved by x_j.

Given a lattice $\mathcal{L}$, its dual lattice $\mathcal{L}^*$ is defined as the set of vectors in $Span(\mathcal{L})$ such that they have an integer inner product with every vector in $\mathcal{L}$. The following theorem of Haviv and Regev [7] shows the existence of a suitably short vector in the dual lattice which defines a unique linearly independent chain in the set of shortest vectors of the lattice.

Theorem 2.3. ([7], **Theorem 4.2).** *Let $\mathcal{L}$ be a lattice of rank k. Let S be the set of shortest vectors in $\mathcal{L}$. Suppose the dimension of $Span(S)$ is k. Then, there exists a vector $v \in \mathcal{L}^*$ that uniquely defines a linearly independent chain of length k in S and satisfies $\|v\| \leq 5k^{17/2} \cdot \lambda_1(\mathcal{L}^*)$.*

3   The $O^*(k^{O(k)})$ Time Algorithm for GGI

We first note that an isomorphism between point sets A and B in $\mathbb{Q}^k$ naturally extends to a linear isometry between the vector spaces spanned by these sets. For simplicity we assume that the point sets A and B contain the zero vector $\bar{0}$.

Lemma 3.1. *Suppose π is a geometric isomorphism between A and B such that $\pi(\bar{0}) = \bar{0}$. Then there is a linear isometry $\mu : Span(A) \to Span(B)$ such that $\mu(x) = \pi(x)$ for all $x \in A$.*

The proof of the above lemma is based on the observations in the next lemma. The proofs involve usual linear algebraic arguments and we omit them in this extended abstract.

Lemma 3.2. *Let π be an isomorphism from A to B such that $\pi(\bar{0}) = \bar{0}$. Let $u_i, u_j \in A$.*

(a) π preserves inner products: i.e. $\langle u_i, u_j \rangle = \langle \pi(u_i), \pi(u_j) \rangle$.
(b) For any linear combination $\alpha u_i + \beta u_j \in A$, $\pi(\alpha u_i + \beta u_j) = \alpha \pi(u_i) + \beta \pi(u_j)$. Similarly, for any linear combination $\alpha v_i + \beta v_j \in B$, $\pi^{-1}(\alpha v_i + \beta v_j) = \alpha \pi^{-1}(v_i) + \beta \pi^{-1}(v_j)$.
(c) $U \subseteq A$ is a basis for $Span(A)$ iff $\pi(U) \subseteq B$ is a basis for $Span(B)$.

We assumed that $\bar{0} \in A, B$ and $\bar{0}$ is fixed by the isometry. We now argue that it suffices to search for such isometries. Observe that the distance of point u_i in set A from the centroid of the points in A is:

$$\|u_i - \frac{1}{n}\sum_{j=i}^{n} u_j\|^2 = \frac{1}{n^2} \cdot \|\sum_{j=1}^{n}(u_i - u_j)\|^2 = \frac{1}{n^2} \cdot \sum_{j=1}^{n}\sum_{k=1}^{n}\langle u_i - u_j, u_i - u_k \rangle$$

$$= \frac{1}{n^2} \cdot \sum_{j=1}^{n}\sum_{k=1}^{n} \frac{1}{2} \cdot (\|u_i - u_j\|^2 + \|u_i - u_k\|^2 - \|u_j - u_k\|^2).$$

Therefore, if sets A and B are isomorphic via a permutation π, the distance of any point u_i from the centroid c_A of set A must be equal to the distance of $\pi(u_i)$ from the centroid c_B of set B. Hence, we can replace A and B by $A \cup \{c_A\}$ and $B \cup \{c_B\}$ respectively, and extend the isomorphism π by mapping the centroid of A to the centroid of B. Clearly, this is an isomorphism between $A \cup \{c_A\}$ and $B \cup \{c_B\}$. Next, we translate the two sets A and B such that their respective centroids are mapped to the zero vector $\bar{0}$. Clearly, A and B are isomorphic if and only if their translations $\tilde{A}$ and $\tilde{B}$, obtained above, are isomorphic via a permutation that maps $\bar{0} \in \tilde{A}$ to $\bar{0} \in \tilde{B}$. Hence, it suffices to solve the following polynomial time equivalent problem: Given point sets $A, B \subset \mathbb{Q}^k$, both with $\bar{0}$ (as centroid), check if there exists an isomorphism mapping A to B that fixes $\bar{0}$.

Algorithm Overview

We first give an overview of a simpler version of the isomorphism algorithm. Consider the *integer lattices* $\mathcal{L}_A$ and $\mathcal{L}_B$ generated by the sets A and B. By Lemma 3.2, any linear isometry μ that maps A bijectively to B also bijectively maps $\mathcal{L}_A$ to $\mathcal{L}_B$. In particular, μ will map the set of shortest vectors of $\mathcal{L}_A$ to the set of shortest vectors of $\mathcal{L}_B$. Also, μ maps the subspace spanned by the shortest vectors of $\mathcal{L}_A$ to the subspace spanned by the shortest vectors of $\mathcal{L}_B$. The algorithm does the following: compute the shortest vector sets of both $\mathcal{L}_A$ and $\mathcal{L}_B$. Fix a maximal linearly independent set of shortest vectors S in lattice $\mathcal{L}_A$. Branch on all possible (injectively mapped) images of S into shortest vector set of $\mathcal{L}_B$. Since any lattice in $\mathbb{Q}^k$ has at most $2^{O(k)}$ many shortest vectors, this branching is bounded by $2^{O(k^2)}$. Next, on each branch the algorithm projects the set A to the orthogonal complement of the subspace spanned by S and B to the orthogonal complement of the subspace spanned by $\pi(S)$. Recursively continue to compute a geometric isomorphism for these projected point sets that are in subspaces of strictly smaller dimension. If A and B are isomorphic then, for one path of choices for the image set of S we can recover an isomorphism. This approach yields a simple $O^*(2^{O(k^2)})$ time algorithm for GGI.

The improvement of the running time to $O^*(k^{O(k)})$ requires the Haviv-Regev result (stated in Theorem 2.3). We now give a brief overview of our $O^*(k^{O(k)})$ time algorithm for computing the canonical form for a given point set $A \subset \mathbb{Q}^k$. As above, the algorithm first computes the shortest vector set $\mathcal{S}_A$ of the lattice $\mathcal{L}_A$. For the overview description, we assume for simplicity that $\mathcal{S}_A$ spans $Span(A)$. The actual algorithm (Algorithm 1) proceeds by projecting $Span(A)$ to the orthogonal complement of $Span(\mathcal{S}_A)$, similar to the $O^*(2^{O(k^2)})$ algorithm sketched above.

We will apply the Haviv-Regev algorithm [7] (Theorem 2.3) to pick the set $\text{short}_A = \{v \in \mathcal{L}_A^* \mid \|v\| \le 5k^{17/2} \cdot \lambda_1(\mathcal{L}_A^*)\}$ of short vectors in the dual lattice $\mathcal{L}_A^*$ which yield a unique linearly independent chain, of length k, in the set of shortest vectors $\mathcal{S}_A$ of $\mathcal{L}_A$. As shown in [7] such vectors in the dual lattice exist and by Theorem 2.1 the set short_A is of size bounded by $k^{O(k)}$ and can be listed in $O^*(k^{O(k)})$ time. Corresponding to each $v \in \text{short}_A$ the linearly independent chain in $\mathcal{S}_A$, of length k, yields a basis for $Span(A)$. There are in total $k^{O(k)}$ such bases thus obtained. For each such basis B we generate a description of the set A as follows: We first compute the Gram matrix $G(B)$ for B. Then for each vector $u_i \in A$, we compute the k-tuple Γ_i of the coordinates of u_i in basis B. The description of A obtained from basis B is the tuple $(G(B), \Gamma_1, \ldots, \Gamma_n)$. We now explain how these descriptions can be used to compute a canonical form for the point set A.

Suppose A_1 and A_2 are isomorphic point sets in $\mathbb{Q}^k$ and $\mu : Span(A_1) \to Span(A_2)$ is the corresponding linear isometry. Then μ is an isometric map between the lattices $\mathcal{L}_{A_1}$ and $\mathcal{L}_{A_2}$ as also between the dual lattices $\mathcal{L}_{A_1}^*$ and $\mathcal{L}_{A_2}^*$. Furthermore, μ maps short_{A_1} to short_{A_2}. More precisely, if $v \in \text{short}_{A_1}$ gives rise to a unique linearly independent chain B in $\mathcal{S}_{A_1}$ then $\mu(v) \in \text{short}_{A_2}$ gives rise to a unique linearly independent chain in $\mathcal{S}_{A_2}$ (which is in fact $\mu(B)$).

Now, crucially, we note that the description for A_1 $(G(B), \Gamma_1, \ldots, \Gamma_n)$ generated using the chain B is *identical* to the description for A_2 generated for $\mu(B)$. This is because the Gram matrices $G(B)$ and $G(\mu(B))$ are equal.

This suggests that the lexicographically least description is a canonical representation for the input point set A, and can be used to generate a canonical form for it. I.e. for each $v \in \text{short}_A$ satisfying the condition of Theorem 2.3, compute the description $(G(B), \Gamma_1, \ldots, \Gamma_n)$ using the corresponding linearly independent chain B. Among these descriptions pick the lexicographically least one $(G(B), \Gamma_1, \ldots, \Gamma_n)$ from which we will recover a canonical form for A. We now formally describe the algorithm.

Input: A set of vectors $A \subset \mathbb{Q}^k$ s.t. $|A| = n$ and $\bar{0} \in A$.
Output: A canonical set of vectors C_A.

1. While $dim(Span(A)) \neq 0$
   (a) Compute the set S_A of shortest vectors in $\mathcal{L}_A$ using Theorem 2.1.
   (b) Define the lattice $\Lambda_1 = \mathcal{L}_A \cap Span(S_A)$.
   (c) Compute the set of vectors W in the dual lattice Λ_1^* which are of length at most $5k^{17/2} \cdot \lambda_1(\Lambda_1^*)$ using Theorem 2.1.
   (d) For each vector in W, check if it defines a linearly independent chain in S_A. If yes, compute the chain. Otherwise, discard w from W.
   (e) Update set A to its component orthogonal to $Span(S_A)$. I.e. replace every $u \in A$ by $u - u_{S_A}$.
2. Let $W_1, \ldots, W_l$ be the sets computed during the l iterations of Step 1(c)–(d). For every tuple $(w_1, \ldots, w_l) \in W_1 \times \cdots \times W_l$,
   (a) Define the basis $B = C_1 \cup \cdots \cup C_l$, where C_i is the unique chain corresponding to vector w_i. Let the set B be $\{r_1, \ldots, r_k\}$.
   (b) Compute the Gram matrix $G(B)$ for the set B.
   (c) For each u_i in the input set A, let $\gamma_1, \ldots, \gamma_k$ be the coefficients such that $u_i = \sum_{j=1}^{k} \gamma_j r_j$. Let Γ_i denote the tuple $(\gamma_1, \ldots, \gamma_k)$, which can be computed by solving a system of linear equations.
   (d) Define the string σ for the tuple $(w_1, \ldots, w_l)$ to be $(G(B), (\Gamma_1, \ldots, \Gamma_n))$.
3. Let Σ be the set of all strings generated in the previous step. Search the lexicographically least string σ_0 in Σ.
4. Given the string $\sigma_0 = (G, (\Gamma_1, \ldots, \Gamma_n))$,
   (a) Let L be the unique lower triangular matrix such that $G = LL^T$.
   (b) Let B_0 be the set of rows of L.
   (c) Compute the set C_A of vectors $\{u_1, \ldots, u_n\}$ where u_i is the Γ_i-linear-combination of B_0.
5. Output C_A as the canonical form for the set A.

Algorithm 1

The following two lemmas show that C_A is indeed a canonical form for the point set A. We omit the detailed proofs in this extended abstract.

Lemma 3.3. *The point sets A and C_A are geometrically isomorphic.*

Lemma 3.4. *Two point sets A and B in $\mathbb{Q}^k$ are geometrically isomorphic if and only if $C_A = C_B$.*

We now formally state the result of this section.

Theorem 3.5. *Given a finite point set $A \subset \mathbb{Q}^k$ of size n as input, there is a deterministic $O^*(k^{O(k)})$ time algorithm that computes a canonizing function $f(A)$. As a consequence, the GGI problem for point sets in $\mathbb{Q}^k$ has a deterministic $O^*(k^{O(k)})$ time algorithm.*

Proof. It follows from Lemmas 3.3 and 3.4 that Algorithm 1 correctly computes a canonical form for any input point set $A \subset \mathbb{Q}^k$. It remains to prove the the running time bound. We first bound the time taken in Step 1 which can execute for at most k iterations. Computing sets S_A and W takes time $O^*(k^{O(k)}) \cdot poly(M)$, as a consequence of Theorem 2.1. The projection operations are routine polynomial time linear algebraic procedures. The bit-complexity of the entries in set A after projection can increase by at most a $poly(k)$ factor, by the properties of Gaussian elimination. Therefore, all the bit-sizes remain bounded by $O(k^k)poly(M)$ during the execution of the algorithm. Overall, the running time complexity of Step 1 is bounded by $O^*(k^{O(k)})$.

Next, we bound the time spent in Step 2. By Lemma 2.2, $|W_i|$ is at most $(25k^{17/2}+1)^k = k^{O(k)}$. The number of tuples examined is at most $|W_1| \cdots \cdots |W_l|$ which is bounded by $k_1^{O(k_1)} \cdot \cdots \cdot k_l^{O(k_l)} \leq k^{O(k)}$. Other operations in Step 2 are polynomial-time computations. Hence, Step 2 takes $O^*(k^{O(k)})$ time. Steps 3-5 are polynomial time solvable.

Remark 3.6. An $O^*(k^{O(k)})$ time FPT algorithm for the isomorphism problem GGI follows from the above theorem: we compute the canonical forms for the input point sets A and B, and accept if and only if $C_A = C_B$.

4   Geometric Isomorphism in Other l_p Metrics

We briefly discuss the GGI problem for other l_p metrics. For the 2-dimensional case we obtain a polynomial-time algorithm. The algorithm works as follows. Given two point sets A and B of size n, we fix three points in set A and branch on their possible images in B under an isomorphism. Using these points, we will construct two colored graphs G and H such that (a) each graph has color class size at most two and (b) the point sets A and B are isomorphic if and only if the graphs G and H are isomorphic via a color-preserving isomorphism. This computation can be performed in polynomial time. The isomorphism problem for color class size two graphs, denoted by $BCGI_2$ is in polynomial time [9] which yields the result.

Theorem 4.1. *Given subsets A and B of $\mathbb{Q}^2$ as input, for any l_p metric, there is a deterministic polynomial-time algorithm for checking if A and B are isomorphic in that metric.*

Proof. We will describe the algorithm for the l_∞ case and then indicate how it can be adapted for other l_p metrics.

Input: Point sets A and B of size n in $\mathbb{Q}^2$ (the l_∞ case).
Output: **Accept** if A and B are isomorphic, and **reject** otherwise.

1. Check if sets A and B are collinear by iterating over all triples and checking whether the three points are collinear.
   - If not collinear: pick the first three non-collinear points $\{a, b, c\}$.
   - If collinear:
     - Construct two colored graphs G and H: The graph G is $(A, \emptyset)$. The color of vertex u_i is the unordered pair $\{d_1, d_2\}$ of the distances of u_i from the two extreme points in the set A. Similarly define H for set B.
     - Return **accept** iff G and H are isomorphic. The isomorphism can be decided using the algorithm of [9].
2. Otherwise, w.l.o.g we have $a, b, c \in A$ (the other case is symmetric). Branch on all possible images of $\{a, b, c\}$ in B, denoted by $\{a', b', c'\}$.
3. First, we compute a coloring of sets A and B. For A, we color a point $u \in A$ by the ordered triple $(d_{u,a}, d_{u,b}, d_{u,c})$ of its distances from a, b, c.
4. Second, we can refine these colorings and ensure that each color class is of size two as follows:
   - If some subset of vertices form a color class of size more than two, they will lie on a line segment parallel to x-axis or y-axis (proof of correctness explains). Each such color class has two extreme points.
   - For each vertex $u \in A, B$, check if it lies in a color class of size more than two. If yes, update the color of u, say C, with the color $(C, \{d_1, d_2\})$ where d_1, d_2 are the distances of u from the extreme points in the color class.
5. Third, we construct weighted colored graphs G' and H' over vertex sets A and B respectively. The graphs G' and H' are complete graphs, and have color classes of size at most two. The coloring of the vertices have been already computed in Step 4. Every edge $\{u, v\}$ in G' or H' is labeled with the weight d_{uv}, the distance between points u and v.
6. Finally, we can easily modify the weighted graphs G' and H' to obtain unweighted graphs G and H such that G is isomorphic to H iff G' is isomorphic to H'. For every pair of color classes C_i and C_j, we can examine the induced graphs $G'[C_i \cup C_j]$ and $H'[C_i \cup C_j]$. By a simple case analysis, we can either (a) claim the graphs G' and H' to be non-isomorphic and **reject**, or (b) replace weighted edges by unweighted edges.
7. Test whether G is isomorphic to H using the algorithm of [9]. If the answer is yes **accept**, else move to the next branch in Step 2. If all branches are exhausted, return **reject**.

Algorithm 2

It is easy to verify that the algorithm works correctly for the case when the sets are collinear. Therefore, we concentrate on the general case. If the above algorithm accepts, clearly the sets are isomorphic. Conversely, suppose there is a isomorphism π from A and B. In Step 2, one of the branches for the image of $\{a, b, c\}$ will coincide with $(\pi(a), \pi(b), \pi(c))$. Furthermore, π must respect the color classes defined by the algorithm based on the distance triples in Step 3. It also respects the color refinements in Step 4 due to the following fact which can be easily verified by induction. A color class of collinear points must map to another class of collinear points in a manner that preserves the order of vertices (therefore, in at most two possible ways). Hence, π respects the colors assigned by the algorithm.

Next, we verify the bound on the color class sizes. The set of points $S_{r,x}$ at l_∞-distance r from a point x is easily seen to be a square in $\mathbb{Q}^2$ centered at x. It has sides of length $2r$ parallel to the coordinate axes. Consider the squares $S_{\alpha,a}$ and $S_{\beta,b}$. Their intersection is one of the following: (a) empty, or (b) at most two points, or (c) a common edge, or (d) two common incident edges. If we consider the third square $S_{\gamma,c}$, its intersection with $S_{\alpha,a} \cap S_{\beta,b}$ is therefore one of these cases: (a) empty or (b) at most two points or (c) a common edge. The last case is ruled our since three squares with non-collinear centres cannot have more than two edges common. Therefore, every color class is bounded by two unless the points in the color class lie on a common edge of three squares. Such a class is refined in Step 4 to have size at most two. Therefore, π must be an isomorphism between the weighted graphs G' and H' since it preserves mutual distances. By construction, the graphs G and H must be isomorphic and therefore, the algorithm accepts in Step 7.

Other l_p metrics. We now briefly explain how the above algorithm can be adapted to solve the 2-dimensional GGI problem for other l_p-metrics.

The set $S_{r,x}$ is a l_p-metric circle of radius r centered at the point x. For $p = 1$, such circles are squares of side $2r$ centered at x which have been rotated by $\pi/4$. The intersection of such squares is similar to the l_∞ case above. Hence, the above algorithm adapts to this case. For the case $p \in (1, \infty)$, it is known that l_p balls are *strictly convex* sets I.e., for any two distinct points u, v on the boundary of such a set, any convex combination $\theta u + (1 - \theta)v$ for $0 < \theta < 1$ is in the interior of the set. For $\mathbb{Q}^2$, this implies that any two l_p circles can intersect in at most two points ([11], Theorem 1). Therefore, any color class can be of size two and therefore, a similar algorithm which reduces the problem to $BCGI_2$ works and we can apply the algorithm of [9].

5   Discussion

In this paper we gave an $O^*(k^{O(k)})$ time FPT algorithm for Geometric Isomorphism in the l_2 metric, which is asymptotically faster than previous algorithms for this problem. A natural open question is to improve the running time. From the point of view of the general Graph Isomorphism problem it would be very interesting to obtain a "geometric" algorithm of running time $2^{O(k)} \cdot poly(nM)$, since the well-known algorithms for this problem are group-theoretic.

As observed in the introduction, Graph Isomorphism (for n-vertex graphs) is polynomial-time reducible to GGI, where, in the reduced instance, the output point sets are contained in $\mathbb{Q}^n$. We note a similar reduction even for hypergraph isomorphism.[2] More precisely, given a pair of hypergraphs (X_1, X_2) on n vertices the reduction outputs a pair of point sets (A, B), where $A, B \subset \mathbb{Q}^{5n}$, such that X_1 and X_2 are isomorphic if and only if A and B are isomorphic.

The current best algorithm for Hypergraph Isomorphism [12] crucially uses a group-theoretic algorithm (for Coset-Intersection of permutation groups) and has running time $O^*(2^{O(n)})$ for n-vertex hypergraphs. However, the only known algorithm for computing canonical forms of hypergraphs is the trivial $O^*(n!)$ time algorithm which picks the lexicographically least hypergraph isomorphic to the input hyerpgraph. Obtaining an $O^*(2^{O(k)})$ algorithm for computing canonical forms of point sets in $\mathbb{Q}^k$ would imply an $O^*(2^{O(n)})$ time (non-group-theoretic) canonization algorithm for hypergraphs, which is a long standing open problem.

Finally, we note that the complexity of GGI for point sets in $\mathbb{Q}^k$ in other l_p metrics is wide open. We do not know if the problem is FPT with k as parameter. One approach to solving GGI for a metric space (X, d) is to try and efficiently embed the given points sets A and B isometrically into a different metric space (X', d') for which we already know an efficient algorithm. For instance, known results about embedding metric spaces imply that there is a reduction from l_1^k-GGI to $l_\infty^{2^k}$-GGI in time $2^k \cdot poly(k, n, M)$, where l_p^k denotes the l_p metric on $\mathbb{Q}^k$. We do not know of a reduction that avoids the blow-up from k to 2^k in dimension.

References

1. Evdokimov, S.A., Ponomarenko, I.N.: On the geometric graph isomorphism problem. Pure Appl. Algebra **117–118**, 253–276 (1997)
2. Akutsu, T.: On determining the congruence of point sets in d dimensions. Comput. Geom. **9**(4), 247–256 (1998)
3. Braß, P., Knauer, C.: Testing the congruence of d-dimensional point sets. In: Symposium on Computational Geometry, pp. 310–314 (2000)
4. Alt, H., Mehlhorn, K., Wagener, H., Welzl, E.: Congruence, similarity, and symmetries of geometric objects. Discrete Comput. Geom. **3**, 237–256 (1988)
5. Papadimitriou, C.H., Safra, S.: The complexity of low-distortion embeddings between point sets. In: Proceedings of the 16th ACM SODA Conference, pp. 112–118 (2005)
6. Micciancio, D., Voulgaris, P.: A deterministic single exponential time algorithm for most lattice problems based on voronoi cell computations. SIAM J. Comput. **42**(3), 1364–1391 (2013)

---

[2] There is a standard reduction that reduces hypergraph isomorphism for n-vertex and m-edge hypergraphs to bipartite graph isomorphism on $n + m$ vertices. However, the point sets thus obtained will be in $\mathbb{Q}^{n+m}$ and m could be much larger than n. The aim is to obtain point sets in as low a dimension as possible.

7. Haviv, I., Regev, O.: On the lattice isomorphism problem. In: Proceedings of the 25th Annual ACM-SIAM Conference, SODA 2014, pp. 391–404 (2014)
8. Babai, L., Luks, E.M.: Canonical labeling of graphs. In: Proceedings of the ACM STOC Conference, pp. 171–183 (1983)
9. Furst, M.L., Hopcroft, J.E., Luks, E.M.: Polynomial-time algorithms for permutation groups. In: Proceedings of the 21st IEEE FOCS Conference, pp. 36–41 (1980)
10. Schrijver, A.: Theory of Integer and Linear Programming. Discrete Mathematics and Optimization. Wiley-Interscience, New York (1998)
11. Corbalan, A.G., Mazon, M., Recio, T.: About voronoi diagrams for strictly convex distances. In: 9th European Workshop on Computational Geometry (1993)
12. Luks, E.M.: Hypergraph isomorphism and structural equivalence of boolean functions. In: Proceedings of the 31st ACM STOC Conference, pp. 652–658 (1999)

The Role of Planarity in Connectivity Problems Parameterized by Treewidth

Julien Baste and Ignasi Sau[(✉)]

AlGCo Project-team, CNRS, LIRMM, Montpellier, France
jbaste@ens-cachan.fr, ignasi.sau@lirmm.fr

Abstract. For some years it was believed that for "connectivity" problems such as HAMILTONIAN CYCLE, algorithms running in time $2^{O(\mathbf{tw})} \cdot n^{O(1)}$ –called *single-exponential*– existed only on planar and other sparse graph classes, where $\mathbf{tw}$ stands for the treewidth of the n-vertex input graph. This was recently disproved by Cygan *et al.* [FOCS 2011], Bodlaender *et al.* [ICALP 2013], and Fomin *et al.* [SODA 2014], who provided single-exponential algorithms on general graphs for most connectivity problems that were known to be solvable in single-exponential time on sparse graphs. In this article we further investigate the role of planarity in connectivity problems parameterized by treewidth, and convey that several problems can indeed be distinguished according to their behavior on planar graphs. Known results from the literature imply that there exist problems, like CYCLE PACKING, that *cannot* be solved in time $2^{o(\mathbf{tw} \log \mathbf{tw})} \cdot n^{O(1)}$ on general graphs but that can be solved in time $2^{O(\mathbf{tw})} \cdot n^{O(1)}$ when restricted to planar graphs. Our main contribution is to show that there exist natural problems that can be solved in time $2^{O(\mathbf{tw} \log \mathbf{tw})} \cdot n^{O(1)}$ on general graphs but that *cannot* be solved in time $2^{o(\mathbf{tw} \log \mathbf{tw})} \cdot n^{O(1)}$ even when restricted to planar graphs. Furthermore, we prove that PLANAR CYCLE PACKING and PLANAR DISJOINT PATHS *cannot* be solved in time $2^{o(\mathbf{tw})} \cdot n^{O(1)}$. The mentioned negative results hold unless the ETH fails. We feel that our results constitute a first step in a subject that can be further exploited.

Keywords: Parameterized complexity · Treewidth · Connectivity problems · Single-exponential algorithms · Planar graphs · Dynamic programming

1 Introduction

Motivation and previous work. Treewidth is a fundamental graph parameter that, loosely speaking, measures the resemblance of a graph to a tree. It was introduced, among other equivalent definitions given by other authors, by Robertson and Seymour in the early stages of their monumental Graph Minors

Research supported by the Languedoc-Roussillon Project "Chercheur d'avenir" KERNEL and by the grant EGOS ANR-12-JS02-002-01.

© Springer International Publishing Switzerland 2014
M. Cygan and P. Heggernes (Eds.): IPEC 2014, LNCS 8894, pp. 63–74, 2014.
DOI: 10.1007/978-3-319-13524-3_6

project [18], and its algorithmic importance was significantly popularized by Courcelle's theorem [3], stating that any graph problem that can be expressed in CMSO logic can be solved in time $f(\mathbf{tw}) \cdot n$ on graphs with n vertices and treewidth $\mathbf{tw}$, where f is some function depending on the problem. Nevertheless, the function $f(\mathbf{tw})$ given by Courcelle's theorem is unavoidably huge [10], so from an algorithmic point of view it is crucial to identify problems for which $f(\mathbf{tw})$ grows *moderately* fast.

Many problems can be solved in time $2^{O(\mathbf{tw} \log \mathbf{tw})} \cdot n^{O(1)}$ when the n-vertex input (general) graph comes equipped with a tree-decomposition of width $\mathbf{tw}$. Intuitively, this is the case of problems that can be solved via dynamic programming on a tree-decomposition by enumerating all *partitions* or *packings* of the vertices in the bags of the tree-decomposition, which are $\mathbf{tw}^{O(\mathbf{tw})} = 2^{O(\mathbf{tw} \log \mathbf{tw})}$ many. In this article we only consider this type of problems and, more precisely, we are interested in which of these problems can be solved in time $2^{O(\mathbf{tw})} \cdot n^{O(1)}$; such a running time is called *single-exponential*. This topic has been object of extensive study during the last decade. Let us briefly overview the main results on this line of research.

It is well known that problems that have *locally checkable certificates*[1], like VERTEX COVER or DOMINATING SET, can be solved in single-exponential time on general graphs. Intuitively, for this problems it is enough to enumerate *subsets* of the bags of a tree-decomposition (rather than partitions or packings), which are $2^{O(\mathbf{tw})}$ many. A natural class of problems that do *not* have locally checkable certificates is the class of so-called *connectivity problems*, which contains for example HAMILTONIAN CYCLE, STEINER TREE, or CONNECTED VERTEX COVER. These problems have the property that the solutions should satisfy a *connectivity* requirement (see [2,4,19] for more details), and using classical dynamic programming techniques it seems that for solving such a problem it is necessary to enumerate partitions or packings of the bags of a tree-decomposition.

A series of articles provided single-exponential algorithms for connectivity problems when the input graphs are restricted to be sparse, namely planar [9], of bounded genus [7,19], or excluding a fixed graph as a minor [8,20]. The common key idea of these works is to use special types of branch-decompositions (which are objects similar to tree-decompositions) with nice combinatorial properties, which strongly rely on the fact that the input graphs are sparse.

Until very recently, it was a common belief that all problems solvable in single-exponential time on general graphs should have locally checkable certificates, specially after Lokshtanov *et al.* [16] proved that one connectivity problem, namely DISJOINT PATHS, cannot be solved in time $2^{o(\mathbf{tw} \log \mathbf{tw})} \cdot n^{O(1)}$ on general graphs unless the Exponential Time Hypothesis (ETH) fails[2]. This

---

[1] That is, certificates consisting of a constant number of bits per vertex that can be checked by a cardinality check and by iteratively looking at the neighborhoods of the input graph.

[2] The ETH states that there exists a positive real number s such that 3-CNF-SAT with n variables and m clauses cannot be solved in time $2^{sn} \cdot (n+m)^{O(1)}$. See [15] for more details.

credence was disproved by Cygan *et al.* [4], who provided single-exponential *randomized* algorithms on general graphs for several connectivity problems, like LONGEST PATH, FEEDBACK VERTEX SET, or CONNECTED VERTEX COVER. More recently, Bodlaender *et al.* [2] presented single-exponential *deterministic* algorithms for basically the same connectivity problems, and an alternative proof based on matroids was given by Fomin *et al.* [11]. These results have been considered a breakthrough, and in particular they imply that most connectivity problems that were known to be solvable in single-exponential time on sparse graph classes [7–9,19,20] are also solvable in single-exponential time on general graphs [2,4].

Our main results. In view of the above discussion, a natural conclusion is that sparsity may not be particularly helpful or relevant for obtaining single-exponential algorithms. However, in this article we convey that sparsity (in particular, planarity) *does* play a role in connectivity problems parameterized by treewidth. To this end, among the problems that can be solved in time $2^{O(\mathbf{tw} \log \mathbf{tw})} \cdot n^{O(1)}$ on general graphs, we distinguish the following three disjoint types:

- **Type 1:** Problems that can be solved in time $2^{O(\mathbf{tw})} \cdot n^{O(1)}$ on general graphs.
- **Type 2:** Problems that *cannot* be solved in time $2^{o(\mathbf{tw} \log \mathbf{tw})} \cdot n^{O(1)}$ on general graphs unless the ETH fails, but that can be solved in time $2^{O(\mathbf{tw})} \cdot n^{O(1)}$ when restricted to planar graphs.
- **Type 3:** Problems that *cannot* be solved in time $2^{o(\mathbf{tw} \log \mathbf{tw})} \cdot n^{O(1)}$ even when restricted to planar graphs, unless the ETH fails.

Problems that have locally checkable certificates are of Type 1 (see also [17] for a logical characterization of such problems). As discussed in Sect. 2, known results [4,13] imply that there exist problems of Type 2, such as CYCLE PACKING. Our main contribution is to show that there exist (natural) problems of Type 3, thus demonstrating that some connectivity problems can indeed be distinguished according to their behavior on planar graphs. More precisely, we prove the following results:

- In Sect. 2 we provide some examples of problems of Type 2. Furthermore, we prove that PLANAR CYCLE PACKING cannot be solved in time $2^{o(\mathbf{tw})} \cdot n^{O(1)}$ unless the ETH fails, and therefore the running time $2^{O(\mathbf{tw})} \cdot n^{O(1)}$ is tight.
- In Sect. 3 we provide an example of problem of Type 3: MONOCHROMATIC DISJOINT PATHS, which is a variant of the DISJOINT PATHS problem on a vertex-colored graph with additional restrictions on the allowed colors for each path. To the best of our knowledge, problems of this type had not been identified before.

In order to obtain our results, for the upper bounds we strongly follow the algorithmic techniques based on *Catalan structures* used in [7–9,19,20], and for some of the lower bounds we use the framework introduced in [16], and that has been also used in [4]. Due to space limitations, the proofs of the results marked with '[⋆]' can be found in the full version of this article [1].

Additional results and further research. We feel that our results about the role of planarity in connectivity problems parameterized by treewidth are just a first step in a subject that can be much exploited, and we think that the following avenues are particularly interesting:

- It is known that DISJOINT PATHS can be solved in time $2^{O(\mathbf{tw} \log \mathbf{tw})} \cdot n^{O(1)}$ on general graphs [21], and that this bound is asymptotically tight under the ETH [16]. The fact whether DISJOINT PATHS belongs to Type 2 or Type 3 (or maybe even to some other type in between) remains an important open problem that we have been unable to solve (it is worth noting that Catalan structures do *not* seem to yield an algorithm in time $2^{O(\mathbf{tw})} \cdot n^{O(1)}$). Towards a possible answer to this question, we prove in [1] that PLANAR DISJOINT PATHS cannot be solved in time $2^{o(\mathbf{tw})} \cdot n^{O(1)}$ unless the ETH fails.
- Lokshtanov *et al.* [14] have proved that for a number of problems such as DOMINATING SET or q-COLORING, the best known constant c in algorithms of the form $c^{\mathbf{tw}} \cdot n^{O(1)}$ on general graphs is best possible unless the Strong ETH fails (similar results have been also given by Cygan *et al.* [4]). Is it possible to provide better constants for these problems on planar graphs? The existence of such algorithms would permit to further refine the problems belonging to Type 1.
- Are there NP-hard problems solvable in time $2^{o(\mathbf{tw})} \cdot n^{O(1)}$?
- Finally, it would be interesting to obtain similar results for problems parameterized by pathwidth, and to extend our algorithms to more general classes of sparse graphs.

Notation and definitions. We use standard graph-theoretic notation, and the reader is referred to [6] for any undefined term. All the graphs we consider are undirected and contain neither loops nor multiple edges. We denote by $V(G)$ the set of vertices of a graph G and by $E(G)$ its set of edges. We use the notation $[k]$ for the set of integers $\{1, \ldots, k\}$. In the set $[k] \times [k]$, a *row* is a set $\{i\} \times [k]$ and a *column* is a set $[k] \times \{i\}$ for some $i \in [k]$. If $\mathbf{P}$ is a problem defined on graphs, we denote by PLANAR $\mathbf{P}$ the restriction of $\mathbf{P}$ to planar input graphs.

A *subgraph* $H = (V_H, E_H)$ of a graph $G = (V, E)$ is a graph such that $V_H \subseteq V$ and $E_H \subseteq E \cap \binom{V_H}{2}$. The *degree* of a vertex v in a graph G, denoted by $\deg_G(v)$, is the number of edges of G containing v. The *grid* $m * k$ is the graph $Gr_{m,k} = (\{a_{i,j} | i \in [m], j \in [k]\}, \{(a_{i,j}, a_{i+1,j}) | i \in [m-1], j \in [k]\} \cup \{(a_{i,j}, a_{i,j+1}) | i \in [m], j \in [k-1]\})$. When $m = k$ we just speak about the *grid of size* k. We say that there is a *path* $s \ldots t$ in a graph G if there exist $m \in \mathbb{N}$ and $x_0, \ldots, x_m$ in $V(G)$ such that $x_0 = s$, $x_m = t$, and for all $i \in [m]$, $(x_{i-1}, x_i) \in E(G)$.

A *tree-decomposition* of width w of a graph $G = (V, E)$ is a pair (T, σ), where T is a tree and $\sigma = \{B_t | B_t \subseteq V, t \in V(T)\}$ such that:

- $\bigcup_{t \in V(T)} B_t = V$;
- For every edge $\{u, v\} \in E$ there is a $t \in V(T)$ such that $\{u, v\} \subseteq B_t$;

- $B_i \cap B_k \subseteq B_j$ for all $\{i, j, k\} \subseteq V(T)$ such that j lies on the path between i and k in T;
- $\max_{t \in V(T)} |B_t| = w + 1$.

The sets B_t are called *bags*. The *treewidth* of G, denoted by $\mathbf{tw}(G)$, is the smallest integer w such that there is a tree-decomposition of G of width w. An *optimal tree-decomposition* is a tree-decomposition of width $\mathbf{tw}(G)$. A *path-decomposition* of a graph $G = (V, E)$ is a tree-decomposition (T, σ) such that T is a path. The *pathwidth* of G, denoted by $\mathbf{pw}(G)$, is the smallest integer w such that there is a path-decomposition of G of width w. Clearly, for any graph G, we have $\mathbf{tw}(G) \leq \mathbf{pw}(G)$.

Throughout the paper, in order to simplify the notation, when the problem and the input graph under consideration are clear, we let n denote the number of vertices of the input graph, $\mathbf{tw}$ its treewidth, and $\mathbf{pw}$ its pathwidth.

2   Problems of Type 2

In this section we deal with CYCLE PACKING. Other problems of Type 2 are discussed in [1].

CYCLE PACKING
Input: An n-vertex graph $G = (V, E)$ and an integer ℓ_0.
Parameter: The treewidth $\mathbf{tw}$ of G.
Question: Does G contain ℓ_0 pairwise vertex-disjoint cycles?

It is proved in [4] that CYCLE PACKING *cannot* be solved in time $2^{o(\mathbf{tw} \log \mathbf{tw})} \cdot n^{O(1)}$ on general graphs unless the ETH fails. On the other hand, a dynamic programming algorithm for PLANAR CYCLE PACKING running in time $2^{O(\mathbf{tw})} \cdot n^{O(1)}$ can be found in [13]. Therefore, it follows that CYCLE PACKING is of Type 2. In Lemma 1 below we provide an alternative algorithm for PLANAR CYCLE PACKING running in time $2^{O(\mathbf{tw})} \cdot n^{O(1)}$, which is a direct application of the techniques based on *Catalan structures* introduced in [9]. We include its proof for completeness, as it yields slightly better constants than [13].

Lemma 1. [⋆] PLANAR CYCLE PACKING *can be solved in time* $2^{O(\mathbf{tw})} \cdot n^{O(1)}$.

It is usually believed that NP-hard problems parameterized by $\mathbf{tw}$ cannot be solved in time $2^{o(\mathbf{tw})} \cdot n^{O(1)}$ under some reasonable complexity assumption. This has been proved in [12] for problems on *general* graphs such as q-COLORABILITY, INDEPENDENT SET, and VERTEX COVER, or in [5] for PLANAR HAMILTONIAN CYCLE, all these results assuming the ETH. Nevertheless, such a result requires an ad-hoc proof for each problem. For instance, to the best of our knowledge, such lower bounds are not known for CYCLE PACKING when the input graph is restricted to be *planar*. We now prove that the running time given by Lemma 1 is asymptotically tight.

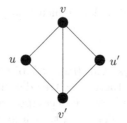

Fig. 1. The choice gadget.

Theorem 1. PLANAR CYCLE PACKING *cannot be solved in time* $2^{o(\sqrt{n})} \cdot n^{O(1)}$ *unless the ETH fails. Therefore,* PLANAR CYCLE PACKING *cannot be solved in time* $2^{o(\mathbf{tw})} \cdot n^{O(1)}$ *unless the ETH fails.*

Proof. To prove this theorem, we reduce from the PLANAR INDEPENDENT SET problem, which consists in finding a maximum-sized set of vertices in a given planar graph that are pairwise nonadjacent. It is shown in [15] that PLANAR INDEPENDENT SET cannot be solved in time $2^{o(\sqrt{n})} \cdot n^{O(1)}$ unless the ETH fails[3], where n stands for the number of vertices of the input graph.

Let G be a planar graph on which we want to solve PLANAR INDEPENDENT SET. We construct from G a graph H as follows. For each vertex a of G, we add to H a cycle of length equal to the degree of a. If the degree of a is smaller than 3, then the added cycle is of size 3. For each edge (a, b) of G, we add the choice gadget depicted in Fig. 1, by identifying vertex u of the choice gadget with one of the vertices of the cycle representing a, and identifying vertex u' of the choice gadget with one of the vertices of the cycle representing b, in such a way that all the choice gadgets are vertex-disjoint. That concludes the construction of H. Note that a planar embedding of H can be obtained in a straightforward way from a planar embedding of G.

Claim 1. G *contains an independent set of size* k *if and only if* H *admits a packing of* $m + k$ *cycles, where* $m = |E(G)|$.

Proof. Assume first that there is an independent set S of size k in G. Then we choose in H the k cycles that represent the k elements of S, plus a cycle of size 3 in each of the m choice gadgets. Note that we always can take a cycle of size 3 in the choice gadgets, as two adjacent vertices in G cannot be in S. This gives $m + k$ cycles in H.

Conversely, assume that H contains a packing $\mathcal{P}$ of $m + k$ vertex-disjoint cycles. We say that a cycle in $\mathcal{P}$ *hits* a choice gadget if it contains vertex v

---

[3] In [15] that PLANAR VERTEX COVER problem is mentioned, which is equivalent to solving PLANAR INDEPENDENT SET, as the complement of a vertex cover is an independent set.

or v', and note that each choice gadget can be hit by at most one cycle, which is necessarily a triangle. We transform $\mathcal{P}$ into a packing $\mathcal{P}'$ with $|\mathcal{P}'| \geq |\mathcal{P}|$ and such that every choice gadget is hit by $\mathcal{P}'$, as follows. Start with $\mathcal{P}' = \mathcal{P}$. Then, for each choice gadget in H, corresponding to an edge (a, b) of G, that is not hit by a cycle in $\mathcal{P}$, we arbitrarily choose one of a or b (say, a), remove from $\mathcal{P}'$ the cycle containing a (if any), and add to $\mathcal{P}'$ the triangle in the choice gadget containing vertex u. As there are only m choice gadgets in H, and as $|\mathcal{P}'| \geq |\mathcal{P}| \geq m + k$, it follows that at least k cycles in $\mathcal{P}'$ do not hit any choice gadget. By construction of H, it clearly means that each of these k cycles is a cycle corresponding to a vertex of G. Let R be the set of these vertices, so we have that $|R| \geq k$. Finally, note that R is an independent set in G, because for each edge $(a, b) \in E(G)$, the corresponding choice gadget is contained in $\mathcal{P}'$, so at most one of a and b can be in R. $\qquad\square$

Thus, if we have an algorithm solving PLANAR CYCLE PACKING in time $2^{o(\sqrt{n})} \cdot n^{O(1)}$, then by Claim 1 (and using that $m = O(n)$ because G is planar) we could solve PLANAR INDEPENDENT SET in time $2^{o(\sqrt{n})} \cdot n^{O(1)}$, which is impossible unless the ETH fails [15]. $\qquad\square$

3   Problems of Type 3

In this section we prove that the MONOCHROMATIC DISJOINT PATHS problem is of Type 3. We first need to introduce some definitions. Let $G = (V, E)$ be a graph, let k be an integer, and let $c : V \rightarrow \{0, \dots, k\}$ be a color function. Two colors c_1 and c_2 in $\{0, \dots, k\}$ are *compatible*, and we denote it by $c_1 \equiv c_2$, if $c_1 = 0$, $c_2 = 0$, or $c_1 = c_2$. A path $P = x_1 \dots x_m$ in G is *monochromatic* if for all $i, j \in [m]$, $i \neq j$, $c(x_i)$ and $c(x_j)$ are two compatible colors. We abuse notation and also define $c(P) = \max_{i \in [m]}(c(x_i))$. We say that P is *colored* x if $x = c(P)$. Two monochromatic paths P and P' are *color-compatible* if $c(P) \equiv c(P')$.

MONOCHROMATIC DISJOINT PATHS
Input: A graph $G = (V, E)$ of treewidth **tw**, a color function $\gamma : V \rightarrow \{0, \dots, \mathbf{tw}\}$, an integer m, and a set $\mathcal{N} = \{\mathcal{N}_i = \{s_i, t_i\} | i \in [m], s_i, t_i \in V\}$.
Parameter: The treewidth **tw** of G.
Question: Does G contain m pairwise vertex-disjoint monochromatic paths from s_i to t_i, for $i \in [m]$?

The proof of the following lemma is inspired from the algorithm given in [21] for the DISJOINT PATHS problem on general graphs.

Lemma 2. [$\star$] MONOCHROMATIC DISJOINT PATHS *can be solved in time* $2^{O(\mathbf{tw} \log \mathbf{tw})} \cdot n^{O(1)}$.

We now proceed to provide a matching lower bound for PLANAR MONOCHROMATIC DISJOINT PATHS. For this, we need to define the $k \times k$-HITTING SET problem, first introduced in [16].

$k \times k$-HITTING SET
Input: A family of sets $S_1, S_2, \ldots, S_m \subseteq [k] \times [k]$, such that each set contains at most one element from each row of $[k] \times [k]$.
Parameter: k.
Question: Is there a set S containing exactly one element from each row such that $S \cap S_i \neq \emptyset$ for any $1 \leq i \leq m$?

Theorem 2. (Lokshtanov) et al. [16]. $k \times k$-HITTING SET *cannot be solved in time* $2^{o(k \log k)} \cdot m^{O(1)}$ *unless the ETH fails.*

We state the following theorem in terms of the pathwidth of the input graph, and as any graph G satisfies $\mathbf{tw}(G) \leq \mathbf{pw}(G)$, it implies the same lower bound in the treewidth.

Theorem 3. PLANAR MONOCHROMATIC DISJOINT PATHS *cannot be solved in time* $2^{o(\mathbf{pw} \log \mathbf{pw})} \cdot n^{O(1)}$ *unless the ETH fails.*

Proof. We reduce from $k \times k$-HITTING SET. Let k be an integer and S_1, $S_2, \ldots, S_m \subseteq [k] \times [k]$ such that each set contains at most one element from each row of $[k] \times [k]$. We will first present an overview of the reduction with all the involved gadgets, and then we will provide a formal definition of the constructed planar graph G.

We construct a gadget for each row $\{r\} \times [k]$, $r \in [k]$, which selects the unique pair p of S in this row. First, for each $r \in [k]$, we introduce two new vertices s_r and t_r, a request $\{s_r, t_r\}$, $m + 1$ vertices $v_{r,i}$, $i \in \{0, \ldots, m\}$, and $m + 2$ edges $\{e_{r,0} = (s_r, v_{r,0})\} \cup \{e_{r,i} = (v_{r,i-1}, v_{r,i}) | i \in [m]\} \cup \{e_{r,m+1} = (v_{r,m}, t_r)\}$. That is, we have a path with $m + 2$ edges between s_r and t_r.

Each edge of these paths, except the last one, will be replaced with an appropriate gadget. Namely, for each $r \in [k]$, we replace the edge $e_{r,0}$ with the gadget depicted in Fig. 2, which we call *color-selection* gadget. In this figure, vertex $u_{r,i}$ is colored i. The color used by the path from s_r to t_r in the color-selection gadget will define the pair of the solution of S in the row $\{r\} \times [k]$.

Now that we have described the gadgets that allow to define S, we need to ensure that $S \cap S_i \neq \emptyset$ for any $i \in [m]$. For this, we need the gadget depicted in Fig. 3, which we call *expel* gadget. Each time we introduce this gadget, we add to $\mathcal{N}$ the request $\{s, t\}$. This new requested path uses either vertex u or vertex v, so only one of these vertices can be used by other paths. For each $i \in [m]$, we replace all the edges $\{e_{r,i} | r \in [k]\}$ with the gadget depicted in Fig. 4, which we call *set* gadget. In this figure, $a_{r,i}$ is such that if $(\{r\} \times [k]) \cap S_i = \{\{r, c_{r,i}\}\}$ then $a_{r,i}$ is colored $c_{r,i}$, and if $(\{r\} \times [k]) \cap S_i = \emptyset$ then vertex $a_{r,i}$ is removed from the gadget.

This completes the construction of the graph G, which is illustrated in Fig. 5. Note that G is indeed planar.

The color function γ of G is defined such that for each $r \in [k]$ and $c \in [k]$, $\gamma(u_{r,c}) = c$, and for each $i \in [m]$ and $(r, c) \in S_i$, $\gamma(a_{r,i}) = c$. For any other vertex $v \in V(G)$, we set $\gamma(v) = 0$. Finally, the input of PLANAR MONOCHROMATIC

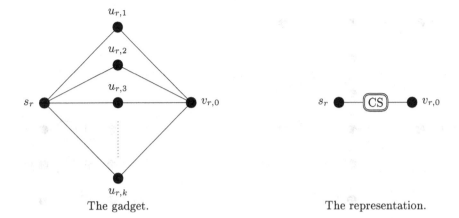

The gadget.   The representation.

Fig. 2. Color-selection gadget, where $u_{r,i}$ is colored c_i for each $i \in [k]$.

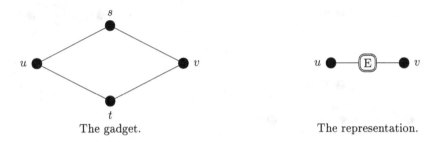

The gadget.   The representation.

Fig. 3. Expel gadget.

DISJOINT PATHS is the planar graph G, the color function γ, and the $k+(k-1)\cdot m$ requests $\mathcal{N} = \{\{s_r, t_r\}|r \in [k]\} \cup \{\{s_{r,i}, t_{r,i}\}|r \in [k-1], i \in [m]\}$, the second set of requests corresponding to the ones introduced by the expel gadgets.

Note that because of the expel gadgets, the request $\{s_r, t_r\}$ forces the existence of a path between $v_{r,i-1}$ and $v_{r,i}$ for each $r \in [k]$. Note also that because of the expel gadgets, at least one of the paths between $v_{r,i-1}$ and $v_{r,i}$ should use an $a_{r,i}$ vertex, as otherwise at least two paths would intersect. Conversely, if one path uses a vertex $a_{r,i}$, then we can find all the desired paths in the corresponding set gadgets by using the vertices $w_{r,i,b}$.

Given a solution of PLANAR MONOCHROMATIC DISJOINT PATHS in G, we can construct a solution of $k \times k$-HITTING SET by letting $S = \{(r, c)|r \in [k]$ such that the path from s_r to t_r is colored with color $c\}$. We have that S contains exactly one element of each row, so we just have to check if $S \cap S_i \neq \emptyset$ for each $i \in [m]$. Because of the property of the set gadgets mentioned above, for each $i \in [m]$, the set gadget labeled i ensures that $S \cap S_i \neq \emptyset$.

Conversely, given a solution S of $k \times k$-HITTING SET, for each $(r, c) \in S$ we color the path from s_r to t_r with color c. We assign an arbitrary coloring

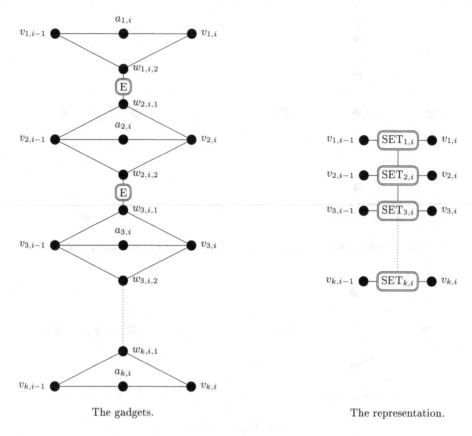

The gadgets.   The representation.

Fig. 4. Set gadgets.

to the other paths. For each $i \in [m]$, we take $(r, c) \in S \cap S_i$ and in the set gadget labeled i, we impose that the path from $v_{r,i-1}$ to $v_{r,i}$ uses vertex $a_{r,i}$. By using the vertices $w_{r,i,b}$ for the other paths, we find the desired $k + (k-1) \cdot m$ monochromatic paths.

Let us now argue about the pathwidth of G. We define for each $r, c \in [k]$ the bag $B_{0,r,c} = \{s_{r'}|r' \in [k]\} \cup \{v_{r',0}|r' \in [k]\} \cup \{u_{r,c}\}$, for each $i \in [m]$, the bag $B_i = \{v_{r,i-1}|r \in [k]\} \cup \{v_{r,i}|r \in [k]\} \cup \{a_{r,i} \in V(G)|r \in [k]\} \cup \{w_{r,i,b} \in V(G)|r \in [k], b \in [2]\} \cup \{s_{r,i}|r \in [k-1]\} \cup \{t_{r,i}|r \in [k-1]\}$, and the bag $B_{m+1} = \{v_{r,m}|r \in [k]\} \cup \{t_r|r \in [k]\}$. Note that for all i in $[m]$, $w_{1,i,1}$ and $w_{k,i,2}$ are not in $V(G)$. The size of each bag is at most $2 \cdot (k-1) + 5 \cdot k - 2 = O(k)$. A path decomposition of G consists of all bags $B_{0,r,c}$, $r, c \in [k]$ and B_i, $i \in [m+1]$ and edges $\{B_i, B_{i+1}\}$ for each $i \in [m]$, $\{B_{0,r,c}, B_{0,r,c+1}\}$ for $r \in [k]$, $c \in [k-1]$, $\{B_{0,r,k}, B_{0,r+1,1}\}$ for $r \in [k]$, and $\{B_{0,k,k}, B_1\}$. Therefore, as we have that $\mathbf{pw}(G) = O(k)$, if one could solve PLANAR MONOCHROMATIC DISJOINT PATHS in time $2^{o(\mathbf{pw} \log \mathbf{pw})} \cdot n^{O(1)}$, then one could also solve $k \times k$-HITTING SET in time $2^{o(k \log k)} \cdot m^{O(1)}$, which is impossible by Theorem 2 unless the ETH fails. $\square$

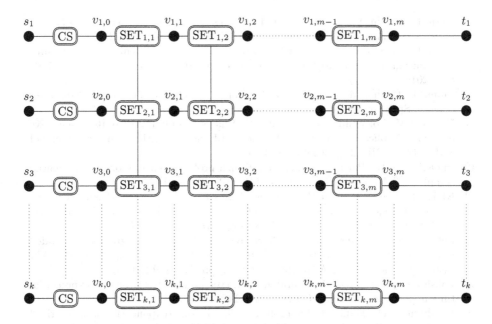

Fig. 5. Final graph G in the reduction of Theorem 3.

Acknowledgement. We would like to thank the anonymous referees for helpful remarks that improved the presentation of the manuscript.

References

1. Baste, J., Sau, I.: The role of planarity in connectivity problems parameterized by treewidth. CoRR, abs/1312.2889 (2013)
2. Bodlaender, H.L., Cygan, M., Kratsch, S., Nederlof, J.: Deterministic single exponential time algorithms for connectivity problems parameterized by treewidth. In: Fomin, F.V., Freivalds, R., Kwiatkowska, M., Peleg, D. (eds.) ICALP 2013, Part I. LNCS, vol. 7965, pp. 196–207. Springer, Heidelberg (2013)
3. Courcelle, B.: The monadic second-order logic of graphs. I. Recognizable sets of finite graphs. Inform. Comput. **85**(1), 12–75 (1990)
4. Cygan, M., Nederlof, J., Pilipczuk, M., Pilipczuk, M., van Rooij, J.M.M., Wojtaszczyk, J.O.: Solving connectivity problems parameterized by treewidth in single exponential time. In: Proceeding of the 52nd Annual IEEE Symposium on Foundations of Computer Science (FOCS), pp. 150–159. IEEE Computer Society (2011)
5. Deineko, V.G., Steiner, G., Xue, Z.: Robotic-cell scheduling: special polynomially solvable cases of the traveling salesman problem on permuted monge matrices. J. Comb. Optim. **9**, 381–399 (2005)
6. Diestel, R.: Graph Theory, 3rd edn. Springer, New York (2005)
7. Dorn, F., Fomin, F.V., Thilikos, D.M.: Fast subexponential algorithm for non-local problems on graphs of bounded genus. In: Arge, L., Freivalds, R. (eds.) SWAT 2006. LNCS, vol. 4059, pp. 172–183. Springer, Heidelberg (2006)

8. Dorn, F., Fomin, F.V., Thilikos, D.M.: Catalan structures and dynamic programming in H-minor-free graphs. J. Syst. Sci. **78**(5), 1606–1622 (2012)
9. Dorn, F., Penninkx, E., Bodlaender, H.L., Fomin, F.V.: Efficient exact algorithms on planar graphs: exploiting sphere cut decompositions. Algorithmica **58**(3), 790–810 (2010)
10. Flum, J., Grohe, M.: Parameterized Complexity Theory. Theoretical Computer Science. Springer, Berlin (2006)
11. Fomin, F.V., Lokshtanov, D., Saurabh, S.: Efficient Computation of Representative Sets with Applications in Parameterized and Exact Algorithms. In: Proceeding of SODA'14. CoRR, abs/1304.4626 (2013)
12. Impagliazzo, R., Paturi, R., Zane, F.: Which problems have strongly exponential complexity? J. Comput. Syst. Sci. **63**(4), 512–530 (2001)
13. Kloks, T., Lee, C.M., Liu, J.: New algorithms for k-face cover, k-feedback vertex set, and k-disjoint cycles on plane and planar graphs. In: Kučera, L. (ed.) WG 2002. LNCS, vol. 2573, pp. 282–295. Springer, Heidelberg (2002)
14. Lokshtanov, D., Marx, D., Saurabh, S.: Known algorithms on graphs of bounded treewidth are probably optimal. In: Proceeding of the 22nd Annual ACM-SIAM Symposium on Discrete Algorithms (SODA), pp. 777–789 (2011)
15. Lokshtanov, D., Marx, D., Saurabh, S.: Lower bounds based on the exponential time hypothesis. Bull. EATCS **105**, 41–72 (2011)
16. Lokshtanov, D., Marx, D., Saurabh, S.: Slightly superexponential parameterized problems. In: Proceeding of the 22nd Annual ACM-SIAM Symposium on Discrete algorithms (SODA), pp. 760–776 (2011)
17. Pilipczuk, M.: Problems parameterized by treewidth tractable in single exponential time: a logical approach. In: Murlak, F., Sankowski, P. (eds.) MFCS 2011. LNCS, vol. 6907, pp. 520–531. Springer, Heidelberg (2011)
18. Robertson, N., Seymour, P.D.: Graph minors. II. Algorithmic aspects of tree-width. J. Algorithms **7**(3), 309–322 (1986)
19. Rué, J., Sau, I., Thilikos, D.M.: Dynamic programming for graphs on surfaces. In: Short Version in the Proceeding of ICALP'10 in ACM Transactions on Algorithms (TALG). CoRR, abs/1104.2486 (2011)
20. Bhattacharya, B., Kameda, T.: A linear time algorithm for computing minmax regret 1-median on a tree. In: Gudmundsson, J., Mestre, J., Viglas, T. (eds.) COCOON 2012. LNCS, vol. 7434, pp. 1–12. Springer, Heidelberg (2012)
21. Scheffler, P.: A practical linear time algorithm for disjoint paths in graphs with bounded tree-width. Fachbereich 3 Mathematik, Technical Report 396/1994, FU Berlin (1994)

Parameterized Inapproximability of Degree Anonymization

Cristina Bazgan[1,3] and André Nichterlein[2(✉)]

[1] PSL, Université Paris-Dauphine, LAMSADE UMR CNRS, 7243 Paris, France
`bazgan@lamsade.dauphine.fr`
[2] Institut für Softwaretechnik und Theoretische Informatik,
TU Berlin, Berlin, Germany
`andre.nichterlein@tu-berlin.de`
[3] Institut Universitaire de France, Paris, France

Abstract. The DEGREE ANONYMITY problem arises in the context of combinatorial graph anonymization. It asks, given a graph G and two integers k and s, whether G can be made k-*anonymous* with at most s modifications. Here, a graph is k-anonymous if the graph contains for every vertex at least $k-1$ other vertices of the same degree. Complementing recent investigations on its computational complexity, we show that this problem is very hard when studied from the viewpoints of approximation as well as parameterized approximation. In particular, for the optimization variant where one wants to minimize the number of either edge or vertex deletions there is no factor-$n^{1-\varepsilon}$ approximation running in polynomial time unless P = NP, for any constant $0 < \varepsilon \le 1$. For the variant where one wants to maximize k and the number s of either edge or vertex deletions is given, there is no factor-$n^{1/2-\varepsilon}$ approximation running in time $f(s) \cdot n^{O(1)}$ unless W[1] = FPT, for any constant $0 < \varepsilon \le 1/2$ and any function f. On the positive side, we classify the general decision version as fixed-parameter tractable with respect to the combined parameter solution size s and maximum degree.

1 Introduction

Releasing social network data without violating the privacy of the users has become an important and active field of research [14]. One model aiming for this goal was introduced by Liu and Terzi [12] who transferred the k-anonymity concept from tabular data in databases [9] to graphs. Herein, Liu and Terzi [12] require that a released graph contains for every vertex at least $k-1$ other vertices with the same degree. The parameter k controls how many individuals are at least linked to one particular degree and thus higher values for k give higher levels of anonymity. We remark that this model has also some weaknesses. Refer to Wu et al. [14] for more details and further anonymization models.

Here, we study the following variant of the model of Liu and Terzi [12].

DEGREE ANONYMITY(ANONYM)

Input: An undirected graph $G = (V, E)$ and two positive integers k and s.

© Springer International Publishing Switzerland 2014
M. Cygan and P. Heggernes (Eds.): IPEC 2014, LNCS 8894, pp. 75–84, 2014.
DOI: 10.1007/978-3-319-13524-3_7

Question: Can G be transformed with at most s modifications into a k-anonymous graph $G' = (V', E')$, that is, for each vertex in G' there are $k - 1$ other vertices of the same degree?

We will use the name scheme ANONYM-{E/V}-{Ins/Del/Edt} to distinguish the different graph modification operations edge/vertex insertion/deletion/editing. Liu and Terzi [12] studied edge insertions (ANONYM-E-INS), but also vertex deletions (ANONYM-V-DEL) [3] and vertex insertions (ANONYM-V-INS) [2,5] have been considered. While the focus of previous work was on experimentally evaluated heuristics and algorithms [10,12] or computational complexity and fixed-parameter algorithms [2,3,11], we study the polynomial-time and parameterized approximability of these problems. To this end, we mostly concentrate on natural optimization variants of the two problems where either edge deletions (ANONYM-E-DEL) or vertex deletions (ANONYM-V-DEL) are allowed. Partially answering an open question of Chester et al. [4], we show strong inapproximable results, even when allowing the running time to be exponential in s. We remark that our results do not transfer to the problem variants allowing to edit up to s edges (ANONYM-E-EDT) and the status of the (parameterized) approximability of the corresponding optimization problems remains unsolved.

Related Work. The basic degree anonymization model was introduced by Liu and Terzi [12] (also see Clarkson et al. [6] for an extended version); they also gave an experimentally evaluated heuristic for ANONYM-E-INS. One of the first theoretical works on this model was done by Chester et al. [4]. They provided polynomial-time algorithms for bipartite graphs and showed NP-hardness of generalizations of ANONYM-E-INS with edge labels. In particular, they asked for effective approximation algorithms for ANONYM-E-INS and generalizations. Hartung et al. [11] proved that ANONYM-E-INS is NP-hard and W[1]-hard with respect to (w.r.t.) the solution size s, even if $k = 2$. On the positive side, using the heuristic of Liu and Terzi [12], they showed fixed-parameter tractability of ANONYM-E-INS w.r.t. the maximum degree in the input graph.

Chester et al. [5] considered a variant of ANONYM-V-INS and gave an approximation algorithm with an additive error of at most k. Bredereck et al. [2] investigated the parameterized complexity of several variants of ANONYM-V-INS which differ in the rules how the inserted vertices can be made adjacent to existing vertices. The ANONYM-V-DEL variant studied by Bredereck et al. [3] turned out to be NP-hard even on very restricted graph classes such as trees, split graphs, or trivially perfect graphs.

Our Results. We investigate the approximability of natural optimization variants of ANONYM-V-DEL and ANONYM-E-DEL: Either the budget s is given and one wants to maximize the level k of anonymity, or k is given and the goal is to minimize the number of modifications s. The optimization problems maximizing k with a given budget s are denoted by MAX ANONYM-V-DEL and MAX ANONYM-E-DEL. The variants minimizing s with given k are denoted by MIN ANONYM-E-DEL and MIN ANONYM-V-DEL.

We show that one cannot approximate MAX ANONYM-E-DEL(MAX ANONYM-V-DEL) within a factor of $n^{1-\varepsilon}$ $(n^{1/2-\varepsilon})$ in $f(s)n^{O(1)}$ time unless FPT = W[1], for any function f and any $0 < \varepsilon \le 1$ $(0 < \varepsilon \le 1/2)$. As the parameter k has size $\Theta(n)$ in all employed gap-reductions, we only manage to exclude *polynomial-time* approximations for the minimization versions. More precisely, both MIN ANONYM-E-DEL and MIN ANONYM-V-DEL cannot be approximated in polynomial time within a factor of $n^{1-\varepsilon}$ unless P = NP.

Complementing the NP-hardness of ANONYM-V-DEL with $k = 2$ on trees [3], we show that ANONYM-E-DEL remains NP-hard on caterpillars (a tree having a dominating path), even if $k = 2$. Extending the fixed-parameter tractability of ANONYM-V-DEL w.r.t. the combined parameter budget and maximum degree (s, Δ), we classify ANONYM (allowing edge and vertex insertion as well as deletion) as fixed-parameter tractable w.r.t. (s, Δ).

Due to the space constraints, some proofs are deferred to a full version.

2   Preliminaries

Graph terminology. We use standard graph-theoretic notation. All graphs studied in this paper are undirected and simple, that is, there are no self-loops and no multi-edges. For a given graph $G = (V, E)$ with vertex set V and edge set E we set $n := |V|$ and $m := |E|$. Furthermore, by $\deg_G(v)$ we denote the degree of a vertex $v \in V$ in G, and Δ_G denotes the maximum degree of any vertex of G. For $0 \le d \le \Delta_G$, let $D_G(d) := \{v \in V \mid \deg_G(v) = d\}$ be the *block* of degree d, that is, the set of all vertices with degree d in G. Thus, being k-anonymous is equivalent to each block being of size either zero or at least k.

The subgraph of G induced by a vertex subset $V' \subseteq V$ is denoted by $G[V']$. For an edge subset $E' \subseteq E$, $V(E')$ denotes the set of all endpoints of edges in E' and $G[E'] := (V(E'), E')$. Furthermore, for a vertex subset $V' \subseteq V$ we set $G - V' := G[V \setminus V']$ and for an edge set $E' \subseteq \binom{V}{2}$ we set $G - E' := (V, E \setminus E')$ and $G + E' = (V, E \cup E')$. A graph G is k-anonymous if for every vertex $v \in V$ there are at least $k - 1$ other vertices in G having the same degree.

A vertex subset $V' \subseteq V$ (an edge subset $E' \subseteq E$) is called k-deletion set if $G - V'$ ($G - E'$, respectively) is k-anonymous. Analogously, for a set E'' of edges with endpoints in a graph G such that $V + E''$ is k-anonymous, we call E'' an for G. We omit subscripts if the graph is clear from the context.

Approximation. Let Σ be a finite alphabet. Given an optimization problem $Q \subseteq \Sigma^*$ and an instance I of Q, we denote by $|I|$ the size of I, by $opt(I)$ the optimum value of I and by $val(I, S)$ the value of a feasible solution S of I. The *performance ratio* of S (or *approximation factor*) is $r(I, S) = \max\left\{\frac{val(I,S)}{opt(I)}, \frac{opt(I)}{val(I,S)}\right\}$. For a function ρ, an algorithm is a $\rho(n)$-*approximation*, if for every instance I of Q, it returns a solution S such that $r(I, S) \le \rho(|I|)$. An optimization problem is $\rho(n)$-*approximable in polynomial time* if there exists a $\rho(n)$-approximation algorithm running in time $|I|^{O(1)}$ for any instance I. A parameterized optimization problem $Q \subseteq \Sigma^* \times \mathbb{N}$ is $\rho(n)$-*approximable in fpt-time w.r.t. the parameter k* if there exists a $\rho(n)$-approximation algorithm running in time $f(k) \cdot |I|^{O(1)}$ for any instance (I, k) and f is a computable function [13]. It is worth pointing that in this case, k is not related to the optimization value.

In this paper we use a *gap*-reduction between a decision problem and a minimization or maximization problem. A decision problem A is called *gap-reducible* to a maximization problem Q with gap $\rho \geq 1$ if there exists a polynomial-time computable function that maps any instance I of A to an instance I' of Q, while satisfying the following properties: (i) if I is a yes-instance, then $opt(I') \geq c\rho$, and (ii) if I is a no-instance, then $opt(I') < c$, where c and ρ are functions of $|I'|$. If A is NP-hard, then Q is not ρ-approximable in polynomial time, unless P = NP. In this paper we also use a variant of this notion, called fpt gap-reduction.

Definition 1 (fpt gap-reduction). *A parameterized (decision) problem A is called* fpt gap-reducible *to a parameterized maximization problem Q with gap $\rho \geq 1$ if any instance (I, k) of A can be mapped to an instance (I', k') of Q in $f(k) \cdot |I|^{O(1)}$ time while satisfying the following properties: (i) $k' \leq g(k)$ for some function g, (ii) if I is a yes-instance, then $opt(I') \geq c\rho$, and (iii) if I is a no-instance, then $opt(I') < c$, where c and ρ are functions of $|I'|$ and k.*

The interest of the fpt gap-reduction is the following result that immediately follows from the previous definition:

Lemma 1. *If a parameterized problem A is $\mathcal{C}$-hard and fpt gap-reducible to a parameterized optimization problem Q with gap ρ, then Q is not ρ-approximable in fpt-time, unless FPT $= \mathcal{C}$, where $\mathcal{C}$ is any class of the parameterized complexity hierarchy.*

3   Inapproximability of Vertex Deletion Versions

In this section we consider the optimization problems associated to ANONYM-V-DEL, that is MIN ANONYM-V-DEL and MAX ANONYM-V-DEL. We prove that MIN ANONYM-V-DEL is not $n^{1-\varepsilon}$-approximable in polynomial time, while MAX ANONYM-V-DEL is not $n^{1/2-\varepsilon}$-approximable in fpt-time w.r.t. parameter s, even on trees.

Theorem 1. MIN ANONYM-V-DEL *is not $n^{1-\varepsilon}$-approximable for any $0 < \varepsilon \leq 1$, unless P = NP.*

Theorem 2. *For every $0 < \varepsilon \leq 1/2$,* MAX ANONYM-V-DEL *is not $n^{1/2-\varepsilon}$-approximable in fpt-time w.r.t. parameter s, even on trees, unless FPT = W[2].*

Proof. Let $0 < \varepsilon \leq 1/2$ be a constant. We provide an fpt gap-reduction from the W[2]-hard SET COVER problem [7] parameterized by the solution size h. SET COVER is defined as follows: given a universe $U = \{e_1, \ldots, e_m\}$, a collection $\mathcal{C} = \{S_1, \ldots, S_n\}$ of sets over U, and $h \in \mathbb{N}$ the task is to decide whether there is a set cover $\mathcal{C}' \subseteq \mathcal{C}$ of size $|\mathcal{C}'| \leq h$, that is $\bigcup_{S \in \mathcal{C}'} S = U$. Let $I = (U, \mathcal{C}, h)$ be an instance of SET COVER. We assume without loss of generality that for each element $e_i \in U$ there exists a set $S_j \in \mathcal{C}$ with $e_i \in S_j$. To reduce the amount of indices in the construction given below we introduce the function $f \colon U \to \mathbb{N}$ that maps an element $e_i \in U$ to $f(e_i) = (h+4)i$. Let t be an integer greater than or equal to $(mn)^{(1-2\varepsilon)/(2\varepsilon)}$. (We will aim for making the constructed graph t-anonymous.)

The instance I' of MAX ANONYM-V-DEL is defined by $s = h$ and on a graph $G = (V, E)$ constructed as follows: For each element $e_i \in U$ add a star $K_{1,f(e_i)}$ with the center vertex v_i^e. Denote with $V_U = \{v_1^e, \dots, v_m^e\}$ the set of all these center vertices. Furthermore, for each element $e_i \in U$ add t stars $K_{1,f(e_i)+1}$.

For each set $S_j \in \mathcal{C}$ add a tree rooted in a vertex v_j^S. The root has $|S_j|t$ child vertices where each element $e_i \in S_j$ corresponds to exactly t of these children, denoted by $v_1^{e_i,S_j}, \dots, v_t^{e_i,S_j}$. Additionally, for each $\ell \in \{1, \dots, t\}$ we add to $v_\ell^{e_i,S_j}$ exactly $f(e_i)$ degree-one neighbors. Hence, the set gadget is a tree of depth two rooted in v_j^S. To ensure that the root v_j^S does not violate the t-anonymous property we add t stars $K_{1,\deg(v_j^S)}$. We denote with $V_{\mathcal{C}} = \{v_1^S, \dots, v_n^S\}$ the set of all root vertices. Finally, to end up with one tree instead of a forest, repeatedly add edges between any degree-one-vertices of different connected components.

We now show that if I is a yes-instance then $opt(I') \geq t$ and if I is a no-instance then $opt(I') = 1$.

Suppose that I has a set cover of size h. Observe that for each element $e_i \in U$ the only vertex of degree $f(e_i)$ is v_i^e, and there are no other vertices violating the t-anonymous property. The key point in the construction is that, in order to get a t-anonymous graph, one has to delete vertices of $V_{\mathcal{C}}$. Indeed, let $e_i \in U$ be an element and v_j^S a root vertex such that $e_i \in S_j$. By construction the child vertices $v_\ell^{e_i,S_j}$ of v_j^S correspond to e_i and therefore have $f(e_i)$ child vertices. Thus, deleting v_j^S lowers the degree of all $v_\ell^{e_i,S_j}$ to $f(e_i)$ and, hence, v_i^e no longer violates the t-anonymous property. Hence, given a set cover of size h one can construct a corresponding t-deletion set for G.

Conversely, we show that if there exists a 2-deletion of size at most h in G, then $(U, \mathcal{C}, h)$ is a yes-instance of SET COVER. Let $S \subseteq V$ be a 2-deletion of size at most h. First, we show how to construct a 2-deletion $S' \subseteq V_{\mathcal{C}}$ such that $|S'| \leq |S|$. To this end, initialize S' as $S' = S \cap V_{\mathcal{C}}$. If S' is a 2-deletion, then the construction of S' is finished. Otherwise, there is a vertex v in $G - S'$ such that there is no other vertex with the same degree as v. Observe that since $S' \subseteq V_{\mathcal{C}}$, it follows that $v \in V_U$, that is $v = v_i^e$ for some $1 \leq i \leq m$. Furthermore, observe that is exactly one vertex in G having a degree d between $f(e_i) - h \leq d \leq f(e_i)$, namely v_i^e. As S is a 2-deletion, it follows that S either contains v_i^e or a vertex u that is adjacent to a vertex w with $\deg_G(w) > \deg(v_i^e)$. In either case, we add to S' a vertex $v_j^S \in V_{\mathcal{C}}$ such that $e_i \in S_j$. By exhaustively applying this procedure, we end up with S' being a 2-deletion. Since the vertices in $V_{\mathcal{C}}$ are the only ones in G that are adjacent to more than one vertex of degree at least three and all vertices in V_U have degree more than three, it follows that $|S'| \leq |S|$.

It remains to show that the set $\mathcal{C}'$ of sets corresponding to the vertices in S' forms a set cover. To this end, assume by contradiction that $\mathcal{C}'$ is not a set cover, that is, there is an element $e_i \notin \bigcup_{S_j \in \mathcal{C}'} S_j$. However, this implies that in $G - S'$ there is exactly one vertex of degree $f(e_i)$, namely v_i^e, implying that S' is not a 2-deletion, a contradiction. As $|\mathcal{C}'| = |S'| \le |S| \le h$, it follows that if G contains a 2-deletion of size h, then $(U, \mathcal{C}, h)$ is a yes-instance. Hence, if $(U, \mathcal{C}, h)$ is a no-instance, then there exist no 2-deletion of size at most h.

Thus, we obtain a fpt gap-reduction with the gap $t = (mn)^{\frac{1-2\varepsilon}{2\varepsilon}} = (t^2 m^2 n^2)^{1/2-\varepsilon} \ge |V|^{1/2-\varepsilon}$ since $|V| < t^2 m^2 n^2$. From Lemma 1 and since SET COVER is W[2]-hard [7], we have that MAX ANONYM-V-DEL is not $n^{1/2-\varepsilon}$-approximable in fpt-time w.r.t. parameter s, even on trees, unless FPT=W[2]. $\qquad\square$

4   Inapproximability of Edge Deletion Versions

In this section, we first show that ANONYM-E-DEL is NP-hard on caterpillars; the corresponding proof is an adaption of the reduction provided in the proof of Theorem 2. A caterpillar is a tree that has a dominating path [1], that is, a caterpillar is a tree such that deleting all leaves results in a path. Then we provide polynomial-time inapproximability results for MIN ANONYM-E-DEL and MAX ANONYM-E-DEL for bounded-degree graphs and parameterized inapproximability results for MAX ANONYM-E-DEL on general graphs.

Theorem 3. ANONYM-E-DEL *is NP-hard on caterpillars, even if* $k = 2$.

Theorem 4. *For every* $0 < \varepsilon \le 1$, MAX ANONYM-E-DEL *is not* $n^{1-\varepsilon}$-*approximable even on bounded-degree graphs, unless* $P = NP$.

Theorem 5. *For every* $0 < \varepsilon \le 1$, MIN ANONYM-E-DEL *is not* $n^{1-\varepsilon}$-*approximable even on bounded-degree graphs, unless* $P = NP$.

Theorem 6. *For every* $0 < \varepsilon \le 1$, MAX ANONYM-E-DEL *is not* $n^{1-\varepsilon}$-*approximable in fpt-time w.r.t. parameter* s, *unless* $FPT=W[1]$.

Proof. We provide an fpt gap-reduction from the W[1]-hard CLIQUE problem [7] parameterized by the solution size h. CLIQUE is defined as follows: given a graph $G = (V, E)$ and an integer $h \in \mathbb{N}$, the task is to decide whether there is a subset $V' \subseteq V$ of at least h pairwise adjacent vertices. Let $I = (G, h)$ be an instance of CLIQUE. Assume w.l.o.g. that $\Delta_G + 2h + 1 \le n$, where $n = |V|$. If this is not the case, then one can add isolated vertices to G until the bound holds.

We construct an instance $I' = (G' = (V', E'), s)$ of MAX ANONYM-E-DEL as follows: First, copy G into G'. Then, add a vertex u and connect it to the n vertices in G'. Next, for each vertex $v \in V$ add to G' degree-one vertices that are adjacent only to v such that $\deg_{G'}(v) = n - h$. This is always possible, since we assumed $\Delta_G + 2h + 1 \le n$. Observe that in this way at most $n(n-h)$ degree-one vertices are added. Now, set $x := \lceil (4n)^{3/\varepsilon} \rceil$ and add cliques with two, $n - 2h + 1$, and $n - h + 1$ vertices such that after adding these cliques the number of degree-d vertices in G', for each $d \in \{1, n - 2h, n - h\}$, is between $x + h$ and $x + h + n$, that is, $x + h \leq |D_{G'}(d)| \leq x + h + n$. After inserting these cliques, the graph consists of four blocks: of degree one, $n - h$, $n - 2h$, and n, where the first three blocks are roughly of the same size (between $x + h$ and $x + h + n$ vertices) and the last block of degree n contains exactly one vertex. To finish the construction, set $s := \binom{h}{2} + h$.

Now we show that if I is a yes-instance, then $opt(I') \geq x$ and if I is a no-instance, then $opt(I') < 2s$.

Suppose that I contains a clique $C \subseteq V$ of size h. Then, deleting the $\binom{h}{2}$ edges within C and the h edges between the vertices in C and u does not exceed the budget s and results in an x-anonymous graph G''. Since h edges incident to u are deleted, it follows that $\deg_{G''}(u) = n - h$. Furthermore, for each clique-vertex $v \in C$ also h incident edges are deleted ($h - 1$ edges to other clique-vertices and the edge to u), thus it follows that $\deg_{G''}(v) = n - 2h$. Since the degree of the remaining vertices remain unchanged, and $|D_{G'}(n - h)| \geq x + h$, it follows that each of the three blocks in G'' has size at least x. Hence, G'' is x-anonymous.

For the reverse direction, suppose that there is a $2s$-deletion set S of size at most s in G'. Since u is the only vertex in G' with degree n, and all other vertices in G' have degree at most $n - h$, it follows that S contains at least h edges that are incident to u. Since $N_{G'}(u) = V$, it follows that the degree of at least h vertices of the block $D_{G'}(n - h)$ is decreased by one. Denote these vertices by C. Since $|S| \leq s$ and h edges incident to u are contained in S, it follows that at most $2s - h + 1$ vertices are incident to an edge in S. Furthermore, since S is a $2s$-deletion set, it follows that the vertices in C are in $G' - S$ either contained in the block of degree one or in the block of degree $n - 2h$. Thus, by deleting the at most $\binom{h}{2}$ remaining edges in S, the degree of each of the h vertices in C is decreased by at least $h - 1$. Hence, these $\binom{h}{2}$ edges in S form a clique on the vertices in C and thus I is a yes-instance. Therefore, it follows that if I is a no-instance, then there is no $2s$-deletion set of size s in G' and hence $opt(I') < 2s$.

Thus we obtain a gap-reduction with the gap at least $\frac{x}{2s}$. Set $n' := |V'|$. By construction we have $3x \leq n' \leq n^2 + 3x + 3h + 3n + 1$. By the choice of x it follows that $x > n'/4$, since

$$\frac{n'}{4} \leq \frac{1}{4}(n^2 + 3x + 3h + 3n + 1) = x + \underbrace{\frac{1}{4}(n^2 + 3h + 3n + 1 - x)}_{<0} < x.$$

Hence the gap is

$$\frac{x}{2s} > \frac{n'^{1-\varepsilon+\varepsilon}}{4(h^2 + h)} \geq n'^{1-\varepsilon} \frac{n'^{\varepsilon}}{8h^2} > n'^{1-\varepsilon} \frac{x^{\varepsilon}}{8n^2} = n'^{1-\varepsilon} \frac{(4n)^{3\varepsilon/\varepsilon}}{8n^2} > n'^{1-\varepsilon}.$$

□

5   Fixed-Parameter Tractability

In previous work, it was shown that ANONYM-E-INS and ANONYM-V-DEL are
both fixed-parameter tractable with respect to the combined parameter budget s
and maximum degree Δ [3,11]. Here we generalize the ideas behind these results
and show fixed-parameter tractability for the general problem variant where one
might insert and delete specified numbers of vertices and edges.

k-DEGREE ANONYMITY EDITING (ANONYM-EDT)

Input: An undirected graph $G = (V, E)$ and five positive integers s_1, s_2, s_3, s_4
   and k.

Question: Is it possible to obtain a graph $G' = (V', E')$ from G using at most s_1
   vertex deletions, s_2 vertex insertions, s_3 edge deletions, and s_4 edge
   insertions, such that G' is k-anonymous?

Observe that here we require that the inserted vertices have degree zero and
we have to "pay" for making the inserted vertices adjacent to the existing ones.
In particular, if $s_4 = 0$, then all inserted vertices are isolated in the target
graph. Note that there are other models where the added vertices can be made
adjacent to an arbitrary number of vertices [2,5]. Our ideas, however, do not
directly transfer to this variant.

For convenience, we set $s := s_1 + s_2 + s_3 + s_4$ to be the number of allowed
graph modifications.

Theorem 7. ANONYM-EDT *is fixed-parameter tractable w.r.t.* (s, Δ).

Proof (sketch). Let $I = (G = (V, E), k, s_1, s_2, s_3, s_4)$ be an instance of ANONYM-
EDT. In the following we give an algorithm finding a solution if existing. Intuitively, the algorithm first guesses a "solution structure" and then checks whether
the graph modifications associated to this solution structure can be performed
in G. A solution structure is a graph S with at most $s(\Delta + 1)$ vertices where

1. each vertex is colored with a color from $\{0, \ldots, \Delta\}$ indicating the degree of
   the vertex in G and
2. each edge and each vertex is marked either as "to be deleted", "to be inserted",
   or "not to be changed" such that:
   (a) all edges incident to a vertex marked as "to be inserted" are also marked
       as "to be inserted",
   (b) at most s_1 vertices and at most s_3 edges are marked as "to be deleted",
       and
   (c) at most s_2 vertices and at most s_4 edges are marked as "to be inserted".

The intuition about this definition is that a solution structure S contains all
graph modifications in a solution *and* the vertices that are affected by the modifications, that is, the vertices whose degree is changed when performing these
modifications. Observe that any solution for I defines such a solution structure
with at most $s(\Delta + 1)$ vertices as each graph modification affects at most $\Delta + 1$

vertices. This bound is tight in the sense that deleting a vertex v affects v and his up to Δ neighbors. Furthermore, observe that once given such a solution structure, we can check in polynomial time whether performing the marked edge/vertex insertions/deletions results in a k-anonymousgraph G', since the coloring of the vertex indicates the degrees of the vertices that are affected by the graph modifications.

Our algorithm works as follows: First it branches into all possibilities for the solution structure S. In each branch it checks whether performing the graph modifications indicated by the marks in S indeed result in a k-anonymous graph. If yes, then the algorithm checks whether the graph modifications associated to S can be performed in G: To this end, all edges and vertices marked as "to be inserted" are removed from S and the marks at the remaining vertices and edges are also removed and the resulting "cleaned" graph is called S'. Finally the algorithm tries to find S' as an induced subgraph of G such that the vertex degrees coincide with the vertex-coloring in S'. This is done by a meta-theorem for bounded local tree-width graphs [8]. If the algorithm succeeds and finds S' as an induced subgraph, then the graph modifications encoded in S can be performed which proves that I is a yes-instance. If the algorithm fails in every branch, then, due to the exhaustive search over all possibilities for S, it follows that I is a no-instance. Thus, the algorithm is indeed correct.

6 Conclusion

We have shown strong inapproximability results for the optimization variants of ANONYM-E-DEL and ANONYM-V-DEL. We leave two major open questions concerning polynomial-time approximability and parameterized approximability: In all our gap reductions the value of k is in the order of n. This leads to the question whether with constant k MIN ANONYM-E-DEL or MIN ANONYM-V-DEL are constant-factor approximable in polynomial time? Second, we failed to transfer the inapproximability results to ANONYM-E-EDTwhere we require that the number of edge insertions plus deletions is at most s. Here, handling the possibility to *revert* already changed degrees seems to be crucial in order to obtain any approximation result (positive or negative) for the optimization variants of ANONYM-E-EDT. This leads to the question whether there are "reasonable" (parameterized) approximation algorithms for the optimization variants of ANONYM-E-EDT?

References

1. Brandstädt, A., Le, V.B., Spinrad, J.P.: Graph Classes: A Survey. Monographs on Discrete Mathematics and Applications. SIAM, Philadelphia (1999)
2. Bredereck, R., Froese, V., Hartung, S., Nichterlein, A., Niedermeier, R., Talmon, N.: The complexity of degree anonymization by vertex addition. In: Gu, Q., Hell, P., Yang, B. (eds.) AAIM 2014. LNCS, vol. 8546, pp. 44–55. Springer, Heidelberg (2014)

3. Bredereck, R., Hartung, S., Nichterlein, A., Woeginger, G.J.: The complexity of finding a large subgraph under anonymity constraints. In: Cai, L., Cheng, S.-W., Lam, T.-W. (eds.) Algorithms and Computation. LNCS, vol. 8283, pp. 152–162. Springer, Heidelberg (2013)

4. Chester, S., Kapron, B., Srivastava, G., Venkatesh, S.: Complexity of social network anonymization. Soc. Netw. Anal. Min. **3**(2), 151–166 (2013)

5. Chester, S., Kapron, B.M., Ramesh, G., Srivastava, G., Thomo, A., Venkatesh, S.: Why Waldo befriended the dummy? k-anonymization of social networks with pseudo-nodes. Soc. Netw. Anal. Min. **3**(3), 381–399 (2013)

6. Clarkson, K.L., Liu, K., Terzi, E.: Towards identity anonymization in social networks. In: Yu, P., Han, J., Faloutsos, C. (eds.) Link Mining: Models, Algorithms, and Applications, pp. 359–385. Springer, New York (2010)

7. Downey, R.G., Fellows, M.R.: Fundamentals of Parameterized Complexity. Springer, London (2013)

8. Frick, M., Grohe, M.: Deciding first-order properties of locally tree-decomposable structures. J. ACM **48**(6), 1184–1206 (2001)

9. Fung, B.C.M., Wang, K., Chen, R., Yu, P.S.: Privacy-preserving data publishing: a survey of recent developments. ACM Comput. Surv. **42**(4), 14:1–14:53 (2010)

10. Hartung, S., Hoffmann, C., Nichterlein, A.: Improved upper and lower bound heuristics for degree anonymization in social networks. In: Gudmundsson, J., Katajainen, J. (eds.) SEA 2014. LNCS, vol. 8504, pp. 376–387. Springer, Heidelberg (2014)

11. Hartung, S., Nichterlein, A., Niedermeier, R., Suchý, O.: A refined complexity analysis of degree anonymization in graphs. In: Fomin, F.V., Freivalds, R., Kwiatkowska, M., Peleg, D. (eds.) ICALP 2013, Part II. LNCS, vol. 7966, pp. 594–606. Springer, Heidelberg (2013)

12. Liu, K., Terzi, E.: Towards identity anonymization on graphs. In: Proceedings of the ACM SIGMOD International Conference on Management of Data (SIGMOD '08), pp. 93–106. ACM (2008)

13. Marx, D.: Parameterized complexity and approximation algorithms. Comput. J. **51**(1), 60–78 (2008)

14. Wu, X., Ying, X., Liu, K., Chen, L.: A survey of privacy-preservation of graphs and social networks. In: Aggarwal, C.C., Wang, H. (eds.) Managing and Mining Graph Data, pp. 421–453. Springer, Berlin (2010)

The k-Distinct Language: Parameterized Automata Constructions

Ran Ben-Basat, Ariel Gabizon, and Meirav Zehavi$^{(\boxtimes)}$

Department of Computer Science, Technion, 32000 Haifa, Israel
{sran,arielga,meizeh}@cs.technion.ac.il

Abstract. In this paper, we pioneer a study of parameterized automata constructions for languages relevant to the design of parameterized algorithms. We focus on the k-DISTINCT language $L_k(\Sigma) \subseteq \Sigma^k$, defined as the set of words of length k whose symbols are all distinct. This language is implicitly related to several breakthrough techniques, developed during the last two decades, to design parameterized algorithms for fundamental problems such as k-PATH and r-DIMENSIONAL k-MATCHING. Building upon the well-known color coding, divide-and-color and narrow sieves techniques, we obtain the following automata constructions for $L_k(\Sigma)$. We develop non-deterministic automata (NFAs) of sizes $4^{k+o(k)}n^{O(1)}$ and $(2e)^{k+o(k)} \cdot n^{O(1)}$, where the latter satisfies a 'bounded ambiguity' property relevant to approximate counting, as well as a non-deterministic xor automaton (NXA) of size $2^k n^{O(1)}$, where $n = |\Sigma|$. We show that our constructions lead to a *unified* approach for the design of both deterministic and randomized algorithms for parameterized problems, considering also their approximate counting variants. To demonstrate our approach, we consider the k-PATH, r-DIMENSIONAL k-MATCHING and MODULE MOTIF problems.

1 Introduction

Parameterized algorithms solve NP-hard problems by confining the combinatorial explosion to a parameter k. More precisely, a problem is *fixed-parameter tractable (FPT)* with respect to a parameter k if it can be solved in time $O^*(f(k))$ for some function f, where O^* hides factors polynomial in the input size.

In this paper, we pioneer a study of parameterized automata constructions for languages relevant to the design of parameterized algorithms. We focus on the k-DISTINCT language, formally defined as follows.

Definition 1. *Let Σ be an alphabet of size n, and k be a positive integer. We define the k-DISTINCT language, denoted $L_k(\Sigma) \subseteq \Sigma^k$, to be the set of words $w_1 \cdots w_k \in \Sigma^k$ such that $w_1, \ldots, w_k$ are all distinct.*

Given integers $k \leq n$, we also use the abbreviated notation $L_k(n)$ to denote the language $L_k([n])$, where $[n] \triangleq \{1, \ldots, n\}$.

The research leading to these results has received funding from the European Community's Seventh Framework Programme (FP7/2007-2013) under grant agreement number 257575.

© Springer International Publishing Switzerland 2014
M. Cygan and P. Heggernes (Eds.): IPEC 2014, LNCS 8894, pp. 85–96, 2014.
DOI: 10.1007/978-3-319-13524-3_8

The language k-DISTINCT is a natural candidate for investigating relations between automata and parameterized problems. Indeed, we show that k-DISTINCT is related to several breakthrough techniques, developed during the last two decades, to design algorithms for classic, fundamental parameterized problems. More precisely, building upon the well-known color coding [3], divide-and-color [9] and narrow sieves [5] techniques, we obtain three different parameterized automata constructions for k-DISTINCT.

We show that our constructions lead to a *unified* approach for the design of parameterized algorithms, which allows to obtain *three* types of such algorithms–deterministic, approximate counting and randomized algorithms–by constructing *only one* problem-specific automaton. We also argue that our approach allows obtaining these algorithms in an easy, natural manner. In particular, for researchers knowledgeable with automata and graphs, our deterministic algorithm for k-PATH can be summarized by saying that, essentially, we just "take the product of the graph with the automaton for $L_k(V)$" (see, e.g., [30]).

To demonstrate our approach, we consider the following three problems.

Weighted k-Path: Given a directed graph $G = (V, E)$, a weight function $w \colon E \to \mathbb{R}$ and a parameter $k \in \mathbb{N}$, determine if G has a *simple k-path* (i.e., an acyclic/simple walk on k vertices), and if so, return such a path of minimal weight.

Weighted r-Dimensional k-Matching: Given disjoint universes $U_1, \ldots, U_r$, think of an element in $U_1 \times \ldots \times U_r$ as a set that contains exactly one element from each universe U_i. Now, given a family $\mathcal{S} \subseteq U_1 \times \ldots \times U_r$, a weight function $w \colon \mathcal{S} \to \mathbb{R}$ and a parameter $k \in \mathbb{N}$, determine if there exists a subfamily of k pairwise-disjoint sets in $\mathcal{S}$, and if so, return such a subfamily of maximal weight.

Weighted Module Motif: Let $G = (V, E)$ be an undirected graph, and let $k \in \mathbb{N}$. A k-*module* is a subset $U \subseteq V$ of k vertices that have the same neighbors outside of U. More formally, for every $u_1, u_2 \in U$ and $r \in V \setminus U$, $(u_1, r) \in E$ iff $(u_2, r) \in E$. Now, given a set C of colors, a function $Col \colon V \to 2^C$ specifying the allowed colors for each vertex, and a weight function $w \colon V \to \mathbb{R}$,[1] determine if there exists a pair of a k-module U in G and a coloring $col \colon U \to C$ such that

- For every $v \in U$, $col(v) \in Col(v)$.
- For every $c \in C$, c is the color of at most one vertex in U (according to col).

If such a pair exists, return such a pair that maximizes the weight $w(U)$ of the module U, defined as $w(U) \triangleq \sum_{v \in U} w(v)$.

We highlight two variants of the above problems, considered in this paper.

1. The *unweighted* variant, where all elements are assumed to have weight zero.
2. The *approximate counting* unweighted variant, where, given an *accuracy parameter* $\delta > 0$, the goal is to return a value in the interval $[S/\delta, \ \delta \cdot S]$, where S is the number of solutions (e.g., the number of simple k-paths).

---

[1] We consider vertex weights, rather than color similarity scores (see [41]), for the sake of clarity. Our algorithms can be easily modified to handle color similarity scores.

1.1   Prior Work

The k-PATH and r-DIMENSIONAL k-MATCHING problems are two of the most well-studied problems in the field of parameterized complexity. Indeed, the k-PATH problem has enjoyed a race towards obtaining the fastest parameterized algorithm for it [1,3,5,6,9,15,16,21,24,31,35,39]. Currently, the best known parameterized algorithm for WEIGHTED k-PATH runs in time $O^*(2.619^k)$ [16,35]. For (unweighted) k-PATH, Williams [39] gave a randomized algorithm running in time $O^*(2^k)$. Restricted to undirected graphs, k-PATH can be solved in randomized time $O^*(1.66^k)$ [5]. We also note that approximately counting the number of simple k-paths in a graph can be performed in time $O^*((2e)^{k+o(k)})$ [2].

The classic decision version of the r-DIMENSIONAL k-MATCHING problem is listed as one of the six fundamental NP-complete problems in Garey and Johnson [17]. A considerable number of papers presented parameterized algorithms for this problem [5,7–11,13,14,19,23–26,28,37,38]. Currently, the best known parameterized algorithm for WEIGHTED r-DIMENSIONAL k-MATCHING runs in time $O^*(2.851^{(r-1)k})$ [19]. For r-DIMENSIONAL k-MATCHING, Björklund et al. [5] gave a randomized algorithm running in time $O^*(2^{(r-2)k})$. We also note that approximately counting the number of 3-dimensional k-matchings can be performed in randomized time $O^*(5.48^{3k})$ [27].

The MODULE MOTIF problem was recently introduced by Rizzi et al. [32], primarily motivated from the analysis of complex biological networks. This problem can be solved in deterministic and randomized times $O^*(4.32^k)$ and $O^*(2^k)$, respectively [41].

1.2   Our Contribution

Building upon previous techniques, we obtain automata constructions for k-DISTINCT. We also give lower bounds on the size of such automata. Furthermore, we show that automata for k-DISTINCT can be used to develop algorithms for the problems described above. While most of our current constructions are not good enough to obtain improved algorithms,[2] we find they provide a unified and 'clean' approach to present and design different types of parameterized algorithms. As our constructions are essentially 'an automata-based view' of well-known parameterized algorithms, improving our lower bounds will show that a large class of algorithms 'cannot do better' (e.g., improving our lower bound on NFA-size described below, would imply a statement of the form 'known color coding-related techniques cannot achieve faster running times'). We describe our results in more detail in the next paragraphs. Unless noted otherwise, we assume that our alphabet is $[n] = \{1, \ldots, n\}$ and use the notation $L_k(n)$ from Definition 1.

NFAs for $L_k(n)$ and Weighted Problems: We show that the divide-and-color technique of [9] can be used to construct an NFA for $L_k(n)$ of size $O^*(4^{k+o(k)})$.

---

[2] We slightly improve the previous *randomized* algorithm that approximately counts only 3-dimensional k-matchings, and deterministic algorithm for MODULE MOTIF.

We also show, on the other hand, that any NFA for $L_k(n)$ must have at least $\Omega(2^k \log n)$ states.[3] Roughly speaking, our lower bound shows that some subset of the states of such an NFA must correspond to elements of an (n, k)-universal set (cf. Definition 5). It is known that such a set must have size $\Omega(2^k \log n)$ [22,34].

Our NFA construction implies deterministic algorithms for the WEIGHTED k-PATH, WEIGHTED r-DIMENSIONAL k-MATCHING and WEIGHTED MODULE MOTIF problems, whose time complexities are $O^*(4^{k+o(k)})$, $O^*(4^{(r-1)k+o((r-1)k)})$ and $O^*(4^{k+o(k)})$, respectively. More generally, our approach implies that an NFA for $L_k(n)$ of size $O^*(s(k))$, that can be constructed in deterministic time $O^*(s(k))$, can be used to obtain deterministic algorithms for these problems with running times $O^*(s(k)), O^*(s((r-1) \cdot k))$ and $O^*(s(k))$, respectively.

Bounded Ambiguity NFAs and Approximate Counting Problems: Roughly speaking, a *bounded ambiguity* NFA is one where the number of accepting paths of any word in its language is roughly the same. Using the balanced hash families construction of [2], we build such an NFA of size $O^*((2e)^k)$ for $L_k(n)$. This NFA allows us to obtain deterministic algorithms for the approximate counting versions of k-PATH, r-DIMENSIONAL k-MATCHING and MODULE MOTIF, running in times $O^*((2e)^{k+o(k)})$, $O^*((2e)^{(r-1)k+o((r-1)k)})$ and $O^*((2e)^{k+o(k)})$, respectively, for any accuracy parameter $\delta > 1 + \frac{1}{poly(k)}$.

NXAs and Randomized Algorithms for Unweighted Problems: Based on the narrow sieves technique of [5] and its interpretation by [1], we construct a *randomized non-deterministic xor automaton (NXA)* of size $O^*(2^k)$ for $L_k(n)$ (see Definitions 6 and 7). We also show a matching lower bound for such an NXA.

Using our construction, we obtain randomized algorithms for k-PATH, r-DIMENSIONAL k-MATCHING and MODULE MOTIF, running in times $O^*(2^k)$, $O^*(2^{(r-1)k})$ and $O^*(2^k)$, respectively. Our lower bound may be viewed as a statement that 'known techniques for (directed) k-PATH based on polynomials over fields of characteristic two cannot do better than $O^*(2^k)$'.

Organization: Due to lack of space, some of the results are omitted, and will appear in [4]. Section 2 presents some preliminaries regarding automata. Sections 3 and 4 present our NFA and NXA constructions for k-DISTINCT. The details of our 'bounded ambguity' automaton are omitted. Sections 5 and 6 present the relation of these constructions to parameterized algorithms.

2   Preliminaries

In this section we recall basic definitions related to automata theory.

2.1   Non-deterministic Finite Automata (NFA)

An NFA, the simplest automaton relevant to this paper, is defined as follows.

---

[3] We note that determining the minimal NFA size, even for a *finite* regular language, is computationally hard [20].

Definition 2. *An NFA M over alphabet Σ is a labeled directed graph $M =< Q, \Delta, q_0, F >$, where*

- *Q is the set of vertices, called* states.
- *Δ is the set of edges, called* transitions, *each labeled by an element of Σ.*
- *q_0 is an element in Q, called the* start state.
- *F is a subset of states from Q, called* accepting states.

We say that M is *acyclic* if it is acyclic as a directed graph. We define the *size* of M as the sum of the number of states and transitions in M, i.e., size(M) $\triangleq$ $|Q| + |\Delta|$. For the sake of clarity, when referring to a transition in Δ, we use the notation $(u \to v)$ rather than (u, v). Moreover, we allow the transitions to be weighted (i.e., we also have a weight function $w : \Delta \to \mathbb{R}$).

For a word $w = w_1 \cdots w_t \in \Sigma^t$, let $M(w) \subseteq Q$ be the *subset of states reached by w.*[4] We now give the standard definition of the language of M.

Definition 3. *The* language *of an NFA M, denoted $L(M)$, is defined by*

$$L(M) \triangleq \{w \in \Sigma^* \mid M(w) \cap F \neq \emptyset\}.$$

2.2   Intersection

To integrate automata for k-Distinct with problem-specific automata, we need the following definition.

Definition 4. *Given NFAs $M_1 =< Q_1, \Delta_1, q_0^1, F_1 >$ and $M_2 =< Q_2, \Delta_2, q_0^2, F_2 >$ over the same alphabet Σ, we define the* intersection NFA

$$M_1 \cap M_2 \triangleq < Q_1 \times Q_2, \Delta, < q_0^1, q_0^2 >, F_1 \times F_2 >$$

over Σ, where the set of transitions Δ is defined as follows. For every pair of transitions $(u_1 \to v_1) \in \Delta_1$ and $(u_2 \to v_2) \in \Delta_2$ that are both labeled by the same element $a \in \Sigma$, we have a transition $(< u_1, u_2 > \to < v_1, v_2 >) \in \Delta$ labeled by a, whose weight is the sum of the weights of $(u_1 \to v_1)$ and $(u_2 \to v_2)$.

Observe that a word is accepted by $M_1 \cap M_2$ iff it is accepted by both M_1 and M_2.

3   NFA Construction for k-Distinct

We now give an explicit construction of an NFA of size $O^*(4^{k+o(k)})$ for k-Distinct, following principles of the divide-and-color technique [9]. To this end, we need the following standard tool for derandomizing parameterized algorithms.

Definition 5. *Let $\mathcal{F}$ be a set of functions $f : [n] \to \{0,1\}$. We say that $\mathcal{F}$ is an (n,t)-universal set if for every subset $I \subseteq [n]$ of size t and a function $f' : I \to \{0,1\}$, there is a function $f \in \mathcal{F}$ such that for all $i \in I$, $f(i) = f'(i)$.*

---

[4] If while reading a word we reach a state where we cannot progress by reading the next symbol, this run is rejected and the state we are at is *not* added to $M(w)$.

The next result asserts that small universal sets can be computed efficiently.

Theorem 1 ([29]). *There is an algorithm that, given a pair of integers (n,t), computes an (n,t)-universal set $\mathcal{F}$ of size $2^{t+O(\log^2 t)}\log n$ in time $O(2^{t+O(\log^2 t)}n\log n)$.*

We now state the main result of this section, followed by a related lower bound.

Theorem 2. *An acyclic NFA M of size $O^*(4^{k+o(k)})$ for $L_k(n)$ can be constructed in time $O^*(4^{k+o(k)})$.*

Proof. Given a subset $S \subseteq [n]$, recall that $L_k(S)$ is the set of words $w \in S^k$ whose symbols are all distinct, i.e., $L_k(S) \triangleq L_k(n) \cap S^k$. For every $1 \leq \ell \leq k$ and subset $S \subseteq [n]$, we construct in time $O^*(4^{\ell+o(\ell)})$, by induction on ℓ, an acyclic NFA $M_{\ell,S}$ for $L_k(S)$ of size $O^*(4^{\ell+o(\ell)})$. Setting $\ell = k$ and $S = [n]$, the correctness of the theorem follows.

Basis: Fix some subset $S \subseteq [n]$. Given a word $w \in [n]$, the NFA $M_{1,S}$ simply checks if w consists of exactly one symbol, and this symbol belongs to S (i.e., $|w| = 1$ and $w \in S$). Thus, $M_{1,S}$ can be constructed in time $O(n)$.

Step: Again, fix some subset $S \subseteq [n]$. Next, assume that we have a construction of an NFA $M_{\ell',S'}$ for every $1 \leq \ell' < \ell$ and $S' \subseteq S$.

Fix disjoint subsets $S_1, S_2 \subseteq S$. Towards constructing $M_{\ell,S}$, we define an acyclic NFA M_{ℓ,S_1,S_2}, which accepts exactly the words $w \in L_\ell(S)$ whose first $\lceil \ell/2 \rceil$ symbols are in S_1, and last $\lfloor \ell/2 \rfloor$ symbols are in S_2. The construction of M_{ℓ,S_1,S_2} simply consists of a copy of $M_{\lceil \ell/2 \rceil,S_1}$ that reads the first $\lceil \ell/2 \rceil$ symbols of w, followed by a copy of $M_{\lfloor \ell/2 \rfloor,S_2}$ that reads the last $\lfloor \ell/2 \rfloor$ symbols of w.

We construct the acyclic NFA $M_{\ell,S}$ as follows. Let $\mathcal{F}$ be an (n, ℓ)-universal set, computed using Theorem 1. Given a function $f \in \mathcal{F}$, let $S[f] = \{a \in S : f(a) = 1\}$. For every function $f \in \mathcal{F}$, we add an ϵ-transition from the start state of $M_{\ell,S}$ to a copy of $M_{\ell,S[f],S\setminus S[f]}$. Thus, $M_{\ell,S}$ accepts a word w iff (at least) one automaton of the form $M_{\ell,S[f],S\setminus S[f]}$ accepts w. Now, using the properties of a universal set, the correctness of $M_{\ell,S}$ can be verified (details on this and the construction time bound are omitted). □

Theorem 3. *Any NFA M over alphabet $[n]$ with $L(M) = L_k(n)$ has $\Omega(2^k \log n)$ states.*

Proof. Let $M =< Q, \Delta, q_0, F >$ be an NFA for $L_k(n)$. Towards showing the claimed lower bound, we associate two sets, $A_q, B_q \subseteq [n]$, with each state $q \in Q$. The set A_q contains letters that belong to some path from q_0 to q, while the set B_q contains letters that belong to some path from q to an accepting state. Since each word in $L_k(n)$ consists of k distinct letters, we get that $A_q \cap B_q = \emptyset$. Now, for each $q \in Q$, define the function $f_q : [n] \to \{0,1\}$ by $f_q(i) = 1$ iff $i \in A_q$. Finally, let $\mathcal{F} \triangleq \{f_q\}_{q\in Q}$ be the set of all the functions f_q.

Now, let S and S' be two disjoint subsets of $[n]$, whose union $I = S \cup S'$ contains exactly k elements. Note that there exists a word in k-DISTINCT whose first $|S|$ letters belong to S, and whose last $|S'|$ letters belong to S'. Therefore, there exists a state $q \in Q$ such that $S \subseteq A_q$ and $S' \subseteq B_q$. For this q, for $i \in I$, $f(i) = 1$ iff $i \in S$. Since we have chosen S and S' arbitrarily, we conclude that $\mathcal{F}$ is an (n, k)-universal set, which is known to contain $\Omega(2^k \log n)$ functions [22,34]; thus, the NFA M contains $\Omega(2^k \log n)$ states. $\qquad\square$

4 Randomized NXA Construction for k-DISTINCT

A *non-deterministic xor automaton (NXA)* is simply an NFA where the acceptance criteria for a word w is that *there is an odd number of accepting paths for w.*[5] We formalize this by defining the *XOR-language* of an NXA.

Definition 6. *Let M be an NXA over an alphabet Σ. We define the XOR-language of M, denoted $L_\oplus(M) \subseteq \Sigma^*$, to be the set of words w that have an odd number of paths to an accept state in M.*

In this section we build upon the algebraic narrow sieves technique [5] (see also [1,24]), and construct an NXA for the language $L_k(n)$. This NXA, together with our two previous NFA constructions, form the infrastructure of our unified approach for the design of parameterized algorithms (see Sects. 5 and 6). We will actually construct a *randomized* NXA, a notion which we now formally define.

Definition 7. *Fix a language $L \subseteq \Sigma^*$, and any $0 < \epsilon < 1$. An ϵ-NXA of size s for L is a randomized algorithm R, which outputs an NXA M with $size(M) \le s$ such that*

- $L_\oplus(M) \subseteq L$ *with probability one, i.e., M never accepts a word outside L.*
- *For any fixed word $w \in L$, $w \in L_\oplus(M)$ with probability at least ϵ.*

It will be convenient to refer to the running time of R as the construction time *of R. By the number of states of R we mean the maximal number of states in an NXA outputted by R with non-zero probability.*

The main purpose of this section is to prove the following theorem.

Theorem 4. *Fix any positive integers $k \le n$. There is a $1/4$-NXA for $L_k(n)$ of size $s = O(2^k \cdot k \cdot n)$, containing $O(2^k \cdot k)$ states, with construction time $O(n \cdot 2^k)$.*

As an intermediate step, we construct an NXA for languages that are a strict subset of $L_k(n)$. Informally, these languages correspond to subsets of columns in a fixed matrix over $\mathbb{F}_2$ that are linearly independent (see below).

Definition 8 (The language L_A). *Fix a $k \times n$ matrix A over $\mathbb{F}_2$. The language $L_A \subseteq L_k(n)$ consists of all $(i_1 \cdots i_k) \in L_k(n)$ such that the columns $\{i_1, \ldots, i_k\}$ are linearly independent in A.*

In the rest of this section sums are always in $\mathbb{F}_2$, i.e., modulo 2. For each non-empty subset $S \subseteq [k]$, define the function $\phi_S : (\{0,1\}^k)^k \to \{0,1\}$ by

---

[5] We assume NXAs do not contain directed cycles of only ϵ-transitions, as their XOR-language is not properly defined in this case.

$$\phi_S(v_1, \ldots, v_k) \triangleq \prod_{i=1}^{k} \sum_{j \in S} v_{i,j}$$

and define $\phi : (\{0,1\}^k)^k \to \{0,1\}$ by

$$\phi(v_1, \ldots, v_k) \triangleq \sum_{\emptyset \neq S \subseteq [k]} \phi_S(v_1, \ldots, v_k).$$

From Ryser's formula for the permanent [33] we know that

Lemma 1. $\phi(v_1, \ldots, v_k)$ *is equal to the determinant of the $k \times k$ matrix over $\mathbb{F}_2$ whose columns are $v_1, \ldots, v_k$.*

Fix a $k \times n$ matrix A over $\mathbb{F}_2$ with columns $v_1, \ldots, v_n \in \{0,1\}^k$. For each non-empty subset $S \subseteq [k]$, we define a function $f_{A,S} : [n]^k \to \{0,1\}$ by $f_{A,S}(i_1, \ldots, i_k) \triangleq \phi_S(v_{i_1}, \ldots, v_{i_k})$. We define $f_A : [n]^k \to \{0,1\}$ by

$$f_A(i_1, \ldots, i_k) \triangleq \phi(v_{i_1}, \ldots, v_{i_k}) = \sum_{\emptyset \neq S \subseteq [k]} \phi_S(v_{i_1}, \ldots, v_{i_k}) = \sum_{\emptyset \neq S \subseteq [k]} f_{A,S}(i_1, \ldots, i_k).$$

Lemma 2. *Fix any $k \times n$ matrix A over $\mathbb{F}_2$ and non-empty $S \subseteq [k]$. There is a deterministic automaton $M_{A,S}$ for $f_{A,S}^{-1}(1)$ with $k+1$ states and at most $k \cdot n$ transitions. $M_{A,S}$ can be constructed in time $O(k \cdot n)$.*

Proof. Let $v_1, \ldots, v_n$ be the columns of A. Let $T \subseteq [n]$ be the set of elements $i \in [n]$ such that

$$\sum_{j \in S} v_{i,j} = 1.$$

Observe that $f_{A,S}(i_1, \ldots, i_k) = 1$ if and only if $i_1, \ldots, i_k$ are all contained in T. This motivates the following construction: $M_{A,S}$ will contain the start state q_0, and the states $q_1, \ldots, q_k$, where q_k will be the only accept state. For each $0 \leq j \leq k-1$, and for every $i \in T$, there will be an edge from q_j to q_{j+1} labeled i. □

Lemma 3. *Fix any positive integers $k \leq n$ and any $k \times n$ matrix A over $\mathbb{F}_2$. There is an NXA M_A over $[n]$ of size $O(2^k \cdot k \cdot n)$, containing $O(k \cdot 2^k)$ states, such that $L_\oplus(M_A) = L_A$. M_A can be constructed in time $O(2^k \cdot k \cdot n)$.*

Proof. For every non-empty $S \subseteq [k]$, M_A will contain a copy of the automaton $M_{A,S}$ as described in Lemma 2. We unite the start state q_0 and accept state q_k of all the automata $M_{A,S}$ to one start state q_0 and accept state q_k of M_A. $L_\oplus(M_A)$ contains exactly the words $(i_1 \cdots i_k)$ that are accepted by an odd number of the automata $M_{A,S}$. Since $L(M_{A,S}) = f_{A,S}^{-1}(1)$, this is exactly $f_A^{-1}(1)$. Now note that $f_A(i_1, \ldots, i_k) = 1$ if and only if the (multi-)set of vectors $\{v_{i_1}, \ldots, v_{i_k}\}$ are linearly independent. In particular, $f_A(i_1, \ldots, i_k) = 1$ only if $\{i_1, \ldots, i_k\}$ are all distinct. Therefore, $L_\oplus(M_A) = f_A^{-1}(1) = L_A$. Finally, as M_A consists of $2^k - 1$ automata of the form $M_{A,S}$ that each have size $O(k \cdot n)$ and can be constructed in time $O(k \cdot n)$, the claim about M_A's size and construction time follows. □

Now, to prove Theorem 4, we choose a random $k \times n$ matrix A and output M_A (the details are omitted).

We next consider the tightness of our randomized NXA construction.

Theorem 5. *Fix positive integers $k \leq n$.*

- *Any NXA M over $[n]$ with $L_\oplus(M) = L_k(n)$ has at least 2^k states.*
- *Any $\frac{1}{4}$-NXA R over $[n]$ with $L_\oplus(M) = L_k(n)$ must have at least $\frac{1}{4} \cdot 2^k$ states.*

Proof. The *Hankel matrix* H_L of a language $L \subseteq [n]^k$ has its rows and columns indexed[6] by the elements of $[n]^k$. The (x, y)-entry of H_L is one if $x \cdot y \in L$. Otherwise the (x, y)-entry of H_L is zero.

Vuillemin and Gama [36] show that the rank, $r(H_L)$, of H_L as a matrix over $\mathbb{F}_2$, is a lower bound on the number of states of any NXA M with $L_\oplus(M) = L$. We first argue that $H_{L_k(n)}$ contains as a submatrix a $2^k \times 2^k$ identity matrix: For every $S = \{i_1, \ldots, i_d\} \subseteq [k]$, let $x_S \in [n]^k$ be the word $x_S = i_1 \cdots i_d$. For every $S \subseteq [k]$, define $y_S = x_{\bar{S}}$. It is clear that for every $S \subseteq [k]$, $x_S \cdot y_S \in L_k(n)$; and for every $S \neq T \subseteq [k]$, $x_S \cdot y_T \notin L_k(n)$. Thus, the set of words $L' \triangleq \{x_S \cdot y_T | S, T \subseteq [k]\}$ corresponds to a $2^k \times 2^k$ identity matrix, that is a submatrix of $H_{L_k(n)}$. This implies that $r(H_{L_k(n)}) \geq 2^k$, and thus the first item of the theorem. The second item follows from the first (the details are omitted). $\square$

5   Problem-Specific Automata

In this section we present problem-specific automata, to be integrated with the automata constructions for k-DISTINCT given in Sect. 6. Interestingly, we need only design one automaton per problem, regardless of the automaton with whom it will next be integrated. That is, using our approach, one can obtain fast deterministic, approximate counting and randomized parameterized algorithms for a given problem, at the price of designing only one automaton.

Intuitively, our problem-specific automata are designed to capture all the restrictions of the problems associated with them, except for a restriction that concerns uniqueness of elements.

k-**Path:** The following definition and straightforward lemma formally convert a graph into an automaton accepting the paths of the graph.

Definition 9. *Let (G, w, k) be an instance of WEIGHTED k-PATH. The NFA*

$$M(G, w) \triangleq < Q = V \cup \{q_0\}, \Delta = E \cup \{(q_0, v) : v \in V\}, q_0, F = V >$$

over alphabet $\Sigma = V$ is defined with the following labeling and weights of transitions. Each transition $(v \rightarrow u)$ is labeled by the target state u and its weight is $w(v, u)$, where, if $v = q_0$, its weight is 0.

Lemma 4. *The language $L(M(G, w))$ is the set of words $w = v_1 \cdots v_m$, where $m \in \mathbb{N}$, such that $v_1 \rightarrow \ldots \rightarrow v_m$ is a path from v_1 to v_m in G. Moreover, M is*

---

[6] Usually the Hankel matrix is defined as an infinite matrix, and we are defining a submatrix of it; for a *finite* language $L \subseteq [n]^k$, that the rank of this submatrix is clearly the same as the rank of the entire matrix, and we are only concerned with its rank.

an unambiguous (i.e., $(1,1)$-ambiguous) NFA, and the weight of a path accepting a word w is the weight of the corresponding path in G.

r-**Dimensional** k-**Matching:** Let $\mathcal{I} = (U_1, \ldots, U_r, \mathcal{S}, w, k)$ be an instance of WEIGHTED r-DIMENSIONAL k-MATCHING. Assume an order $<$ on $\mathcal{S}$, and given a subfamily $\mathcal{S}' \subseteq \mathcal{S}$, let $\mathcal{S}'[i]$ be the ith set in $\mathcal{S}'$ (according to $<$), for all $1 \leq i \leq |\mathcal{S}'|$. Given a word $w = u_1 \cdots u_{(r-1)m}$, where $m \in \mathbb{N}$, denote

$$\mathcal{I}(w) = \{\mathcal{S}' \subseteq \mathcal{S} : |(\bigcup \mathcal{S}') \cap U_1| = |\mathcal{S}'| = m,$$
$$[\forall i \in \{1, \ldots, m\}, j \in \{2, \ldots, r\} : \mathcal{S}'[i] \cap U_j = \{u_{(r-1)(i-1)+(j-1)}\}]\}.$$

Informally, $\mathcal{I}(w)$ includes every subfamily of $\mathcal{S}$ that contains sets with unique elements from U_1, such that the set of all of their elements together, excluding those in U_1, is the set of symbols in w, under a restriction related to the order $<$.

We design a simple automaton $M(\mathcal{I})$ in time $O^*(1)$, proving the next result.

Lemma 5. *The language $L(M(\mathcal{I}))$ is the set of words $w = u_1 \cdots u_{(r-1)m}$, where $m \in \mathbb{N}$, such that $\mathcal{I}(w) \neq \emptyset$. Moreover, a word w accepted by M is accepted by exactly $|\mathcal{I}(w)|$ paths, whose total weight is the total weight of the families in $\mathcal{I}(w)$.*

The details concerning this automaton, and a similar construction and lemma related to WEIGHTED MODULE MOTIF are omitted.

6 Applications for Parameterized Algorithms

We next show that, by merely intersecting automata obtained using Theorem 2 with problem-specific automata, we obtain problem-specific parameterized algorithms. Assume that the weights of the transitions of an automaton obtained using Theorem 2 are all 0.

Let (G, w, k) be an instance of WEIGHTED k-PATH. Now, let $M_{k,V}$ be an acyclic automaton whose language is $L_k(V)$, obtained using Theorem 2. Definition 4 and Lemma 4 directly imply that the (acyclic) intersection automaton $M_{k,V} \cap M(G, w)$ contains a path of weight W that accepts a word $w = v_1 \cdots v_m$ iff $v_1 \to \ldots \to v_m$ is a simple k-path in G of weight W. Thus, by computing a minimal weight accepting path in $M_{k,V} \cap M(G, w)$ in time $O(\text{size}(M_{k,V} \cap M(G, w)))$ (using, e.g., DFS), we solve WEIGHTED k-PATH in time $O^*(4^{k+o(k)})$.

Given an instance $\mathcal{I} = (U_1, \ldots, U_r, \mathcal{S}, w, k)$ of WEIGHTED r-DIMENSIONAL k-MATCHING, we simply consider the intersection automaton $M_{(r-1)k, U_2 \cup \ldots \cup U_r} \cap M(\mathcal{I})$, rather than $M_{k,V} \cap M(G, w)$. Similarly, given an instance $\mathcal{J} = (G, C, Col, w, k)$ of MODULE MOTIF, we simply consider the intersection automaton $M_{k,C} \cap M(\mathcal{J})$, rather than $M_{k,V} \cap M(G, w)$.

We thus obtain the following result.

Theorem 6. *The weighted variants of k-PATH, r-DIMENSIONAL k-MATCHING and MODULE MOTIF can be solved in times $O^*(4^{k+o(k)})$, $O^*(4^{(r-1)k+o((r-1)k)})$ and $O^*(4^{k+o(k)})$, respectively.*

Our approximate counting algorithms can be developed in a similar manner, while our randomized algorithms also use a fast randomized procedure that checks the emptiness of an NXA (the details are omitted).

Acknowledgement. We thank Hasan Abasi, Nader Bshouty, Michael Forbes and Amir Shpilka for helpful conversations.

References

1. Abasi, H., Bshouty, N.: A simple algorithm for undirected hamiltonicity. Electron. Colloquium Comput. Complex. (ECCC) **20**, 12 (2013)
2. Alon, N., Gutner, S.: Balanced hashing, color coding and approximate counting. In: Chen, J., Fomin, F.V. (eds.) IWPEC 2009. LNCS, vol. 5917, pp. 1–16. Springer, Heidelberg (2009)
3. Alon, N., Yuster, R., Zwick, U.: Color coding. J. Assoc. Comput. Mach. **42**(4), 844–856 (1995)
4. Ben-Basat, R.: M.Sc. thesis. Technical reports and theses, Technion (2015)
5. Björklund, A., Husfeldt, T., Kaski, P., Koivisto, M.: Narrow sieves for parameterized paths and packings. CoRR abs/1007.1161 (2010)
6. Bodlaender, H.L.: On linear time minor tests with depth-first search. J. Algorithms **14**(1), 1–23 (1993)
7. Chen, J., Feng, Q., Liu, Y., Lu, S., Wang, J.: Improved deterministic algorithms for weighted matching and packing problems. Theor. Comput. Sci. **412**(23), 2503–2512 (2011)
8. Chen, J., Friesen, D., Jia, W., Kanj, I.: Using nondeterminism to design effcient deterministic algorithms. Algorithmica **40**(2), 83–97 (2004)
9. Chen, J., Kneis, J., Lu, S., Molle, D., Richter, S., Rossmanith, P., Sze, S.H., Zhang, F.: Randomized divide-and-conquer: improved path, matching, and packing algorithms. SIAM J. Comput. **38**(6), 2526–2547 (2009)
10. Chen, J., Liu, Y., Lu, S., Sze, S.H., Zhang, F.: Iterative expansion and color coding: an improved algorithm for 3D-matching. ACM Trans. Algorithms **8**(1), 6 (2012)
11. Chen, S., Chen, Z.: Faster deterministic algorithms for packing, matching and t-dominating set problems. CoRR abs/1306.360 (2013)
12. Downey, R.G., Fellows, M.R.: Parameterized Complexity. Springer, New York (1999)
13. Downey, R.G., Fellows, M.R., Koblitz, M.: Techniques for exponential parameterized reductions in vertex set problems (Unpublished, reported in [12], Sect. 8.3)
14. Fellows, M.R., Knauer, C., Nishimura, N., Ragde, P., Rosamond, F.A., Stege, U., Thilikos, D.M., Whitesides, S.: Faster fixed-parameter tractable algorithms for matching and packing problems. Algorithmica **52**(2), 167–176 (2008)
15. Fomin, F., Lokshtanov, D., Saurabh, S.: Efficient computation of representative sets with applications in parameterized and exact agorithms. In: SODA, pp. 142–151 (2014)
16. Fomin, F.V., Lokshtanov, D., Panolan, F., Saurabh, S.: Representative sets of product families. In: ESA (2014, to appear)
17. Garey, M.R., Johnson, D.S.: Computers and Intractability. Freeman, San Francisco (1979)
18. Goyal, P., Misra, N., Panolan, F.: Faster deterministic algorithms for r-dimensional matching using representative sets. In: FSTTCS, pp. 237–248 (2013)
19. Goyal, P., Misra, N., Panolan, F., Zehavi, M.: Faster deterministic algorithms for matching and packing problems. Unpublished, results reported in [18, 40] (2014)
20. Gruber, H., Holzer, M.: Computational complexity of NFA minimization for finite and unary languages. In: LATA, pp. 261–272 (2007)
21. Hüffner, F., Wernicke, S., Zichner, T.: Algorithm engineering for color-coding with applications to signaling pathway detection. Algorithmica **52**(2), 114–132 (2008)

22. Kleitman, D.J., Spencer, J.: Families of k-independent sets. Discrete Math. **6**(3), 255–262 (1972)
23. Koutis, I.: A faster parameterized algorithm for set packing. Inf. Proc. Lett. **94**, 7–9 (2005)
24. Koutis, I.: Faster algebraic algorithms for path and packing problems. In: Aceto, L., Damgård, I., Goldberg, L.A., Halldórsson, M.M., Ingólfsdóttir, A., Walukiewicz, I. (eds.) ICALP 2008, Part I. LNCS, vol. 5125, pp. 575–586. Springer, Heidelberg (2008)
25. Koutis, I., Williams, R.: Limits and applications of group algebras for parameterized problems. In: Albers, S., Marchetti-Spaccamela, A., Matias, Y., Nikoletseas, S., Thomas, W. (eds.) ICALP 2009, Part I. LNCS, vol. 5555, pp. 653–664. Springer, Heidelberg (2009)
26. Liu, Y., Chen, J., Wang, J.: On efficient FPT algorithms for weighted matching and packing problems. In: TAMC, pp. 575–586 (2007)
27. Liu, Y., Chen, J., Wang, J.: A randomized approximation algorithm for parameterized 3-d matching counting problem. In: Lin, G. (ed.) COCOON 2007. LNCS, vol. 4598, pp. 349–359. Springer, Heidelberg (2007)
28. Liu, Y., Lu, S., Chen, J., Sze, S.-H.: Greedy localization and color-coding: improved matching and packing algorithms. In: Bodlaender, H.L., Langston, M.A. (eds.) IWPEC 2006. LNCS, vol. 4169, pp. 84–95. Springer, Heidelberg (2006)
29. M. Naor, L.J.S., Srinivasan, A.: Splitters and near-optimal derandomization. In: FOCS, pp. 182–191 (1995)
30. Mendelzon, A.O., Wood, P.T.: Finding regular simple paths in graph databases. SIAM J. Comput. **24**(6), 1235–1258 (1995)
31. Monien, B.: How to find long paths efficiently. Ann. Discrete Math. **25**, 239–254 (1985)
32. Rizzi, R., Sikora, F.: Some results on more flexible versions of graph motif. In: Hirsch, E.A., Karhumäki, J., Lepistö, A., Prilutskii, M. (eds.) CSR 2012. LNCS, vol. 7353, pp. 278–289. Springer, Heidelberg (2012)
33. Ryser, H.J.: Combinatorial Mathematics. The Carus Mathematical Monographs. The Mathematical Association of America, Washington (1963)
34. Seroussi, G., Bshouty, N.H.: Vector sets for exhaustive testing of logic circuits. IEEE Trans. Inf. Theory **34**(3), 513–522 (1988)
35. Shachnai, H., Zehavi, M.: Representative families: a unified tradeoff-based approach. In: Schulz, A.S., Wagner, D. (eds.) ESA 2014. LNCS, vol. 8737, pp. 786–797. Springer, Heidelberg (2014)
36. Vuillemin, J., Gama, N.: Compact normal form for regular languages as xor automata. In: Maneth, S. (ed.) CIAA 2009. LNCS, vol. 5642, pp. 24–33. Springer, Heidelberg (2009)
37. Wang, J., Feng, Q.: Improved parameterized algorithms for weighted 3-set packing. In: Hu, X., Wang, J. (eds.) COCOON 2008. LNCS, vol. 5092, pp. 130–139. Springer, Heidelberg (2008)
38. Wang, J., Feng, Q.: An $O^*(3.523^k)$ parameterized algorithm for 3-set packing. In: Agrawal, M., Du, D.-Z., Duan, Z., Li, A. (eds.) TAMC 2008. LNCS, vol. 4978, pp. 82–93. Springer, Heidelberg (2008)
39. Williams, R.: Finding paths of length k in $O^*(2^k)$ time. Inf. Proc. Let. **109**(6), 315–318 (2009)
40. Zehavi, M.: Deterministic parameterized algorithms for matching and packing problems. CoRR abs/1311.0484 (2013)
41. Zehavi, M.: Parameterized algorithms for module motif. In: Chatterjee, K., Sgall, J. (eds.) MFCS 2013. LNCS, vol. 8087, pp. 825–836. Springer, Heidelberg (2013)

A 14k-Kernel for Planar Feedback Vertex Set via Region Decomposition

Marthe Bonamy[1] and Łukasz Kowalik[2]($^{(\boxtimes)}$)

[1] LIRMM, Montpellier, France
[2] University of Warsaw, Warsaw, Poland
Kowalik@mimuw.edu.pl

Abstract. We show a kernel of at most 14k vertices for the PLANAR FEEDBACK VERTEX SET problem. This improves over the previous kernel of size bounded by 97k. Our algorithm has a few new reduction rules. However, our main contribution is an application of the region decomposition technique in the analysis of the kernel size.

1 Introduction

A *feedback vertex set* in a graph $G = (V, E)$ is a set of vertices $S \subseteq V$ such that $G - S$ is a forest. In the FEEDBACK VERTEX SET problem, given a graph G and integer k one has to decide whether G has a feedback vertex set of size k. This is one of the fundamental NP-complete problems, in particular it is among the 21 problems considered by Karp [8]. It has applications e.g. in operating systems (see [9]), VLSI design, synchronous systems and artificial intelligence (see [6]).

In this paper we study kernelization algorithms, i.e., polynomial-time algorithms which, for an input instance (G, k) either conclude that G has no feedback vertex set of size k or return an equivalent instance (G', k'), called *kernel*. In this paper, by the size of the kernel we mean the number of vertices of G'. Burrage et al. [5] showed that FEEDBACK VERTEX SET has a kernel of size $O(k^{11})$, which was next improved to $O(k^3)$ by Bodlaender [3] and to $4k^2$ by Thomassé [10].

In this paper we study PLANAR FEEDBACK VERTEX SET problem, i.e., FEEDBACK VERTEX SET restricted to planar graphs. Then one can get a kernel of size $O(k)$ by general tools based on the bidimensionality theory [7]. However, in this general algorithm the constants hidden in the O notation are very large. Bodlaender and Penninkx [4] gave a simple (to implement) algorithm which outputs a kernel of size at most 112k. This was improved recently by Abu-Khzam and Khuzam [1] to 97k. In this paper we improve this bound substantially, to 14k. More precisely, we show the following result.

Theorem 1.1. *There is an algorithm that, given an instance (G, k) of* PLANAR FEEDBACK VERTEX SET, *either reports that G has no feedback vertex set of*

Work partially supported by the ANR Grant EGOS (2012–2015) 12 JS02 002 01 (MB) and by the National Science Centre of Poland, grant number UMO-2013/09/B/ST6/03136 (ŁK).

© Springer International Publishing Switzerland 2014
M. Cygan and P. Heggernes (Eds.): IPEC 2014, LNCS 8894, pp. 97–109, 2014.
DOI: 10.1007/978-3-319-13524-3_9

size k or produces an equivalent instance with at most $14k - 24$ vertices. The algorithm runs in $O(kn)$ time, where n is the number of vertices of G.

To obtain Theorem 1.1, we extend the algorithms in the previous works [1,4] by a few new reduction rules. However, our main contribution is an application of the region decomposition technique in the analysis of the kernel size. Region decomposition was developed for the DOMINATING SET problem by Alber et al. [2]. Roughly, in this method the plane instance is decomposed into $O(k)$ *regions* (i.e. subsets of the plane) such that every region contains $O(1)$ vertices of the graph. We apply this approach in a slightly relaxed way: the regions are the faces of a k-vertex plane graph and the number of vertices in each region is linear in the length of the corresponding face. In [1,4] kernel size was bounded using different methods, e.g., using bounds on the number of edges in general and bipartite planar graphs. In our opinion region decomposition gives tighter bounds. In particular, we present a tight example, i.e., an example of a family of graphs which can be returned by our algorithm and have $14k - O(1)$ vertices.

Organization of the paper. In Sect. 2 we present a kernelization algorithm which is obtained from the algorithms in [1,4] by generalizing a few reduction rules, and adding some completely new rules. In Sect. 3 we present an analysis of the size of the kernel obtained by our algorithm. In the analysis we assume that in the reduced graph, for every induced path with ℓ internal vertices, the internal vertices have at least three neighbors outside the path. Based on this, we get the bound of $(2\ell + 4)k - (4\ell + 6)$ for the number of vertices in the kernel. In Sect. 2 we present reduction rules which guarantee that in the kernel $\ell \leq 6$, resulting in the kernel size bound of $16k - 30$. To get the claimed bound of $14k - 24$ vertices we use a complex set of reduction rules, which allow us to conclude that $\ell \leq 5$. Due to space limitations these additional rules are deferred to the journal version.

Notation. In this paper we deal with multigraphs, though for simplicity we refer to them as graphs. (Even if the input graph is simple, our algorithm may introduce multiple edges.) Recall that by the *degree* of a vertex x in a multigraph G, denoted by $\deg_G(x)$, we mean the number of edges incident to x in G. By $N_G(x)$, or shortly $N(x)$, we denote the set of neighbors of x. Note that in a multigraph $|N_G(x)| \leq \deg_G(x)$, but the equality does not need to hold. The neighborhood of a set of vertices S is defined as $N(S) = (\bigcup_{v \in S} N(v)) \setminus S$. For a face f in a plane graph, a *facial walk* of f is the shortest closed walk induced by all edges incident with f. The *length* of f, denoted by $d(f)$ is the length of its facial walk.

2   Our Kernelization Algorithm

In this section we describe our algorithm which outputs a $16k$-kernel for PLANAR FEEDBACK VERTEX SET. The algorithm exhaustively applies reduction rules. Each reduction rule is a subroutine which finds in polynomial time a certain structure in the graph and replaces it by another structure, so that the resulting

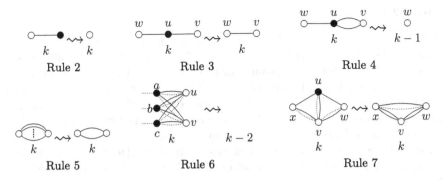

Fig. 1. Reduction rules 1–7. Dashed edges are optional. We draw in black the vertices whose incident edges are all already drawn (as solid or dashed edges), in white the vertices which might be incident to other edges. Regardless of their color, vertices in the figures may not coincide.

instance is equivalent to the original one. More precisely, we say that a reduction rule for parameterized graph problem P is *correct* when for every instance (G, k) of P it returns an instance (G', k') such that:

(a) (G', k') is an instance of P,
(b) (G, k) is a yes-instance of P iff (G', k') is a yes-instance of P, and
(c) $k' \leq k$.

Below we state the rules we use. The rules are applied in the given order, i.e., in each rule we assume that the earlier rules do not apply. We begin with some rules used in the previous works [1,4] (Fig. 1).

Rule 1. If there is a loop at a vertex v, remove v and decrease k by one.
Rule 2. Delete vertices of degree at most one.
Rule 3. If a vertex u is of degree two, with incident edges uv and uw, then delete u and add the edge vw. (Note that if $v = w$ then a loop is added.)
Rule 4. If a vertex u has exactly two neighbors v and w, edge uv is double, and edge uw is single, then delete v and u and decrease k by one.
Rule 5. If there are at least three edges between a pair of vertices, remove all but two of the edges.
Rule 6. Assume that there are five vertices a, b, c, v, w such that 1) both v and w are neighbors of each of a, b, c and 2) each vertex $x \in \{a, b, c\}$ is incident with at most one edge xy such that $y \notin \{v, w\}$. Then remove all the five vertices and decrease k by two.

The correctness of the above reduction rules was proven in [1]. (In [1], Rule 6 is formulated in a slightly less general way which forbids multiplicity of some edges, but the correctness proof stays the same.) Now we introduce a few new rules.

Rule 7. If a vertex u has exactly three neighbors v, w and x, v is also adjacent to w and x, and both edges uw and ux are single, then contract uv and add an edge wx (increasing its multiplicity if it already exists). If edge uv was not single, add a loop at v.

Lemma 2.1. *Rule 7 is correct.*

Proof. Let G' be the graph obtained from a graph G by a single application of Rule 7. Let S be a feedback vertex set of size k in G'. We claim S is a feedback vertex set in G too. Assume for a contradiction that there is a cycle C in $G - S$. Then $u \in V(C)$, for otherwise $C \subseteq G'$. If $v \in S$ then $\{wu, ux\} \subseteq C$ and $C - \{wu, ux\} + \{wx\}$ is a cycle in G', a contradiction. If $v \notin S$, then $w, x \in S$ and hence v is the only neighbor of u in $G - S$, so C is the 2-cycle uvu. But then $G' - S$ contains a loop at v, a contradiction.

Let S be a feedback vertex set of size k in G. If $|\{u, v\} \cap S| = 2$, then $S \setminus \{u\} \cup \{w\}$ is a feedback vertex set of size k in G'. Assume $|\{u, v\} \cap S| = 1$. Then we can assume $v \in S$ for otherwise we replace S by $S \setminus \{u\} \cup \{v\}$, which is also a feedback vertex set in G. If there is a cycle C in $G' - S$, then $wx \in E(C)$, for otherwise $C \subseteq G - S$. But then $C - \{wx\} + \{wu, ux\}$ is a cycle in G, a contradiction. Finally, if $|\{u, v\} \cap S| = 0$ then both w and x are in S, so S is also a feedback vertex set in G'. $\qquad\square$

Rule 8. Let $A \subseteq V(G)$ and let w_1 and w_2 be two vertices in G, $w_1, w_2 \notin A$. If (i) no cycle in $G \setminus \{w_1, w_2\}$ intersects A, and (ii) there is a subgraph $Q \subseteq G[A \cup \{w_1, w_2\}]$ with $|V(Q)| \geq 2$ such that for every vertex $x \in V(Q) \setminus \{w_1\}$, we have $\deg_Q(x) \leq |E(Q)| - |A| - 1$, then remove w_1 and decrease k by 1.

Lemma 2.2. *Rule 8 is correct.*

Proof. Let G' be the graph obtained from a graph G by a single application of Rule 8, i.e., $G' = G - w_1$. Let S be a feedback vertex set of size $k - 1$ in G'. Then every cycle in $G - S$ contains w_1, so $S \cup \{w_1\}$ is a feedback vertex set of size k in G.

Let S be a feedback vertex set of size k in G. If $w_1 \in S$, then clearly $S \setminus \{w_1\}$ is a solution of the instance $(G', k - 1)$. Hence assume $w_1 \notin S$. We claim that $|S \cap V(Q)| \geq 2$. Assume the contrary, i.e., $|S \cap V(Q)| \leq 1$. Since $Q - S$ is a forest,

$$|E(Q - S)| \leq |V(Q - S)| - 1 = |V(Q)| - |S \cap V(Q)| - 1 \leq |A| + 1 - |S \cap V(Q)|. \quad (1)$$

On the other hand, by the degree bound, and because $w_1 \notin S$ and $|S \cap V(Q)| \leq 1$,

$$|E(Q - S)| \geq |E(Q)| - (|E(Q)| - |A| - 1)|S \cap V(Q)|. \quad (2)$$

By (1) and (2), $|A| + 1 \geq |E(Q)| - (|E(Q)| - |A| - 2)|S \cap V(Q)|$. Since $|S \cap V(Q)| \leq 1$ this implies $|A| + 1 \geq |E(Q)| - (|E(Q)| - |A| - 2) = |A| + 2$, a contradiction. It follows that $|S \cap V(Q)| \geq 2$. Then $S' = S \setminus \{u, v_1, v_2, v\} \cup \{w_1, w_2\}$ is of size at most k. Moreover, S' is a feedback vertex set in G, since S is a feedback vertex set and by (i). Again, this implies that $S' \setminus \{w_1\}$ is a solution of the instance $(G', k - 1)$, as required. $\qquad\square$

Rule 8 is not used directly in our algorithm, because it seems impossible to detect it in $O(n)$ time. However, to get the claimed kernel size we need just a few special cases of Rule 8, which are stated in Lemmas 2.3, 2.4 and 2.5 below (Fig. 2).

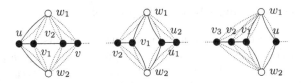

Fig. 2. Configurations in Lemmas 2.3, 2.4, and 2.5.

Lemma 2.3. *Assume there is an induced path uv_1v_2v such that for some vertices w_1, w_2 outside the path we have $N(u) = \{v_1, w_1, w_2\}$, $N(\{v_1, v_2\}) \setminus \{u, v\} \subseteq \{w_1, w_2\}$, and there is at most one edge incident to v and a vertex outside $\{w_1, w_2, v_2\}$. Then Rule 8 applies.*

Proof. It is easy to see that condition (i) of Rule 8 is satisfied. We proceed to condition (ii). By symmetry we can assume $|N(w_1) \cap \{v, v_1, v_2\}| \geq |N(w_2) \cap \{v, v_1, v_2\}|$. Let $A = \{u, v_1, v_2, v\}$. We build $E(Q)$ as follows. First, we put the five edges uw_1, uw_2, uv_1, v_1v_2, v_2v in Q. Since Rule 3 does not apply, there are no vertices of degree two in G and all of v, v_1 and v_2 are adjacent to w_1 or w_2 (or to both). For every $y \in \{v, v_1, v_2\}$, if $yw_1 \in E$, then we add edge yw_1 to Q, and otherwise we add yw_2 to Q. Thus $|E(Q)| = 8$. Moreover, since $|N(w_1) \cap \{v, v_1, v_2\}| \geq |N(w_2) \cap \{v, v_1, v_2\}|$, in this last step at least two edges added to Q are incident with w_1, and at most one to w_2. Hence, for every $x \in V(Q) \setminus \{w_1\}$ we have $\deg_Q(x) \leq 3 = |E(Q)| - |A| - 1$. □

Lemma 2.4. *Assume there are six vertices v_1, v_2, u_1, u_2, w_1, w_2 such that $N(v_1) = \{w_1, w_2, v_2\}$, $N(u_1) = \{w_1, w_2, u_2\}$, there is at most one edge incident to v_2 and a vertex outside $\{w_1, w_2, v_1\}$ and at most one edge incident to u_2 and a vertex outside $\{w_1, w_2, u_1\}$. Moreover, assume that the edges v_1v_2 and u_1u_2 are simple. Then Rule 8 applies.*

Proof. It is easy to see that condition (i) of Rule 8 is satisfied. It is easy to see that condition (i) of Rule 8 is satisfied. We proceed to condition (ii). By symmetry we can assume $|N(w_1) \cap \{v_2, v_3\}| \geq |N(w_2) \cap \{v_2, v_3\}|$. Let $A = \{u, v_1, v_2, v_3\}$. We build $E(Q)$ as follows. First, we put the six edges v_1v_2, v_1w_1, v_1w_2, v_2v_3, uw_1, and uw_2 in Q. Since Rule 3 does not apply, there are no vertices of degree two in G and both v_2 and v_3 are adjacent to w_1 or w_2 (or to both). For every $y \in \{v_2, v_3\}$, if $yw_1 \in E$, then we add edge yw_1 to Q, and otherwise we add yw_2 to Q. Thus $|E(Q)| = 8$. Moreover, since $|N(w_1) \cap \{v_2, v_3\}| \geq |N(w_2) \cap \{v_2, v_3\}|$, in this last step at least one edge added to Q is incident with w_1, and at most one to w_2. Hence, for every $x \in V(Q) \setminus \{w_1\}$ we have $\deg_Q(x) \leq 3 = |E(Q)| - |A| - 1$. □

Lemma 2.5. *Assume there are six vertices v_1, v_2, v_3, u, w_1, w_2 such that $N(v_1) = \{w_1, w_2, v_2\}$, $\{v_1, v_3\} \subseteq N(v_2) \subseteq \{w_1, w_2, v_1, v_3\}$, there is at most one edge incident to v_3 and a vertex outside $\{w_1, w_2, v_2\}$ and at most one edge incident to u and a vertex outside $\{w_1, w_2\}$. Moreover, the edges v_1v_2 and v_2v_3 are simple. Then Rule 8 applies.*

Proof. We proceed very similarly as in the proof of Lemma 2.4.   □

Rule 9. Assume there is an induced path with endpoints u and v and with six internal vertices $v_1, \ldots, v_6$ such that for some vertices w_1, w_2 outside the path $N(\{v_1, \ldots, v_6\}) \setminus \{u, v\} = \{w_1, w_2\}$. If $|N(w_1) \cap \{v_1, \ldots, v_6\}| \geq |N(w_2) \cap \{v_1, \ldots, v_6\}|$, then remove w_1 and decrease k by one.

The correctness of Rule 9 is shown in [1]. In [1] it was assumed that when Rule 9 described above is applied, G does not contain an induced path $v_1, \ldots, v_5$ such that for some vertex w, we have $N(v_2, v_3, v_4) \setminus \{v_1, v_5\} = \{w\}$. In our algorithm this is guaranteed by Rule 7 (slightly more general than their Rule 6).

To complete the analysis we need a final rejecting rule which is applied when the resulting graph is too big. In Sect. 3 we prove that Rule 10 is correct.

Rule 10. If the graph has more than $16k - 30$ vertices, return a trivial no-instance (conclude that there is no feedback vertex set of size k in G).

We are able to extend Rule 9 as follows.

Lemma 2.6. *Assume there is an induced path with endpoints u and v and with five internal vertices $v_1, \ldots, v_5$ such that for some vertices w_1, w_2 outside the path $N(\{v_1, \ldots, v_5\}) \setminus \{u, v\} = \{w_1, w_2\}$. Then there is an instance (G', k') with $|V(G')| < |V(G)|$ such that (G, k) is a yes-instance iff (G', k') is a yes-instance and $k' \leq k$.*

The proof of Lemma 2.6 contains many cases and is thus deferred to a journal version due to space limitations. We stress here that even without Lemma 2.6, in this paper we give a self-contained kernelization algorithm which returns a kernel of size at most $16k$. If one aims at a $14k$-kernel, beside adding the reduction rule from Lemma 2.6, the bound in Rule 10 should be replaced by $14k - 26$.

Running time. It is easy to verify that the whole algorithm works in $O(kn)$ time (details deferred to the journal version).

3   The Size Bound

In this section we prove the following theorem.

Theorem 3.1. *Let G be a planar graph such that rules 1–7 do not apply and G does not contain the configurations described in Lemmas 2.3, 2.4 and 2.5. Assume also that for every induced path P with endpoints u and v and with ℓ internal vertices $v_1, \ldots, v_\ell$ the internal vertices have at least three neighbors outside the path, i.e., $|N(\{v_1, \ldots, v_\ell\}) \setminus \{u, v\}| \geq 3$. If there is a feedback vertex set of size k in G, then $|V(G)| \leq (2\ell + 4)k - (4\ell + 6)$.*

Let S be a feedback vertex set of size k in G (i.e., a "solution"), and let F be the forest induced by $V(G) \setminus S$. Denote the set of vertices of F by $V_F = V(G) \setminus S$. We call the vertices in S *solution vertices* and the vertices in V_F *forest vertices*.

A partition of V_F. Now we define some subsets of V_F. Let $I_2, I_{3+} \subseteq V_F$ denote the vertices whose degree in F is two or at least three, respectively. The leaves of F are further partitioned into two subsets. Let L_2 and L_{3+} be the leaves of F that have two or at least three solution neighbors, respectively. By rules 2 and 3 all the vertices in G have degree at least 3. Hence, if a leaf of F has fewer than two solution neighbors, Rule 4 or 5 applies. It follows that every leaf of F belongs to $L_2 \cup L_{3+}$. This proves claim (i) of Lemma 3.2 below.

Lemma 3.2. *Graph G satisfies the following properties.*

 (i) *The sets I_2, I_{3+}, L_2, L_{3+} form a partition of V_F.*
 (ii) *For every pair u, v of solution vertices there are at most two vertices $x, y \in L_2$ such that $N(x) \cap S = N(y) \cap S = \{u, v\}$.*
(iii) *Every vertex of G is of degree at least three.*
(iv) *Every face of G is of length at least two.*

Claim (ii) follows from the fact that Rule 6 does not apply to G. Claim (iii) follows because rules 2 and 3 do not apply to G and Claim (iv) by Rule 1.

The inner forest. Let F_I be the forest on the vertex set $I_{3+} \cup L_{3+}$ such that $uv \in E(F_I)$ iff for some integer $i \geq 0$, there is a path $ux_1 \cdots x_i v$ in forest F such that $u, v \in I_{3+} \cup L_{3+}$ and for every $j = 1, \ldots, i$, vertex x_i belongs to I_2.

Three sets of short chains. A path in F consisting of vertices from $I_2 \cup L_2$ will be called a *chain*. A chain is maximal if it is not contained in a bigger chain. In what follows we introduce three sets of (not necessarily maximal) chains, denoted by CL_2, C_{2-} and C_{3+}. We will do it so that each vertex in I_2 belongs to *at least one* chain from these sets of chains.

For every vertex $x \in L_2$, we consider the maximal chain $(y_1, \ldots, y_p)$ of degree 2 vertices in F such that y_1 is adjacent to x and no y_i has a solution neighbor outside $N_G(x) \cap S$. Then the chain $(x, y_1, \ldots, y_p)$ is an element of CL_2. Note that $L_2 \subseteq V(CL_2)$.

Chains of C_{2-} and C_{3+} are defined using the following algorithm. We consider maximal chains in F, one by one (note that all maximal chains are vertex-disjoint). Let $c = (x_1, x_2, \ldots, x_p)$ be a maximal chain. The vertices of c are ordered so that if $\{x_1, x_p\} \cap L_2 \neq \emptyset$, then $x_p \in L_2$. Using vertices of c we form disjoint bounded length chains and put them in the sets C_{2-} and C_{3+} as follows. Assume that for some $i < p$ the vertices of a prefix $(x_1, x_2, \ldots, x_i)$ have been already partitioned into such chains (in particular $i = 0$ if we begin to process c). There are three cases to consider.

Consider a shortest chain $c_i = (x_{i+1}, \ldots, x_j)$ such that the vertices of c_i have at least three solution neighbors, i.e., $|S \cap N(\{x_{i+1}, \ldots, x_j\})| \geq 3$. If the chain c_i exists, we put it in C_{3+}, and we proceed to the next vertices of c. Otherwise we consider the chain $c'_i = (x_{i+1}, \ldots, x_p)$. Note that vertices of c'_i have at most two solution neighbors.

If $x_p \in I_2$, then we add the chain c'_i to C_{2-} and we finish processing c. Note that then x_p is adjacent to a vertex $u \in L_{3+} \cup I_{3+}$ (otherwise c is not maximal, as we can extend it by a vertex in L_2). Moreover, because of the order of the vertices in c, we know that $x_1 \notin L_2$. It follows that x_1 is also adjacent to a vertex $v \in L_{3+} \cup I_{3+}$. Hence, $uv \in E(F_I)$. We assign chain c_i' to edge uv.

If $x_p \in L_2$, then we do not form a new chain and we finish processing c. Note, however, that the vertices $\{x_{i+1}, \ldots, x_p\} \cap I_2$ belong to a chain in CL_2.

Note also that some vertices of the first chain c_0 can belong to two chains, one in C_{3+} and one in CL_2.

Let us summarize the main properties of the construction.

Lemma 3.3. *The following properties hold:*

(i) Every vertex from I_2 belongs to a chain in CL_2, C_{2-} or C_{3+}.

(ii) Every chain in $CL_2 \cup C_{2-}$ has at most two solution neighbors.

(iii) Every chain in C_{3+} has at least three solution neighbors.

(iv) Every chain in C_{2-} is assigned to a different edge of inner forest F_I.

(v) Every chain in $C_{2-} \cup CL_2$ has at most $\ell - 1$ vertices.

(vi) Every chain in C_{3+} has at most ℓ vertices.

A solution graph H_S. Let us introduce a new plane multigraph $H_S = (S, E_S)$. Since the vertices of H_S are the solution vertices we call it a *solution graph*. From now on, we fix a plane embedding of G. The vertices of H_S are embedded in the plane exactly in the same points as in G. The edge multiset E_S is defined as follows. For every triple (u, x, v) such that $u, v \in S$, $x \in L_2$ and there is a path uxv in G, we put an edge uv in E_S. Moreover, the edge uv is embedded in the plane exactly as one of the corresponding paths uxv (note that there can be up to four such paths if some edges are double). Note that by Lemma 3.2(ii), every edge of H_S has multiplicity at most two.

The set of faces of H_S is denoted by F_S. By $F_{S,2}$ we denote its subset with the faces of length two, while $F_{S,3+}$ are the remaining faces. Note that there are no faces of length 1 in H_S.

Lemma 3.4. *We have $|V(CL_2)| \leq 3(|E_S| - |F_{S,2}|)$.*

Proof. By the definition, for every vertex $x \in L_2$ there is a corresponding edge $uv \in E_S$, where $N_G(x) \cap S = \{u, v\}$. Also, for every chain c in CL_2 there is a corresponding vertex $x \in L_2$, and thus a corresponding edge $uv \in E_S$. We *assign* x, c and the vertices of c to the pair $\{u, v\}$.

Consider an arbitrary pair u, v such that $uv \in E_S$. Note that there are exactly $|E_S| - |F_{S,2}|$ such pairs. We claim that there are at most three elements in $V(CL_2)$ assigned to the pair $\{u, v\}$. Indeed, by Lemma 3.2(ii), there are at most two vertices in L_2 assigned to $\{u, v\}$. If there are no such vertices, no chain in CL_2 is assigned to $\{u, v\}$, so the claim holds. If there is exactly one vertex $x \in L_2$ assigned, there is exactly one chain $c \in CL_2$ assigned. By Lemma 2.3, chain c has at most three vertices, so the claim holds. Finally, if there are exactly two vertices $x, y \in L_2$ assigned, there are exactly two chains c_x and c_y assigned. By Lemmas 2.4 and 2.5 we have $|V(c_x)| + |V(c_y)| \leq 3$. This concludes the proof. $\qquad\square$

Maximality. In what follows we assume that graph G is *maximal*, meaning that one can add neither an edge to $E(G)$ nor a vertex to L_2 obtaining a graph G' such that S is still a feedback vertex set of G' and all the claims of Lemmas 3.2, 3.3 and 3.4 hold. Note that the number of L_2-vertices which can be added to G is bounded, since each such vertex corresponds to an edge in H_S, and H_S has at most $6|S|$ edges as a plane multigraph with edge multiplicity at most two. Similarly, once the set of L_2-vertices is maximal, and hence the vertex set of G is fixed, the number of edges which can be added to G is bounded by $6|V(G)|$. It follows that such a maximal supergraph of G exists. Clearly, it is sufficient to prove Theorem 3.1 only in the case when G is maximal.

Lemma 3.5. *The planar graph H_S is connected.*

Proof. Assume now for contradiction that there is a partition $S = S_1 \cup S_2$ such that there is no edge in H_S between a vertex of S_1 and a vertex of S_2.

Every face of G is incident to at least one vertex of S, for otherwise the boundary of the face does not contain a cycle, a contradiction. Assume that a face f of G contains a solution vertex u_1 in S_1 and a solution vertex u_2 in S_2. Then we can add a vertex x, two edges xu_1 and two edges xu_2. Note that S is still a feedback vertex set in the new graph; in particular now $x \in L_2$. In the new graph there are no more vertices in L_2 adjacent to both u_1 and u_2 because of our assumption that S_1 and S_2 are not connected by an edge in H_S, so Lemma 3.2(*ii*) holds. Moreover, $|V(CL_2)|$ was increased by one and $|E_S| - |F_{S,2}|$ was also increased by one, so Lemma 3.4 holds. The other claims of Lemmas 3.2 and 3.3 trivially hold, so F is not maximal, a contradiction.

Let $\mathcal{F}_1$ and $\mathcal{F}_2$ be the collections of faces of G containing a vertex in S_1, or in S_2, respectively. We have shown above that $\mathcal{F}_1 \cup \mathcal{F}_2$ is a partition of the set of all the faces of G. Let V_1 and V_2 denote the sets of vertices incident to a face in $\mathcal{F}_1$, or in $\mathcal{F}_2$, respectively. Note that $V_1 \cap V_2 \neq \emptyset$, since there must be two neighboring faces, one in $\mathcal{F}_1$ and the other in $\mathcal{F}_2$. Let $x \in V_1 \cap V_2$. Since faces of G are of length at least two, x has in G at least two neighbors in $V_1 \cap V_2$. It follows that $G[V_1 \cap V_2]$ has minimum degree two, so $G[V_1 \cap V_2]$ contains a cycle. However, $(V_1 \cap V_2) \cap S = \emptyset$, since $\mathcal{F}_1$ and $\mathcal{F}_2$ are disjoint. Hence $V_1 \cap V_2 \subseteq F$, a contradiction. $\qquad\square$

Bounding the number of forest vertices in a face of H_S. For a face f of H_S and a set of vertices $A \subseteq V(G)$ we define A^f as the subset of A of vertices which are embedded in f or belong to the boundary of f. Note that all vertices of every chain belong to the same face f of H_S. When C is a set of chains, by C^f we denote the subset of chains of C which lie in f, i.e., $C^f = \{c \in C : V(c) \subseteq V(G)^f\}$.

Lemma 3.6. *For every face f of H_S, it holds that $|L^f_{3+}| + |I^f_{3+}| + |C^f_{3+}| \leq d(f) - 2$.*

Proof. First we note that the forest F^f is in fact a tree. Indeed, if F^f has more than one component, we can add an edge between two solution vertices on the boundary of f preserving planarity, what contradicts the assumed maximality.

Consider a plane subgraph A of G induced by $V(G)^f$, i.e., we take the plane embedding of G and we remove the vertices outside $V(G)^f$. Then we can define graph A_S, analogously to H_S. We treat f as a face of A_S. Let $u_1u_2\cdots u_{d(f)}u_1$ be the facial walk of f.

Consider an arbitrary vertex x of I_{3+}^f. Let $T_1,\ldots,T_r$ be the r trees obtained from the tree T in F containing x after removing r from T. Then $r \geq 3$ since x has at least three neighbors in T. By planarity, there are $2r$ indices $b_1, e_1, b_2, e_2, \ldots, b_r, e_r$ such that for every $i = 1,\ldots,r$

$$\{u_{b_i}, u_{e_i}\} \subseteq N(V(T_t)) \cap \{u_1,\ldots,u_{d(f)}\} \subseteq \{u_{b_i}, u_{b_i+1},\ldots,u_{e_i}\}.$$

Then, for every $j \in \{b_1,b_2,\ldots,b_r\}$ there is an edge xu_j, for otherwise we can add it in the current plane embedding, contradicting the maximality of G. This means that every vertex in I_{3+}^f has at least three neighbors in $\{u_1, u_2, \ldots, u_{d(f)}\}$.

We further define B as the plane graph obtained from A by (1) replacing every triple (u, x, v) where $x \in L_2$, $u, v \in S$ and uxv forms a path by a single edge uv, (2) removing vertices of $V(CL_2)$, (3) contracting every chain from C_{3+} into a single vertex, and (4) contracting every chain from C_{2-} into a single edge. By (4) we mean that every maximal chain $d = x_1,\ldots,x_i$ of I_2 vertices which is contained in a chain from C_{2-}, is replaced by the edge yz where y and z are the forest neighbors (in $L_{3+} \cup I_{3+}$) of x_1 and x_i outside the chain d. Let us call the vertices of B that are not on the boundary of f as *inner vertices*.

Note that the set of inner vertices is in a bijection with $L_{3+}^f \cup I_{3+}^f \cup C_{3+}^f$. Moreover, I forms a tree, since F^f is a tree. Also, each inner vertex has at least three neighbors in $\{u_1, u_2, \ldots, u_{d(f)}\}$. We show that $|I| \leq d(f) - 2$ by the induction on $d(f)$. When $d(f) = 2$ the claim follows since each inner vertex has at least three neighbors on the boundary of f. Now assume $d(f) > 2$. Let x be leaf in the tree I. Then the edges from x to the boundary of face f split F into at least three different faces. The subtree $I - x$ lies in one of these faces, say face bounded by the cycle $xu_iu_{i+1}\cdots u_jx$. We remove x and vertices $u_{j+1},\ldots,u_{i-1}$ (there is at least one of them) and we add edge u_iu_j. The outer face of the resulting graph is of length at most $d(f) - 1$, so we can apply induction and the claim follows. $\square$

Lemma 3.7. *For every face f in H_S of length at least three,*

$$|V_F^f \setminus V(CL_2^f)| \leq \ell \cdot (d(f) - 2) - (\ell - 1).$$

Proof. We have

$$|V_F^f \setminus V(CL_2^f)| \leq |L_{3+}^f| + |I_{3+}^f| + |V(C_{3+}^f)| + |V(C_{2-}^f)|.$$

By Lemma 3.3(v) we get

$$|V_F^f \setminus V(CL_2^f)| \leq |L_{3+}^f| + |I_{3+}^f| + \ell|C_{3+}^f| + (\ell - 1)|C_{2-}^f|. \tag{3}$$

By Lemma 3.3(iv), $|C_{2-}^f|$ is bounded by the number of edges of the inner forest F_I. Hence, $|C_{2-}^f| \leq |L_{3+}^f| + |I_{3+}^f| - 1$ when $|L_{3+}^f| + |I_{3+}^f| > 0$ and $|C_{2-}^f| = 0$

otherwise. In the prior case, by (3) we get that

$$|V_F^f \setminus V(CL_2^f)| \le \ell(|L_{3+}^f| + |I_{3+}^f| + |C_{3+}^f|) - (\ell - 1),$$

and the result then follows from Lemma 3.6. Hence it suffices to prove the claim when $|L_{3+}^f| = |I_{3+}^f| = |C_{2-}^f| = 0$. Then the forest F^f is a non-empty collection of paths, each with both endpoints in L_2. Let c be such a path on p vertices $x_1, \ldots, x_p$. Then $x_1 \in L_2$ and x_1 has exactly two neighbors u, v in S. Let i be the largest such that $N(\{x_1, \ldots, x_i\}) \cap S = \{u, v\}$. By definition, $(x_1, \ldots, x_i)$ is a chain in CL_2^f. We infer that if $i = p$ for every such path, then $|V_F^f \setminus V(CL_2^f)| = 0$ and the claim follows. Hence we can assume that $i < p$, i.e., x_{i+1} has a neighbor in $S \setminus \{u, v\}$. Then, by definition, $(x_1, \ldots, x_{i+1})$ is a chain in C_{3+}^f. Since $(x_1, \ldots, x_i) \in CL_2^f$, we get $|\{x_1, \ldots, x_{i+1}\} \setminus V(CL_2^f)| = 1$. Hence,

$$|V_F^f \setminus V(CL_2^f)| \le 1 + \ell(|C_{3+}^f| - 1),$$

what, by Lemma 3.6, is bounded by $1 + \ell \cdot (d(f) - 3) = \ell \cdot (d(f) - 2) - (\ell - 1)$, as required. $\square$

Lemma 3.8. *For every face f in H_S of length two, $V_F^f \subseteq V(CL_2^f)$.*

Proof. Since the boundary of f has only two solution vertices, F^f contains no vertices of L_{3+}^f, $V(C_{3+})^f$ or I_{3+}^f. Then by Lemma 3.3(iv), C_{2-}^f is also empty. The claim follows. $\square$

Now we proceed to the bound of Theorem 3.1. By Lemmas 3.7 and 3.8 we have

$$|V_F| \le |V(CL_2)| + \sum_{f \in F_{S,3+}} (\ell(d(f) - 2) - (\ell - 1))$$

By Lemma 3.4 we get

$$|V_F| \le 3(|E_S| - |F_{S,2}|) + \sum_{f \in F_{S,3+}} (\ell(d(f) - 2) - (\ell - 1))$$

$$= 3(|E_S| - |F_{S,2}|) + \sum_{f \in F_S} (\ell(d(f) - 2) - (\ell - 1)) + (\ell - 1)|F_{2,S}|$$

$$= (2\ell + 3)|E_S| - (3\ell - 1)|F_S| + (\ell - 4)|F_{2,S}|$$

$$= (2\ell + 3)|E_S| - (2\ell + 3)|F_S| - (\ell - 4)|F_S| + (\ell - 4)|F_{2,S}|$$

$$\le (2\ell + 3)(|E_S| - |F_S|).$$

By Lemma 3.5 graph H_S is connected, so we can apply Euler's formula $|S| - |E_S| + |F_S| = 2$. Thus,

$$|V(G)| = |V_F| + |S| \le (2\ell + 3)(|S| - 2) + |S|,$$
$$= (2\ell + 4)k - (4\ell + 6).$$

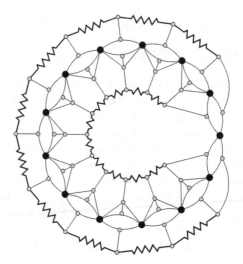

Fig. 3. A tight example. The big black vertices are solution vertices, the small grey ones are forest vertices. The zigzag edges represent paths of $\ell - 1$ forest vertices, each adjacent to the two available solution vertices. Asymptotically for larger cycles, we have $2\ell + 3$ forest vertices for each solution vertex.

This concludes the proof of Theorem 3.1. For $\ell = 6$, we get $|V(G)| \leq 16k - 30$. If we use Lemma 2.6, we can put $\ell = 5$, which results in $|V(G)| \leq 14k - 24$. In Fig. 3 we show an example of a graph, where our reduction rules do not apply and our analysis is tight (up to a constant additive term).

Note. We have learned that very recently Xiao [11] obtained independently a $29k$-kernel for PLANAR FEEDBACK VERTEX SET.

References

1. Abu-Khzam, F.N., Bou Khuzam, M.: An improved kernel for the undirected planar feedback vertex set problem. In: Thilikos, D.M., Woeginger, G.J. (eds.) IPEC 2012. LNCS, vol. 7535, pp. 264–273. Springer, Heidelberg (2012)
2. Alber, J., Fellows, M.R., Niedermeier, R.: Polynomial-time data reduction for dominating set. J. ACM **51**(3), 363–384 (2004)
3. Bodlaender, H.L.: A cubic kernel for feedback vertex set. In: Thomas, W., Weil, P. (eds.) STACS 2007. LNCS, vol. 4393, pp. 320–331. Springer, Heidelberg (2007)
4. Bodlaender, H.L., Penninkx, E.: A linear kernel for planar feedback vertex set. In: Grohe, M., Niedermeier, R. (eds.) IWPEC 2008. LNCS, vol. 5018, pp. 160–171. Springer, Heidelberg (2008)
5. Burrage, K., Estivill-Castro, V., Fellows, M.R., Langston, M.A., Mac, S., Rosamond, F.A.: The undirected feedback vertex set problem has a poly(k) kernel. In: Bodlaender, H.L., Langston, M.A. (eds.) IWPEC 2006. LNCS, vol. 4169, pp. 192–202. Springer, Heidelberg (2006)

6. Festa, P., Pardalos, P., Resende, M.: Feedback Set Problems. Encyclopedia of Optimization, pp. 1005–1016. Springer, New York (2009)
7. Fomin, F.V., Lokshtanov, D., Saurabh, S., Thilikos, D.M.: Bidimensionality and kernels. In: SODA, pp. 503–510 (2010)
8. Karp, R.: Reducibility among combinatorial problems. In: Miller, R., Thatcher, J. (eds.) Complexity of Computer Computations, pp. 85–103. Plenum Press, New York (1972)
9. Silberschatz, A., Galvin, P.B., Gagne, G.: Operating System Concepts, 8th edn. Wiley Publishing, New York (2008)
10. Thomassé, S.: A $4k^2$ kernel for feedback vertex set. ACM Trans. Algorithms **6**(2), 1–8 (2010)
11. Xiao, M.: A new linear kernel for undirected planar feedback vertex set: smaller and simpler. In: Gu, Q., Hell, P., Yang, B. (eds.) AAIM 2014. LNCS, vol. 8546, pp. 288–298. Springer, Heidelberg (2014)

The Complexity of Bounded Length Graph Recoloring and CSP Reconfiguration

Paul Bonsma[1]($\boxtimes$), Amer E. Mouawad[2], Naomi Nishimura[2], and Venkatesh Raman[3]

[1] Faculty of EEMCS, University of Twente, Enschede, The Netherlands
p.s.bonsma@ewi.utwente.nl
[2] David R. Cheriton School of Computer Science, University of Waterloo, Waterloo, Canada
{aabdomou,nishi}@uwaterloo.ca
[3] The Institute of Mathematical Sciences, Chennai, India
vraman@imsc.res.in

Abstract. In the first part of this work we study the following question: Given two k-colorings α and β of a graph G on n vertices and an integer ℓ, can α be modified into β by recoloring vertices one at a time, while maintaining a k-coloring throughout and using at most ℓ such recoloring steps? This problem is weakly PSPACE-hard for every constant $k \geq 4$. We show that the problem is also strongly NP-hard for every constant $k \geq 4$ and W[1]-hard (but in XP) when parameterized only by ℓ. On the positive side, we show that the problem is fixed-parameter tractable when parameterized by $k + \ell$. In fact, we show that the more general problem of ℓ-length bounded reconfiguration of constraint satisfaction problems (CSPs) is fixed-parameter tractable parameterized by $k + \ell + r$, where r is the maximum constraint arity and k is the maximum domain size. We show that for parameter ℓ, the latter problem is W[2]-hard, even for $k = 2$. Finally, if p denotes the number of variables with different values in the two given assignments, we show that the problem is W[2]-hard when parameterized by $\ell - p$, even for $k = 2$ and $r = 3$.

1 Introduction

For any graph G and integer k, the *k-Color Graph* $\mathcal{C}_k(G)$ has as vertex set all (proper) k-colorings of G, where two colorings are adjacent if and only if they differ on exactly one vertex. Given an integer k and two k-colorings α and β of G, the *Coloring Reachability* problem asks if there exists a path in $\mathcal{C}_k(G)$ from α to β. This is a well-studied problem, which is known to be solvable in polynomial time for $k \leq 3$ [7], and PSPACE-complete for every constant $k \geq 4$, even for bipartite graphs [3]. For any $k \geq 4$, examples have been explicitly constructed where any path from α to β has exponential length [3]. On the other hand, for $k \leq 3$, the diameter of components of $\mathcal{C}_k(G)$ is known to be polynomial [7].

Amer E. Mouawad, Naomi Nishimura— Research supported by the Natural Science and Engineering Research Council of Canada.

M. Cygan and P. Heggernes (Eds.): IPEC 2014, LNCS 8894, pp. 110–121, 2014.
DOI: 10.1007/978-3-319-13524-3_10

Similar questions can be formulated for almost any search problem: After defining a symmetric adjacency relation between solutions, the *reconfiguration graph* for a problem instance has as vertex set all solutions, with undirected edges defined by the adjacency relation. Such reconfiguration questions have received considerable attention in recent literature; see e.g. the survey by Van den Heuvel [13]. The most well-studied questions are related to the complexity of the *reachability problem*: Given two solutions α and β, does there exist a path from α to β in the reconfiguration graph? In most cases, the reachability problem is PSPACE-hard in general, although polynomial-time solvable restricted cases can be identified. For PSPACE-hard cases, it is not surprising that shortest paths between solutions can have exponential length. More surprisingly, for most known polynomial-time solvable cases, shortest paths between solutions have been shown to have polynomial length. Results of this kind have for instance been obtained e.g. for the reachability of independent sets [4,17], vertex covers [19], shortest paths [1,2,16], or Boolean satisfiability (SAT) assignments [12].

There are various motivations for studying reconfiguration problems [13], and for studying Coloring Reachability in particular (see [6,13,14]). For example, reconfiguration problems model dynamic situations in which we seek to transform a solution into a more desirable one, maintaining feasibility during the process (see [14] for such an application of Coloring Reachability). However, in many applications of reconfiguration problems, the existence of a path between two solutions is irrelevant if every such path has exponential length. So the more important question is in fact: Does there exist a path between two solutions of length at most ℓ, for some integer ℓ? Results on such *length-bounded* reachability questions have been obtained in [2,12,16,19,20]. In some cases where the existence of paths between solutions can be decided efficiently, one can in fact find *shortest* paths efficiently [2,12]. On the other hand, NP-hard cases have also been identified [16,19]. If we wish to obtain a more detailed picture of the complexity of length-bounded reachability, the framework of parameterized complexity [9,10] is very useful, where we choose ℓ as parameter. We refer to [9,10] for an introduction to parameterized complexity and fixed parameter tractable (FPT) algorithms. A systematic study of the parameterized complexity of reachability problems was initiated by Mouawad et al. [20]. However, in [20], only negative results were obtained for length-bounded reachability: various problems were identified where the problem was not only NP-hard, but also W[1]-hard, when parameterized by ℓ (or even when parameterized by $k + \ell$, where k is another problem parameter). In this paper, we give a first example of a length-bounded reachability problem that is NP-hard, but admits an FPT algorithm. Another example, namely Length-Bounded Vertex Cover Reachability on graphs of bounded degree, was very recently obtained by Mouawad et al. in [19].

Our Results. We first study the *Length-Bounded Coloring Reachability (LBCR)* problem: Given is a graph G on n vertices, nonnegative integers k and ℓ, and two k-colorings α and β of G. The question is whether $\mathcal{C}_k(G)$ contains a path from α to β of length at most ℓ. We fully explore how the complexity of the above problem depends on the problem parameters k and ℓ (when viewed as *input variables* or *constants/parameters*). Using a reduction from Coloring Reachability [3], LBCR is easily observed to be PSPACE-hard in general, for any constant $k \geq 4$: Since there are at most k^n different k-colorings of a graph on n vertices, a path from α to β exists if and only if there exists one of length at most k^n. Nevertheless, this only establishes *weak* PSPACE-hardness, since the chosen value of $\ell = k^n$ is exponential in the instance size. In other words, if we require that all integers are encoded in unary, then this is not a polynomial reduction. And indeed, the complexity status of the problem changes under that requirement; in that case, LBCR is easily observed to be in NP. In Sect. 3, we show that LBCR is in fact NP-complete when ℓ is encoded in unary, or in other words, it is *strongly NP-hard*. On the positive side, in Sect. 4, we show that the problem can be solved in time $\mathcal{O}(2^{k(\ell+1)} \cdot \ell^\ell \cdot \mathrm{poly}(n))$. This establishes that LBCR is *fixed parameter tractable (FPT)* when parameterized by $k + \ell$. (We remark that this result was also obtained independently by Johnson et al. [15]. The algorithm in [15] is very different however.) One may ask whether the problem is still FPT when only parameterized by ℓ. In Sect. 3 we show that this is not the case (unless W[1]=FPT), by showing that LBCR is W[1]-hard when only parameterized by ℓ. We observe however that a straightforward branching algorithm can solve the problem in time $n^{\mathcal{O}(\ell)}$, hence in polynomial time for any constant ℓ. In other words, LBCR is in XP, parameterized by ℓ.

Our algorithmic results hold in fact for a much larger class of problems: In a *constraint satisfaction problem (CSP)*, we are given a set X of n variables, which all can take on at most k different values. In addition, a set $\mathcal{C}$ of *constraints* is given, all of arity at most r. Every constraint consists of a subset $T \subseteq X$ of variables with $|T| \leq r$, and a set of allowed value combinations for these variables. A k-coloring can be seen as a CSP solution, where the edges correspond to binary constraints, stating that the two incident vertices/variables cannot have the same color/value. The *Length-Bounded CSP Reachability (LBCSPR)* problem asks, given two satisfying variable assignments α and β for a CSP instance $(X, k, \mathcal{C})$, whether there exists a path from α to β of length at most ℓ. (Two solutions are adjacent if they differ in one variable. See Sect. 4 for precise definitions.) In Sect. 4, we give our main result: an FPT algorithm for LBCSPR, parameterized by $\ell + k + r$. This result has many implications, besides the aforementioned result for LBCR: For instance, it follows that Length-Bounded Boolean SAT Reachability is FPT, parameterized by $\ell + r$. In addition, it implies that Length-Bounded Shortest Path Reachability is FPT, parameterized by $\ell + k$, where k is an upper bound on the number of vertices in one distance layer (See [12] resp. [1,2,16] for more details on these problems). This result prompts two further questions: Firstly, is it possible to also obtain an FPT algorithm for LBCSPR for parameter $\ell + k$? Secondly, clearly any reconfiguration sequence from α to β has length at least p, where $p = |\{x \in X \mid \alpha(x) \neq \beta(x)\}|$. Is it also possible to obtain an FPT algorithm for LBCSPR for parameter $(\ell - p) + k + r$? (This is an *above-guarantee* parameterization). In Sect. 5, we give two W[2]-hardness results that show that the answer to these questions is negative (unless FPT = W[2]). These W[2]-hardness results hold in fact for the restricted case of Boolean SAT instances with only Horn clauses. Together, these hardness results show that

Table 1. Complexity of LBCSPR for different parameterizations

Parameter:	Complexity:
$k + \ell + r$	FPT
$k + r$	para-NP-complete (ℓ unary) / para-PSPACE-complete (ℓ binary) (already for $k = 4$, $r = 2$; Coloring instances)
$k + \ell$	W[2]-hard (already for $k = 2$; Horn SAT instances), in XP
$r + \ell$	W[1]-hard (already for $r = 2$; Coloring instances), in XP
$k + r + \ell - p$	W[2]-hard (already for $k = 2$, $r = 3$; Horn 3SAT instances)

our FPT result for LBCSPR is tight (assuming $FPT \neq W[1]$): to obtain an FPT algorithm, all three variables ℓ, k, and r need to be part of the parameter. See also Table 1, which summarizes our results, and the complexity status of LBCSPR for all different parameterizations in terms of ℓ, k, r and p. (Omitted parameter combinations follow directly from the given rows.)

2   Preliminaries

For general graph theoretic definitions, we refer the reader to the book of Diestel [8]. Let u and v be vertices in a graph G. A *pseudowalk* from u to v of length ℓ is a sequence $w_0, \ldots, w_\ell$ of vertices in G with $w_0 = u$, $w_\ell = v$, such that for every $i \in \{0, \ldots, \ell - 1\}$, either $w_i = w_{i+1}$ or $w_i w_{i+1} \in E(G)$. A k-*coloring* for a graph G is a function $\alpha : V(G) \to \{1, \ldots, k\}$ that assigns *colors* to the vertices of G, such that for all $uv \in E(G)$, $\alpha(u) \neq \alpha(v)$. A graph that admits a k-coloring is called k-*colorable*. Pseudowalks in $\mathcal{C}_k(G)$ from α to β are also called k-*recoloring sequences* from α to β. If there exists an integer k such that $\alpha_0, \ldots, \alpha_m$ is a k-recoloring sequence, then this is called a *recoloring sequence* from α_0 to α_m.

A k-*color list assignment* for a graph G is a mapping L that assigns a *color list* $L(v) \subseteq \{1, \ldots, k\}$ to each vertex $v \in V(G)$. A k-coloring α of G is an *L-coloring* if $\alpha(v) \in L(v)$ for all v. By $\mathcal{C}(G, L)$ we denote the subgraph of $\mathcal{C}_k(G)$ induced by all L-colorings of G, and pseudowalks in $\mathcal{C}(G, L)$ are called L-*recoloring sequences*. The *Length-Bounded L-Coloring Reachability (LB L-CR)* problem asks, given G, L, α, β, and ℓ, where α and β are L-colorings of G, whether there exists an L-recoloring sequence from α to β of length at most ℓ.

For a positive integer $k \geq 1$, we let $[k] = \{1, \ldots, k\}$. For a function $f : D \to I$ and subset $D' \subseteq D$, we denote by $f|_{D'}$ the restriction of f to the domain D'. The (unique) trivial function with empty domain is denoted by $f^\emptyset$. Note that for any function g, $g|_\emptyset = f^\emptyset$. We use $\text{poly}(x_1, \ldots, x_p)$ to denote a polynomial function on variables $x_1, \ldots, x_p$.

3   Hardness Results for Coloring Reachability

To prove W[1]-hardness for LBCR parameterized by ℓ, we give a reduction from the t-*Independent Set (t-IS)* problem. Given a graph G and a positive integer t, t-IS asks whether G has an independent set of size at least t.

The t-IS problem is known to be W[1]-hard [9,10] when parameterized by t. We will also use the following result, which was shown independently by Cereceda [5], Marcotte and Hansen [18] and Jacob [14]: For every pair of k-colorings α and β of a graph G, there exists a path from α to β in $\mathcal{C}_{2k-1}(G)$, and there are examples where at least $2k-1$ colors are necessary. The graphs constructed in [5,14,18] to prove the latter result are in fact very similar. We will use these graphs for our reduction. For any integer $k \geq 1$, the graph B_k has vertex set $V(B_k) = \{b_j^i \mid i,j \in \{1, \ldots, k\}\}$, and two vertices b_j^i and $b_{j'}^{i'}$ are adjacent if and only if $i \neq i'$ and $j \neq j'$. Define two k-colorings α^k and β^k for B_k by setting $\alpha^k(b_j^i) = i$ and $\beta^k(b_j^i) = j$ for all vertices b_j^i.

Theorem 1 ([5],*)[1]. *Let B_k, α^k and β^k be as defined above (for $k \geq 1$). Then (i) every recoloring sequence from α^k to β^k contains a coloring that uses at least $2k-1$ different colors, and (ii) there is a $(2k-1)$-recoloring sequence of length at most $2k^2$ from α^k to β^k.*

Theorem 2 (*). *LBCR is W[1]-hard when parameterized by ℓ.*

Proof sketch: For ease of presentation, we give a reduction from the $(t-1)$-IS problem, which remains W[1]-hard. Given an instance $(G, t-1)$ of $(t-1)$-IS, where $G = (V, E)$ and $V = \{v_1, \ldots, v_n\}$, we construct a graph G' in time polynomial in $|V(G)|$ as follows. (We will use $n + t + 1$ colors.)

G' contains a copy of G and a copy of B_t with all edges between them. In addition, G' contains $n + t + 1$ independent sets C_1, ..., C_{n+t+1}, each of size $2t + 2t^2$ and disjoint from the copies of G and B_t. We say that C_i (for $1 \leq i \leq n+t+1$) is a *color-guard set*, as it will be used to enforce some coloring constraints; in the colorings we define, and all colorings reachable from them using at most $|C_i| - 1$ recolorings, C_i will contain at least one vertex of color i. We let $V_G = \{g_1, \ldots, g_n\}$, $V_B = \{b_j^i \mid i,j \in \{1, \ldots, t\}\}$, $V_C = C_1 \cup \ldots \cup C_{n+t+1}$, and hence $V(G') = V_G \cup V_B \cup V_C$. The total number of vertices in G' is therefore $n + t^2 + (n+t+1)(2t+2t^2)$. For every vertex $g_i \in V_G$, we add all edges between g_i and the vertices in $V_C \setminus (C_i \cup C_{n+t+1})$. Similarly, for every vertex $b \in V_B$, we add all edges between b and the vertices in C_{n+t+1}. We define α as follows. For every vertex $g_i \in V_G$, $1 \leq i \leq n$, we set $\alpha(g_i) = i$. For every $i \in \{1, \ldots, n+t+1\}$ and every vertex $c \in C_i$, we set $\alpha(c) = i$. For every vertex $b_j^i \in V_B$, we choose $\alpha(b_j^i) = n+i$. Considering α and the color guard sets, which all have size $2t+2t^2$, we conclude that for all recoloring sequences $\gamma_0, \ldots, \gamma_p$ with $p \leq 2t + 2t^2$ and $\gamma_0 = \alpha$, for every i and j it holds that $\gamma_j(g_i) \in \{i, n+t+1\}$, and for all $b \in V_B$ and j it holds that $\gamma_j(b) \neq n+t+1$. Finally, we define the target coloring β. For every vertex $v \in V_G \cup V_C$ we set $\beta(v) = \alpha(v)$. For every vertex $b_j^i \in V_B$ (with $i,j \in \{1, \ldots, t\}$), we choose $\beta(b_j^i) = n + j$. So the goal is to change from a 'row coloring' to a 'column coloring' for V_B, while maintaining the same coloring for vertices in $V_G \cup V_C$.

---

[1] A star indicates that (additional) proof details will be given in the full version of the paper.

It can be shown that $\mathcal{C}_k(G')$ contains a path from α to β of length at most $\ell = 2t + t^2$ if and only if G has an independent set S at size at least $t - 1$: If there exists such a set S, then these vertices can be recolored to color $n + t + 1$, which makes $t - 1$ colors available to recolor V_B from a row coloring to a column coloring. That is, the $(2t - 1)$-recoloring sequence of length at most $2t^2$ from Theorem 1 can be applied. Next, the vertices in G are recolored to their original color again. This procedure yields β and uses at most $2t + 2t^2$ recoloring steps in total. If there exists a recoloring sequence from α to β, then this contains a coloring γ that assigns at least $2t - 1$ different colors to V_B (Theorem 1). This includes at least $t - 1$ colors that originally appeared in V_G, on a vertex set S. As observed above, these vertices are then all colored with color $n + t + 1$ in γ, so they form an independent set with $|S| \geq t - 1$. □

Next, we show that the LBCR problem is strongly NP-hard for every fixed constant $k \geq 4$. We give a reduction from the *Planar Graph 3-Colorability (P3C)* problem, which is known to be NP-complete [11]. Given a planar graph G, P3C asks whether G is 3-colorable. In fact we construct an instance of the LB L-CR problem. It was observed in [3] that an instance $(G, L, \alpha, \beta, \ell)$ of the LB L-CR problem with $L(v) \subseteq \{1, \ldots, 4\}$ for all v is easily transformed to an instance $(G', \alpha, \beta, \ell)$ of LBCR, for any $k \geq 4$, by adding one complete graph on k vertices x_i with $i \in \{1, \ldots, k\}$ and $\alpha(x_i) = \beta(x_i) = i$, and edges vx_i for every vertex $v \in V(G)$ and $i \notin L(v)$.

The proof of Lemma 3 makes heavy use of the notion of (a, b)-forbidding paths and their properties, which were introduced in [3]. Informally, these are paths that can be added between any pair of vertices u and v (provided that $L(u), L(v) \neq \{1, \ldots, 4\}$), that function as a special type of edge, which only excludes the color combination (a, b) for u and v respectively, but allows (recoloring to) any other color combination. For any combination of a, b and $L(u), L(v) \neq \{1, \ldots, 4\}$, there exists such a path, of length six, with all color lists in $\{1, \ldots, 4\}$.

Lemma 3 (*). *There exists a graph H (on $\mathcal{O}(1)$ vertices) with color lists L and vertices $u, v, z \in V(H)$ with $L(u) = L(v) = \{1, 2, 3\}$ and $L(z) = \{1, 2, 4\}$, and L-coloring α of H with $\alpha(u) = \alpha(v) = 1$ and $\alpha(z) = 4$, such that the following properties hold:*

- *For every L-coloring γ of H, it holds that $\gamma(z) = 4$ or $\gamma(u) \neq \gamma(v)$.*
- *For any combination of colors $a \in L(u)$, $b \in L(v)$ with $a \neq b$, there exists an L-recoloring sequence from α to an L-coloring γ with $\gamma(u) = a$, $\gamma(v) = b$ and $\gamma(z) \neq 4$, of length at most $|V(H)|$.*

Theorem 4. *For any constant $k \geq 4$, the problem LBCR, with ℓ encoded in unary, is NP-complete.*

Proof: Given an instance G of P3C, we construct an instance $(G', L, \ell, \alpha, \beta)$ of LB L-CR as follows. Start with the vertex set $V(G)$. All of these vertices $u \in V(G)$ receive color $\alpha(u) = 1$ and $L(u) = \{1, 2, 3\}$. For every edge $uv \in E(G)$, add a copy of the graph H from Lemma 3, where the u-vertex and v-vertex from H are identified with u and v, respectively. Note that there is no edge between u

and v in G'. For each $uv \in E(G)$, the z-vertex of the corresponding copy of H is denoted by z_{uv}, and we let $Z = \{z_{uv} \mid uv \in E(G)\}$. For these H-subgraphs, the L-coloring α is as given in Lemma 3. Next, we add a triangle on vertices a, b, c to G', with the following colors and lists: $\alpha(a) = 1$, $\alpha(b) = 2$, $\alpha(c) = 3$, $L(a) = \{1, 2, 3\}$, $L(b) = \{1, 2\}$, and $L(c) = \{3, 4\}$. Add edges from all vertices in Z to c. This yields the graph G'. Finally, we define the target coloring β. For all vertices $v \in V(G') \setminus \{a, b\}$, set $\beta(v) = \alpha(v)$. We set $\beta(a) = 2$ and $\beta(b) = 1$, so the goal is to reverse the colors of these two vertices.

We now argue that G is 3-colorable if and only if there exists an L-recoloring sequence for G' from α to β of length $\mathcal{O}(m)$, where $m = |E(G)|$. Suppose that there exists such an L-recoloring sequence. Considering the vertices a, b, and c, we see that this must contain a coloring γ with $\gamma(c) = 4$. This implies that for every $z_{uv} \in Z$, $\gamma(z_{uv}) \in \{1, 2\}$. By Lemma 3, this implies that for every $uv \in E(G)$, $\gamma(u) \neq \gamma(v)$. Hence γ restricted to $V(G)$ is a 3-coloring of G. On the other hand, if G is 3-colorable, then we can recolor the vertices of G to such a 3-coloring, which allows recoloring all vertices z_{uv} to a color different from 4, using $\mathcal{O}(1)$ recoloring steps for each H-subgraph, and thus $\mathcal{O}(m)$ recoloring steps in total. This makes it possible to recolor the vertices a, b, and c to their target color in $\mathcal{O}(1)$ steps, and subsequently the other recoloring steps can be reversed, which gives $\mathcal{O}(m)$ steps in total.

Combining this reduction with the fact that we can easily transform the LB L-CR instance to an LBCR instance, and the NP-hardness of P3C, shows that LBCR is *strongly* NP-hard. (This uses the fact that ℓ is polynomial in m.)   □

4   An FPT Algorithm for CSP Reachability

We will consider sets of *variables* B, which all can take on the values $D = [k]$. The set D is called the *domain* of the variables. A function $f : B \to D$.[2] is called a *value assignment* from B to D. A set U of value assignments from B to D is called a *VA-set* from B to D. Below, we will consider a fixed set X of variables, and consider VA-sets U for many different subsets $B \subseteq X$, but always for the same domain D, so we will omit D from the terminology and simply call U a *VA-set for B*, and elements of U *value assignments for B*.

An instance $(X, k, \mathcal{C})$ of the *Constraint Satisfaction Problem (CSP)* consists of a set X of *variables*, which all have *domain* $D = [k]$, and a set $\mathcal{C}$ of *constraints*. Every constraint $C \in \mathcal{C}$ is a tuple (T, R), where $T \subseteq X$, and R is a VA-set for T. The VA-set R is interpreted as the set of all value combinations that are allowed for the variables in T. A value assignment $f : X \to D$ is said to *satisfy* constraint $C = (T, R)$ if and only if $f|_T \in R$. If f satisfies all constraints in $\mathcal{C}$, f is called *valid* (for $\mathcal{C}$). *CSP* is a decision problem where the question is whether there exists a valid value assignment.

---

[2] Considering the function f, it is perhaps a little confusing to call D the domain, but this conforms with the terminology used in the context of CSPs.

We remark that for many problems that can be formulated as CSPs, the constraints $(T, R) \in \mathcal{C}$ are not explicitly given, since R would usually be prohibitively (exponentially) large. Instead, a simple and efficient algorithm is given that can verify whether the constraint is satisfied. The factor $g(\mathcal{C})$ in our complexity bounds accounts for this.

In order to study reconfiguration questions for CSPs, we define two distinct value assignments $\alpha : X \to D$ and $\beta : X \to D$ to be *adjacent* if they differ on exactly one variable $v \in X$ (so, expressed differently: if there exists a $v \in X$ such that $\alpha|_{X \setminus \{v\}} = \beta|_{X \setminus \{v\}}$). For a CSP instance $(X, k, \mathcal{C})$, the solution graph $\mathrm{CSP}_k(X, \mathcal{C})$ has as vertex set all value assignments from X to $[k]$ that are valid for $\mathcal{C}$, with adjacency as defined above. Pseudowalks in $\mathrm{CSP}_k(X, \mathcal{C})$ are called *CSP sequences* for $(X, k, \mathcal{C})$. We consider the following problem.

Length-Bounded CSP Reachability (LBCSPR):
INSTANCE: A CSP instance $(X, k, \mathcal{C})$, two valid value assignments α and β for X and $[k]$, and an integer ℓ.
QUESTION: Does $\mathrm{CSP}_k(X, \mathcal{C})$ contain a path from α to β of length at most ℓ?

For every constant ℓ, the LBCSPR problem can be solved in polynomial time, using the following simple branching algorithm. Denote the given instance by $(X, k, \mathcal{C}, \alpha, \beta, \ell)$, with $|X| = n$. Start with the initial value assignment α. For every value assignment generated by the algorithm, consider all adjacent value assignments in $\mathrm{CSP}_k(X, \mathcal{C})$. Recurse on these choices, up to a recursion depth of at most ℓ. Return yes if and only if in one of the recursion branches, the target value assignment β is obtained. Clearly, this algorithm yields the correct answer. One value assignment has at most kn neighbors, so branching with depth ℓ shows that at most $\mathcal{O}((kn)^\ell)$ value assignments will be considered. This proves the claim, or in other words: for parameter ℓ, the problem is in XP.

We let $S = \{x \in X \mid \alpha(x) \neq \beta(x)\}$. Clearly, when $|S| > \ell$ we have a no-instance and when $|S| = 0$ we have a trivial yes-instance. To obtain an FPT algorithm, the main challenge that we need to overcome is that the number of variables that potentially need to be reassigned cannot easily be bounded by a function of ℓ. However, once we know the set B of variables which will change at least once, the problem can be solved using a branching algorithm similar to the one above. Let $\mathcal{S} = \gamma_0, \ldots, \gamma_\ell$ be a CSP sequence for a CSP instance $(X, k, \mathcal{C})$. For a set $B \subseteq X$, the *set of B-variable combinations used by* $\mathcal{S}$ is $\mathrm{USED}(\mathcal{S}, B) = \{\gamma_i|_B : i \in \{0, \ldots, \ell\}\}$. Let U be a VA-set for B. We say that $\mathcal{S}$ *follows* U if $\mathrm{USED}(\mathcal{S}, B) \subseteq U$. A branching algorithm can be given for the following variant of LBCSPR, which is restricted by choices of B and U.

Lemma 5 (*). *Let $(X, k, \mathcal{C}, \alpha, \beta, \ell)$ be an LBCSPR instance, and let $g(\mathcal{C})$ be the complexity of deciding whether a given value assignment for X satisfies $\mathcal{C}$. Let $B \subseteq X$, and U be a VA-set for B. Let $L(x) = \{f(x) \mid f \in U\}$ for all $x \in B$, and $p = \sum_{x \in B}(|L(x)| - 1)$. Then there exists an algorithm LISTCSPRECONFIG with complexity $\mathcal{O}(p^\ell \cdot g(\mathcal{C}) \cdot poly(|U|, |X|))$, that decides whether there exists a CSP sequence $\mathcal{S}$ for $(X, k, \mathcal{C})$ from α to β of length at most ℓ in which only variables in B are changed, with $\mathrm{USED}(\mathcal{S}, B) \subseteq U$.*

Algorithm 1. CSPRECONFIG$(X, k, \mathcal{C}, \alpha, \beta, \ell)$

Input: A variable set $X = \{x_1, \ldots, x_n\}$ with domains $[k]$, a set $\mathcal{C}$ of constraints on X, valid value assignments $\alpha : X \to [k]$ and $\beta : X \to [k]$, and integer $\ell \geq 0$.
Output: "YES" if and only if there exists a CSP sequence of length at most ℓ from α to β.

1: $S := \{x \in X \mid \alpha(x) \neq \beta(x)\}$
2: **if** $|S| > \ell$ **then return** NO
3: **if** $|S| = 0$ **then return** YES
4: **return** Recurse$(\emptyset, \{f^0\}, \{f^0\})$

  Subroutine RECURSE(B, U, L):

5: **if** $\sum_{v \in B}(|L(v)| - 1) > \ell$ **then return** NO
6: **if** $S \subseteq B$ and there are no critical constraints for U, B and α **then**
7:     **return** LISTCSPRECONFIG$(X, k, \mathcal{C}, \alpha, \beta, \ell, B, U)$.
8: **if** not $S \subseteq B$ **then**
9:     Let i be the lowest index such that $x_i \in S \setminus B$
10:    NewVar $:= \{x_i\}$
11: **else**
12:    choose a critical constraint $(T, R) \in \mathcal{C}$ for U, B and α.
13:    NewVar $:= T \setminus B$
14: **for all** $x \in$ NewVar:
15:    $B' := B \cup \{x\}$
16:    **for all** VA-sets U' for B' that extend U, with $|U'| \leq \ell$ and $\{\alpha|_{B'}, \beta|_{B'}\} \subseteq U'$:
17:       $L(x) := \{f(x) \mid f \in U'\}$
18:       **if** $|L(x)| \geq 2$ **then**
19:          **if** Recurse(B', U', L)=YES **then return** YES
20: **return** NO

It remains to give a branching algorithm that, if there exists a CSP sequence $\mathcal{S}$ of length at most ℓ, can determine a proper guess for the sets B of variables that are changed in $\mathcal{S}$, and $U = \text{USED}(\mathcal{S}, B)$. Clearly, $S \subseteq B$ should hold, so we start with $B = S$, and we first consider all possible VA-sets U for this B. We will say that a constraint $C = (T, R)$ is *critical* for B, U and α if there exists an $f \in U$ such that the (unique) value assignment $g : X \to D$ that satisfies $g|_B = f$ and $g|_{X \setminus B} = \alpha|_{X \setminus B}$ does not satisfy C. Note that in this case, if we assume that the combination of values f occurs at some point during the reconfiguration, then for at least one variable in $T \setminus B$, the value must change before this point, so one such variable should be added to B, which yields a new set B'. Let $B \subseteq B' \subseteq X$, and let U and U' be VA-sets for B and B', respectively. We say that U' *extends* U if $U = \{f|_B : f \in U'\}$. In other words, if U and U' are interpreted as guesses of value combinations that will occur during the reconfiguration, then these guesses are consistent with each other.

For given $B \subseteq X$ and VA-set U for B, we let $L(x) = \{f(x) \mid f \in U\}$ for all $x \in B$. If $\sum_{x \in B}(|L(x)| - 1) > \ell$ then the set U cannot correspond to the set $\text{USED}(\mathcal{S}, B)$ for a CSP sequence $\mathcal{S}$ of length at most ℓ, so this guess can be safely ignored. On the other hand, if a guess of B and U is reached where $\sum_{x \in B}(|L(x)| - 1) \leq \ell$ and there are no critical constraints, then the aforementioned LISTCSPRECONFIG algorithm can be used to test whether there exists a corresponding CSP sequence. Using these observations, it can be shown that Algorithm 1 correctly decides the LBCSPR problem.

It is relatively easy to see that the total number of recursive calls made by this algorithm is bounded by some function of ℓ, k and r, where $r = \max_{(T,R) \in \mathcal{C}} |T|$. Indeed, Line 18 guarantees that for every recursive call, the quantity $\sum_{v \in B} (|L(v)| - 1)$ increases by at least one, so the recursion depth is at most $\ell + 1$ (see Line 5). The number of iterations of the for-loops in Lines 14 and 16 is bounded by $r - 1$, and by some function of ℓ and k, respectively. This shows that Algorithm 1 is an FPT algorithm for parameter $k + \ell + r$. Using a sophisticated analysis, one can prove the following bound on the complexity.

Theorem 6 (*). *Let $(X, k, \mathcal{C}, \alpha, \beta, \ell)$ be an LBCSPR instance. Then in time $\mathcal{O}\big((r-1)^{\ell} \cdot k^{\ell(\ell+1)} \cdot \ell^{\ell} \cdot g(\mathcal{C}) \cdot poly(k, \ell, n)\big)$, it can be decided whether there exists a CSP sequence from α to β of length at most ℓ, where $r = \max_{(T,R) \in \mathcal{C}} |T|$ and $n = |X|$, and where $g(\mathcal{C})$ denotes the time to find a constraint in $\mathcal{C}$ that is not satisfied by a given value assignment, if such a constraint exists.*

This result implies e.g. FPT algorithms for LBCR (for parameter $k + \ell$), and Length-Bounded Boolean SAT Reachability (for parameter $\ell + r$). In fact, for CSP problems with binary constraints such as LBCR, the complexity can be improved, since it suffices to guess only the lists $L(x)$ for each vertex/variable x, instead of all value combinations U.

Theorem 7 (*). *Let $G, k, \alpha, \beta, \ell$ be a LBCR instance, with $n = |V(G)|$. There is an algorithm with complexity $\mathcal{O}(2^{k(\ell+1)} \cdot \ell^{\ell} \cdot poly(n))$ that decides whether there exists a k-recoloring sequence from α to β for G of length at most ℓ.*

5 Hardness Results for CSP Reachability

We give two W[2]-hardness results. These hold in fact for very restricted types of CSP instances. A CSP instance $(X, k, \mathcal{C})$ is called a *Horn-SAT* instance if $k = 2$, and every constraint in $\mathcal{C}$ can be formulated as a Boolean clause that uses at most one positive literal. (As is customary in Boolean satisfiability, we assume in this case that the variables can take on the values 0 and 1 instead.) The *Length-Bounded Horn-SAT Reachability* problem is the LBCSPR problem restricted to Horn-SAT instances. The even more restricted problem where all clauses have three variables is called *Length-Bounded Horn-3SAT Reachability*.

In both proofs, we will give reductions from the W[2]-hard p-Hitting Set problem. A p-Hitting Set instance $(\mathcal{U}, \mathcal{F}, p)$ consists of a finite universe $\mathcal{U}$, a family of sets $\mathcal{F} \subseteq 2^{\mathcal{U}}$, and a positive integer p. The question is whether there exists a subset $U \subseteq \mathcal{U}$ of size at most p such that for every set $F \in \mathcal{F}$ we have $F \cap U \neq \emptyset$. We say that such a set U is a *hitting set* of $\mathcal{F}$. This problem is W[2]-hard when parameterized by p [9].

Theorem 8 (*). *Length-Bounded Horn-SAT Reachability is W[2]-hard when parameterized by ℓ.*

Proof sketch: Given an instance $(\mathcal{U}, \mathcal{F}, p)$ of p-Hitting Set, we create a variable x_u for each element $u \in \mathcal{U}$ and two additional variables y_1 and y_2, for a total of $|\mathcal{U}| + 2$ variables. For each set $\{u_1, u_2, \ldots u_t\} \in \mathcal{F}$, we create a Horn clause $(y_1 \vee \overline{y_2} \vee \overline{x_{u_1}}, \overline{x_{u_2}}, \ldots \overline{x_{u_t}})$. Finally, we add an additional clause $(y_2 \vee \overline{y_1})$. These clauses constitute a Horn formula $\mathcal{H}$ with $|\mathcal{F}| + 1$ clauses. Let α be the satisfying assignment for $\mathcal{H}$ that sets all its variables to 1, and β be the satisfying assignment for $\mathcal{H}$ that sets $y_1 = y_2 = 0$ and all other variables to 1.

Observe that before we can set y_2 to 0, y_1 has to be set to 0. Moreover, before y_1 can be set to 0, some of the x variables (i.e. variables corresponding to elements of the universe $\mathcal{U}$) have to be set to 0 to satisfy all the clauses corresponding to the sets. Using the previous two observations, it can be shown that $\mathcal{F}$ has a hitting set of size at most p if and only there is a CSP sequence of length at most $2p + 2$ from α to β. □

Theorem 8 implies in particular that for LBCSPR, there is no FPT algorithm when parameterized only by $k + \ell$, unless FPT=W[2]. Next, we consider the "above-guarantee" version of LBCSPR. Given two valid value assignments α and β for X and $[k]$, we let $S = \{x \in X \mid \alpha(x) \neq \beta(x)\}$. Clearly, the length of any CSP sequence from α to β is at least $|S|$. Hence, in the above-guarantee version of the problem, instead of allowing the running time to depend on the full length ℓ of a CSP sequence, we let $\bar{\ell} = \ell - |S|$ and allow the running time to depend on $\bar{\ell}$ only. However, the next theorem implies that no FPT algorithm for LBCSPR exists, when parameterized by $\bar{\ell} + k + r$, unless $W[2] = FPT$.

Theorem 9 (*). *Length-Bounded Horn-3SAT Reachability is W[2]-hard when parameterized by $\bar{\ell} = \ell - |S|$, where $S = \{x \in X \mid \alpha(x) \neq \beta(x)\}$.*

Proof sketch: Starting from a p-Hitting Set instance $(\mathcal{U}, \mathcal{F}, p)$, we first create a variable x_u for every $u \in \mathcal{U}$. We let $\mathcal{F} = \{F_1, F_2, \ldots F_m\}$ and $\{u_1, u_2, \ldots u_r\}$ be a set in $\mathcal{F}$. For each such set in $\mathcal{F}$, we create r new variables $y_1, y_2, \ldots y_r$ and the clauses $(y_1 \vee \overline{x_{u_1}} \vee \overline{y_2}), (y_2 \vee \overline{x_{u_2}} \vee \overline{y_3}), \ldots, (y_r \vee \overline{x_{u_r}} \vee \overline{y_1})$. We let α be the satisfying assignment for the formula with all variables set to 1, and let β be the satisfying assignment with all the x_u, $u \in \mathcal{U}$, variables set to 1 and the rest set to 0.

Consider the clauses corresponding to a set $\{u_1, u_2, \ldots u_r\}$ in $\mathcal{F}$, with variables $y_1, \ldots, y_r$. None of the y variables can be set to 0 before we flip at least one x variable to 0. Moreover, after flipping any x variable to 0, we can in fact flip all y variables to 0, provided this is done in the proper order. Combining the previous observations with the fact that $|S| = \sum_{i=1}^{m} |F_i|$, it can be shown that $\mathcal{F}$ has a hitting set of size at most p if and only there is a CSP sequence of length at most $\sum_{i=1}^{m} |F_i| + 2p$ from α to β. □

References

1. Bonsma, P.: Rerouting shortest paths in planar graphs. In: FSTTCS 2012. LIPIcs, vol. 18, pp. 337–349. Schloss Dagstuhl - Leibniz-Zentrum fuer Informatik (2012)
2. Bonsma, P.: The complexity of rerouting shortest paths. Theor. Comput. Sci. **510**, 1–12 (2013)
3. Bonsma, P., Cereceda, L.: Finding paths between graph colourings: PSPACE-completeness and superpolynomial distances. Theor. Comput. Sci. **410**(50), 5215–5226 (2009)
4. Bonsma, P., Kamiński, M., Wrochna, M.: Reconfiguring independent sets in claw-free graphs. In: Ravi, R., Gørtz, I.L. (eds.) SWAT 2014. LNCS, vol. 8503, pp. 86–97. Springer, Heidelberg (2014)
5. Cereceda, L.: Mixing graph colourings. Ph.D. thesis, London School of Economics (2007)
6. Cereceda, L., van den Heuvel, J., Johnson, M.: Connectedness of the graph of vertex-colourings. Discrete Math. **308**(56), 913–919 (2008)
7. Cereceda, L., van den Heuvel, J., Johnson, M.: Finding paths between 3-colorings. J. Graph Theor. **67**(1), 69–82 (2011)
8. Diestel, R.: Graph Theory, Electronic Edition. Springer, Heidelberg (2005)
9. Downey, R.G., Fellow, M.R.: Parameterized Complexity. Springer, New York (1997)
10. Flum, J., Grohe, M.: Parameterized Complexity Theory. Springer, Berlin (2006)
11. Garey, M.R., Johnson, D.S., Stockmeyer, L.: Some simplified NP-complete graph problems. Theor. Comput. Sci. **1**(3), 237–267 (1976)
12. Gopalan, P., Kolaitis, P.G., Maneva, E.N., Papadimitriou, C.H.: The connectivity of boolean satisfiability: computational and structural dichotomies. SIAM J. Comput. **38**(6), 2330–2355 (2009)
13. van den Heuvel, J.: The complexity of change. Surv. Combinatorics **2013**, 127–160 (2013)
14. Jacob, R.: Standortplanung mit blick auf online-strategien. Graduate thesis, Universität Würzburg (1997)
15. Johnson, M., Kratsch, D., Kratsch, S., Patel, V., Paulusma, D.: Finding shortest paths between graph colourings. In: Proceedings of IPEC 2014(2014)
16. Kamiński, M., Medvedev, P., Milanič, M.: Shortest paths between shortest paths. Theor. Comput. Sci. **412**(39), 5205–5210 (2011)
17. Kamiński, M., Medvedev, P., Milanič, M.: Complexity of independent set reconfigurability problems. Theor. Comput. Sci. **439**, 9–15 (2012)
18. Marcotte, O., Hansen, P.: The height and length of colour switching. In: Graph Colouring and Applications, pp. 101–110. AMS, Providence (1999)
19. Mouawad, A.E., Nishimura, N., Raman, V.: Vertex cover reconfiguration and beyond. arXiv:1403.0359. Accepted for ISAAC 2014 (2014)
20. Mouawad, A.E., Nishimura, N., Raman, V., Simjour, N., Suzuki, A.: On the parameterized complexity of reconfiguration problems. In: Gutin, G., Szeider, S. (eds.) IPEC 2013. LNCS, vol. 8246, pp. 281–294. Springer, Heidelberg (2013)

Quantified Conjunctive Queries on Partially Ordered Sets

Simone Bova[✉], Robert Ganian, and Stefan Szeider

Vienna University of Technology, Vienna, Austria
{simone.bova,robert.ganian,stefan.szeider}@tuwien.ac.at

Abstract. We study the computational problem of checking whether a quantified conjunctive query (a first-order sentence built using only conjunction as Boolean connective) is true in a finite poset (a reflexive, antisymmetric, and transitive directed graph). We prove that the problem is already NP-hard on a certain fixed poset, and investigate structural properties of posets yielding fixed-parameter tractability when the problem is parameterized by the query. Our main algorithmic result is that model checking quantified conjunctive queries on posets of bounded width is fixed-parameter tractable (the width of a poset is the maximum size of a subset of pairwise incomparable elements). We complement our algorithmic result by complexity results with respect to classes of finite posets in a hierarchy of natural poset invariants, establishing its tightness in this sense.

Keywords: Quantified conjunctive queries · Posets · Parameterized complexity

1 Introduction

Motivation. The *model checking* problem for first-order logic is the problem of deciding whether a given first-order sentence is true in a given finite structure; it encompasses a wide range of fundamental combinatorial problems. The problem is trivially decidable in $O(n^k)$ time, where n is the size of the structure and k is the size of the sentence, but it is not polynomial-time decidable or even *fixed-parameter tractable* when parameterized by k (under complexity assumptions in classical and parameterized complexity, respectively).

Restrictions of the model checking problem to fixed classes of structures or sentences have been intensively investigated from the perspective of parameterized algorithms and complexity [5,10,11]. In particular, starting from seminal work by Courcelle [6] and Seese [15], structural properties of *graphs* sufficient for fixed-parameter tractability of model checking have been identified. An important outcome of this research is the understanding of the interplay between structural properties of graphs and the expressive power of first-order logic, most notably the interplay between sparsity and locality, culminating in the

© Springer International Publishing Switzerland 2014
M. Cygan and P. Heggernes (Eds.): IPEC 2014, LNCS 8894, pp. 122–134, 2014.
DOI: 10.1007/978-3-319-13524-3_11

recent result by Grohe, Kreutzer, and Siebertz that model checking first-order logic on classes of *nowhere dense* graphs is fixed-parameter tractable [12,13]. On graph classes closed under subgraphs the result is known to be tight; at the same time, there are classes of *somewhere dense* graphs (not closed under subgraphs) with fixed parameter tractable first-order (and even monadic second-order) logic model checking; the prominent examples are graph classes of bounded clique-width solved by Courcelle et al. [7].

In this paper, we investigate *posets* (short for *partially ordered sets*). Posets form a fundamental class of combinatorial objects [9] and may be viewed as reflexive, antisymmetric, and transitive directed graphs. Besides their naturality, our motivation towards posets is that they challenge our current model checking knowledge; indeed, posets are somewhere dense (but not closed under substructures) and have unbounded clique-width [1, Proposition 5]. Therefore, not only are they not covered by the aforementioned results [7,12], but most importantly, it seems likely that new structural ideas and algorithmic techniques are needed to understand and conquer first-order logic on posets.

In recent work, we started the investigation of first-order logic model checking on finite posets, and obtained a parameterized complexity classification of *existential* and *universal* logic (first-order sentences in prefix form built using only existential or only universal quantifiers) with respect to classes of posets in a hierarchy generated by basic poset invariants, including for instance width and depth [1].[1] In particular, as articulated more precisely in [1], a complete understanding of the first-order case reduces to understanding the parameterized complexity of model checking first-order logic on bounded width posets (the *width* of a poset is the maximum size of a subset of pairwise incomparable elements); these classes are hindered by the same obstructions as general posets, since already posets of width 2 have unbounded clique-width [1, Proposition 5].

Contribution. In this paper we push the tractability frontier traced in [1] closer towards full first-order logic, by proving that model checking *(quantified) conjunctive positive* logic (first-order sentences built using only conjunction as Boolean connective) is tractable on bounded width posets.[2] The problem of model checking conjunctive positive logic on finite structures, also known as the *quantified constraint satisfaction* problem, has been previously studied with various motivations in various settings [3,5]; somehow surprisingly, conjunctive logic is also capable of expressing rather interesting poset properties (as sampled in Proposition 2).

More precisely, our contribution is twofold. First, we identify conjunctive positive logic as a minimal syntactic fragment of first-order logic that allows for full quantification, and has computationally hard expression complexity on posets; namely, we prove that *there exists a finite poset where model checking (quantified) conjunctive positive logic is NP-hard* (Theorem 1). Next, as our

---

[1] Existential and universal logic are maximal syntactic fragments properly contained in first-order logic.

[2] Conjunctive positive logic and existential (respectively, universal) logic are incomparable syntactic fragments of first-order logic.

main algorithmic result, we establish that *model checking conjunctive positive logic on finite posets, parameterized by the width of the poset and the size of the sentence, is fixed-parameter tractable* with an elementary parameter dependence (Theorem 2). The aforementioned fact that model checking conjunctive positive logic is already NP-hard on a fixed poset justifies the relaxation to fixed-parameter tractability by showing that, if we insist on polynomial-time algorithms, any structural property of posets (captured by the boundedness of a numeric invariant) is negligible.

Informally, the idea of our algorithm is the following. First, given a poset $\mathbf{P}$ and a sentence ϕ, we rewrite the sentence in a simplified form (which we call a *reduced* form), equisatisfiable on $\mathbf{P}$ (Proposition 1). Next, using the properties of reduced forms, we define a syntactic notion of "depth" of a variable in ϕ and a semantic notion of "depth" of a subset of $\mathbf{P}$, and we prove that $\mathbf{P} \models \phi$ if and only if $\mathbf{P}$ verifies ϕ upon "relativizing" variables to subsets of matching depth (Lemmas 1 and 2). The key fact is that the size of the subsets of $\mathbf{P}$ used to relativize the variables of ϕ is bounded above by the width of $\mathbf{P}$ and the size of ϕ (Lemma 3), from which the main result follows (Theorem 2). We remark that the approach outlined above differs significantly from the algebraic approach used in [1]; moreover, both stages make essential use of the restriction that conjunction is the only Boolean connective allowed in the sentences.

It follows immediately that *model checking conjunctive positive logic on classes of finite posets of bounded width, parameterized by the size of the sentence, is fixed-parameter tractable* (Corollary 1). On the other hand, there exist classes of finite posets of bounded depth (the *depth* of a poset is the maximum size of a subset of pairwise comparable elements) and classes of finite posets of bounded cover-degree (the *cover-degree* of a poset is the degree of its cover relation) where model checking conjunctive positive logic is shown to be coW[2]-hard and hence not fixed parameter tractable, unless the exponential time hypothesis [8] fails, see Proposition 3. Combined with the algorithm by Seese [15], these facts complete the parameterized complexity classification of the investigated poset invariants, as depicted in Fig. 1.

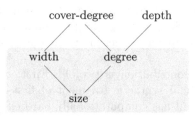

Fig. 1. On all classes of posets bounded under invariants in the gray region, model checking conjunctive positive logic is fixed-parameter tractable; on some classes of posets bounded under the remaining invariants, the problem is not fixed-parameter tractable unless FPT = coW[2].

The classification of conjunctive positive logic in this paper matches the classification of existential logic in [1], and further emphasizes the quest for a classification of full first-order logic on bounded width posets. We believe that the work presented in this paper and [1] enlightens the spectrum of phenomena that a fixed-parameter tractable algorithm for model checking the full first-order logic on bounded width posets, if it exists, has to capture.

Throughout the paper, we mark with ⋆ all statements whose proofs are omitted; we refer to [2] for a full version.

2   Preliminaries

For all integers $k \geq 1$, we let $[k]$ denote the set $\{1, \ldots, k\}$. We focus on relational first-order logic. A *vocabulary* σ is a set of *constant symbols* and *relation symbols*; each relation symbol is associated to a natural number called its *arity*; we let $\mathrm{ar}(R)$ denote the arity of $R \in \sigma$. *All vocabularies considered in this paper are finite.*

An *atom* α (over vocabulary σ) is an equality $t = t'$ or an application of a predicate $Rt_1 \ldots t_{\mathrm{ar}(R)}$, where $t, t', t_1, \ldots, t_{\mathrm{ar}(R)}$ are variable symbols (in a fixed countable set) or constant symbols, and $R \in \sigma$. We let $\mathcal{FO}$ denote the class of first-order sentences.

A *structure* $\mathbf{A}$ (over σ) is specified by a nonempty set A, called the *universe* of the structure, an element $c^{\mathbf{A}} \in A$ for each constant symbol $c \in \sigma$, and a relation $R^{\mathbf{A}} \subseteq A^{\mathrm{ar}(R)}$ for each relation symbol $R \in \sigma$. Given a structure $\mathbf{A}$ and $B \subseteq A$ such that $\{c^{\mathbf{A}} \mid c \in \sigma\} \subseteq B$, we denote by $\mathbf{A}|_B$ the substructure of $\mathbf{A}$ induced by B, defined as follows: the universe of $\mathbf{A}|_B$ is B, $c^{\mathbf{A}|_B} = c^{\mathbf{A}}$ for each $c \in \sigma$, and $R^{\mathbf{A}|_B} = R^{\mathbf{A}} \cap B^{\mathrm{ar}(R)}$ for all $R \in \sigma$. A structure is *finite* if its universe is finite and *trivial* if its universe is a singleton. *All structures considered in this paper are finite and nontrivial.*

For a structure $\mathbf{A}$ and a sentence ϕ over the same vocabulary, we write $\mathbf{A} \models \phi$ if the sentence ϕ is *true* in the structure $\mathbf{A}$. When $\mathbf{A}$ is a structure, f is a mapping from the variables to the universe of $\mathbf{A}$, and $\psi(x_1, \ldots, x_n)$ is a formula over the vocabulary of $\mathbf{A}$, we write $\mathbf{A}, f \models \psi$ or (liberally) $\mathbf{A} \models \psi(f(x_1), \ldots, f(x_n))$ to indicate that ψ is satisfied in $\mathbf{A}$ under f.

We refer the reader to [8] for the standard algorithmic setup of the model checking problem, and for standard notions in parameterized complexity theory. As for notation, the model checking problem for a class of σ-structures $\mathcal{C}$ and a class of σ-sentences $\mathcal{L} \subseteq \mathcal{FO}$ is denoted by $\mathrm{MC}(\mathcal{C}, \mathcal{L})$; it is the problem of deciding, given $(\mathbf{A}, \phi) \in \mathcal{C} \times \mathcal{L}$, whether $\mathbf{A} \models \phi$. We let $\|(\mathbf{A}, \phi)\|$, $\|\mathbf{A}\|$, and $\|\phi\|$ denote, respectively, the size of the (encoding of the) instance $(\mathbf{A}, \phi)$, the structure $\mathbf{A}$, and the sentence ϕ. The parameterization of an instance $(\mathbf{A}, \phi)$ returns $\|\phi\|$.

Conjunctive Positive Logic. In this paper, we study the *(quantified) conjunctive positive* fragment of first-order logic, in symbols $\mathcal{FO}(\forall, \exists, \wedge)$, containing first-order sentences built using only logical symbols in $\{\forall, \exists, \wedge\}$.

A conjunctive positive sentence is in *alternating prefix form* if it has the form

$$\phi = \forall x_1 \exists y_1 \ldots \forall x_l \exists y_l C(x_1, y_1, \ldots, x_l, y_l), \qquad (1)$$

where $l \geq 0$ and $C(x_1, y_1, \ldots, x_l, y_l)$ is a conjunction of atoms whose variables are contained in $\{x_1, y_1, \ldots, x_l, y_l\}$; it is possible to reduce any conjunctive positive sentence to a logically equivalent conjunctive positive sentence of form (1) in polynomial time. For a simpler exposition, *every conjunctive positive sentence considered in this paper is assumed to be given in alternating prefix form (or is implicitly reduced to that form if required by the context).*

Let σ be a relational vocabulary. Let $\mathbf{A}$ be a σ-structure and let ϕ be a conjunctive positive σ-sentence as in (1). It is well known that the truth of ϕ in $\mathbf{A}$ can be characterized in terms of the *Hintikka (or model checking) game* on $\mathbf{A}$ and ϕ. The game is played by two players, Abelard (male, the *universal* player) and Eloise (female, the *existential* player), as follows. For increasing values of i from 1 to l, Abelard assigns x_i to an element $a_i \in A$, and Eloise assigns y_i to an element $b_i \in A$; the sequence $(a_1, b_1, \ldots, a_l, b_l)$ is called a *play* on $\mathbf{A}$ and ϕ, where $(a_1, \ldots, a_l)$ and $(b_1, \ldots, b_l)$ are the plays by Abelard and Eloise respectively; Eloise wins if and only if $\mathbf{A} \models C(a_1, b_1, \ldots, a_l, b_l)$.

A *strategy for Eloise* (in the Hintikka game on $\mathbf{A}$ and ϕ) is a sequence $(g_1, \ldots, g_l)$ of functions of the form $g_i \colon A^i \to A$, for all $i \in [l]$; it *beats* a play $f \colon \{x_1, \ldots, x_l\} \to A$ by Abelard if $\mathbf{A} \models C(f(x_1), g_1(f(x_1)), \ldots, f(x_i), g_i(f(x_1), \ldots, f(x_i)), \ldots)$, where $i \in [l]$. A strategy for Eloise is *winning* (in the Hintikka game on $\mathbf{A}$ and ϕ) if it beats all Abelard plays. It is well known (and easily verified) that $\mathbf{A} \models \phi$ if and only if Eloise has a winning strategy (in the Hintikka game on $\mathbf{A}$ and ϕ).

For $X_1, Y_1, \ldots, X_l, Y_l \subseteq A$, we denote by

$$\phi' = (\forall x_1 \in X_1)(\exists y_1 \in Y_1) \ldots (\forall x_l \in X_l)(\exists y_l \in Y_l) C(x_1, y_1, \ldots, x_l, y_l), \qquad (2)$$

the relativization in ϕ of variable x_i to X_i and y_i to Y_i for all $i \in [l]$, and liberally write $\mathbf{A} \models \phi'$ meaning that ϕ' is satisfied in the intended expansion of $\mathbf{A}$. It is readily verified that, if ϕ' is as in (2), then $\mathbf{A} \models \phi'$ if and only if, in the Hintikka game on $\mathbf{A}$ and ϕ, Eloise has a strategy of the form $g_i \colon X_1 \times \cdots \times X_i \to Y_i$ for all $i \in [l]$, beating all plays f by Abelard such that $f(x_i) \in X_i$ for all $i \in [l]$.

Partially Ordered Sets. We refer the reader to [4] for the few standard notions in order theory used in the paper but not defined below.

A structure $\mathbf{G} = (G, E^{\mathbf{G}})$ with $\mathrm{ar}(E) = 2$ is called a *digraph*. Two digraphs $\mathbf{G}$ and $\mathbf{H}$ are *isomorphic* if there exists a bijection $f \colon G \to H$ such that for all $g, g' \in G$ it holds that $(g, g') \in E^{\mathbf{G}}$ if and only if $(f(g), f(g')) \in E^{\mathbf{H}}$. The *degree* of $g \in G$, in symbols degree(g), is equal to $|\{(g', g) \in E^{\mathbf{G}} \mid g' \in G\} \cup \{(g, g') \in E^{\mathbf{G}} \mid g' \in G\}|$, and the *degree* of $\mathbf{G}$, in symbols degree($\mathbf{G}$), is the maximum degree attained by the elements of $\mathbf{G}$.

A digraph $\mathbf{P} = (P, \leq^{\mathbf{P}})$ is a *partially ordered set* (in short, a *poset*) if $\leq^{\mathbf{P}}$ is a reflexive, antisymmetric, and transitive relation over P. For all $Q \subseteq P$, we let $\min^{\mathbf{P}}(Q)$ and $\max^{\mathbf{P}}(Q)$ denote, respectively, the set of minimal and maximal elements in the substructure of $\mathbf{P}$ induced by Q; we also write $\min(\mathbf{P})$ instead of $\min^{\mathbf{P}}(P)$, and $\max(\mathbf{P})$ instead of $\max^{\mathbf{P}}(P)$. For all $Q \subseteq P$, we let $(Q]^{\mathbf{P}}$, respectively $[Q)^{\mathbf{P}}$, denote the downset, respectively upset, of $\mathbf{P}$ induced by Q. Let $\mathbf{P}$ be a poset and let $p, q \in P$. We write $p \prec^{\mathbf{P}} q$ if q covers p in $\mathbf{P}$, and $p \parallel^{\mathbf{P}} q$ if p and q are incomparable in $\mathbf{P}$. If $\mathcal{P}$ is a class of posets, we let $\mathrm{cover}(\mathcal{P}) = \{\mathrm{cover}(\mathbf{P}) \mid \mathbf{P} \in \mathcal{P}\}$, where $\mathrm{cover}(\mathbf{P}) = \{(\mathrm{p}, \mathrm{q}) \mid \mathrm{p} \prec^{\mathbf{P}} \mathrm{q}\}$.

We introduce a family of poset invariants. Let $\mathbf{P}$ be a poset. The *size* of $\mathbf{P}$ is $|P|$. The *depth* of $\mathbf{P}$, $\mathrm{depth}(\mathbf{P})$, is the maximum size of a chain in $\mathbf{P}$. The *width* of $\mathbf{P}$, $\mathrm{width}(\mathbf{P})$, is the maximum size of an antichain in $\mathbf{P}$. The *degree* of $\mathbf{P}$, $\mathrm{degree}(\mathbf{P})$, is the degree of $\mathbf{P}$ as a digraph. The *cover-degree* of $\mathbf{P}$, $\mathrm{cover-degree}(\mathbf{P})$, is the degree of the cover relation of $\mathbf{P}$, that is, $\mathrm{degree}(\mathrm{cover}(\mathbf{P}))$. We say that a class of posets $\mathcal{P}$ is *bounded* w.r.t. the poset invariant inv if there exists $b \in \mathbb{N}$ such that $\mathrm{inv}(\mathcal{P}) \leq \mathrm{b}$ for all $\mathbf{P} \in \mathcal{P}$. The above poset invariants are ordered as in Fig. 1, where $\mathrm{inv} \leq \mathrm{inv}'$ if and only if: $\mathcal{P}$ is bounded w.r.t. inv implies $\mathcal{P}$ is bounded w.r.t. inv' for every class of posets $\mathcal{P}$ [1, Proposition 3].

3   Expression Hardness

In this section we prove that conjunctive positive logic on posets is NP-hard in expression complexity. Let $\mathbf{B} = (B, \leq^{\mathbf{B}})$ be the *bowtie* poset defined by the universe $B = \{0, 1, 2, 3\}$ and the covers $0, 2 \prec^{\mathbf{B}} 1, 3$; see Fig. 2.

Theorem 1. $\mathrm{MC}(\{\mathbf{B}\}, \mathcal{FO}(\forall, \exists, \wedge))$ *is NP-hard.*

Proof. Let $\tau = \{\leq\}$ and $\sigma = \tau \cup \{c_0, c_1, c_2, c_3\}$ be vocabularies where $\leq$ is a binary relation symbol and c_i is a constant symbol ($i \in B$). Let $\mathcal{FO}_\sigma(\exists, \wedge)$ contain first-order sentences built using only logical symbols in $\{\exists, \wedge\}$ and nonlogical symbols in σ; $\mathcal{FO}_\tau(\forall, \exists, \wedge)$ is described similarly. Let $\mathbf{B}^*$ be the σ-structure such that $B^* = B$, $(B^*, \leq^{\mathbf{B}^*})$ is isomorphic to $\mathbf{B}$ under the identity mapping, and $c_i^{\mathbf{B}^*} = i$ for all $i \in B$.

By [14, Theorem 2, Case $n = 2$], the problem $\mathrm{MC}(\{\mathbf{B}^*\}, \mathcal{FO}_\sigma(\exists, \wedge))$ is NP-hard. It is therefore sufficient to give a polynomial-time many-one reduction from $\mathrm{MC}(\{\mathbf{B}^*\}, \mathcal{FO}_\sigma(\exists, \wedge))$ to $\mathrm{MC}(\{\mathbf{B}\}, \mathcal{FO}_\tau(\forall, \exists, \wedge))$. The idea of the reduction is to simulate the constants in σ by universal quantification and additional variables; the details follow.

Let ψ be an instance of $\mathrm{MC}(\{\mathbf{B}^*\}, \mathcal{FO}_\sigma(\exists, \wedge))$, and let $\{x_i, y_i, w_i \mid i \in B\}$ be a set of 12 fresh variables (not occurring in ψ). Let ψ' be the $\mathcal{FO}_\tau(\exists, \wedge)$-sentence obtained from ψ by replacing atoms of the form $c_i \leq u$ and $u \leq c_i$, respectively, by atoms of the form $w_i \leq u$ and $u \leq w_i$ (where c_i is a constant in σ and u, w_i are variables). Let α be the conjunction of atoms defined by (see Fig. 2)

$$\{w_0, w_2\} \leq \{w_1, w_3\} \wedge \bigwedge_{j \in \{0,2\}} \{x_j\} \leq \{y_j, w_j\} \wedge \bigwedge_{j \in \{1,3\}} \{y_j, w_j\} \leq \{x_j\},$$

where, for sets of variables S and S', the notation $S \leq S'$ denotes the conjunction of atoms of the form $s \leq s'$ for all $(s, s') \in S \times S'$.

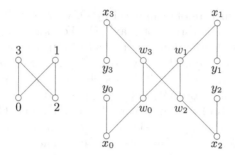

Fig. 2. The Hasse diagrams of the bowtie poset **B** (left) and of the representation $\mathbf{M}_\alpha$ of the formula α (right, see Sect. 4 for the interpretation of $\mathbf{M}_\alpha$) used in Theorem 1. The idea of the reduction is to simulate the constant c_i in $\psi \in \mathcal{FO}_v(\exists, \wedge)$, interpreted on the element $i \in B$, by the variable w_i in $\phi \in \mathcal{FO}_\tau(\forall, \exists, \wedge)$, where $i \in \{0, 1, 2, 3\}$.

We finally define the $\mathcal{FO}_\tau(\forall, \exists, \wedge)$-sentence ϕ by putting $\phi = \forall y_0 \ldots \forall y_3 \exists x_0 \ldots \exists x_3 \exists w_0 \ldots \exists w_3 (\alpha \wedge \psi')$. The reduction is clearly feasible in polynomial time; we now prove that the reduction is correct, that is, $\mathbf{B}^* \models \psi$ if and only if $\mathbf{B} \models \phi$.

An assignment $f \colon \{y_0, y_1, y_2, y_3\} \to B$ is said to be *nontrivial* if $\{f(y_0), f(y_2)\} = \{0, 2\}$ and $\{f(y_1), f(y_3)\} = \{1, 3\}$, and *trivial* otherwise; in particular, nontrivial assignments are bijective.

Claim 1. $(\star)$ $\mathbf{B}, f \models \exists x_0 \ldots x_3 w_0 \ldots w_3 (\alpha \wedge \psi')$ *for all trivial assignments* f.

Claim 2. $(\star)$ *Let* f *be a nontrivial assignment. The following are equivalent.*

(i) $\mathbf{B}, f \models \exists x_0 \ldots x_3 w_0 \ldots w_3 (\alpha \wedge \psi')$.
(ii) $\mathbf{B}^* \models \psi$.

We conclude the proof by showing that $\mathbf{B}^* \models \psi$ if and only if $\mathbf{B} \models \phi$. If $\mathbf{B} \not\models \phi$, then there exists an assignment f such that $\mathbf{B}, f \not\models \exists x_0 \ldots \exists x_3 \exists w_0 \ldots \exists w_3 (\alpha \wedge \psi')$; by Claim 1, f is nontrivial. Then $\mathbf{B}^* \not\models \psi$ by Claim 2. Conversely, if $\mathbf{B} \models \phi$, then in particular $\mathbf{B}, f \models \exists x_0 \ldots \exists x_3 \exists w_0 \ldots \exists w_3 (\alpha \wedge \psi')$ for all nontrivial assignments f, and hence $\mathbf{B}^* \models \psi$ by Claim 2.

4   Reduced Forms

In this section, we introduce *reduced* forms for conjunctive positive sentences on posets and prove that, given a poset **P** and a sentence ϕ, a reduced form for ϕ is easy to compute and equivalent to ϕ on **P**.

In the rest of this section, $\sigma = \{\leq\}$ is the vocabulary of posets, and ϕ is a conjunctive positive σ-sentence as in (1). Since ϕ will be evaluated on posets, where the formulas $x \leq y \wedge y \leq x$ and $x = y$ are equivalent, we assume that no atom of the form $x = y$ occurs in ϕ; otherwise, such an atom can be replaced by the formula $x \leq y \wedge y \leq x$ maintaining logical equivalence.

We represent ϕ by the pair $(\mathbf{Q}_\phi, \mathbf{M}_\phi)$, where $\mathbf{Q}_\phi = (Q_\phi, E^{\mathbf{Q}_\phi})$ and $\mathbf{M}_\phi = (M_\phi, E^{\mathbf{M}_\phi})$ are digraphs encoding the *prefix* and the *matrix* of ϕ respectively, as follows. The universes are $Q_\phi = M_\phi = \{x_1, y_1, \ldots, x_l, y_l\}$; we let $M_\phi^\forall = \{x_1, \ldots, x_l\}$ and $M_\phi^\exists = \{y_1, \ldots, y_l\}$ denote, respectively, the set of *universal* and *existential* variables in ϕ. The structure $\mathbf{Q}_\phi$ is a chain with cover relation $x_1 \prec^{\mathbf{Q}_\phi} y_1 \prec^{\mathbf{Q}_\phi} \cdots \prec^{\mathbf{Q}_\phi} x_l \prec^{\mathbf{Q}_\phi} y_l$. The structure $\mathbf{M}_\phi$ is defined by the edge relation $E^{\mathbf{M}_\phi} = \{(x,y) x \leq y \text{ is an atom of } \phi\}$. We say that ϕ is in *reduced form* if:

(i) $\mathbf{M}_\phi$ is a poset;

(ii) the substructure of $\mathbf{M}_\phi$ induced by $M_\phi^\forall$ is an antichain;

(iii) for all distinct x and x' in $M_\phi^\forall$, it holds that $[x)^{\mathbf{M}_\phi} \cap [x')^{\mathbf{M}_\phi} = (x]^{\mathbf{M}_\phi} \cap (x']^{\mathbf{M}_\phi} = \emptyset$;

(iv) for all $x \in M_\phi^\forall$ and all $y \in M_\phi^\exists \cap ((x]^{\mathbf{M}_\phi} \cup [x)^{\mathbf{M}_\phi})$, it holds that $x <^{\mathbf{Q}_\phi} y$.

Let $\phi \in \mathcal{FO}(\forall, \exists, \wedge)$. For all $Z \subseteq M_\phi$, we let $\phi|_Z$ denote the conjunctive positive sentence represented by $(\mathbf{Q}_\phi|_Z, \mathbf{M}_\phi|_Z)$. It is readily observed that, for all $Z \subseteq M_\phi$, it holds that $\phi \models \phi|_Z$.

Proposition 1. $(\star)$ *Let $\mathcal{P}$ be a class of posets. There exists a polynomial-time algorithm that, given an instance $(\mathbf{P}, \phi)$ of $\mathrm{MC}(\mathcal{P}, \mathcal{FO}(\forall, \exists, \wedge))$, either correctly rejects, or returns a sentence $\phi' \in \mathcal{FO}(\forall, \exists, \wedge)$ in reduced form such that $\mathbf{P} \models \phi'$ if and only if $\mathbf{P} \models \phi$.*

5 Fixed-Parameter Tractability

In this section, we prove that model checking conjunctive positive logic is fixed-parameter tractable parameterized by the size of the sentence *and* the width of the poset; it follows, in particular, that model checking conjunctive positive logic is fixed-parameter tractable (parameterized by the size of the sentence) on classes of posets of bounded width. We refer the reader to the introduction for an informal outline of the proof idea.

In the rest of this section, $\sigma = \{\leq\}$ is the vocabulary of posets, $\mathbf{P}$ is a poset and $\phi = (\mathbf{Q}_\phi, \mathbf{M}_\phi)$ is a conjunctive positive σ-sentence as in (1) satisfying clauses (i) and (ii) of the definition of reduced form.

5.1 Depth in the Sentence

Using the fact that ϕ is in reduced form, we define the following. For all $y \in M_\phi^\exists$: lower-depth$(y) = \mathrm{depth}(\mathbf{M}_\phi|_{(y]^{\mathbf{M}_\phi}})$; upper-depth$(y) = \mathrm{depth}(\mathbf{M}_\phi|_{[y)^{\mathbf{M}_\phi}})$. In words, lower-depth$(y)$ is the size of the largest chain in the substructure of $\mathbf{M}_\phi$ induced by the downset of y in $\mathbf{M}_\phi$, and upper-depth(y) is the size of the largest chain in the substructure of $\mathbf{M}_\phi$ induced by the upset of y in $\mathbf{M}_\phi$.

Next, we define a partition of $M_\phi^\exists$ into two blocks L_ϕ and U_ϕ, the *lower* and *upper* variables respectively, as follows. For all $y \in M_\phi^\exists$ let $y \in L_\phi$ if and only if there either exists $x \in M_\phi^\forall$ such that $y \leq^{\mathbf{M}_\phi} x$, or $y \parallel^{\mathbf{M}_\phi} x$ for all

$x \in M_\phi^\forall$ and lower-depth$(y) \leq$ upper-depth(y). Similarly, $y \in U_\phi$ if and only if there either exists $x \in M_\phi^\forall$ such that $y \geq^{M_\phi} x$, or $y \parallel^{M_\phi} x$ for all $x \in M_\phi^\forall$ and lower-depth$(y) >$ upper-depth(y). In words, an existential variable y in ϕ is lower if and only if it is below a universal variable in the matrix of ϕ, or is incomparable to all universal variables in the matrix of ϕ but "closer" to the bottom of the matrix of ϕ in that lower-depth$(y) \leq$ upper-depth(y); a similar idea drives the definition of upper variables.

Finally we define, for all $y \in M_\phi^\exists$: depth$(y) =$ lower-depth(y) if $y \in L_\phi$, and depth$(y) =$ upper-depth(y) if $y \in U_\phi$; in words, the depth of a lower variable is its "distance" from the bottom as measured by lower-depth(y), and similarly for upper variables.

5.2   Depth in the Structure

Relative to the poset $\mathbf{P}$, we define, for all $i \geq 0$, the set P_i as follows.

- $L_0 = \min(\mathbf{P})$, $U_0 = \max(\mathbf{P}) \setminus L_0$, and $P_0 = L_0 \cup U_0$.
- Let $i \geq 1$, and let $R \subseteq P_{i-1}$ be such that $R \cap L_{i-1}$ is downward closed in $\mathbf{P}|_{L_{i-1}}$ (that is, for all $l, l' \in L_{i-1}$, if $l \in R \cap L_{i-1}$ and $l' \leq^\mathbf{P} l$, then $l' \in R$) and $R \cap U_{i-1}$ is upward closed in $\mathbf{P}|_{U_{i-1}}$ (that is, for all $u, u' \in U_{i-1}$, if $u \in R \cap U_{i-1}$ and $u \leq^\mathbf{P} u'$, then $u' \in R$). Let

$$P_{i-1,R} = \left\{ p \in P \ \middle| \ \begin{array}{l} \text{for all } l \in L_{i-1}, l \leq^\mathbf{P} p \text{ if and only if } l \in R, \\ \text{for all } u \in U_{i-1}, p \leq^\mathbf{P} u \text{ if and only if } u \in R \end{array} \right\};$$

in words, $p \in P_{i-1,R}$ if and only if the elements in L_{i-1} below p are exactly those in $R \cap L_{i-1}$ (and the elements in $L_{i-1} \setminus R$ are incomparable to p) and the elements in U_{i-1} above p are exactly those in $R \cap U_{i-1}$ (and the elements in $U_{i-1} \setminus R$ are incomparable to p). We now define $P_i = L_i \cup U_i$ where L_i and U_i are as follows:

$$L_i = L_{i-1} \cup \bigcup_{R \subseteq P_{i-1}} \min^\mathbf{P}(P_{i-1,R}), \quad U_i = \left(U_{i-1} \cup \bigcup_{R \subseteq P_{i-1}} \max^\mathbf{P}(P_{i-1,R}) \right) \setminus L_i.$$

Let $p \in P$. Let $i \geq 0$ be minimum such that $p \in P_i$ (note that for every $p \in P$ such minimum i exists, and $L_i \cap U_i = \emptyset$ by construction). If $p \in L_i$, then $p \in L_\mathbf{P}$ and lower-depth$(p) = i$ and if $p \in U_i$, then $p \in U_\mathbf{P}$ and upper-depth$(p) = i$. Note that $L_\mathbf{P}$ and $U_\mathbf{P}$ partition P into two blocks containing the *lower* and *upper* elements respectively. Finally we define, for all $p \in P$: depth$(p) =$ lower-depth(p), if $p \in L_\mathbf{P}$, and depth$(p) =$ upper-depth(p), if $p \in U_\mathbf{P}$.

5.3   Depth Restricted Game

We now establish and formalize the relation between the depth in ϕ and the depth in $\mathbf{P}$ (see Lemma 1); this is the key combinatorial fact underlying the model checking algorithm.

Relative to the Hintikka game on $\mathbf{P}$ and ϕ, we define the following. A pair $(y, p) \in M_\phi^\exists \times P$ is *depth respecting* if $(y, p) \in (L_\phi \times L_{\mathbf{P}}) \cup (U_\phi \times U_{\mathbf{P}})$ and depth$(p) \leq$ depth(y). A strategy $(g_1, \ldots, g_l)$ for Eloise is *depth respecting* if, for all $i \in [l]$ and all plays $f \colon \{x_1, \ldots, x_l\} \to P$ by Abelard, the pair $(y_i, g_i(f(x_1), \ldots, f(x_i)))$ is depth respecting.

Let $b \geq 0$ be the maximum depth of a variable in ϕ. A play $f \colon \{x_1, \ldots, x_l\} \to P$ by Abelard is *bounded depth* if, for all $i \in [l]$, it holds that $f(x_i) \in P_{b+1}$.

Lemma 1. *The following are equivalent (w.r.t. the Hintikka game on $\mathbf{P}$ and ϕ).*

(i) Eloise has a winning strategy.

(ii) Eloise has a depth respecting winning strategy.

(iii) Eloise has a depth respecting strategy beating all bounded depth Abelard plays.

Proof. $(ii) \Rightarrow (iii)$ is trivial. We prove $(i) \Rightarrow (ii)$ and $(iii) \Rightarrow (i)$.

$(i) \Rightarrow (ii)$: Let $\mathbf{g} = (g_1, \ldots, g_l)$ be a winning strategy for Eloise. Let the Abelard play $f \colon \{x_1, \ldots, x_l\} \to P$ and the existential variable $y_j \in M_\phi^\exists$ be a minimal witness that the above winning strategy for Eloise is not depth respecting, in the following sense: $(y_j, g_j(f(x_1), \ldots, f(x_j)))$ is not depth respecting, but for all $f' \colon \{x_1, \ldots, x_l\} \to P$ and all $y_{j'} \in M_\phi^\exists$ such that either $y_j, y_{j'} \in L_\phi$ and lower-depth$(y_{j'}) <$ lower-depth(y_j), or $y_j, y_{j'} \in U_\phi$ and upper-depth$(y_{j'}) <$ upper-depth(y_j), it holds that $(y_{j'}, g_j(f'(x_1), \ldots, f'(x_j)))$ is depth respecting.

We define a strategy $\mathbf{g}' = (g_1, \ldots, g_{j-1}, g_j', g_{j+1}, \ldots, g_l)$ for Eloise such that g_j' restricted to $P^j \setminus \{(f(x_1), \ldots, f(x_j))\}$ is equal to g_j (in other words, g_j' differs from g_j only in the move after $f \colon \{x_1, \ldots, x_l\} \to P$), and $(y_j, g_j'(f(x_1), \ldots, f(x_j)))$ is depth respecting. There are two cases to consider, depending on whether $y_j \in L_\phi$ or $y_j \in U_\phi$. We prove the statement in the former case; the argument is symmetric in the latter case.

So, assume $y_j \in L_\phi$. Let $g_j(f(x_1), \ldots, f(x_j)) = p$ and $i =$ depth(y_j). Let $R \subseteq P_{i-1}$ (with $R \cap L_{i-1}$ downward closed in $\mathbf{P}|_{L_{i-1}}$ and $R \cap U_{i-1}$ upward closed in $\mathbf{P}|_{U_{i-1}}$) be such that, for all $l \in L_{i-1}$ and $u \in U_{i-1}$, it holds that $l \leq^{\mathbf{P}} p$ if and only if $l \in R$ and $p \leq^{\mathbf{P}} u$ if and only if $u \in R$. Hence $p \in P_{i-1,R}$. Then there exists $m \in \min^{\mathbf{P}}(P_{i-1,R})$ such that $m \leq^{\mathbf{P}} p$. By construction we have depth$(m) = i$. Let $g_j' \colon P^j \to P$ be exactly as g_j with the exception that $g_j'(f(x_1), \ldots, f(x_j)) = m$; note that the pair (y_j, m) is depth respecting.

Claim 3. $(\star)$ *Let f' be any play by Abelard. Then $\mathbf{g}' = (g_1, \ldots, g_j', \ldots, g_l)$ beats f' in the Hintikka game on $\mathbf{P}$ and ϕ.*

We obtain a depth respecting winning strategy for Eloise by iterating the above argument thanks to Claim 5.

$(iii) \Rightarrow (i)$: Let $b \geq 0$ be the maximum depth of a variable in ϕ, and let $\mathbf{g} = (g_1, \ldots, g_l)$ be a depth respecting strategy for Eloise beating all bounded depth plays by Abelard. We define a strategy $\mathbf{g}' = (g_1', \ldots, g_l')$ for Eloise, as follows.

Let $f: \{x_1, \ldots, x_l\} \to P$ be a play by Abelard, say $f(x_i) = p_i$ for all $i \in [l]$. Let $i \in [l]$ and let $R_i \subseteq P_b$ (with $R_i \cap L_{b-1}$ downward closed in $\mathbf{P}|_{L_{b-1}}$ and $R_i \cap U_{b-1}$ upward closed in $\mathbf{P}|_{U_{b-1}}$) be such that for all $l \in L_b$, it holds that $l \leq^{\mathbf{P}} p_i$ if and only if $l \in R_i$ and for all $u \in U_b$, it holds that $p_i \leq^{\mathbf{P}} u$ if and only if $u \in R_i$. By construction, there exists $r_i \in P_{b+1}$ such that for all $l \in L_b$, it holds that $l \leq^{\mathbf{P}} r_i$ if and only if $l \leq^{\mathbf{P}} p_i$ and for all $u \in U_b$, it holds that $r_i \leq^{\mathbf{P}} u$ if and only if $p_i \leq^{\mathbf{P}} u$. Let $f': \{x_1, \ldots, x_l\} \to P$ be the bounded depth play by Abelard defined by $f'(x_i) = r_i$ for all $i \in [l]$. Finally define, for all $i \in [l]$, $g_i'(f(x_1), \ldots, f(x_i)) = g_i(f'(x_1), \ldots, f'(x_i))$.

Claim 4. $(\star)$ $\mathbf{g}' = (g_1', \ldots, g_l')$ *is a winning strategy for Eloise.*

This concludes the proof of the lemma. $\qquad\square$

5.4 Fixed-Parameter Tractability

The following two lemmas allow to establish the correctness (Lemma 2, relying on Lemma 1) and the tractability (Lemma 3) of the presented model checking algorithm, respectively.

Lemma 2. $(\star)$ *Let $b \geq 0$ be the maximum depth of a variable in ϕ. Let $D = P_{b+1}$ and, for all $i \in [l]$, let*

$$D_i = \begin{cases} L_{\text{depth}(y_i)}, & if\, y_i \in L_\phi, \\ U_{\text{depth}(y_i)}, & if\, y_i \in U_\phi. \end{cases}$$

Then, $\mathbf{P} \models \phi$ if and only if $\mathbf{P} \models (\forall x_1 \in D)(\exists y_1 \in D_1) \ldots (\forall x_l \in D)(\exists y_l \in D_l)C$.

Lemma 3. $(\star)$ *Let $w = \text{width}(\mathbf{P})$ and let $k \geq 0$. Then, $|P_k| \leq 2w^{(3w)^k}$.*

We are now ready to describe the announced algorithm. The underlying idea is that the characterization in Lemma 2 is checkable in fixed-parameter tractable time since $|D_i| \leq |D|$ for all $i \in [l]$, and $|D|$ is bounded above by a computable function of $\text{width}(\mathbf{P})$ and $\|\phi\|$.

Theorem 2. $(\star)$ *There exists an algorithm that, given a poset $\mathbf{P}$ and a sentence $\phi \in \mathcal{FO}(\forall, \exists, \wedge)$, decides whether $\mathbf{P} \models \phi$ in*

$$\exp_w^4(O(k)) \cdot n^{O(1)}$$

time, where $w = \text{width}(\mathbf{P})$, $k = \|\phi\|$, and $n = \|(\mathbf{P}, \phi)\|$.

Corollary 1. *Let $\mathcal{P}$ be a class of posets of bounded width. Then, the problem $\text{MC}(\mathcal{P}, \mathcal{FO}(\forall, \exists, \wedge))$ is fixed-parameter tractable.*

6   Fixed-Parameter Intractability

In this section, we prove that model checking conjunctive positive logic on classes of bounded depth and bounded cover-degree posets is coW[2]-hard, and hence unlikely to be fixed-parameter tractable [8].

We first observe the following. Let ϕ_k be the $\mathcal{FO}(\forall, \exists, \wedge)$-sentence $(k \geq 1)$

$$\forall x_1 \ldots \forall x_k \exists y_1 \ldots \exists y_k \exists w \left(\bigwedge_{i \in [k]} y_i \leq x_i \wedge \bigwedge_{i \in [k]} y_i \leq w \right). \tag{3}$$

Proposition 2. $(\star)$ *For every poset* $\mathbf{P}$ *and* $k \geq 1$, $\mathbf{P} \models \phi_k$ *iff for every* k *elements* $p_1, \ldots, p_k \in \min(\mathbf{P})$, *there exists* $u \in P$ *such that* $p_1, \ldots, p_k \leq^{\mathbf{P}} u$.

We now describe the reductions. Let $\mathcal{H}$ be the class of hypergraphs (a *hypergraph* is a σ-structure $\mathbf{H}$ such that $U^{\mathbf{H}} \neq \emptyset$ for all U in a unary vocabulary σ). For the depth invariant, we define a function d from $\mathcal{H}$ to a class of posets of depth at most 2 where $d(\mathbf{H}) = \mathbf{P}$ such that: $\min(\mathbf{P}) = H$; $\max(\mathbf{P}) = \sigma$; $h \prec^{\mathbf{P}} U$ for all $h \in \min(\mathbf{P})$ and $U \in \max(\mathbf{P})$ such that $h \notin U^{\mathbf{H}}$. For the cover-degree invariant, we similarly define a function c from $\mathcal{H}$ to a class of posets with cover graphs of degree at most 3 (see [2] for details). We then use Proposition 2 to obtain:

Proposition 3. $(\star)$ *Let* $r \in \{c, d\}$. *Then,* $\mathrm{MC}(\{r(\mathbf{H}) \mid \mathbf{H} \in \mathcal{H}\}, \mathcal{FO}(\forall, \exists, \wedge))$ *is coW[2]-hard.*

7   Conclusion

We provided a parameterized complexity classification of the problem of model checking quantified conjunctive queries on posets with respect to the invariants in Fig. 1; in particular, we push the tractability frontier of the model checking problem on bounded width posets closer towards the full first-order logic. The question of whether first-order logic is fixed-parameter tractable on bounded width posets remains open.

Acknowledgments. This research was supported by the European Research Council (Complex Reason, 239962) and the FWF Austrian Science Fund (Parameterized Compilation, P26200 and X-TRACT, P26696).

References

1. Bova, S., Ganian, R., Szeider, S.: Model checking existential logic on partially ordered sets. In: CSL-LICS. Preprint in CoRR. abs/1405.2891 (2014)
2. Bova, S., Ganian, R., Szeider, S.: Quantified conjunctive queries on partially ordered sets. Preprint in CoRR, abs/1408.4263 (2014)
3. Börner, F., Bulatov, A., Chen, H., Jeavons, P., Krokhin, A.: The complexity of constraint satisfaction games and QCSP. Inform. Comput. **207**(9), 923–944 (2009)

4. Caspard, N., Leclerc, B., Monjardet, B.: Finite Ordered Sets. Cambridge University Press, Cambridge (2012)
5. Chen, H., Dalmau, V.: Decomposing quantified conjunctive (or Disjunctive) formulas. In: LICS (2012)
6. Courcelle, B.: The monadic second-order logic of graphs I: recognizable sets of finite graphs. Inform. Comput. **85**(1), 12–75 (1990)
7. Courcelle, B., Makowsky, J.A., Rotics, U.: Linear time solvable optimization problems on graphs of bounded clique-width. Theor. Comput. Syst. **33**(2), 125–150 (2000)
8. Flum, J., Grohe, M.: Parameterized Complexity Theory. Springer, Berlin (2006)
9. Graham, R.L., Grötschel, M., Lovász, L. (eds.): Handbook of Combinatorics, vol. 1. MIT Press, Cambridge (1995)
10. Grohe, M.: The complexity of homomorphism and constraint satisfaction problems seen from the other side. J. ACM **54**(1), 1–24 (2007)
11. Grohe, M., Kreutzer, S.: Methods for algorithmic meta theorems. In: Model Theoretic Methods in Finite Combinatorics, pp. 181–206. AMS, Providence (2011)
12. Grohe, M., Kreutzer, S., Siebertz, S.: Deciding first-order properties of nowhere dense graphs. In: STOC, 2014. Preprint in CoRR. abs/131.3899 (2013)
13. Nešetřil, J., de Mendez, P.O.: Sparsity. Springer, Berlin (2012)
14. Pratt, V.R., Tiuryn, J.: Satisfiability of inequalities in a poset. Fund. Inform. **28** (1–2), 165–182 (1996)
15. Seese, D.: Linear time computable problems and first-order descriptions. Math. Struct. Comp. Sci. **6**(6), 505–526 (1996)

Graph Isomorphism Parameterized
by Elimination Distance to Bounded Degree

Jannis Bulian[✉] and Anuj Dawar[✉]

University of Cambridge Computer Laboratory, Cambridge, UK
{jannis.bulian,anuj.dawar}@cl.cam.ac.uk

Abstract. A commonly studied means of parameterizing graph problems is the deletion distance from triviality [10], which counts vertices
that need to be deleted from a graph to place it in some class for which
efficient algorithms are known. In the context of graph isomorphism,
we define triviality to mean a graph with maximum degree bounded by
a constant, as such graph classes admit polynomial-time isomorphism
tests. We generalise deletion distance to a measure we call elimination
distance to triviality, based on elimination trees or tree-depth decompositions. We establish that graph canonisation, and thus graph isomorphism, is FPT when parameterized by elimination distance to bounded
degree, generalising results of Bouland et al. [2] on isomorphism parameterized by tree-depth.

1 Introduction

The *graph isomorphism problem* (GI) is the problem of determining, given a
pair of graphs G and H, whether they are isomorphic. This problem has an
unusual status in complexity theory as it is neither known to be in P nor known
to be NP-complete, one of the few natural problems for which this is the case.
Polynomial-time algorithms are known for a variety of special classes of graphs.
Many of these lead to natural parameterizations of GI by means of structural
parameters of the graphs. For instance, it is known that GI is in XP parameterized
by the genus of the graph [7,17], by maximum degree [1,14] and by the size
of the smallest excluded minor [19], or more generally, the smallest excluded
topological minor [9]. For each of these parameters, it remains an open question
whether the problem is FPT. On the other hand, GI has been shown to be FPT
when parameterized by eigenvalue multiplicity [5], tree distance width [22], the
maximum size of a simplicial component [20,21] and minimum feedback vertex
set [11]. Bouland et al. [2] showed that the problem is FPT when parameterized
by the tree depth of a graph and in a recent advance on this, Lokshtanov et al. [13]
have announced that it is also FPT parameterized by *tree width*.

Our main result extends the results of Bouland et al. and is incomparable with
that of Lokshtanov et al. We show that graph canonisation is FPT parameterized

Research supported in part by EPSRC grant EP/H026835, DAAD grant A/13/05456,
and DFG project Logik, Struktur und das Graphenisomorphieproblem.

M. Cygan and P. Heggernes (Eds.): IPEC 2014, LNCS 8894, pp. 135–146, 2014.
DOI: 10.1007/978-3-319-13524-3_12

by *elimination distance to degree d*, for any constant d. The structural graph parameter we introduce is an instance of what Guo et al. [10] call *distance to triviality* and it may be of interest in the context of other graph problems.

To put this parameter in context, consider the simplest notion of distance to triviality for a graph G: the number k of vertices of G that must be deleted to obtain a graph with no edges. This is, of course, just the size of a minimal vertex cover in G and is a parameter that has been much studied (see for instance [6]). Indeed, it is also quite straightforward to see that GI is FPT when parameterized by vertex cover number. Consider two ways this observation might be strengthened. The first is to relax the notion of what we consider to be "trivial". For instance, as there is for each d a polynomial time algorithm deciding GI among graphs with maximum degree d, we may take this as our trivial base case. We then parameterize G by the number k of vertices that must be deleted to obtain a subgraph of G with maximum degree d. This yields the parameter *deletion distance to bounded degree*, which we consider in Sect. 3 below. Alternatively, we relax the notion of "distance" so that rather than considering the sequential deletion of k vertices, we consider the recursive deletion of vertices in a tree-like fashion. To be precise, say that a graph G has *elimination distance* $k + 1$ from triviality if, in each connected component of G we can delete a vertex so that the resulting graph has distance k to triviality. If triviality is understood to mean the empty graph, this just yields a definition of the tree depth of G. In our main result, we combine these two approaches by parameterizing G by the elimination distance to triviality, where a graph is trivial if it has maximum degree d. We show that, for any fixed d, this gives a structural parameter on graphs for which graph canonisation is FPT. Along the way, we establish a number of characterisations of the parameter that may be interesting in themselves. The key idea in the proof is the separation of any graph of elimination distance k to degree d into two subgraphs, one of which has degree bounded by d and the other tree-depth bounded by a function of k and d, in a canonical way.

A central technique used in the proof is to construct, from a graph G, a term (or equivalently a labelled, ordered tree) T_G that is an isomorphism invariant of the graph G. It should be noted that this general method is widely deployed in practical isomorphism tests such as McKay's graph isomorphism testing program "nauty" [15,16]. The recent advance by Lokshtanov et al. [13] is also based on such an approach.

In Sect. 2 we recall some definitions from graph theory and parameterized complexity theory. Section 3 introduces the notion of deletion distance to bounded degree and presents a kernelisation procedure that allows us to decide isomorphism. In Sect. 4 we introduce the main parameter of our paper, elimination distance to bounded degree, and establish its key properties. The main result on FPT graph canonisation is established in Sect. 5. Due to limitations of space, proofs of several key lemmas are deferred to the full version of this paper, which may be found at `arxiv:1406.4718`.

2   Preliminaries

Parameterized complexity theory is a two-dimensional approach to the study of the complexity of computational problems. A *language* (or *problem*) L is a set of strings $L \subseteq \Sigma^*$ over a finite alphabet Σ. A *parameterization* is a function $\kappa : \Sigma^* \to \mathbb{N}$. We say that L is *fixed-parameter tractable* with respect to κ if we can decide whether an input $x \in \Sigma^*$ is in L in time $O(f(\kappa(x)) \cdot |x|^c)$, where c is a constant and f is some computable function. For a thorough discussion of the subject we refer to the books by Downey and Fellows [4], Flum and Grohe [8] and Niedermeier [18].

A *graph* G is a set of vertices $V(G)$ and a set of edges $E(G) \subseteq V(G) \times V(G)$. We will usually assume that graphs are loop-free and undirected, i.e. that E is irreflexive and symmetric. If E is not symmetric, we call G a *directed graph*. We mostly follow the notation in Diestel's book [3].

If $v \in G$ and $S \subseteq V(G)$, we write $E_G(v, S)$ for the set of edges $\{vw \mid w \in S\}$ between v and S.

The *neighbourhood* of a vertex v is $N_G(v) := \{w \in V(G) \mid vw \in E(G)\}$. The *degree* of a vertex v is the size of its neighbourhood $\deg_G(v) := |N_G(v)|$. For a set of vertices $S \subseteq V(G)$ its neighbourhood is defined to be $N_G(S) := \bigcup_{v \in S} N_G(v)$. The *degree* of a graph G is the maximal degree of its vertices $\Delta(G) := \max\{\deg_G(v) \mid v \in V(G)\}$. If it is clear from the context what the graph is, we will sometimes omit the subscript.

Let H be a subgraph of G and $v, w \in V(G)$. A *path through H from w to v* is a path P from w to v in G with all vertices, except possibly the endpoints, in $V(H)$, i.e. $(V(P) \setminus \{v, w\}) \subseteq V(H)$.

A *(k-)colouring* of a graph G is a map $c : V(G) \to \{1, \ldots, k\}$ for some $k \in \mathbb{N}$. We call a graph together with a colouring a *coloured* graph. Two coloured graphs G, G' with respective colourings $c : V(G) \to \{1, \ldots, k\}, c' : V(G') \to \{1, \ldots, k\}$ are *isomorphic* if there is a bijection $\phi : V(G) \to V(G')$
such that:

- for all $v, w \in V(G)$ we have that $vw \in E(G)$ if and only if $\phi(v)\phi(w) \in E(G')$;
- for all $v \in V(G)$, we have that $c(v) = c'(\phi(v))$.

Note that we require the colour classes to match exactly, and do not allow a permutation of the colour classes.

Let C be a class of (coloured) graphs closed under isomorphism. A *canonical form for* C is a function $F : \mathsf{C} \to \mathsf{C}$ such that

- for all $G \in \mathsf{C}$, we have that $F(G) \cong G$;
- for all $G, H \in \mathsf{C}$, we have that $G \cong H$ if, and only if, $F(G) = F(H)$.

A *partial order* is a binary relation $\leq$ on a set S which is reflexive, antisymmetric and transitive. If $\leq$ is a partial order on S, and for each element $a \in S$, the set $\{b \in S \mid b \leq a\}$ is totally ordered by $\leq$, we say $\leq$ is a *tree order*. (Note that the covering relation of a tree order is not necessarily a tree, but may be a forest.)

Definition 2.1. *An* elimination order ≤ *is a* tree order *on the vertices of a graph* G, *such that for each edge* $uv \in E(G)$ *we have either* $u \leq v$ *or* $v \leq u$.

We say that an order has *height* k if the length of the longest chain in it is k.

Isomorphism on (coloured) bounded degree graphs. Luks [14] proved that isomorphism of bounded degree graphs can be decided in polynomial time, and Babai and Luks [1] give a polynomial-time canonisation algorithm for bounded degree graphs. These results can be extended to coloured graphs by means of an easy reduction (which can be found in the full version of this paper). Thus, we have:

Theorem 2.2. *Let* C *be a class of (coloured) bounded degree graphs closed under isomorphism. Then there is a canonical form* F *for* C *that allows us to compute* $F(G)$ *in polynomial time.*

3   Deletion Distance to Bounded Degree

We first study the notion of deletion distance to bounded degree and establish in this section that graph isomorphism is FPT with this parameter. Though the result in this section is subsumed by the more general one in Sect. 5, it provides a useful warm-up and a tighter, polynomial kernel. In the present warm-up we only give an algorithm for the graph isomorphism problem, though the result easily holds for canonisation as well (and this follows from the more general result in Sect. 5). The notion of deletion distance to bounded degree is a particular instance of the general notion of distance to triviality introduced by Guo et al. [10]. In the context of graph isomorphism, we have chosen triviality to mean graphs of bounded degree.

Definition 3.1. *A graph* G *has* deletion distance k *to degree* d *if there are* k *vertices* $v_1, \ldots, v_k \in V(G)$ *such that* $G \setminus \{v_1, \ldots, v_k\}$ *has degree* d. *We call the set* $\{v_1, \ldots, v_k\}$ *a* d-deletion set.

Remark 3.2. To say that G has deletion distance 0 from degree d is just to say that G has maximum degree d. Also note that if $d = 0$, then the d-deletion set is just a vertex cover and the minimum deletion distance the vertex cover number of G.

We show that isomorphism are fixed-parameter tractable on such graphs parameterized by k with fixed degree d; in particular we give a procedure that computes a polynomial kernel in linear time.

Theorem 3.3. *For any graph* G *and integers* $d, k > 0$, *we can identify in linear time a subgraph* G' *of* G, *a set of vertices* $U \subseteq V(G')$ *with* $|U| = O(k(k + d)^2)$ *and a* $k' \leq k$ *such that: G has deletion distance k to degree d if and only if G' has deletion distance k' to d and, moreover, if G' has deletion distance at most k', then any minimum size d-deletion set for G' is contained in U.*

Proof. Let $H := \{v \in V(G) \mid \deg(v) > k + d\}$. Now, if R is a minimum size d-deletion set for G and G has deletion distance at most k to degree d, then $|R| \leq k$ and the vertices in $V(G \setminus R)$ have degree at most $k + d$ in G. So $H \subseteq R$. This means that if $|H| > k$, then G must have deletion distance greater than k to degree d and in that case we let $G' := G, k' := k$ and $U = \emptyset$.

Otherwise let $G' := G \setminus H$ and $k' := k - |H|$. We have shown that every d-deletion set of size at most k must contain H. Thus G has deletion distance k to degree d if and only if G' has deletion distance k' to degree d.

Let $S := \{v \in V(G') \mid \deg_{G'}(v) > d\}$ and $U := S \cup N_{G'}(S)$. Let $R' \subseteq V(G')$ be a minimum size d-deletion set for G'. We show that $R' \subseteq U$. Let $v \notin U$. Then by the definition of U we know that $\deg_{G'}(v) \leq d$ and all of the neighbours of v have degree at most d in G'. So if $v \in R'$, then $G \setminus (R' \setminus \{v\})$ also has maximal degree d, which violates that R' is of minimum size. Thus $v \notin R'$.

Note that the vertices in $G' \setminus (R' \cup N(R'))$ have the same degree in G' as in G and thus all have degree at most d. So $S \subseteq R' \cup N(R')$ and thus $|U| \leq k' + k'(k + d) + k'(k + d)^2 = O(k(k + d)^2)$.

Finally, the sets H and U defined as above can be found in linear time, and G', k' can be computed from H in linear time. $\qquad\square$

Remark 3.4. Note that if $U = \emptyset$ and $k' > 0$, then there are no d-deletion sets of size at most k'.

Next we see how the kernel U can be used to determine whether two graphs with deletion distance k to degree d are isomorphic by reducing the problem to isomorphism of coloured graphs of degree at most d.

Suppose we are given two graphs G and H with d-deletion sets $S = \{v_1, \ldots, v_k\}$ and $T = \{w_1, \ldots, w_k\}$ respectively. Further suppose that the map $v_i \mapsto w_i$ is an isomorphism on the induced subgraphs $G[S]$ and $H[T]$. We can then test if this map can be extended to an isomorphism from G to H using Theorem 2.2. To be precise, we define the coloured graphs G' and H' which are obtained from $G \setminus S$ and $H \setminus T$ respectively, by colouring vertices. A vertex $u \in V(G')$ gets the colour $\{i \mid v_i \in N_G(u)\}$, i.e. the set of indices of its neighbours in S. Vertices in H' are similarly coloured by the sets of indices of their neighbours in T. It is clear that G' and H' are isomorphic if, and only if, there is an isomorphism between G and H, extending the fixed map between S and T. The coloured graphs G' and H' have degree bounded by d, so Theorem 2.2 gives us a polynomial-time isomorphism test on these graphs.

Now, given a pair of graphs G and H which have deletion distance k to degree d, let A and B be the sets of vertices of degree greater than $k + d$ in the two graphs respectively. Also, let U and V be the two kernels in the graphs obtained from Theorem 3.3. Thus, any d-deletion set in G contains A and is contained in $A \cup U$ and similarly, any d-deletion set for H contains B and is contained in $B \cup V$. Therefore to test G and H for isomorphism, it suffices to consider all k-element subsets S of $A \cup U$ containing A and all k-element subsets T of $B \cup V$ containing B, and if they are d-deletion sets for G and H, check for all $k!$ maps between them whether the map can be extended to an isomorphism from G to H. As d is constant this takes time $O^* \left(\binom{k^3}{k}^2 \cdot k! \right)$, which is $O^* \left(2^{7k \log k} \right)$.

4   Elimination Distance to Bounded Degree

In this section we introduce a new structural parameter for graphs. We generalise the idea of deletion distance to triviality by recursively allowing deletions from each component of the graph. This generalises the idea of elimination height or tree-depth, and is equivalent to it when the notion of triviality is the empty graph. In the context of graph isomorphism and canonisation we again define triviality to mean bounded degree, so we look at the elimination distance to bounded degree.

Definition 4.1. *The* elimination distance to degree d *of a graph G is defined as follows:*

$$ed_d(G) := \begin{cases} 0, & \text{if } \Delta(G) \le d; \\ 1 + \min\{ed_d(G \setminus v) \mid v \in V(G)\}, & \text{if } \Delta(G) > d \text{ and } G \text{ is connected}; \\ \max\{ed_d(H) \mid H \text{ a connected component of } G\}, & \text{otherwise.} \end{cases}$$

We first introduce other equivalent characterisations of this parameter. (The proofs of equivalence are omitted due to lack of space and can be found in the full version of the paper.) If G is a graph that has elimination distance k to degree d, then we can associate a certain tree order $\le$ with it:

Definition 4.2. *A tree order $\le$ on $V(G)$ is an* elimination order to degree d *for G if for each $v \in V(G)$ the set*

$$S_v := \{u \in V(G) \mid uv \in E(G) \text{ and } u \not\le v \text{ and } v \not\le u\}$$

satisfies either:

- $S_v = \emptyset$; *or*
- v *is $\le$-maximal, $|S_v| \le d$, and for all $u \in S_v$, we have $\{w \mid w < u\} = \{w \mid w < v\}$.*

Remark 4.3. Note that if $S_v = \emptyset$ for all $v \in V(G)$, then an elimination order to degree d is just an elimination order, in the sense of Definition 2.1.

Proposition 4.4. *A graph G has $ed_d(G) \le k$ if, and only if, there is an elimination order $\le$ to degree d of height k for G.*

We can split a graph with an elimination order to degree d in two parts: one of low degree, and one with an elimination order defined on it. So if G is a graph that has elimination distance k to degree d, we can associate an elimination order $\le$ for a subgraph H of G of height k with G, so that each component of $G \setminus V(H)$ has degree at most d and is connected to H in a certain way:

Proposition 4.5. *Let G be a graph and $\leq$ an elimination order to degree d for G of height k. If A is the set of vertices in $V(G)$ that are not $\leq$-maximal, then:*

1. *$\leq$ restricted to A is an elimination order of height k of $G[A]$; and*
2. *$G \setminus A$ has degree at most d;*
3. *if C is the vertex set of a component of $G \setminus A$, and $u, v \in A$ are $\leq$-incomparable, then either $E(u, C) = \emptyset$ or $E(v, C) = \emptyset$.*

We also have a converse to the above in the following sense.

Proposition 4.6. *Suppose G is a graph with $A \subseteq V(G)$ a set of vertices and $\leq_A$ an elimination order of $G[A]$ of height k, such that:*

1. *$G \setminus A$ has degree at most d;*
2. *if C is the vertex set of a component of $G \setminus A$, and $u, v \in A$ are incomparable, then either $E(u, C) = \emptyset$ or $E(v, C) = \emptyset$.*

Then, $\leq_A$ can be extended to an elimination order to degree d for G of height $k + 1$.

Remark 4.7. In the following, given a graph G and an elimination order to degree d, $\leq$, we call the subgraph of G induced by the non-maximal elements of the order $\leq$ the *non-maximal subgraph of G under $\leq$.*

These alternative characterisations are very useful. In the following series of lemmas, we use them to construct a *canonical* elimination order to degree d of G, based on an elimination order of a graph we call the *torso* of G, which consists of the high-degree vertices of G, along with some additional edges. (Due to the lack of space we again omit proofs. These proofs can be found in the full version of the paper.)

Lemma 4.8. *Let G be a graph with maximal degree $\Delta(G) \leq k + d$. Let $\leq$ be an elimination order to degree d of height k of G with non-maximal subgraph H.*
Then $H' = G[V(H) \cup \{v \in V(G) \mid \deg_G(v) > d\}]$ has an elimination order $\sqsubseteq$ of height at most $k(k + d + 1)$ that can be extended to an elimination order to degree d of G.

Lemma 4.9. *Let G be a graph. Let $\leq$ be an elimination order to degree d of G of height k with non-maximal subgraph H, such that H contains all vertices of degree greater than d.*
Then $H' = G[\{v \in V(H) \mid \deg_G(v) > d\}]$ has an elimination order $\sqsubseteq$ of height at most $k((k + 1)d)^{2^k}$ that can be extended to an elimination order to degree d of G of height $k((k + 1)d)^{2^k} + 1$.

Next we introduce the notion of d-degree torso and prove that it captures the properties that we require of an elimination tree to degree d.

Definition 4.10. *Let G be a graph, let $d > 0$ and let H be the induced subgraph of G containing the vertices of degree larger than d. The d-degree torso of G is the graph C obtained from H by adding an edge between two vertices $u, v \in H$ if there is a path through $G \setminus V(H)$ from u to v in G.*

Lemma 4.11. *Let G be a graph and let C be the d-degree torso of G. Let $H = G[V(C)]$ and let $\leq$ be an elimination order for H. Then $\leq$ is an elimination order for C of height h if, and only if, $\leq$ can be extended to an elimination order to degree d for G of height $h + 1$.*

Lemma 4.12. *Let G be a graph with elimination distance k to degree d and maximum degree $\Delta(G) \leq k + d$. Let C be the d-degree torso of G and let $\leq$ be a minimum height elimination order for C. Then $\leq$ has height at most*

$$k(k + d + 1)((k(k + d + 1) + 1)d)^{2^{k(k+d+1)}}.$$

We are ready to prove the main result now:

Theorem 4.13. *Let G be a graph that has elimination distance k to degree d. Let $\leq$ be a minimum height elimination order of the d-degree torso G. Then $\leq$ can be extended to an elimination order to degree d of G of height*

$$k((k + 1)(k + d))^{2^k} + k(1 + k + d)(k(1 + k + 2d))^{2^{k(1+k+d)}} + 1.$$

Proof. We show that the d-degree torso of G has an elimination order of height at most $k((k + 1)(k + d))^{2^k} + k(1 + k + d)(k(1 + k + 2d))^{2^{k(1+k+d)}}$. The Theorem then follows by Lemma 4.11.

Let C be the $(k + d)$-degree torso of G. We first show that the tree-depth of C is bounded by $k((k+1)(k+d))^{2^k}$. To see this, let $\sqsubseteq$ be an elimination order to degree d of G of minimum height with non-maximal subgraph H. Note that H contains all vertices of degree greater than $k + d$, because vertices in $G \setminus V(H)$ are adjacent to at most k vertices in H.

Let $H' = G[\{v \in V(H) \mid \deg_G(v) > k+d\}]$. By Lemma 4.9, the subgraph H' has an elimination order $\preceq$ of depth at most $h := k((k + 1)(k + d))^{2^k}$ that can be extended to an elimination order to degree $k + d$ of G of height $h + 1$. Note that $V(H') = V(C)$, so by Lemma 4.11, the order $\preceq$ is an elimination order for C. Let $\preceq'$ denote its extension to G.

Let Z be a component of $G \setminus V(H')$ and let C_Z be the d-degree torso of Z. By Lemma 4.12, there is an elimination order $\preceq_Z$ for C_Z of height at most $k(k + d + 1)((k(k + d + 1) + 1)d)^{2^{k(k+d+1)}}$. Let v_Z be the $\preceq$-maximal element in C such that there is a $w \in C_Z$ with $v_Z \preceq' w$. Define

$$\leq' := \preceq \cup \bigcup_Z \preceq_Z$$

$$\cup \bigcup_Z \{(v, w) \mid v \preceq' v_Z, w \in C_Z\}$$

$$\cup \bigcup_Z \{(v, w) \mid v \preceq' v_Z \text{ or } v \in C_Z, w \in \hat{C}, \hat{C} \text{ a component of } Z \setminus V(C_Z), E(v, \hat{C}) \neq \emptyset\}.$$

Observe that $C \cup \bigcup_Z C_Z$ is a subgraph of the d-degree torso of G. Thus $\leq'$ is an elimination order for the d-degree torso of G. The height of $\leq'$ is bounded by

$$td(C) + \max\{td(C_Z)\}_Z \leq k((k + 1)(k + d))^{2^k} + k(1 + k + d)(k(1 + k + 2d))^{2^{k(1+k+d)}}.$$

$\square$

5 Canonisation Parameterized by Elimination Distance to Bounded Degree

In this section we show that graph canonisation, and thus graph isomorphism, is FPT parameterized by elimination distance to bounded degree. The main idea is to construct a labelled directed tree T_G from a graph G (of elimination distance k to degree d) that is an isomorphism invariant for G. From the labelled tree T_G we obtain a canonical labelled tree using the tree canonisation algorithm from Lindell [12]. In the last step we construct a canonical form of G from the canonical labelled tree.

The tree T_G is obtained from G by taking a tree-depth decomposition of the d-degree torso of G and labelling the nodes with the isomorphism types of the low-degree components that attach to them. The tree-depth decomposition of a graph is just the elimination order in tree form. We formally define it as follows:

Definition 5.1. *Given a graph H and an elimination order $\leq$ on H, the tree-depth decomposition associated with $\leq$ is the directed tree with nodes $V(H)$ and an arc $a \to b$ if, and only if, $a < b$ and there is no c such that $a < c < b$.*

Remark 5.2. The tree-depth decomposition corresponding to an elimination order is what, in the language of partial orders, is known as its covering relation.

Note that, in general, the tree-depth decomposition of a graph that is not connected may be a forest. By results of Bouland *et al.* [2], we can construct a canonical tree-depth decomposition of an n-vertex graph of tree-depth k in time $f(k) \cdot n^c$ for some comutable f and constant c.

Before defining T_G formally, we need one piece of terminology.

Definition 5.3. *Let G be a graph and let $\leq$ be a tree order for G. The level of a vertex $v \in V(G)$ is the length of the chain $\{w \in V(G) \mid w \leq v\}$. We denote the level of v by $level_{\leq}(v)$.*

Given a graph G of elimination distance k to degree d, let C be the d-degree torso of G, let T be a canonical tree-depth decomposition of C and $\leq$ the corresponding elimination order. Let Z be a component of $G \backslash C$. We let Z^C denote the coloured graph that is obtained by colouring each vertex v in Z by the colour $\{i \mid uv \in E(G)$ for some $u \in C$ with $level_{\leq}(u) = i\}$. We write $F(Z^C)$ for the canonical form of this coloured graph given by Theorem 2.2. Note that, by the definition of elimination distance, there is, for each Z and i at most one vertex $u \in C$ with $level_{\leq}(u) = i$ which is in $N_G(Z)$.

We are now ready to define the labelled tree T_G. The nodes of T_G are the nodes of T together with a new node r, and the arcs are the arcs of T along with new arcs from r to the root of each tree in T. Define, for each node u of T_G, $\mathcal{Z}_u$ to be the set $\{Z \mid Z$ is a component of $G \backslash C$ with $u \leq$-maximal in $C \cap N_G(Z)\}$ (if $u \neq r$) and $\{Z \mid Z$ is a component of $G \backslash C$ with $C \cap N_G(Z) = \emptyset\}$ (if $u = r$). Each node u in T carries a label consisting of two parts:

– $L_w := \{level(w) \mid w < u$ and $uw \in E(G)\}$; and
– the multiset $\{F(Z^C) \mid Z \in \mathcal{Z}_u\}$.

Proposition 5.4. *For any graphs G and G', T_G and $T_{G'}$ are isomorphic labelled trees if, and only if, $G \cong G'$.*

Proof. If $G \cong G'$ then, by construction, their d-degree torsos induce isomorphic graphs. The canonical tree-depth decomposition of Bouland et al. then produces isomorphic directed trees and the isomorphism must preserve the labels that encode the rest of the graphs G and G' respectively.

For the converse direction, suppose we have an isomorphism ϕ between the labelled trees T_G and $T_{G'}$. Since the label L_u of any node u encodes all ancestors of u which are neighbours, ϕ must preserve all edges and non-edges in the d-degree torso C of G. To extend ϕ to all of G, for each node u in T_G, let β_u be a bijection from $\mathcal{Z}_u$ to the corresponding set $\mathcal{Z}_{\phi(u)}$ of components of $G' \setminus C'$, such that $F(Z^C) = F(\beta_u(Z)^{C'})$ (such a bijection exists as u and $\phi(u)$ carry the same label). Thus, in particular, there is an isomorphism between Z^C and $\beta_u(Z)^{C'}$, since they have the same canonical form. We define, for each $v \in V(G) \setminus C$, $\phi(v)$ to be the image of v under the isomorphism taking the component Z containing v to $\beta_u(Z)$. Note that this gives a well-defined function on $V(G)$, because for each such v, there is exactly one node u of T_G such that the component containing v is in $\mathcal{Z}_u$. We claim that ϕ is now an isomorphism from G to G'. Let vw be an edge of G. If both v and w are in C, then either $v < w$ or $w < v$. Assume, without loss of generality, that it is the former. Then, $level(v) \in L_w$ in the label of w in T_G and since φ is a label-preserving isomorphism from T_G to $T_{G'}$, $\varphi(v)\varphi(w)$ is an edge in G'. If both v and w are in $G \setminus C$, then there is some component Z of $G \setminus C$ that contains them both. Since φ maps Z to an isomorphic component of $G' \setminus C'$, $\varphi(v)\varphi(w) \in E(G')$. Finally, suppose v is in C and w in $G \setminus C$ and let Z be the component containing w. Then $i := level(v)$ is part of the colour of w in Z^C and hence part of the colour of $\varphi(w)$ in the corresponding component of $G' \setminus C'$. Moreover, if u is the $\leq$-maximal element in $C \cap N_G(Z)$, then we must have $v \leq u$. Thus $\varphi(v)$ is the unique element of level i in $C' \cap N_{G'}(\beta_u(Z))$ and we conclude that $\varphi(v)\varphi(w) \in E(G')$. By a symmetric argument, we have that for any edge $vw \in E(G')$, $\phi^{-1}(v)\phi^{-1}(w) \in E(G)$ and we conclude that ϕ is an isomorphism. □

With this, we are able to establish our main result.

Theorem 5.5. *Graph Canonisation is* FPT *parameterized by elimination distance to bounded degree.*

Proof. Suppose we are given a graph G with $|V(G)| = n$. We first compute the d-degree torso C of G in $O(n^4)$ time. Using the result from Bouland *et. al.* [2, Theorem 11], we can find a canonical tree-depth decomposition for C in time $O(h(k)n^3 log(n))$ for some computable function h. To compute the labels of the nodes in the trees (and hence obtain) T_G, we determine, for each $u \in C$, the set $\{level(w) \mid w < u$ and $uw \in E(G)\}$. This can be done in time $O(n^2)$.

Then, we find the components of $G \setminus C$, and colour the vertices with the levels of their neighbours in C. This can be done in $O(n^2)$ time. Finally, we compute for each coloured component Z^C the canonical representative $F(Z^C)$ which, by Theorem 2.2 can be done in polynomial time (where the degree of the polynomial depends on d).

Having obtained T_G, we compute the canonical form T'_G in linear time using Lindell's canonisation algorithm [12]. Using the labels of T'_G one can, in linear time, construct a graph G' such that $T(G') = T'_G$. By Proposition 5.4, this is a canonical form G' of G. $\qquad\square$

Corollary 5.6. *Graph Isomorphism is* FPT *parameterized by elimination distance to bounded degree.*

6 Conclusion

We introduce a new way of parameterizing graphs by their distance to triviality, i.e. by elimination distance. In the particular case of graph canonisation, and thus also graph isomorphism, taking triviality to mean graphs of bounded degree, we show that the problem is FPT.

A natural question that arises is what happens when we take other classes of graphs for which graph isomorphism is known to be tractable as our "trivial" classes. For instance, what can we say about GI when parameterized by elimination distance to planar graphs? Unfortunately techniques such as those deployed in the present paper are unlikely to work in this case. Our techniques rely on identifying a canonical subgraph which defines an elimination tree into the trivial class. In the case of planar graphs, consider graphs which are subdivisions of K_5, each of which is deletion distance 1 away from planarity. However the deletion of *any* vertex yields a planar graph and it is therefore not possible to identify a canonical such vertex.

More generally, the notion of elimination distance to triviality seems to offer promise for defining tractable parameterizations for many graph problems other than isomorphism. This is a direction that bears further investigation.

It is easy to see that if a class of graphs C is characterised by a finite set of excluded minors, that the class $\hat{\mathsf{C}}$ of graphs with bounded elimination distance to C is characterised by a finite set of excluded minors as well. An interesting question is whether we can, given the set of excluded minors for C, compute the excluded minors for $\hat{\mathsf{C}}$ as well?

References

1. Babai, L., Luks, E.M.: Canonical labeling of graphs. In: Proceedings of 15th ACM Symposium. Theory of Computing, pp. 171–183. ACM, New York (1983)
2. Bouland, A., Dawar, A., Kopczyński, E.: On tractable parameterizations of graph isomorphism. In: Thilikos, D.M., Woeginger, G.J. (eds.) IPEC 2012. LNCS, vol. 7535, pp. 218–230. Springer, Heidelberg (2012)

3. Diestel, R.: Graph Theory. Springer, New York (2000)
4. Downey, R.G., Fellows, M.R.: Parameterized Complexity. Springer, Heidelberg (2012)
5. Evdokimov, S., Ponomarenko, I.: Isomorphism of coloured graphs with slowly increasing multiplicity of Jordan blocks. Combinatorica **19**(3), 321–333 (1999)
6. Fellows, M.R., Lokshtanov, D., Misra, N., Rosamond, F.A., Saurabh, S.: Graph layout problems parameterized by vertex cover. In: Proceedings 19th International Symposium Algorithms and Computation, pp. 294–305 (2008)
7. Filotti, I.S., Mayer, J.N.: A polynomial-time algorithm for determining the isomorphism of graphs of fixed genus. In: STOC '80: Proceedings of the Twelfth Annual ACM Symposium on Theory of Computing, ACM Request Permissions, April 1980
8. Flum, J., Grohe, M.: Parameterized Complexity Theory. Springer, Berlin (2006)
9. Grohe, M., Marx, D.: Structure theorem and isomorphism test for graphs with excluded topological subgraphs. In: Proceedings 44th Symposium on Theory of Computing, pp. 173–192 (2012)
10. Guo, J., Hüffner, F., Niedermeier, R.: A structural view on parameterizing problems: distance from triviality. In: Downey, R.G., Fellows, M.R., Dehne, F. (eds.) IWPEC 2004. LNCS, vol. 3162, pp. 162–173. Springer, Heidelberg (2004)
11. Kratsch, S., Schweitzer, P.: Isomorphism for graphs of bounded feedback vertex set number. In: Kaplan, H. (ed.) SWAT 2010. LNCS, vol. 6139, pp. 81–92. Springer, Heidelberg (2010)
12. Lindell, S.: A logspace algorithm for tree canonization (extended abstract). In: STOC '92: Proceedings of the Twenty-fourth Annual ACM Symposium on Theory of Computing, ACM Request Permissions, July 1992
13. Lokshtanov, D., Pilipczuk, M., Pilipczuk, M., Saurabh, S.: Fixed-parameter tractable canonization and isomorphism test for graphs of bounded treewidth. arxiv:1404.0818 [cs.DS] (2014)
14. Luks, E.M.: Isomorphism of graphs of bounded valence can be tested in polynomial time. J. Comput. Syst. Sci. **25**(1), 42–65 (1982)
15. McKay, B.D.: Practical graph isomorphism. Congressus Numerantium **30**, 45–87 (1981)
16. McKay, B.D., Piperno, A.: Practical graph isomorphism. II. J. Symb. Comput. **60**, 94–112 (2014)
17. Miller, G.: Isomorphism testing for graphs of bounded genus. In: STOC '80: Proceedings 12th ACM Symposium Theory of Computing. ACM (1980)
18. Niedermeier, R.: Invitation to Fixed-Parameter Algorithms. Oxford University Press, Oxford (2006)
19. Ponomarenko, I.N.: The isomorphism problem for classes of graphs closed under contraction. J. Sov. Math. **55**(2), 1621–1643 (1991)
20. Toda, S.: Computing automorphism groups of chordal graphs whose simplicial components are of small size. IEICE Trans. Inf. Syst. **E89-D**(8), 2388–2401 (2006)
21. Uehara, R., Toda, S., Nagoya, T.: Graph isomorphism completeness for chordal bipartite graphs and strongly chordal graphs. Discrete Appl. Math. **145**(3), 479–482 (2005)
22. Yamazaki, K., Bodlaender, H.L., De Fluiter, B., Thilikos, D.M.: Isomorphism for. In: Bongiovanni, G., Bovet, D.P., Di Battista, G. (eds.) CIAC 1997. LNCS, vol. 1203, pp. 276–287. Springer, Heidelberg (1997)

Exact Exponential Algorithms to Find
a Tropical Connected Set of Minimum Size

Mathieu Chapelle[1], Manfred Cochefert[2], Dieter Kratsch[2]([✉]),
Romain Letourneur[3], and Mathieu Liedloff[3]

[1] Université Libre de Bruxelles, CP 212, 1050 Bruxelles, Belgium
mathieu.chapelle@ulb.ac.be
[2] LITA,Université de Lorraine, 57045 Metz Cedex 01, France
{manfred.cochefert,dieter.kratsch}@univ-lorraine.fr
[3] LIFO, Université d'Orléans, 45067 Orléans Cedex 2, France
{romain.letourneur,mathieu.liedloff}@univ-orleans.fr

Abstract. The input of the *Tropical Connected Set* problem is a vertex-colored graph $G = (V, E)$ and the task is to find a connected subset $S \subseteq V$ of minimum size such that each color of G appears in S. This problem is known to be NP-complete, even when restricted to trees of height at most three. We show that *Tropical Connected Set* on trees has no subexponential-time algorithm unless the Exponential Time Hypothesis fails. This motivates the study of exact exponential algorithms to solve *Tropical Connected Set*. We present an $\mathcal{O}^*(1.5359^n)$ time algorithm for general graphs and an $\mathcal{O}^*(1.2721^n)$ time algorithm for trees.

1 Introduction

Problems on vertex-colored graphs have been widely studied, notably the *Graph Motif* problem which was introduced in 1994 by McMorris et al. [17]. This problem is motivated by applications in biology and metabolic networks [16,19]. *Graph Motif* is a decision problem asking whether a given vertex-colored graph $G = (V, E)$ has a connected subset S of vertices such that there is bijection between S and a multiset of colors; the latter being part of the input. Equivalently, the question is whether a vertex-colored graph given with a vector of multiplicities of the colors of G has a connected vertex set S such that each color appears in S with its required multiplicity. As an immediate consequence, the size of a solution S is given by the input.

Fellows et al. proved that *Graph Motif* is NP-complete, even if the multiset of colors is actually a set and if the graph is a tree of maximum degree three [9]. They also proved that this problem is NP-complete even if the multiset contains only two colors and if the graph is bipartite of maximum degree four [9]. Many variants of the *Graph Motif* problem have been studied; typical contributions being NP-hardness proofs and fixed-parameter tractable algorithms [4,9,10,14,16]. To the best of our knowledge, the unique paper to study exact exponential time algorithms for a variant of *Graph Motif* is the one by Dondi et al. [8].

© Springer International Publishing Switzerland 2014
M. Cygan and P. Heggernes (Eds.): IPEC 2014, LNCS 8894, pp. 147–158, 2014.
DOI: 10.1007/978-3-319-13524-3_13

In this paper we study the exact complexity of another variant of the *Graph Motif* problem, i.e. the optimization problem *Tropical Connected Set*. Let $G = (V, E)$ be a graph and let c be a (not necessarily proper) vertex coloring assigning to each vertex of G a color, i.e. a positive integer. By C we denote the set of colors of the vertices of G. Now $S \subseteq V$ is a *tropical set* of the vertex-colored graph G if all colors of G appear in S. The problem *Tropical Connected Set* takes as input a vertex-colored graph and the task is to find a tropical connected subset of vertices S of minimum size. The study of tropical connected sets was initiated by Angles d'Auriac et al. [3]. With a reduction from the well-known NP-complete problem Dominating Set, Angles d'Auriac et al. have shown that finding a minimum tropical connected set is NP-complete, even when restricted to trees of height three [3].

Using their reduction, we first show that *Tropical Connected Set* on trees has no subexponential-time algorithm unless the Exponential Time Hypothesis fails (Sect. 3); thus the existence of such a subexponential-time algorithm is considered unlikely. This provides the following motivation for studying the exact complexity of *Tropical Connected Set*. Many NP-hard graph problems when restricted to trees are polynomial-time solvable; often they are fixed-parameter tractable when parameterized by treewidth, based on [6]. Also NP-hard graph problems on planar graphs and even graphs of bounded genus often can be solved by subexponential-time parameterized algorithms [7,13]. Thus, if a problem is not subexponential-time solvable on trees, what kind of fast exponential algorithm on trees, or even general graphs can we expect; a $\mathcal{O}^*(\alpha^n)$ time algorithm with $\alpha \geq 1$ much smaller than 2, or even α close to 1? Our main contributions are two fast exact exponential algorithms solving the NP-hard problem *Tropical Connected Set*. The algorithm for general graphs has running time $\mathcal{O}^*(1.5359^n)$ and uses algorithms for the NP-hard problems CONNECTED RED-BLUE DOMINATING SET and STEINER TREE as subroutines (Sect. 4). The branching algorithm for trees has running time $\mathcal{O}^*(1.2721^n)$ and heavily exploits structural properties of trees (Sect. 5).

Proofs not given in this version will be provided in the full version of this paper.

2 Preliminaries

Throughout this paper, we denote by $G = (V, E)$ an undirected graph, and by $T = (V, E)$ an undirected tree with vertex set V and edge set E. We adopt the convention $n = |V|$ and $m = |E|$. For a subset $X \subseteq V$ of vertices, we denote by $G[X]$ the subgraph of G induced by X. For a vertex $v \in V$ of G, we denote by $N(v)$ the set of all neighbors of v; and we let $N[v] = N(v) \cup \{v\}$. For every $X \subseteq V$, we denote by $N[X] = \bigcup_{x \in X} N[x]$ the closed neighbourhood of X, and by $N(X) = N[X] \setminus X$ the open neighbourhood of X. A vertex set $S \subseteq V$ of G is connected if the subgraph $G[S]$ is connected. Let $G = (V, E)$ be a graph, and let $c : V \to \mathbb{N}$ be a (not necessarily proper) coloring of G. Then we call (G, c) a vertex-colored graph and $C = \{c(v) : v \in V\}$ the set of colors of G. For a subset of vertices S of a vertex-colored graph (G, c), we denote by $c(S) = \{c(v) : v \in S\}$ the set of colors of S, and we call S tropical if $c(S) = C$.

We denote by $l_1(G)$ the number of colors appearing exactly once in a vertex-colored graph (G, c), and by $l_2(G)$ the number of colors appearing at least twice in (G, c). A connected component $G[B]$ of a disconnected graph $G = (V, E)$ has a tropical connected set if and only if all colors of the graph G appear in $B \subseteq V$. Hence (G, c) has no tropical connected set if and only if none of its components contains all colors of G. Thus we may consider only connected graphs.

Throughout our paper all graphs and trees are vertex-colored and often denoted by G and T instead of (G, c) and (T, c). In this paper, we focus on the TROPICAL CONNECTED SET problem:

TROPICAL CONNECTED SET
Input: Graph $G = (V, E)$ with a coloring $c : V \to \mathbb{N}$ and set of colors C.
Question: Find a minimum size subset $S \subseteq V$ such that $G[S]$ is connected, and S contains at least one vertex of each color in C.

3   No Subexponential-Time Algorithm for Trees

We show that TROPICAL CONNECTED SET on trees of height at most 3 admits no subexponential-time algorithm unless the Exponential Time Hypothesis (short ETH) fails. ETH has been defined by Impagliazzo et al. [15]. On one hand it is considered unlikely that ETH fails, and on the other hand a proof of ETH would have many important consequences in complexity theory. The non existence of a subexponential-time algorithm under the assumption of ETH stresses the significance of designing an exact algorithm running in time $\mathcal{O}^*(\alpha^n)$ (for some fixed constant α much smaller than 2). The result is achieved by combining a reduction from DOMINATING SET to TROPICAL CONNECTED SET described in [3], and a proof that DOMINATING SET admits no subexponential-time algorithm on graphs with maximum degree 6 given in [12].

Theorem 1. *The* TROPICAL CONNECTED SET *problem on trees of height at most 3 admits no subexponential-time algorithm, unless* SNP $\subseteq$ SUBEXP, *which would imply that the Exponential Time Hypothesis fails.*

Proof. Angles d'Auriac et al. [3] proved that TROPICAL CONNECTED SET is NP-complete even on trees of height at most 3, by a reduction from DOMINATING SET. For the sake of completeness, we briefly recall the construction used in their proof.

Given an input graph $G = (V, E)$, instance for DOMINATING SET, we construct a vertex-colored tree (T, c) as follows. Consider an arbitrary ordering $\sigma : V \to \mathbb{N}$ on the vertices of G. Initially, T contains a unique vertex r colored $R \notin \{0, 1, \ldots, |V|\}$. Now, for every vertex $v \in V(G)$, add to T a star whose center is colored $\sigma(v)$, with one leaf colored $\sigma(u)$ for every neighbor u of v in G, and an extra leaf colored 0 linked to the vertex r of T. See Fig. 1.

Now, let us refer to Fomin et al. [12] who proved that DOMINATING SET admits no subexponential-time algorithm on graphs with maximum degree 6,

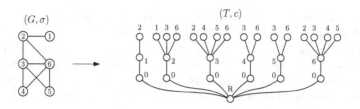

Fig. 1. The reduction from DOMINATING SET to TROPICAL CONNECTED SET [3].

unless SNP $\subseteq$ SUBEXP. Observe that given an input graph $G = (V, E)$ of maximum degree 6, instance for DOMINATING SET, the reduction from DOMINATING SET to TROPICAL CONNECTED SET described above yields a tree T, instance for TROPICAL CONNECTED SET, of height 3 and containing at most $8n + 1 = \mathcal{O}(n)$ vertices. Thus a subexponential-time algorithm for TROPICAL CONNECTED SET on trees of maximum height 3 would yield a subexponential-time algorithm for DOMINATING SET on graphs with maximum degree 6, hence implying that SNP $\subseteq$ SUBEXP; and the latter implies that ETH fails [15]. $\square$

The fact that there is no subexponential-time algorithm for TROPICAL CONNECTED SET on trees unless ETH fails, triggered our interest in constructing fast exponential time algorithms for TROPICAL CONNECTED SET on trees as well as on general graphs.

4   An Exact Exponential Algorithm for General Graphs

This section is devoted to the design and analysis of an exact algorithm for TROPICAL CONNECTED SET. A naive brute-force algorithm would solve this problem in $\mathcal{O}^*(2^n)$ time. Using reductions to the NP-hard problems CONNECTED RED-BLUE DOMINATING SET and STEINER TREE and by balancing the corresponding algorithms, we design an algorithm computing a minimum tropical connected set in time $\mathcal{O}^*(1.5359^n)$.

Let us recall the definition of these two problems:

STEINER TREE
Input: Graph $G = (V, E)$, weight function $w : E \to \mathbb{N}$, set of terminals $K \subseteq V$.
Question: Find a connected subtree $T = (V', E')$ of G, with $V' \subseteq V$ and $E' \subseteq E$, such that $K \subseteq V'$ and $\sum_{e \in E'} w(e)$ is minimum.

CONNECTED RED-BLUE DOMINATING SET
Input: Graph $G = (R \cup B, E)$ where vertices are colored either red (vertices in R) or blue (vertices in B).
Question: Find the smallest subset $S \subseteq R$ of red vertices such that $G[S]$ is connected, and every vertex in B has at least one neighbor in S, that is $B \subseteq N(S)$.

The currently best known exact algorithms (for our purposes) solving STEINER TREE and CONNECTED RED-BLUE DOMINATING SET are due to Nederlof [18] and

Abu-Khzam et al. [1] respectively. Their running times are $\mathcal{O}^*(W \cdot 2^{|K|})$ (where K is the set of terminals and W the maximum weight of the input) and $\mathcal{O}^*(1.36443^n)$.

Now we define a construction that will be used by our algorithms to reduce TROPICAL CONNECTED SET to STEINER TREE and CONNECTED RED-BLUE DOMINATING SET. Let $G = (V, E)$ be a graph, let c be a coloring of G and C the set of colors of G. First the graph $G' = (R' \cup B', E')$ to be used to reduce TROPICAL CONNECTED SET to STEINER TREE (see Lemma 1) is defined as follows: $R' = \{v' \mid v \in V\}$, $B' = \{r_i \mid i \in C\}$, $E' = \{u'v' \mid uv \in E\} \cup \{v'r_i \mid v \text{ is of color } i \text{ in } G\}$.

Secondly we construct the graph $G'' = (R'' \cup B'', E'')$ to be used to reduce TROPICAL CONNECTED SET to CONNECTED RED-BLUE DOMINATING SET (see Lemma 2). Initially, set $R'' = R'$, $B'' = B'$ and $E'' = E'$. For every vertex $v \in V$ whose color appears exactly once in G, move the corresponding vertex v' from R'' to B'', and remove the vertex r_i from B'', where $c(v) = i$ in G. After this step, let $B_1, \ldots, B_p$ be the components of the subgraph induced in $G''[B'']$ by those vertices that had been moved to B''. For each $i = 1, 2, \ldots p$, contract the component B_i in $G''[B'']$ so that it remains only one vertex and call this vertex b_i. Note that after all contractions, B'' is now an independent set of G''. Finally for all b_i, $1 \leq i \leq p$, turn the neighborhood $N_{G''}(b_i) \subseteq R''$ into a clique by adding corresponding edges. It is worth mentioning that this construction can be done in polynomial time.

Now we describe three ways to solve TROPICAL CONNECTED SET: by brute-force, via STEINER TREE and via CONNECTED RED-BLUE DOMINATING SET.

Brute-force. Let (G, c) be an instance of TROPICAL CONNECTED SET. Our brute-force algorithm first computes the set U of all vertices having a color appearing only once in G (forced to be in any tropical set). Then for every $A \subseteq V \setminus U$ verify in polynomial time whether $U \cup A$ is a tropical connected set and compute the minimum size of (such) a tropical connected set. This algorithm runs in time $\mathcal{O}^*\left(2^{n-l_1(G)}\right)$.

Using Steiner tree. TROPICAL CONNECTED SET can easily be reduced to STEINER TREE as follows. Let (G, c) be an instance of TROPICAL CONNECTED SET. We construct an instance (G', w, K) for STEINER TREE, where $G' = (R' \cup B', E')$ is the graph constructed above and the terminal set $K = B'$. Note that $|K| = |B'| = |C|$. The weight function is defined as follows: $w(e) = n = |V|$ for every $e \in E(G')$ incident to a vertex of B', and $w(e) = 1$ for every edge $e = u'v'$ with $u', v' \in R'$.

Lemma 1. *The vertex-colored graph (G, c) admits a tropical connected set of size k if and only if (G', w, B') admits a Steiner tree of weight $k' = k - 1 + |C|n$.*

Using the exact algorithm of Nederlof [18] to solve STEINER TREE in $\mathcal{O}^*(W2^k)$ time (with k the number of terminals and maximum weight $W = n$), this reduction yields an algorithm solving TROPICAL CONNECTED SET in $\mathcal{O}^*\left(2^{|C|}\right)$ time.

Using Connected red-blue dominating set. TROPICAL CONNECTED SET can be reduced to CONNECTED RED-BLUE DOMINATING SET as follows.

Let (G, c) be an instance of TROPICAL CONNECTED SET. We construct an instance $G'' = (R'' \cup B'', E'')$ of CONNECTED RED-BLUE DOMINATING SET where G'' is the graph constructed above, R'' set of red and B'' set of blue vertices.

Lemma 2. (G, c) *admits a tropical connected set of size* k *if and only if* $G'' = (R'' \cup B'', E'')$ *admits a connected red-blue dominating set of size* $k' = k - l_1(G)$.

Proof. In the proof we identify a vertex v in G and its copy v' in G''.

Let S be a tropical connected set of G of size k. Let U be the (possibly empty) set of vertices having a color appearing only once in G. Clearly $U \subseteq S$. We claim that $D = S \setminus U$ is a connected red-blue dominating set of G''. By construction, $B_1, B_2, \ldots, B_p$ are the connected components of $G[U]$, and for each $i \in \{1, 2, \ldots, p\}$, the connected set B_i of G is contracted to b_i in G''. Since S is connected in G, $(S \setminus U) \cup \{b_1, b_2, \ldots, b_p\}$ is connected in G'' since we contract connected sets in G. Now when removing $\{b_1, b_2, \ldots, b_p\}$ from $(S \setminus U) \cup \{b_1, b_2, \ldots, b_p\}$ we obtain D. Consider the induced subgraphs in G''. Since by construction of G'' for every b_i the neighbours of b_i form a clique in G'', removal of a vertex b_i cannot make the set D disconnected. Hence D is connected in G''. Finally by construction, $D \subseteq R''$ and $N(D) \cap B''$ contains all vertices $r_i \in B''$ with a color i appearing at least twice in G. Since S is connected in G and $\{b_1, b_2, \ldots, b_p\} \subseteq B''$ is an independent set of G'', every vertex of $\{b_1, b_2, \ldots, b_p\}$ has a neighbour in D. Hence D is red-blue dominating in G''. Summarizing, D is a connected red-blue dominating set of G'' such that $|D| = |S| - |U| = k - l_1(G) = k'$.

Conversely, let $D \subseteq R''$ be a red-blue dominating set of G'' such that $|D| = k' = k - l_1(G)$. By construction of G'' and since D is a red-blue dominating set, for every $i = 1, 2, \ldots, p$, the set D contains at least one vertex of $N(b_i) \cap R''$ in G''. Consequently $D \cup \bigcup_{i=1}^{p} B_i$ is a connected set in G', and also in G. To see this recall that for every i, the vertex b_i is obtained by contracting the connected set B_i. Furthermore $D \cup \bigcup_{i=1}^{p} B_i$ is tropical since it contains all vertices of G with a color appearing once in G, and D dominates $B'' \setminus \bigcup_{i=1}^{p} B_i$, i.e. D contains a vertex of each color appearing at least twice. Hence $D \cup \bigcup_{i=1}^{p} B_i$ is a tropical connected set in G of size $k' + l_1(G) = k$. □

The exact algorithm due to Abu-Khzam et al. [1] solves CONNECTED RED-BLUE DOMINATING SET in time $\mathcal{O}^*(1.36443^n)$. Hence our reduction yields an algorithm solving TROPICAL CONNECTED SET on (G, c) in $\mathcal{O}^*(1.36443^{n+l_2(G)})$ time.

Balancing three algorithms. We are now ready to describe our exact algorithm for TROPICAL CONNECTED SET on general graphs, by balancing between STEINER TREE, CONNECTED RED-BLUE DOMINATING SET, and a brute-force algorithm depending on the value of $l_1(G)$ (see Fig. 2). Let (G, c) be the instance of TROPICAL CONNECTED SET.

- If $l_1(G) < 0.23814 \cdot n$, then reduce to STEINER TREE and use the algorithm of Nederlof [18] on (G', w, K); this yields an algorithm solving TROPICAL CONNECTED SET in $\mathcal{O}^*(2^{|C|})$ time. Notice that in this case, $|C| = l_1(G) + l_2(G) < 0.23814 \cdot n + \frac{1 - 0.23814}{2} \cdot n = 0.61907 \cdot n$.

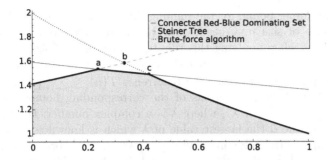

Fig. 2. Representation of the algorithms running time by a dark curve. The x-axis corresponds to the ratio $\alpha = \frac{l_1(G)}{n}$, while the y-axis corresponds to the constant c in the running time $\mathcal{O}^*(c^n)$. Point **a** corresponds to $\alpha = 0.23814$; point **b** to $\alpha = \frac{1}{3}$; point **c** to $\alpha = 0.42218$. Our algorithms running time $\mathcal{O}^*(1.5359^n)$ is achieved for a $= 0.23814$.

- If $0.23814 \cdot n \leq l_1(G) \leq 0.42218 \cdot n$, then reduce to CONNECTED RED-BLUE DOMINATING SET and use the algorithm of Abu-Khzam et al. [1] on G''; this yields an algorithm solving TROPICAL CONNECTED SET in time $\mathcal{O}^*(1.36443^{n+l_2(G)})$. Notice that in this case, $n+l_2(G) \leq n + \frac{1}{2} \cdot n - \frac{1}{2} \cdot l_1(G) \leq 1.38093 \cdot n$.
- If $0.42218 \cdot n < l_1(G)$, then use the brute-force algorithm on (G,c); this yields an algorithm solving TROPICAL CONNECTED SET in $\mathcal{O}^*(2^{n-l_1(G)})$ time. Notice that $n - l_1(G) \leq (1 - 0.42218) \cdot n = 0.57782 \cdot n$.

We establish an exact exponential algorithm which needs only polynomial space since all three algorithms used need only polynomial space. It is worth mentioning that Brute-force is not needed to achieve the claimed running time.

Theorem 2. *The* TROPICAL CONNECTED SET *problem on general graphs can be solved in time* $\mathcal{O}^*(1.5359^n)$ *and polynomial space.*

5   An Exact Exponential Algorithm for Trees

The problem of finding a minimum tropical connected set on trees of height at most three is NP-hard [3], and has no subexponential-time algorithm unless ETH fails, as shown in Sect. 3. The main result of this section is a branching algorithm to compute a minimum tropical connected set on trees in time $\mathcal{O}^*(1.2721^n)$. Branching algorithms are a main tool to design exact exponential algorithms. For an introduction to branching algorithms we refer to [11]. We shortly summarize the main ideas in the design and analysis of branching algorithms. Firstly, such an algorithm solves instances of a problem by recursively branching into instances of subproblems of smaller sizes; thereby using branching and reduction rules. The execution of a branching algorithm can be illustrated by a search tree. Secondly, we shortly summarize how time analysis of such algorithms is done.

Time analysis.   The running time of a branching algorithm is usually obtained by upper bounding the maximum number of vertices in a search tree. To do this, let $T(n)$ be an upper bound for the running time of the algorithm when applied to any instance of size n. Consider any branching rule. Assume the sizes of the subproblems when applying this rule to any instance of size n are at most $n - t_1, \ldots, n - t_b$. Then we say that the rule has branching vector $(t_1, \ldots, t_b)$. The running time $T(n)$ satisfies the recurrence $T(n) \leq \sum_{i=1}^{b} T(n - t_i)$. It is well-known that all basic solutions of the corresponding homogeneous linear recurrence are of the form λ^n, where λ is a complex number. For the running time analysis we need the largest value of λ which is known to be the unique positive real root of the polynomial $\lambda^n - \sum_{i=1}^{b} \lambda^{n-t_i}$, called branching number of $(t_1, \ldots, t_b)$; often denoted by $\tau(t_1, \ldots, t_b)$. If different branching rules are applied and different branching vectors are involved at different steps of an algorithm, then the branching vector with the largest branching number, say $\lambda_{\max}$, implies an upper bound of $O^*((\lambda_{\max})^n)$ for the running time of the algorithm. In our presentation we shall carefully analyze the branching vectors, provide them in Fig. 3, and give their branching numbers without details on calculations.

Notation. Let $T = (V, E)$ be a tree and (T, c) an instance of TROPICAL CONNECTED SET. In the rest of this section, we consider T as a tree rooted at a distinguished vertex $r \in V$. For a given vertex v, we denote by $T(v)$ the subtree of T rooted at v, and by $|T(v)|$ its number of vertices. We denote by $path(v, r) = (x_0, x_1, \ldots, x_k)$ the vertex set of the unique path in the tree T from the vertex $v = x_0$ to the root $r = x_k$ in T. By $d(v, w)$ we denote the length of the unique and thus shortest path from v to w in T. Let $S \subseteq V$ be a connected subset of T and $v \in V$, then we denote by $d(v, S)$ the distance from v to S in T, i.e. $d(v, S) = \min_{s \in S} d(v, s)$ the shortest length of a path from v to any vertex s of S. Observe that $d(v, S) = 0$ for all $v \in S$, and that $d(v, S) = 1$ for all $v \in N(S)$. For a vertex $v \in (V \setminus \{r\})$, we denote by $p(v)$ the parent of v in the rooted tree T. A vertex $v \in V$ is called a leaf of the rooted tree T if $|T(v)| = 1$, otherwise v is called an internal of T. Finally, for an internal $v \in V$ we denote by $s_1(v), \ldots, s_k(v)$, $k \geq 1$, the children of v in T.

Instances and subproblems. As already mentioned branching algorithms recursively solve subproblems by applying reduction and branching rules. Given a (rooted) tree $T = (V, E)$ with a coloring c and set of colors C as an input of TROPICAL CONNECTED SET, T, c and C are global for our recursive algorithm. Now we define an instance of a subproblem as a 3-partition (S, F, D) of V. For such an instance (S, F, D) the task is to find a tropical connected set S^* of (T, c) such that $S \subseteq S^*$, $S^* \cap D = \emptyset$ and the size of S^* is minimum. We call such a vertex set S^* a solution of (S, F, D). Finally such an instance (S, F, D) is defined as follows.

- S is the set of *selected* vertices, i.e. those vertices that have already been chosen as being a subset of any solution.
- F is the set of *free* vertices, i.e. no decision has been made whether they are in S or D.
- D is the set of *discarded* vertices, i.e. those vertices that cannot belong to any solution.

As will become clear later, the construction of our algorithm implies that an instance (S, F, D) always satisfies the following properties: *(i)* the root r of T belongs to S, *(ii)* S and $S \cup F$ are connected sets of T, and *(iii)* for every $v \in D$ all vertices of $T(v)$ also belong to D. Finally for any instance (S, F, D), we denote by T' the subtree induced by $S \cup F$ and we denote by $C' \subseteq C$ the set of those colors that do not already appear in S: $C' = C \setminus c(S)$.

Description of algorithm. The following two procedures are used in the reduction and branching rules of our algorithm when applied to any instance (S, F, D) to obtain new instances and subproblems. Let $v \in F$. Note that selecting v implies to also add all free vertices of $path(v, r)$ to S, since otherwise no superset set of $S \cup \{v\}$ will be connected, and that for the same reason, discarding v implies that all vertices of $T(v)$ are also moved to D.

ADD(v) adds all free vertices of $path(v, r)$ to S, also removes them from F, and removes all colors of $path(v, r)$ from C'.

RMV(v) removes all free vertices of $T(v)$ from F, and adds them to D, while C' remains unchanged.

Let $X \subseteq F$. To ease notation, we write ADD(X) to denote applying ADD(v) for all $v \in X$ (order does not matter). Similarly, we write RMV(X) to denote applying RMV(v) for all $v \in X$ (again order does not matter).

Now we describe our branching algorithm to compute a minimum tropical connected set on trees which runs in time $O^*(1.2721^n)$. Let $T = (V, E)$ be the input tree, c its vertex coloring and C the set of colors of T. For each $v \in V$, we consider the tree T rooted at v and we apply the branching algorithm to the instance $(S, F, D) = (\{v\}, V \setminus \{v\}, \emptyset)$. Note that the root always belongs to S. Therefore a minimum tropical connected set of (T, c) is a solution of $(\{v\}, V \setminus \{v\}, \emptyset)$ of minimum size, taken over all $v \in V$. To describe our branching algorithm, let (S, F, D) be any instance: an initial one with $F = V \setminus \{v\}$ or one obtained by a sequence of recursive calls. In Fig. 3 a listing of the reduction and branching rules of our algorithm is given. Also we assign to each branching rule its branching vector. The details of the case analysis are given in Theorem 3. As is common, the rules are listed in the order in which they have to be applied, i.e. a rule can only be applied to an instance if all previous ones cannot be applied.

Analysis of algorithm. In Lemma 3 and Theorem 3, we prove the correctness of the reduction and branching rules given in Fig. 3.

Lemma 3. *The reduction rules of the branching algorithm TCS-Tree are safe.*

Proof. Let (S, F, D) be any instance of a subproblem and $C' = C \setminus c(S)$. We denote by S^* a solution of (S, F, D). Hence S^* is a tropical connected set satisfying $S \subseteq S^*$ and $D \cap S^* = \emptyset$ and being of minimum size among all such sets, if there is one.

Algorithm TCS-Tree(S, F, D)

Reduction rules.

R0.1. If $C' = \emptyset$, then STOP: S is a tropical connected set of T.

R0.2. If $F = \emptyset$, then STOP: there is no solution for this instance.

R1. If there is $v \in F$, v a leaf of T', such that $c(v) \notin C'$, then RMV(v).

R2. If there is $v \in F$ such that $c(v) \in C'$ and for every $u \in F \setminus \{v\}$, $c(v) \neq c(u)$, then ADD(v).

R3. If there is $v \in F$, v a leaf of T', such that there exists $u \in F \setminus \{v\}$, such that $c(u) = c(v)$ and $d(u, S) = 1$, then RMV(v).

R4. If there is $v \in F$, v a leaf of T', such that there exists $u \in F \setminus \{v\}$, with $c(u) = c(v)$ such that $d(u, path(v, r)) \leq 1$, then RMV(v).

R5. If there is $u \in F$, such that there is a leaf v of T', $v \in S$, with $c(u) = c(v)$, then RMV(u).

R6. If all vertices in F are leaves in T', then ADD(v), for any $v \in F$.

Branching rules. Each of the branching rules creates two subproblems. We write $\langle \mathcal{O}_1 \parallel \mathcal{O}_2 \rangle$ to express that the algorithm branches into 2 subinstances, where the set of operations $\mathcal{O}_i$ is applied to instance (S, F, D) to obtain the ith subinstance.

B1. If there is a free vertex $v \in F$, v leaf of T' with $d(v, S) \geq 4$, then the algorithm distinguishes three cases.

(a) If there are two internals v' and v'' in F satisfying $c(v) = c(v') = c(v'')$: $\langle \{\text{ADD}(v), \text{RMV}(\{v', v''\})\} \parallel \text{RMV}(v)\rangle$ $(\mathbf{7}, \mathbf{1})$

(b) If there is one leaf $v' \in F$ of T' satisfying $c(v) = c(v')$: $\langle \{\text{ADD}(p(v)), \text{RMV}(v')\} \parallel \text{RMV}(p(v))\rangle$ $(\mathbf{4}, \mathbf{2})$

(c) If there is a unique internal $v' \in F$ of T' satisfying $c(v) = c(v')$: $\langle \{\text{ADD}(v), \text{RMV}(v')\} \parallel \{\text{RMV}(v), \text{ADD}(v')\}\rangle$ $(\mathbf{6}, \mathbf{3})$

B2. If there is a vertex $v \in F$ such that v is a leaf of T' and $d(v, S) = 3$, then the algorithm distinguishes four cases.

(a) If there are two leaves $v', v'' \in F$ of T' satisfying $c(v) = c(v') = c(v'')$: $\langle \{\text{ADD}(p(v)), \text{RMV}(\{v', v''\})\} \parallel \text{RMV}(p(v))\rangle$ $(\mathbf{4}, \mathbf{2})$

(b) If there are two internals $v', v'' \in F$ of T' satisfying $c(v) = c(v') = c(v'')$: $\langle \{\text{ADD}(v), \text{RMV}(\{v', v''\})\} \parallel \text{RMV}(v)\rangle$ $(\mathbf{7}, \mathbf{1})$

(c) If there is a leaf $v' \in F$ and an internal $v'' \in F$ of T' satisfying $c(v) = c(v') = c(v'')$: $\langle \{\text{ADD}(v''), \text{RMV}(\{v, v'\})\} \parallel \text{RMV}(v'')\rangle$ $(\mathbf{4}, \mathbf{2})$

(d) If there is a unique vertex $v' \in F$ of T' satisfying $c(v) = c(v')$: $\langle \{\text{ADD}(v), \text{RMV}(v')\} \parallel \{\text{RMV}(v), \text{ADD}(v')\}\rangle$ $(\mathbf{4}, \mathbf{3})$

B3. If there is a free vertex $v \in F$, v leaf of T' with $d(v, S) = 2$, then the algorithm distinguishes two cases.

(a) If there are $k \geq 2$ vertices $v_1, \ldots, v_k \in F$ of T' satisfying $c(v) = c(v_1) = \ldots = c(v_k)$: $\langle \{\text{ADD}(p(v), v), \text{RMV}(\{v_1, \ldots, v_k\})\} \parallel \text{RMV}(p(v))\rangle$ $(\mathbf{4}, \mathbf{2})$

(b) If there is a unique vertex $v' \in F$ of T' satisfying $c(v) = c(v')$: $\langle \{\text{ADD}(\{p(v), v\}), \text{RMV}(v')\} \parallel \{\text{RMV}(p(v)), \text{ADD}(v')\}\rangle$ $(\mathbf{3}, \mathbf{4})$

Fig. 3. Reduction rules, branching rules and branching vectors. The largest branching number is $\tau(4, 2) \leq 1.2721$.

R0.1. Let $C' = \emptyset$. Due to the properties of instances, $G[S]$ is connected, and as $C' = \emptyset$, S is tropical. Hence S is a tropical connected set of T and S is a solution of (S, F, D).

R0.2. Let $C' \neq \emptyset$ and $F = \emptyset$. By the properties of instances, $G[S]$ is a connected set of T. However, since $C' \neq \emptyset$, S is not tropical. Since S cannot be extended, (S, F, D) has no solution.

R1. Let $v \in F$ be a leaf of $T' = T[S \cup F]$ such that $c(v) \notin C'$. Let $v' \in S$ with $c(v) = c(v')$. Suppose S^* is a solution of (S, F, D). Assume that $v \in S^*$. Since $S \subseteq S^*$ and $v' \in S$, we also have $v' \in S^*$. Clearly $S^* \setminus \{v\}$ is tropical and a connected set. Hence S^* is not a solution of (S, F, D), contradiction. Thus we may safely discard v and apply $\texttt{RMV}(v)$.

R2. Let $v \in F$ be the unique vertex in F of color $c(v) \in C'$. Hence any solution S^* of (S, F, D) must contain v to be tropical. Thus we may safely apply $\texttt{ADD}(v)$.

R3. Let $v \in F$ be a leaf of $T' = T[S \cup F]$, and let $u \in F$ such that $c(u) = c(v)$ and $d(u, S) = 1$. Assume that $v \in S^*$ for a solution S^* of (S, F, D). If S^* contains u, then S^* is not minimum, since $S^* \setminus \{v\}$ is tropical and connected. If S^* does not contain u, then $(S^* \setminus \{v\}) \cup \{u\}$ is tropical and connected too. Hence, if there is a solution containing v, then there is also a solution not containing v. Thus it is safe to apply $\texttt{RMV}(v)$.

R4. Let $v \in F$ be a leaf of T' and let $u \in F$ with $c(u) = c(v)$ and $d(u, path(v, r)) \leq 1$. Let S^* be a solution containing v. If $u \in path(v, r)$, then $S^* \setminus \{v\}$ is tropical connected, and so S^* is not a solution, contradiction. If $u \notin path(v, r)$, then there is an $x \in path(v, r)$, $x \neq v$, such that $xu \in E$. Moreover, since S^* is connected, $v \in S^*$ implies $x \in S^*$. Thus $(S^* \setminus \{v\}) \cup \{u\}$ is tropical and connected. Hence there is a solution not containing v and we safely apply $\texttt{RMV}(v)$.

R5. Let $u \in F$ and v be a leaf of T', and $v \in S$ with $c(u) = c(v)$. Let S^* be any solution of (S, F, D) containing a vertex $x \in T'(u)$. Observe that $u \in path(x, r)$. Now $r, x \in S^*$ and S^* is connected implies $u \in S^*$. Clearly, $v \in S$ implies $v \in S^*$ and together with $u \in S^*$ we obtain $S^* \setminus \{v\}$ is tropical and connected, contradicting $v \in S$. Hence $u \notin S^*$ and we may safely apply $\texttt{RMV}(u)$.

R6. When this rule applies, then any subset of leaves of T' with pairwise different colors can be added to S to obtain a solution S^*. Hence adding any leaf of T' to S is safe. $\qquad\square$

The correctness of the branching rules and the time analysis including the determination of the branching vectors will be given in the full version. Combined with Lemma 3 we obtain

Theorem 3. *Our branching algorithm* TCS-Tree *computes a minimum tropical connected set of a tree in time* $\mathcal{O}^*(1.2721^n)$.

References

1. Abu-Khzam, F.N., Mouawad, A.E., Liedloff, M.: An exact algorithm for connected red-blue dominating set. J. Discret. Algorithm. **9**(3), 252–262 (2011)
2. Ambalath, A.M., Balasundaram, R., Rao H., C., Koppula, V., Misra, N., Philip, G., Ramanujan, M.S.: On the kernelization complexity of colorful motifs. In: Raman, V., Saurabh, S. (eds.) IPEC 2010. LNCS, vol. 6478, pp. 14–25. Springer, Heidelberg (2010)
3. Angles D'Auriac, J.-A., Cohen, N., El Maftouhi, A., Harutyunyan, A., Legay, S., Manoussakis, Y.: Connected tropical subgraphs in vertex-colored graphs. http:// people.maths.ox.ac.uk/harutyunyan/Tropical%20sets.pdf
4. Betzler, N., Van Bevern, R., Fellows, M.R., Komusiewicz, C., Niedermeier, R.: Parameterized algorithmics for finding connected motifs in biological networks. IEEE/ACM Trans. Comput. Biol. Bioinf. **8**(5), 1296–1308 (2011)
5. Bruckner, S., Hüffner, F., Karp, R.M., Shamir, R., Sharan, R.: Topology-free querying of protein interaction networks. J. Comput. Biol. **17**(3), 237–252 (2010)
6. Courcelle, B.: The monadic second-order logic of graphs. I. Recognizable sets of finite graphs. Inf. Comput. **85**, 12–75 (1990)
7. Demaine, E.D., Hajiaghayi, M.T.: The bidimensionality theory and its algorithmic applications. Comput. J. **51**, 292–302 (2008)
8. Dondi, R., Fertin, G., Vialette, S.: Complexity issues in vertex-colored graph pattern matching. J. Discret. Algorithm. **9**(1), 82–99 (2011)
9. Fellows, M.R., Fertin, G., Hermelin, D., Vialette, S.: Sharp tractability borderlines for finding connected motifs in vertex-colored graphs. In: Arge, L., Cachin, C., Jurdziński, T., Tarlecki, A. (eds.) ICALP 2007. LNCS, vol. 4596, pp. 340–351. Springer, Heidelberg (2007)
10. Fellows, M.R., Fertin, G., Hermelin, D., Vialette, S.: Upper and lower bounds for finding connected motifs in vertex-colored graphs. J. Comput. Syst. Sci. **77**(4), 799–811 (2011)
11. Fomin, F.V., Kratsch, D.: Exact Exponential Algorithms. Springer, Heidelberg (2010)
12. Fomin, F.V., Kratsch, D., Woeginger, G.J.: Exact (Exponential) algorithms for the dominating set problem. In: Hromkovič, J., Nagl, M., Westfechtel, B. (eds.) WG 2004. LNCS, vol. 3353, pp. 245–256. Springer, Heidelberg (2004)
13. Fomin, F.V., Thilikos, D.M.: A simple and fast approach for solving problems on planar graphs. In: Diekert, V., Habib, M. (eds.) STACS 2004. LNCS, vol. 2996, pp. 56–67. Springer, Heidelberg (2004)
14. Guillemot, S., Sikora, F.: Finding and counting vertex-colored subtrees. Algorithmica **65**(4), 828–844 (2013)
15. Impagliazzo, R., Paturi, R.: On the complexity of k-SAT. J. Comput/Syst. Sci. **62**, 367–375 (2001)
16. Lacroix, V., Fernandes, C.G., Sagot, M.-F.: Motif search in graphs: application to metabolic networks. IEEE/ACM Trans. Comput. Biol. Bioinf. **3**(4), 360–368 (2006)
17. McMorris, F., Warnow, T., Wimer, T.: Triangulating vertex-colored graphs. SIAM J. Discret. Math. **7**, 296–306 (1994)
18. Nederlof, J.: Fast polynomial-space algorithms using Inclusion-Exclusion. Algorithmica **65**, 868–884 (2013)
19. Scott, J., Ideker, T., Karp, R.M., Sharan, R.: Efficient algorithms for detecting signaling pathways in protein interaction networks. J. Comput. Biol. **13**, 133–144 (2006)

A Tight Algorithm for Strongly Connected Steiner Subgraph on Two Terminals with Demands (Extended Abstract)

Rajesh Hemant Chitnis[1]([✉]), Hossein Esfandiari[1],
MohammadTaghi Hajiaghayi[1], Rohit Khandekar[2], Guy Kortsarz[3],
and Saeed Seddighin[1]

[1] Department of Computer Science, University of Maryland at College Park,
College Park, USA
{rchitnis,hossein,hajiagha}@cs.umd.edu
[2] KCG Holdings Inc., New York, USA
rkhandekar@gmail.com
[3] Department of Computer Science, Rutgers University-Camden, Camden, USA
guyk@camden.rutgers.edu

Abstract. Given an edge-weighted directed graph $G = (V, E)$ on n vertices and a set $T = \{t_1, t_2, \ldots t_p\}$ of p terminals, the objective of the STRONGLY CONNECTED STEINER SUBGRAPH (SCSS) problem is to find an edge set $H \subseteq E$ of minimum weight such that $G[H]$ contains a $t_i \rightarrow t_j$ path for each $1 \leq i \neq j \leq p$. The problem is NP-hard, but Feldman and Ruhl [FOCS '99; SICOMP '06] gave a novel $n^{O(p)}$ algorithm for the p-SCSS problem.

In this paper, we investigate the computational complexity of a variant of 2-SCSS where we have demands for the number of paths between each terminal pair. Formally, the 2-SCSS-(k_1, k_2) problem is defined as follows: given an edge-weighted directed graph $G = (V, E)$ with weight function $\omega : E \rightarrow \mathbb{R}_{\geq 0}$, two terminal vertices s, t, and integers k_1, k_2 ; the objective is to find a set of k_1 paths $F_1, F_2, \ldots, F_{k_1}$ from $s \rightsquigarrow t$ and k_2 paths $B_1, B_2, \ldots, B_{k_2}$ from $t \rightsquigarrow s$ such that $\sum_{e \in E} \omega(e) \cdot \phi(e)$ is minimized, where $\phi(e) = \max \left\{ |\{i : i \in [k_1], e \in F_i\}| \; ; \; |\{j : j \in [k_2], e \in B_j\}| \right\}$. For each $k \geq 1$, we show the following:
- The 2-SCSS-$(k, 1)$ problem can be solved in $n^{O(k)}$ time.
- A matching lower bound for our algorithm: the 2-SCSS-$(k, 1)$ problem does not have an $f(k) \cdot n^{o(k)}$ algorithm for any computable function f, unless the Exponential Time Hypothesis (ETH) fails.

Our algorithm for 2-SCSS-$(k, 1)$ relies on a structural result regarding the optimal solution followed by using the idea of a "token game" similar to that of Feldman and Ruhl. We show with an example that the structural

---

A full version of the paper is available on arXiv.org.

Chitnis, Esfandiari, Hajiaghayi, and Seddighin—Supported in part by NSF CAREER award 1053605, NSF grant CCF-1161626, ONR YIP award N000141110662, DARPA/AFOSR grant FA9550-12-1-0423.

Guy Kortsarz—Supported by NSF grant 1218620.

M. Cygan and P. Heggernes (Eds.): IPEC 2014, LNCS 8894, pp. 159–171, 2014.
DOI: 10.1007/978-3-319-13524-3_14

result does not hold for the 2-SCSS-(k_1, k_2) problem if $\min\{k_1, k_2\} \geq 2$. Therefore 2-SCSS-$(k, 1)$ is the most general problem one can attempt to solve with our techniques. To obtain the lower bound matching the algorithm, we reduce from a special variant of the GRID TILING problem introduced by Marx [FOCS '07; ICALP '12].

1 Introduction

The STEINER TREE (ST) problem is one of the earliest and most fundamental problems in combinatorial optimization: given an undirected edge-weighted graph $G = (V, E)$ with edge weights $c : E \to \mathbb{R}^+$ and a set $T \subseteq V$ of terminals, the objective is to find a tree S of minimum cost $c(S) := \sum_{e \in S} c(e)$ which spans all the terminals. The STEINER TREE problem is believed to have been first formally defined by Gauss in a letter in 1836. In the directed version of the ST problem, called DIRECTED STEINER TREE (DST), we are also given a root vertex r and the objective is to find a minimum size arborescence in the directed graph which connects the root r to each terminal from T. An easy reduction from SET COVER shows that the DST problem is also NP-complete.

Steiner-type of problems arise in the design of networks. Since many networks are symmetric, the directed versions of Steiner type of problems were mostly of theoretical interest. However in recent years, it has been observed [13] that the connection cost in various networks such as satellite or radio networks are not symmetric. Therefore, directed graphs form the most suitable model for such networks. In addition, Ramanathan [13] also used the DST problem to find low-cost multicast trees, which have applications in point-to-multipoint communication in high bandwidth networks. A generalization of the DST problem is the STRONGLY CONNECTED STEINER SUBGRAPH (SCSS) problem. In the p-SCSS problem, given a directed graph $G = (V, E)$ and a set $T = \{t_1, t_2, \ldots, t_p\}$ of p terminals the objective is to find a set $S \subseteq V$ such that $G[S]$ contains a $t_i \to t_j$ path for each $1 \leq i \neq j \leq p$. The best known approximation ratio in polynomial time for SCSS is p^ϵ for any $\epsilon > 0$ [2]. A result of Halperin and Krauthgamer [8] implies SCSS has no $\Omega(\log^{2-\epsilon} n)$-approximation for any $\epsilon > 0$, unless NP has quasi-polynomial Las Vegas algorithms.

The 2-SCSS-(k_1, k_2) Problem: We define the following generalization of the 2-SCSS problem:

2-**SCSS**-(k_1, k_2)

Input: An edge-weighted digraph $G = (V, E)$ with weight function $\omega : E \to \mathbb{R}_{\geq 0}$, two terminal vertices s, t, and integers k_1, k_2

Question: Find a set of k_1 paths $F_1, F_2, \ldots, F_{k_1}$ from $s \leadsto t$ and k_2 paths $B_1, B_2, \ldots, B_{k_2}$ from $t \leadsto s$ such that $\sum_{e \in E} \omega(e) \cdot \phi(e)$ is minimized where $\phi(e) = \max\Big\{ |\{i : 1 \leq i \leq k_1, e \in F_i\}| \; ; \; |\{j : 1 \leq j \leq k_2, e \in B_j\}| \Big\}$.

Observe that 2-SCSS-$(1, 1)$ is the same as the 2-SCSS problem. The definition of the 2-SCSS-(k_1, k_2) problem allows us to potentially choose the same edge multiple times, but we have to pay for each time we use it in a path between a given terminal pair. This can be thought of as **"buying disjointness"** by adding parallel edges. In large real-world networks, it might be more feasible to modify the network by adding some parallel edges to create disjoint paths than finding disjoint paths in the existing network. Teixira et al. [14,15] model path diversity in Interner Service Provider (ISP) networks and the Sprint network by disjoint paths between two hosts. There have been several patents [7,12] attempting to design multiple paths between the components of Google Data Centers.

The 2-SCSS-(k_1, k_2) problem is a special case of the DIRECTED SURVIVABLE NETWORK DESIGN (DIR-CAP-SNDP) problem [6] in which we are given an directed multigraph with costs and capacities on the edges, and the question is to find a minimum cost subset of edges that satisfies all pairwise minimum-cut requirements. In the 2-SCSS-(k_1, k_2) problem, we do not require disjoint paths. As observed in Chakrabarty et al. [1] and Goemans et al. [6], the DIR-CAP-SNDP problem becomes much easier to approximate if we allow taking multiple copies of each edge.

1.1 Our Results and Techniques

In this paper, we consider the 2-SCSS-$(k, 1)$ problem parameterized by k, which is the sum of all the demands. To the best of our knowledge, we are unaware of any non-trivial exact algorithms for a version of the SCSS problem with demands between the terminal pairs. Our main algorithmic result is the following:

Theorem 1. *The 2-SCSS-$(k, 1)$ problem can be solved in $n^{O(k)}$ time.*

Our algorithm proceeds as follows: In Sect. 2.1 we first show that there is an optimal solution for the 2-SCSS-$(k, 1)$ problem which satisfies a structural property which we call as **reverse-compatibility**. Then in Sect. 2.2 we introduce a "Token Game" (similar to that of Feldman and Ruhl [5], and show that it can be solved in $n^{O(k)}$. Finally in Sect. 2.3, using the existence of an optimal solution satisfying reverse-compatibility, we give a reduction from the 2-SCSS-$(k, 1)$ problem to the Token Game which gives an $n^{O(k)}$ algorithm for the 2-SCSS-$(k, 1)$ problem. This algorithm also generalizes the result of Feldman and Ruhl [5] for 2-SCSS, since 2-SCSS is equivalent to 2-SCSS-$(1, 1)$. In Sect. 2.4, we show with an example (see Fig. 3) that the structural result does not hold for the 2-SCSS-(k_1, k_2) problem if $\min\{k_1, k_2\} \geq 2$. Therefore, 2-SCSS-$(k, 1)$ is the most general problem one can attempt to solve with our technique.

Theorem 1 does not rule out the possibility that the 2-SCSS-$(k, 1)$ problem is actually solvable in polynomial time. Our main hardness result rules out this possibility by showing that our algorithm is *tight* in the sense that the exponent of $O(k)$ is best possible.

Theorem 2. *The 2-SCSS-$(k, 1)$ problem is W[1]-hard parameterized by k. Moroever, under the ETH, the 2-SCSS-$(k, 1)$ problem cannot be solved in $f(k) \cdot n^{o(k)}$ time for any function f where n is the number of vertices in the graph.*

We reduce from the GRID TILING problem formulated in the pioneering work of Marx [9]:

$k \times k$ **Grid Tiling**

Input: Integers k, n, and k^2 non-empty sets $S_{i,j} \subseteq [n] \times [n]$ where $i, j \in [k]$

Question: For each $1 \leq i, j \leq k$ does there exist a value $s_{i,j} \in S_{i,j}$ such that

- If $s_{i,j} = (x, y)$ and $s_{i,j+1} = (x', y')$ then $x = x'$.
- If $s_{i,j} = (x, y)$ and $s_{i+1,j} = (x', y')$ then $y = y'$.

The GRID TILING problem has turned to be a convenient starting point for parameterized reductions for planar problems, and has been used recently in various W[1]-hardness proofs on planar graphs [4,10,11]. Under the ETH, Chen et al. [3] showed that k-CLIQUE[1] does not admit an algorithm running in time $f(k) \cdot n^{o(k)}$ for any function f. Marx [9] gave a reduction from k-CLIQUE to $k \times k$ GRID TILING. In Sect. 3, we give a reduction from $k \times k$ GRID TILING to 2-SCSS-$(k, 1)$. Since the parameter blowup is linear, the $f(k) \cdot n^{o(k)}$ lower bound for GRID TILING from [9] transfers to 2-SCSS-$(k, 1)$. In fact, the reduction in [9] from k-CLIQUE to $k \times k$ GRID TILING actually shows the hardness of a special case of the GRID TILING problem where the sets are constructed as follows: given a graph $G = (V, E)$ for the k-CLIQUE problem with $V = \{v_1, v_2, \dots, v_n\}$ we set $S_{i,i} = \{(j, j) : 1 \leq j \leq [n]\}$ for each $i \in [k]$ and $S_{i,f} = \{(j, \ell) : 1 \leq j \neq \ell \leq n, (v_j, v_\ell) \in E\}$ for each $1 \leq i \neq f \leq k$. We call this as the GRID TILING* problem and actually give a reduction from this problem to 2-SCSS-$(k, 1)$. To the best of our knowledge, this is the first use of the special structure of GRID TILING* in a W[1]-hardness proof.

In the full version of the paper, we show that the edge-weighted and the vertex-weighted variants of 2-SCSS-(k_1, k_2) are computationally equivalent. Therefore, henceforth we consider only the edge-weighted version of 2-SCSS-(k_1, k_2).

2 An $n^{O(k)}$ Algorithm for 2-SCSS-$(k, 1)$

In this section we describe an algorithm for the 2-SCSS-$(k, 1)$ problem running in $n^{O(k)}$ time where n is the number of vertices in the graph. First in Sect. 2.1 we present a structural property called as *reverse compatibility* for one optimal solution of this problem. Next we define a Token Game in Sect. 2.2 and provide an $n^{O(k)}$ algorithm to solve the game. Finally, in Subsect. 2.3 we present an algorithm that finds the optimum solution of 2-SCSS-$(k, 1)$ in time $n^{O(k)}$ via a reduction to the Token Game problem.

2.1 Structural Lemma for 2-SCSS-$(k, 1)$

For simplicity, we replace each edge e of the input graph G with k copies $e_1, e_2, \dots, e_k$, each having the same cost as that of e. Let the new graph constructed in this way be G'. In G, different $s \rightsquigarrow t$ paths must pay each time they

---

[1] The k-CLIQUE problem asks whether there is a clique of size $\geq k$?

Fig. 1. Let F be a $s \leadsto t$ path given by $s \to u \to v \to w \to y \to z \to t$ and B be a $t \leadsto s$ path given by $t \to y \to z \to u \to v \to s$. The two paths P_1 and P_2 shown in blue are the maximal common sub-paths between F and B. From Definition 1, it follows that F and B are *path-reverse-compatible* since B first sees P_2 and then P_1 (colour figure online).

use different copies of the same edge. We can alternately view this as the $s \leadsto t$ paths in G' being **edge-disjoint**.

Definition 1 (path-reverse-compatible). *Let F be a $s \leadsto t$ path and B be a $t \leadsto s$ path. Let $\{P_1, P_2, \ldots, P_d\}$ be the set of maximal sub-paths that F and B share and for all $j \in [d]$, P_j is the j-th sub-path as seen while traversing F. We say the pair (F, B) is path-reverse-compatible if for all $j \in [d]$, P_j is the $(d-j+1)$-th sub-path that is seen while traversing B, i.e., P_j is the j-th sub-path that is seen while traversing B backward.*

See Fig. 1 for an illustration of path-reverse-compatibility.

Definition 2 (reverse-compatible). *Let $\boldsymbol{F} = \{F_1, F_2, \ldots, F_d\}$ be a set of $s \leadsto t$ paths and b be a $t \leadsto s$ path. We say $(\boldsymbol{F}, B)$ is reverse-compatible, if for all $1 \le i \le d$ the pair (F_i, B) is path-reverse-compatible.*

The next lemma shows that there exists an optimum solution for 2-SCSS-$(k, 1)$ which is reverse-compatible.

Lemma 1 (structurallemma). *There exists an optimum solution for 2-SCSS-$(k, 1)$ which is reverse-compatible.*

Proof In order to prove this lemma, we first introduce the notion of rank of a solution for 2-SCSS-$(k, 1)$. Later, we show that an optimum solution of 2-SCSS-$(k, 1)$ with the minimum rank is reverse-compatible.

Definition 3. *Let $\boldsymbol{F} = \{F_1, F_2, \ldots, F_k\}$ be a set of paths form $s \leadsto t$, and B be a path from $t \leadsto s$. For each $i \in [k]$, let d_i be the number of maximal sub-paths that B and F_i share. The rank of $(\boldsymbol{F}, B)$ is given by*

$$\mathcal{R}(\boldsymbol{F}, B) = \sum_{i=1}^{k} d_i$$

Let $(\mathbf{F}, B)$ be an optimum solution of 2-SCSS-$(k, 1)$ with the minimum rank. Assume for the sake of contradiction that $(\mathbf{F}, B)$ is not reverse-compatible, i.e., there exists some $F_i \in \mathbf{F}$ such that (F_i, B) is not path-reverse-compatible. From Definition 1, this means that F_i and B share two maximal sub-paths $u \to v$

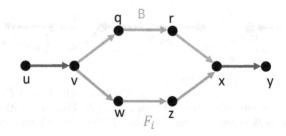

Fig. 2. Let the $u \rightsquigarrow y$ sub-path of F_i be a $u \rightsquigarrow v \rightsquigarrow w \rightsquigarrow z \rightsquigarrow x \rightsquigarrow y$ and the $u \rightsquigarrow y$ sub-path of B be $u \rightsquigarrow v \rightsquigarrow q \rightsquigarrow r \rightsquigarrow x \rightsquigarrow y$. From Definition 1, it follows that F_i and B are not *path-reverse-compatible* since they both first see $u \rightsquigarrow v$ and then see $x \rightsquigarrow y$.

and $x \rightarrow y$, and at the same time F_i and B both contain $u \rightarrow y$ sub-paths (see Fig. 2).

We replace the $u \rightarrow y$ sub-path of B by the $u \rightarrow y$ sub-path of F_i. On one hand, B shares all of the $u \rightarrow y$ sub-path with F_i. Thus, this change does not increase the cost of the network, therefore it remains an optimum solution. On the other hand, by this change, the sub-paths $u \rightarrow v$ and $x \rightarrow y$ join. Hence, d_i decreases by 1. Also, since the forward paths are edge-disjoint, after the change all other d_j's remain same (for $i \neq j$ since B shares the whole $u \rightarrow y$ sub-path with only F_i. Therefore, this change strictly decreases the rank of the solution. Existence of an optimum solution with a smaller rank contradicts the selection of $(\mathbf{F}, B)$ and completes the proof.   □

2.2   The Token Game

In the token game, we are given a graph G, a set of tokens $\mathcal{T}$, vertices s and t, a set of moves $\mathcal{M}$, and a cost function $\hat{C} : \mathcal{M} \to \mathbb{R}$. Each move $m \in \mathcal{M}$ consists of a set of triples (t_i, u_i, v_i) where $t_i \in \mathcal{T}$ is a token, and u_i and v_i are vertices of the graph. In order to apply a move $m = \{(t_1, u_1, v_1), (t_2, u_2, v_2), \ldots, (t_d, u_d, v_d)\}$ to a state of the game, each token t_i should be on vertex u_i for all $1 \leq i \leq d$ and after applying this move, for every triple $(t_i, u_i, v_i) \in m$ token t_i will be transported to the vertex v_i. For each $m \in \mathcal{M}$, $\hat{C}(m)$ specifies the cost of applying m to the game. Initially, all of the tokens are placed on vertex s. In each step, we apply a move $m \in \mathcal{M}$ to the game with cost $\hat{C}(m)$ and the goal is to transport all of the tokens to the vertex t with minimum cost.

In the following, we present an algorithm to solve an instance $\langle G, s, t, \mathcal{T}, \mathcal{M}, \hat{C} \rangle$ of the Token game in time $\mathcal{O}(n^{|\mathcal{T}|} \cdot |\mathcal{M}| \cdot \log(n^{|\mathcal{T}|}))$, where n is the number of the vertices of G.

Lemma 2 [⋆] [2] **(algorithm for Token Game).** *There exists an algorithm which solves the Token game in time* $\mathcal{O}(n^{|\mathcal{T}|}|\mathcal{M}|\log(n^{|\mathcal{T}|}))$.

_____
[2] The proofs of the results labeled with ⋆ are available in the full version on arXiv.

2.3   Reduction to the Token Game

Here, we provide a reduction from the 2-SCSS-$(k, 1)$ problem to the Token game. As a consequence, we show that one can use the presented algorithm in Subsect. 2.2 to solve 2-SCSS-$(k, 1)$ in time $\mathcal{O}(n^{\mathcal{O}(k)})$.

Let $I = \langle G, s, t \rangle$ be an instance of the 2-SCSS-$(k, 1)$. We reduce I to an instance $\text{Cor}(I) = \langle G', s', t', \mathcal{T}, \mathcal{M}, \hat{C} \rangle$ of the Token Game problem where $G = G'$, $s = s'$, $t = t'$ and $\mathcal{T}$ is a set of $k+1$ tokens $\{\mathcal{F}_1, \mathcal{F}_2, \ldots, \mathcal{F}_k, \mathcal{B}\}$. Furthermore, $\mathcal{M}$ and $\hat{C}$ are constructed in the following way:

- For every edge $(u, v) \in E(G)$, we add k moves $\{(\mathcal{F}_i, u, v)\}$ to $\mathcal{M}$ for all $1 \leq i \leq k$. Cost of each move is equal to the length of its corresponding edge in G.
- For every edge $(u, v) \in E(G)$ with weight w, we add a move $\{(\mathcal{B}, v, u)\}$ to $\mathcal{M}$ with cost w.
- For every pair of vertices u and v in G, we add k moves $\{(\mathcal{F}_i, u, v), (\mathcal{B}, v, u)\}$ to $\mathcal{M}$ for all $1 \leq i \leq k$. Cost of each move is equal to the distance of vertex v from vertex u in G.

Next we show that $\text{OPT}(I) = \text{OPT}(\text{Cor}(I))$, where $\text{OPT}(I)$ and $\text{OPT}(\text{Cor}(I))$ stand for the optimum solutions of I and $\text{Cor}(I)$ respectively. We do this by the following two lemmas:

Lemma 3. [⋆] *For a given instance I of the 2-SCSS-$(k, 1)$ we have $OPT(I) \geq OPT(Cor(I))$.*

Lemma 4. [⋆] *For a given instance I of the 2-SCSS-$(k, 1)$ we have $OPT(I) \leq OPT(Cor(I))$.*

Theorem 3. [⋆] *There exists an algorithm that solves the 2-SCSS-$(k, 1)$ in time $\mathcal{O}(n^{\mathcal{O}(k)})$.*

2.4   Structural Lemma Fails for 2-SCSS-(k_1, k_2) if $\min\{k_1, k_2\} \geq 2$

In the 2-SCSS-(k_1, k_2) problem we want k_1 paths from $s \rightsquigarrow t$ and k_2 paths from $t \rightsquigarrow s$. So, we define a natural generalization of Definition 1 to reverse-compatibility of a set of forward paths and a set of backward paths as follows.

Definition 4. *Let $\boldsymbol{F} = \{F_1, F_2, \ldots, F_{k_1}\}$ be a set of $s \rightsquigarrow t$ paths and $\boldsymbol{B} = \{B_1, B_2, \ldots, B_{k_2}\}$ be a set of $t \rightsquigarrow s$ paths. We say $(\boldsymbol{F}, \boldsymbol{B})$ is reverse-compatible, if for all $1 \leq i \leq k_2$, $(\boldsymbol{F}, B_i)$ is reverse-compatible.*

Figure 3 gives an instance of 2-SCSS-$(2, 2)$ where no optimum solution is reverse-compatible. The following lemma shows that Lemma 1 does not hold for the 2-SCSS-(k_1, k_2) problem when $\min\{k_1, k_2\} \geq 2$, i.e., Lemma 1 is in its most general form.

Lemma 5. *The instance of 2-SCSS-$(2, 2)$ shown in Fig. 3 has no reverse-compatible optimum solution.*

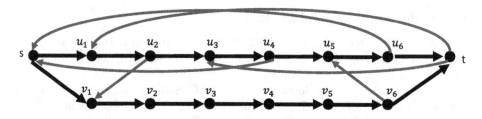

Fig. 3. Each black edge has weight 1 and each red edge has weight 0. Let the $u \rightsquigarrow y$ sub-path of F_i be a $u \rightsquigarrow v \rightsquigarrow w \rightsquigarrow z \rightsquigarrow x \rightsquigarrow y$ and the $u \rightsquigarrow y$ sub-path of B be $u \rightsquigarrow v \rightsquigarrow q \rightsquigarrow r \rightsquigarrow x \rightsquigarrow y$. From Definition 1, it follows that F_i and B are not *path-reverse-compatible* since they both first see $u \rightsquigarrow v$ and then see $x \rightsquigarrow y$ (colour figure online).

Proof. In Fig. 3, each black has weight 1 and each red edge has weight 0. Note that the length of each shortest path from s to t is 7, and there are exactly two such paths viz. $P_1 := s \rightarrow v_1 \rightarrow v_2 \rightarrow v_3 \rightarrow v_4 \rightarrow v_5 \rightarrow v_6 \rightarrow t$ and $P_2 := s \rightarrow u_1 \rightarrow u_2 \rightarrow u_3 \rightarrow u_4 \rightarrow u_5 \rightarrow u_6 \rightarrow t$. Thus, any optimum solution to 2-SCSS-$(2, 2)$ has cost at least 14. In addition, if we select both P_1 and P_2 then we can select two paths $Q_1 := t \rightarrow u_3 \rightarrow u_4 \rightarrow s$ and $Q_2 = t \rightarrow u_1 \rightarrow u_2 \rightarrow v_1 \rightarrow v_2 \rightarrow v_3 \rightarrow v_4 \rightarrow v_5 \rightarrow v_6 \rightarrow u_5 \rightarrow u_6 \rightarrow s$ from t to s without any cost. Therefore, $(\{P_1, P_2\}, \{Q_1, Q_2\})$ is an optimum solution to 2-SCSS-$(2, 2)$ with cost 14.

If we select paths P_1 and P_2 as forward paths, (Q_1, Q_2) is the only pair of backward paths which is free. On the other hand, if we select one of P_1 or P_2 twice, then it is easy to see that there is no free backward path. Thus, $(\{P_1, P_2\}, \{Q_1, Q_2\})$ is the unique optimum solution to 2-SCSS-$(2, 2)$. However, one can see that paths P_2 and Q_2 are not reverse-compatible since they see the common maximal sub-paths $u_1 \rightarrow u_2$ and $u_5 \rightarrow u_6$ in the same order.

3 $f(k) \cdot n^{o(k)}$ Hardness for 2-SCSS-$(k, 1)$

In this section we prove Theorem 2. We reduce from the GRID TILING problem (see Sect. 1.1 for definition). Chen et al. [3] showed that for any function f an $f(k) \cdot n^{o(k)}$ algorithm for CLIQUE implies ETH fails. Marx [9] gave the following reduction which transforms the problem of finding a k-CLIQUE into an instance of $k \times k$ GRID TILING as follows: For a graph $G = (V, E)$ with $V = \{v_1, v_2, \ldots, v_n\}$ we build an instance I_G of GRID TILING

- For each $1 \leq i \leq k$, we have $(j, \ell) \in S_{i,i}$ if and only if $j = \ell$.
- For any $1 \leq i \neq j \leq k$, we have $(\ell, r) \in S_{i,j}$ if and only if $\{v_\ell, v_r\} \in E$.

It is easy to show that G has a clique of size k if and only if the instance I_G of GRID TILING has a solution. Therefore, assuming ETH, the following special case of $k \times k$ GRID TILING also cannot be solved in time $f(k) \cdot n^{o(k)}$ for any computable function f.

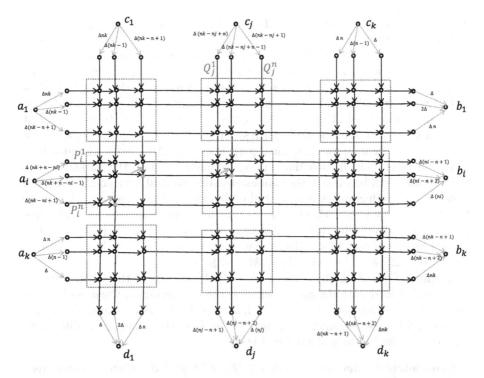

Fig. 4. The instance of 2-SCSS-$(k, 1)$ created from an instance of Grid Tiling*.

$k \times k$ Grid Tiling*

Input: Integers k, n, and k^2 non-empty sets $S_{i,j} \subseteq [n] \times [n]$ where $1 \leq i, j \leq k$ such that for each $1 \leq i \leq k$, we have $(j, \ell) \in S_{i,i}$ if and only if $j = \ell$

Question: For each $1 \leq i, j \leq k$ does there exist a value $\gamma_{i,j} \in S_{i,j}$ such that

- If $\gamma_{i,j} = (x, y)$ and $\gamma_{i,j+1} = (x', y')$ then $x = x'$.
- If $\gamma_{i,j} = (x, y)$ and $\gamma_{i+1,j} = (x', y')$ then $y = y'$.

Consider an instance of GRID TILING*. We now build an instance of edge-weighted 2-SCSS-$(2k - 1, 1)$ as shown in Fig. 4. We consider $4k$ special vertices: (a_i, b_i, c_i, d_i) for each $i \in [k]$. We introduce k^2 red gadgets where each gadget is an $n \times n$ grid. Let weight of each black edge be 4.

Definition 5. *For each $1 \leq i \leq k$, an $a_i \rightsquigarrow b_i$ canonical path is a path from a_i to b_i which starts with a blue edge coming out of a_i, then follows a horizontal path of black edges and finally ends with a blue edge going into b_i. Similarly an $c_j \rightsquigarrow d_j$ canonical path is a path from c_j to d_j which starts with a blue edge coming out of c_j, then follows a vertically downward path of black edges and finally ends with a blue edge going into d_j.*

For each $1 \leq i \leq k$, there are n edge-disjoint $a_i \rightsquigarrow b_i$ canonical paths: let us call them $P_i^1, P_i^2, \ldots, P_i^n$ as viewed from top to bottom. They are named using magenta color in Fig. 4. Similarly we call the canonical paths from c_j to d_j as $Q_j^1, Q_j^2, \ldots, Q_j^n$ when viewed from left to right. For each $i \in [k]$ and $\ell \in [n]$ we assign a weight of $\Delta(nk - ni + n + 1 - \ell), \Delta(ni - n + \ell)$ to the first, last edges of P_i^ℓ (which are colored blue) respectively. Similarly for each $j \in [k]$ and $\ell \in [n]$ we assign a weight of $\Delta(nk - nj + n + 1 - \ell), \Delta(nj - n + \ell)$ to the first, last edges of Q_j^ℓ (which are colored blue) respectively. Thus the total weight of first and last blue edges on any canonical path is exactly $\Delta(nk + 1)$. The idea is to choose Δ large enough such that in any optimum solution the paths between the terminals will be exactly the canonical paths. We will see that $\Delta = 7n^6$ will suffice for our reduction. Any canonical path uses two blue edges (which sum up to $\Delta(nk + 1)$), $(k + 1)$ black edges not inside the gadgets and $(n - 1)$ black edges inside each gadget. Since the number of gadgets each canonical path visits is k and the weight of each black edge is 4, we have the total weight of any canonical path is $\alpha = \Delta(nk + 1) + 4(k + 1) + 4k(n - 1)$.

Intuitively the k^2 gadgets correspond to the k^2 sets in the GRID TILING* instance. Let us denote the gadget which is the intersection of the $a_i \rightsquigarrow b_i$ paths and $c_j \rightsquigarrow d_j$ paths by $G^{i,j}$. If $i = j$, then we call $G^{i,j}$ as a symmetric gadget; else we call it as a asymmetric gadget. We perform the following modifications on the edges inside the gadget: (see Fig. 4)

- **Symmetric Gadgets:** For each $i \in [k]$, if $(x, y) \in S_{i,i}$ then we color green the vertex in the gadget $G_{i,i}$ which is the unique intersection of the canonical paths P_i^x and Q_i^y. Then we add a shortcut as shown in Fig. 5. The idea is if both the $a_i \rightsquigarrow b_i$ path and $c_i \rightsquigarrow d_i$ path pass through the green vertex then the $a_i \rightsquigarrow b_i$ path can save a weight of 2 by using the green edge and a vertical

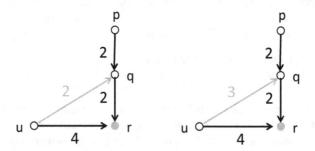

Fig. 5. Let u, r be two consecutive vertices on the canonical path say P_i^ℓ. Let r be on the canonical path $Q_j^{\ell'}$ and let p be the vertex preceding it on this path. If r is a green (respectively orange) vertex then we subdivide the edge (p, r) by introducing a new vertex q and adding two edges (p, q) and (q, r) of weight 2. We also add an edge (u, q) of weight 2 (respectively 3). The idea is if both the edges (p, r) and (u, r) were being used initially then now we can save a weight of 2 (respectively 1) by making the horizontal path choose (u, q) and then we get (q, r) for free, as it is already being used by the vertical canonical path (colour figure online).

downward edge ((which is already being used by $c_j \leadsto d_j$ canonical path)) to reach the green vertex, instead of paying a weight of 4 to use the horizontal edge reaching the green vertex.

- **Aymmetric Gadgets:** For each $i \neq j \in [k]$, if $(x, y) \in S_{i,j}$ then we color orange the vertex in the gadget $G_{i,i}$ which is the unique intersection of the canonical paths P_i^x and Q_i^y. Then we add a shortcut as shown in Fig. 5. The idea is if both the $a_i \leadsto b_i$ path and $c_j \leadsto d_j$ path pass through the green vertex then the $a_i \leadsto b_i$ path can save a weight of 1 by using the orange edge of weight 3 followed by a vertical downward edge (which is already being used by $c_j \leadsto d_j$ canonical path) to reach the orange vertex, instead of paying a weight of 4 to use the horizontal edge reaching the green vertex.

From Fig. 4, it is easy to see that each canonical path has weight equal to α.

3.1   Vertices and Edges Not Shown in Fig. 4

The following vertices and edges are not shown in Fig. 4 for sake of presentation:

- Add two vertices s and t.
- For each $1 \leq i \leq k$, add an edge (s, c_i) of weight 0.
- For each $1 \leq i \leq k$, add an edge (d_i, t) of weight 0.
- Add edges (t, a_k) and (b_1, s) of weight 0.
- For each $2 \leq i \leq k$, introduce two new vertices e_i and f_i. We call these $2k - 2$ vertices as bridge vertices.
- For each $2 \leq i \leq k$, add a path $b_i \to e_i \to f_i \to a_{i-1}$. Set the weights of (b_i, e_i) and (f_i, a_{i-1}) to be zero.
- For each $2 \leq i \leq k$, set the weight of the edge (e_i, f_i) to be W. We call these edges as **connector** edges. The idea is that we will choose W large enough so that each connector edge is used exactly once in an optimum solution for 2-SCSS-$(k, 1)$. We will see later that $W = 53n^9$ suffices for our reduction.

We need a small technical modification: add one dummy row and column to the GRID TILING* instance. Essentially, we now have a dummy index 1. So neither the first row nor the first column of any $S_{i,j}$ has any elements in the GRID TILING* instance. That is, no green vertex or orange vertex can be in the first row or first column of any gadget. We now state two theorems which together give a reduction from GRID TILING* to 2-SCSS-$(k, 1)$. Let

$$\beta = 2k \cdot \alpha + W(k - 1) - (k^2 + k) \tag{1}$$

Theorem 4. [⋆] GRID TILING* *has a solution if and only if OPT for 2-SCSS-*$(k, 1)$ *is at most β.*

3.2   Proof of Theorem 2

Proof. Theorem 4 implies the W[1]-hardness by giving a reduction which transforms the problem of $k \times k$ GRID TILING* into an instance of 2-SCSS-$(k, 1)$ where we want to find k paths from $s \leadsto t$ and one path from $t \leadsto s$.

Chen et al. [3] showed for any function f an $f(k)n^{o(k)}$ algorithm for CLIQUE implies ETH fails. Composing the reduction of [9] from CLIQUE to GRID TILING*, along with our reduction from GRID TILING* to 2-SCSS-$(k, 1)$, we obtain under ETH there is no $f(k)n^{o(k)}$ algorithm for 2-SCSS-$(k, 1)$ for any function f. This shows that the $n^{\mathcal{O}(k)}$ algorithm for 2-SCSS-$(k, 1)$ given in Sect. 2 is optimal.   □

4   Conclusions

In this paper, we studied the 2-SCSS-$(k, 1)$ problem and presented an algorithm which finds an optimum solution in time $n^{\mathcal{O}(k)}$, and that is asymptotically optimal under the ETH. This algorithm was based on the fact that there always exists an optimal solution for 2-SCSS-$(k, 1)$ that has the reverse-compatibility property. However, we show in Sect. 2.4 that the 2-SCSS-(k_1, k_2) problem need not always have a solution which satisfies the reverse-compatibility property. Therefore, it remains an important challenging problem to find a similar structure and generalize our method to solve the 2-SCSS-(k_1, k_2) problem.

References

1. Chakrabarty, D., Chekuri, C., Khanna, S., Korula, N.: Approximability of Capacitated Network Design. In: Günlük, O., Woeginger, G.J. (eds.) IPCO 2011. LNCS, vol. 6655, pp. 78–91. Springer, Heidelberg (2011)
2. Charikar, M., Chekuri, C., Cheung, T.Y., Goel, A., Guha, S., Li, M.: Approximation algorithms for directed Steiner problems. J. Algorithm. **33**(1), 73–91 (1999)
3. Chen, J., Huang, X., Kanj, I.A., Xia, G.: Strong computational lower bounds via parameterized complexity. J. Comput. Syst. Sci. **72**(8), 1346–1367 (2006)
4. Chitnis, R.H., Hajiaghayi, M., Marx, D.: Tight bounds for planar strongly connected Steiner subgraph with fixed number of terminals (and extensions). In: SODA, pp. 1782–1801 (2014)
5. Feldman, J., Ruhl, M.: The directed Steiner network problem is tractable for a constant number of terminals. SIAM J. Comput. **36**(2), 543–561 (2006)
6. Goemans, M.X., Goldberg, A.V., Plotkin, S.A., Shmoys, D.B., Tardos, É., Williamson, D.P.: Improved approximation algorithms for network design problems. In: SODA, pp. 223–232 (1994)
7. Guo, C., Lu, G., Li, D., Wu, H., Shi, Y., Zhang, D., Zhang, Y., Lu, S.: Hybrid butterfly cube architecture for modular data centers (Nov 22 2011). US patent 8,065,433. http://www.google.com/patents/US8065433
8. Halperin, E., Krauthgamer, R.: Polylogarithmic inapproximability. In: STOC '03 (2003)
9. Marx, D.: On optimality of planar & geometric approximation schemes. In: FOCS'07 (2007)
10. Marx, D.: A Tight Lower Bound for Planar Multiway Cut with Fixed Number of Terminals. In: Czumaj, A., Mehlhorn, K., Pitts, A., Wattenhofer, R. (eds.) ICALP 2012, Part I. LNCS, vol. 7391, pp. 677–688. Springer, Heidelberg (2012)
11. Marx, D., Pilipczuk, M.: Everything you always wanted to know about the parameterized complexity of subgraph isomorphism (but were afraid to ask). In: STACS, pp. 542–553 (2014)

12. Ramachandran, K., Kokku, R., Mahindra, R., Rangarajan, S.: Wireless network connectivity in data centers. US patent App. 12/499, 906. http://www.google.com/patents/US20100172292. Accessed 8 Jul 2010
13. Ramanathan, S.: Multicast tree generation in networks with asymmetric links. IEEE/ACM Trans. Netw. $4(4)$, 558–568 (1996)
14. Teixeira, R., Marzullo, K., Savage, S., Voelker, G.M.: Characterizing and measuring path diversity of internet topologies. In: SIGMETRICS, pp. 304–305 (2003)
15. Teixeira, R., Marzullo, K., Savage, S., Voelker, G.M.: In search of path diversity in ISP networks. In: Internet Measurement Conference, pp. 313–318 (2003)

The Firefighter Problem: A Structural Analysis

Janka Chlebíková[1] and Morgan Chopin[2(✉)]

[1] University of Portsmouth, School of Computing, Portsmouth, UK
`janka.chlebikova@port.ac.uk`
[2] Institut für Optimierung und Operations Research, Universität Ulm, Ulm, Germany
`morgan.chopin@uni-ulm.de`

Abstract. We consider the complexity of the firefighter problem where a budget of $b \geq 1$ firefighters are available at each time step. This problem is proved to be NP-complete even on trees of degree at most three and $b = 1$ [10] and on trees of bounded degree $b+3$ for any fixed $b \geq 2$ [3]. In this paper, we provide further insight into the complexity landscape of the problem by showing a complexity dichotomy result with respect to the parameters pathwidth and maximum degree of the input graph. More precisely, we first prove that the problem is NP-complete even on trees of pathwidth at most three for any $b \geq 1$. Then we show that the problem turns out to be fixed parameter-tractable with respect to the combined parameter "pathwidth" and "maximum degree" of the input graph. Finally, we show that the problem remains NP-complete on very dense graphs, namely co-bipartite graphs, but is fixed-parameter tractable with respect to the parameter "cluster vertex deletion".

1 Introduction

The firefighter problem was introduced by Hartnell [13] and received considerable attention in a series of papers [1,5,7,8,10,14–16,18,19]. In its original version, a fire breaks out at some vertex of a given graph. At each time step, one vertex can be protected by a firefighter and then the fire spreads to all unprotected neighbors of the vertices on fire. The process ends when the fire can no longer spread. At the end all vertices that are not on fire are considered as saved. The objective is at each time step to choose a vertex which will be protected by a firefighter such that a maximum number of vertices in the graph is saved at the end of the process. In this paper, we consider a more general version which allows us to protect $b \geq 1$ vertices at each step (the value b is called *budget*).

The original firefighter problem was proved to be NP-hard for bipartite graphs [18], cubic graphs [16] and unit disk graphs [11]. Finbow et al. [10] showed that the problem is NP-hard even on trees. More precisely, they proved the following dichotomy theorem: the problem is NP-hard even for trees of maximum degree three and it is solvable in polynomial-time for graphs with maximum degree three, provided that the fire breaks out at a vertex of degree at most two.

Morgan Chopin—A major part of this work was done during a three-month visit of the University of Portsmouth supported by the ERASMUS program.

© Springer International Publishing Switzerland 2014
M. Cygan and P. Heggernes (Eds.): IPEC 2014, LNCS 8894, pp. 172–183, 2014.
DOI: 10.1007/978-3-319-13524-3_15

Furthermore, the problem is polynomial-time solvable for caterpillars and so-called P-trees [18]. Later, Bazgan et al. [3] extended the previous results by showing that the general firefighter problem is NP-hard even for trees of maximum degree $b + 3$ for any fixed budget $b \geq 2$ and polynomial-time solvable on k-caterpillars. From the approximation point of view, the problem is $\frac{e}{e-1}$-approximable on trees ($\frac{e}{e-1} \approx 1.5819$) [5] and it is not $n^{1-\varepsilon}$-approximable on general graphs for any $\varepsilon > 0$ unless $P = NP$ [1]. Moreover for trees in which each non-leaf vertex has at most three children, the firefighter problem is 1.3997-approximable [15]. Very recently, Costa et al. [7] extended the $\frac{e}{e-1}$-approximation algorithm on trees to the case where the fire breaks out at $f > 1$ vertices and $b > 1$ firefighters are available at each step. From a parameterized perspective, the problem is W[1]-hard with respect to the natural parameters "number of saved vertices" and "number of burned vertices" [2]. Furthermore, it admits an $O(2^\tau k\tau)$-size kernel where τ is the minimum vertex cover of the input graph and k the number of burned vertices [2]. Cai et al. [5] gave first fixed-parameter tractable algorithms and polynomial-size kernels for trees for each of the following parameters: "number of saved vertices", "number of saved leaves", "number of burned vertices", and "number of protected vertices".

In this paper, we provide a complexity dichotomy result of the problem with respect to the parameters maximum degree and pathwidth of the input graph. In Sect. 2, we first provide the formal definition of the problem as well as some preliminaries. In Sect. 3, we complete the hardness picture of the problem on trees by proving that it is also NP-complete on trees of pathwidth three. We note that the given proof is also a simpler proof of the NP-completeness of the problem on trees. In Sect. 4, we devise a parameterized algorithm with respect to the combined parameter "pathwidth" and "maximum degree" of the input graph. In Sect. 5, we show that the problem is also NP-hard on co-bipartite graphs which are very dense graphs but fixed-parameter tractable with respect parameter "cluster vertex deletion" (cvd). This last result strengthen the previous $O(2^\tau k\tau)$-size kernel as it suppresses the dependence with k and the cvd number is smaller than the vertex cover number. The conclusion is given in Sect. 6. Due to space limitation, some proofs are deferred to a full version.

2   Preliminaries

Graph terminology. Let $G = (V, E)$ be an *undirected graph* of order n. For a subset $S \subseteq V$, $G[S]$ is the induced subgraph of G. The *neighborhood* of a vertex $v \in V$, denoted by $N(v)$, is the set of all neighbors of v. We denote by $N^k(v)$ the set of vertices which are at distance at most k from v. The *degree* of a vertex v is denoted by $\deg_G(v)$ and the *maximum degree* of the graph G is denoted by $\Delta(G)$.

A *linear layout* of G is a bijection $\pi : V \to \{1, \ldots, n\}$. For convenience, we express π by the list $L = (v_1, \ldots, v_n)$ where $\pi(v_i) = i$. Given a linear layout L, we denote the distance between two vertices in L by $d_L(v_i, v_j) = j - i$.

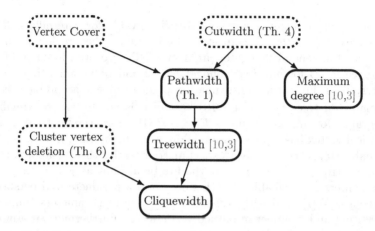

Fig. 1. The parameterized complexity of the FIREFIGHTER problem with respect to some structural graph parameters. An arc from a parameter k_2 to a parameter k_1 means that there exists some function h such that $k_1 \leq h(k_2)$. For any fixed budget, a dotted rectangle means fixed-parameter tractability for this parameter and a thick rectangle means NP-hardness even for constant values of this parameter.

The *cutwidth* $\mathrm{cw}(G)$ of G is the minimum $k \in \mathbb{N}$ such that the vertices of G can be arranged in a linear layout $L = (v_1, \ldots, v_n)$ in such a way that, for every $i \in \{1, \ldots, n-1\}$, there are at most k edges between $\{v_1, \ldots, v_i\}$ and $\{v_{i+1}, \ldots, v_n\}$.

The *bandwidth* $\mathrm{bw}(G)$ of G is the minimum $k \in \mathbb{N}$ such that the vertices of G can be arranged in a linear layout $L = (v_1, \ldots, v_n)$ so that $|d_L(v_i, v_j)| \leq k$ for every edge $v_i v_j$ of G.

A path decomposition $\mathcal{P}$ of G is a pair $(P, \mathcal{H})$ where P is a path with node set X and $\mathcal{H} = \{H_x : x \in X\}$ is a family of subsets of V such that the following conditions are met

1. $\bigcup_{x \in X} H_x = V$.
2. For each $uv \in E$ there is an $x \in X$ with $u, v \in H_x$.
3. For each $v \in V$, the set of nodes $\{x : x \in X \text{ and } v \in H_x\}$ induces a subpath of P.

The width of a path decomposition $\mathcal{P}$ is $\max_{x \in X} |H_x| - 1$. The pathwidth $\mathrm{pw}(G)$ of a graph G is the minimum width over all possible path decompositions of G.

We may skip the argument of $\mathrm{pw}(G)$, $\mathrm{cw}(G)$, $\mathrm{bw}(G)$ and $\Delta(G)$ if the graph G is clear from the context.

A *star* is a tree consisting of one vertex, called the *center* of the star, adjacent to all the other vertices.

Problem definition. We start with an informal explanation of the propagation process for the firefighter problem. Let $G = (V, E)$ be a graph of order n with a vertex $s \in V$, let $b \in \mathbb{N}$ be a *budget*. At step $t = 0$, a fire breaks out at vertex s

and s starts burning. At any subsequent step $t > 0$ the following two phases are performed in sequence:

1. *Protection phase*: The firefighter protects at most b vertices not yet on fire.
2. *Spreading phase*: Every unprotected vertex which is adjacent to a burned vertex starts burning.

Burned and protected vertices remain burned and protected until the propagation process stops, respectively. The propagation process stops when in a next step no new vertex can be burned. We call a vertex saved if it is either protected or if all paths from any burned vertex to it contain at least one protected vertex. Notice that, until the propagation process stops, there is at least one new burned vertex at each step. This leads to the following obvious lemma.

Lemma 1. *The number of steps before the propagation process stops is less or equal to the total number of burned vertices.*

A *protection strategy* (or simply *strategy*) Φ indicates which vertices to protect at each step until the propagation process stops. Since there can be at most n burned vertices, it follows from Lemma 1 that the propagation unfolds in at most n steps. We are now in position to give the formal definition of the investigated problem.

The FIREFIGHTER problem:
Input: A graph $G = (V, E)$, a vertex $s \in V$, and positive integers b and k.
Question: Is there a strategy for an instance (G, s, b, k) with respect to budget b such that at most k vertices are burned if a fire breaks out at s?

When dealing with trees, we use the following observation which is a straightforward adaptation of the one by MacGillivray and Wang for the case $b > 1$ [18, Sect. 4.1].

Lemma 2. *Among the strategies that maximize the number of saved vertices (or equivalently minimize the number of burned vertices) for a tree, there exists one that protects vertices adjacent to a burned vertex at each time step.*

Throughout the paper, we assume all graphs to be connected since otherwise we can simply consider the component where the initial burned vertex s belongs to.

3   Firefighting on Path-Like Graphs

Finbow et al. [10] showed that the problem is NP-complete even on trees of degree at most three. However, the constructed tree in the proof has an unbounded pathwidth. In this section, we show that the FIREFIGHTER problem is NP-complete even on trees of pathwidth three. For that purpose we use the following problem.

The CUBIC MONOTONE 1-IN-3-SAT problem:
Input: A CNF formula with no negative literals in which every clause contains exactly three variables and every variable appears in exactly three clauses.
Question: Is there a 1-perfect satisfying assignment (a truth assignment such that each clause has exactly one true literal) for the formula?

The NP-completeness of the above problem is due to its equivalence with the NP-complete EXACT COVER BY 3-SETS problem [12].

Theorem 1. *For any budget $b \geq 1$, the* FIREFIGHTER *problem is NP-complete even on trees of pathwidth three.*

Proof. Clearly, FIREFIGHTER belongs to NP. Now we provide a polynomial-time reduction from CUBIC MONOTONE 1-IN-3-SAT. We start with the case where $b = 1$ and later explain how to extend the proof for larger values of b.

In the proof, a *guard-vertex* is a star with k leaves where the center is adjacent to a vertex of a graph. It is clear that if at most k vertices can be burned, then the guard-vertex has to be saved.

Let ϕ be a formula of CUBIC MONOTONE 1-IN-3-SAT with n variables $\{x_1, \ldots, x_n\}$ and m initial clauses $\{c_1, \ldots, c_m\}$. Notice that we have $n = m$ since there is a total of $3n = 3m$ literals in ϕ. First, we extend ϕ into a new formula ϕ' by adding m new clauses as follows. For each clause c_j we add the clause $\bar{c}_j$ by taking negation of each variable of c_j. A perfect satisfying assignment for ϕ' is then a truth assignment such that each clause c_j has exactly one true literal (1-perfect) and each clause $\bar{c}_j$ has exactly two true literals (2-perfect). Clearly, we have that ϕ has a 1-perfect satisfying assignment if and only if ϕ' has a perfect one. To see this, observe that a clause c_j has exactly one true literal if and only if $\bar{c}_j$ has two true literals.

Now we construct an instance $I' = (T, s, 1, k)$ of FIREFIGHTER from ϕ' as follows (see Fig. 2). We start with the construction of the tree T, the value of k will be specified later.

– Start with a vertex set $\{s = u_1, u_2, \ldots, u_p\}$ and edges of $\{su_2, u_2u_3, \ldots, u_{p-1}u_p\}$ where $p = 2n - 1$ and add two degree-one vertices v_{x_i} and $v_{\bar{x}_i}$ adjacent to u_{2i-1} for every $i \in \{1, \ldots, n\}$.

Then for each $i \in \{1, \ldots, n\}$ in two steps:

– Add a guard-vertex g_i (resp. $\bar{g}_i$) adjacent to v_{x_i} (resp. $v_{\bar{x}_i}$).
– At each vertex v_{x_i} (resp. $v_{\bar{x}_i}$) root a path of length $2 \cdot (n - i)$ at v_{x_i} (resp. $v_{\bar{x}_i}$) in which the endpoint is adjacent to three degree-one vertices (called literal-vertices) denoted by $\ell_1^{x_i}, \ell_2^{x_i}$, and $\ell_3^{x_i}$ (resp. $\ell_1^{\bar{x}_i}, \ell_2^{\bar{x}_i}$, and $\ell_3^{\bar{x}_i}$). Each literal-vertex corresponds to an occurence of the variable x_i in an initial clause of ϕ. Analogously, the literal-vertices $\ell_1^{\bar{x}_i}, \ell_2^{\bar{x}_i}$, and $\ell_3^{\bar{x}_i}$ represent the negative literal $\bar{x}_i$ that appears in the new clauses of ϕ'.

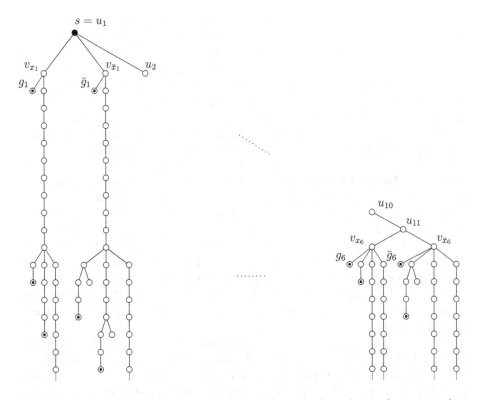

Fig. 2. An example of part of a tree constructed from the formula $\phi = (x_1 \vee x_3 \vee x_6) \wedge (x_1 \vee x_2 \vee x_3) \wedge (x_3 \vee x_4 \vee x_5) \wedge (x_2 \vee x_4 \vee x_5) \wedge (x_1 \vee x_4 \vee x_6) \wedge (x_2 \vee x_6 \vee x_5)$. Guard vertices are represented by a dot within a circle.

Notice that each literal-vertex is at distance exactly $p + 1$ from s.

– For each variable x_i (resp. $\bar{x}_i$), $i \in \{1, \dots, n\}$, there are exactly three clauses containing x_i (resp. $\bar{x}_i$). Let c_j (resp. $\bar{c}_j$), $j \in \{1, \dots, m\}$, be the first one of them. Then root a path $Q_j^{x_i}$ (resp. $Q_j^{\bar{x}_i}$) of length $3 \cdot (j-1)$ at $\ell_1^{x_i}$ (resp. $\ell_1^{\bar{x}_i}$), and add a guard-vertex $g_j^{x_i}$ adjacent to the endpoint of $Q_j^{x_i}$. To the endpoint of $Q_j^{\bar{x}_i}$ (i) add a degree-one vertex $d^{\bar{x}_i}$ (a dummy-vertex) and (ii) root a path $D_j^{\bar{x}_i}$ of length 3 where the last vertex of the path is a guard vertex $g_j^{\bar{x}_i}$. Repeat the same for the two other clauses with x_i (resp. $\bar{x}_i$) and $\ell_2^{x_i}$, $\ell_3^{x_i}$ (resp. $\ell_2^{\bar{x}_i}$, $\ell_3^{\bar{x}_i}$).

To finish the construction, set $k = p + \frac{n}{2}(11n + 7)$.

In what follows, we use Lemma 2 and thus we only consider strategies that protect a vertex adjacent to a burned vertex at each time step. Recall that the budget is set to one in the instance I'. Now we show that there is a perfect satisfying assignment for ϕ' if and only if there exists a strategy for I' such that at most k vertices in T are burned.

"⇒" : Suppose that there is a perfect satisfying assignment τ for ϕ'. We define the following strategy Φ_τ from τ. At each step t from 1 to $p+1$, if t is odd then protect $v_{\bar{x}_{\lceil t/2 \rceil}}$ if $x_{\lceil t/2 \rceil}$ is true, otherwise protect $v_{x_{\lceil t/2 \rceil}}$. If t is even, then protect the guard-vertex $g_{\lceil t/2 \rceil}$ if $v_{\bar{x}_{\lceil t/2 \rceil}}$ has been protected, otherwise protect $\bar{g}_{\lceil t/2 \rceil}$. At the end of time step $p+1$, the number of burned vertices is exactly $p + \sum_{i=1}^{n}(3 + 2(n-i) + 1) = p + 3n + n^2$. Moreover, the literal-vertices that are burned in T correspond to the true literals in ϕ'. Thus, by construction and since τ statisfies ϕ', the vertices adjacent to a burning vertex are exactly one guard-vertex $g_1^{x_a}$, two dummy vertices $d^{\bar{x}_b}, d^{\bar{x}_c}$ and $3n-1$ other vertices where $x_a \vee x_b \vee x_c$ is the first clause, $a, b, c \in \{1, \ldots, n\}$. At step $p+2$, we must protect the guard vertex $g_1^{x_a}$. During the steps $p+3$ and $p+4$, the strategy must protect one vertex lying on the path $D_1^{\bar{x}_b}$ and $D_1^{\bar{x}_c}$, respectively. Thus $3(3n-3) + 5 = 9n - 4$ more vertices are burned at the end of step $p+4$. More generally, from time step $p + 3(j-1) + 2$ to $p + 3(j-1) + 4$, for some $j \in \{1, \ldots, m\}$, the strategy Φ_τ must protect a guard-vertex $g_j^{x_a}$ and one vertex of each path $D_j^{\bar{x}_b}$ and $D_j^{\bar{x}_c}$, where x_a, x_b, x_c appear in the clause c_j, $a, b, c \in \{1, \ldots, n\}$. Thus $9(n-(j-1))-4$ vertices get burned. It follows that the number of burned vertices from step $p+2$ to $p+3m+1$ is $\sum_{j=1}^{m}[9(n-(j-1))-4] = \frac{9}{2}m(m+1) - 4m$. Putting all together, we arrive at a total of $p + 3n + n^2 + \frac{9}{2}m(m+1) - 4m = p + \frac{n}{2}(11n + 7) = k$ burned vertices.

"⇐": Conversely, assume that there is no perfect satisfying assignment for ϕ'. Observe first that any strategy Φ for I' protects either v_{x_i} or $v_{\bar{x}_i}$ for each $i \in \{1, \ldots, n\}$. As a contradiction, suppose that there exists $i \in \{1, \ldots, n\}$ such that Φ does not protect v_{x_i} and $v_{\bar{x}_i}$. Then in some time step both v_{x_i} and $v_{\bar{x}_i}$ get burned. Hence, it is not possible to protect both g_i and $\bar{g}_i$, and at least one will burn implying that more than k vertices would burn, a contradiction. Furthermore, v_{x_i} and $v_{\bar{x}_i}$ cannot be both protected, otherwise we would have protected a vertex not adjacent to a burned vertex at some step. Now consider the situation at the end of step $p+1$. By the previous observation, the literal-vertices that are burned in T can be interpreted as being the literals in ϕ' set to true. As previously, the number of burned vertices so far is exactly $p + \sum_{i=1}^{n}(3 + 2(n-i) + 1) = p + 3n + n^2$. Let n_g and n_d be the number of guard-vertices and dummy-vertices adjacent to a burned vertex, respectively. As it follows from the previous construction, we know that $n_g = 3 - n_d$ with $0 \leq n_g \leq 3$ and $0 \leq n_d \leq 3$. We have the following possible cases:

(1) $n_g > 1$. In this case, a guard-vertex gets burned, and hence more than k vertices would burn.

(2) $n_g = 1$. Let $g_1^{x_a}$ be that guard-vertex and let $d^{\bar{x}_b}, d^{\bar{x}_c}$ be the $n_d = 3 - n_g = 2$ dummy-vertices where x_a, x_b, x_c are variables of the first clause. At time step $p+2$, we must protect $g_1^{x_a}$. Furthermore, during the step $p+3$ (resp. $p+4$), any strategy must protect a vertex lying on the path $D_1^{\bar{x}_b}$ (resp. $D_1^{\bar{x}_c}$). Indeed, if a strategy does otherwise, then at least one guard-vertex $g_1^{\bar{x}_b}$ or $g_1^{\bar{x}_c}$ gets burned. Thus 2 dummy-vertices are burned.

(3) $n_g = 0$. Hence we have exactly $n_d = 3 - n_g = 3$ dummy-vertices $d^{\bar{x}_a}, d^{\bar{x}_b}, d^{\bar{x}_c}$ adjacent to burned vertices. Using a similar argument as before, we know

that during the step $p + 2$ (resp. $p + 3$, $p + 4$), a strategy must protect a vertex lying on the path $D_1^{\bar{x}_a}$ (resp. $D_1^{\bar{x}_b}$, $D_1^{\bar{x}_c}$). Thus 3 dummy-vertices are burned.

Notice that at step $p + 5$, we end up with a similar situation as in step $p + 2$. Now consider an assignment for ϕ'. Since ϕ' is not perfect satisfiable, therefore ϕ is not 1-perfect satisfiable as well. There are two possibilities:

- There exists a clause c_j in ϕ with more than one true literal. Thus, we end up with case (1), and there is no strategy for I' such that at most k vertices are burned.
- There is a clause c_j in ϕ with only false literals. This corresponds to case (3), and the number of burned vertices would be at least $1 + p + \frac{n}{2}(11n + 7)$ (at least one extra dummy-vertex gets burned) giving us a total of at least $k+1$ burned vertices. Hence there is no strategy for I' where at most k vertices are burned.

It remains to prove that the pathwidth of T is at most three. To see this, observe that any subtree rooted at v_{x_i} or $v_{\bar{x}_i}$ has pathwidth two. Let P_{x_i} and $P_{\bar{x}_i}$ be the paths of the path-decompositions of these subtrees, respectively. We construct the path-decomposition for T as follows. For every $i \in \{1, \ldots, n - 1\}$, define the node $B_i = \{u_{2i-1}, u_{2i}, u_{2i+1}\}$. Extend all nodes of the paths P_{x_i} and $P_{\bar{x}_i}$ to P'_{x_i} and $P'_{\bar{x}_i}$ by adding the vertex u_{2i-1} inside it. Finally, connect the paths P'_{x_1}, $P'_{\bar{x}_1}$ and the node B_1 to form a path and continue in this way with P'_{x_2}, $P'_{\bar{x}_2}$, B_2, P'_{x_3}, $P'_{\bar{x}_3}$, B_3, $\ldots$, B_{n-1}, P'_{x_n}, $P'_{\bar{x}_n}$.

Finally, we consider the case where $b > 1$. We start from the above reduction and alter the tree T as follows. Let w_1 be the vertex s (corresponding also to u_1). Add a path $\{w_1w_2, w_2w_3, \ldots, w_{5n}w_{5n+1}\}$ to T together with $b - 1$ guard-vertices added to each w_i. First, one can easily check that the pathwidth remains unchanged, since the added component has pathwidth two and is only connected to the root s. Second, it can be seen that at each time step, only one firefighter can be placed "freely", as the other $b - 1$ firefighters must protect $b - 1$ guard-vertices. It follows that we end up with a similar proof as above. This completes the proof. □

As a side result, we also obtain the following.

Proposition 1. *For any budget $b \geq 1$, the* FIREFIGHTER *problem is NP-complete even on line graphs.*

4   Path-Like Graphs of Bounded Degree

As previously shown, for any fixed budget $b \geq 1$, the FIREFIGHTER problem is NP-complete on trees of bounded degree $b + 3$ [3,10] and on trees of bounded pathwidth three (Theorem 1). It is thus natural to ask for the complexity of the problem when both the degree and the pathwidth of the input graph are bounded. In what follows, we answer this question positively. A first step toward this goal is to use the following combinatorial characterization of the number of burned vertices in a graph.

Theorem 2. *Consider a graph of pathwidth* pw *and maximum degree* Δ. *If the number of initially burned vertices is bounded by* $f_1(\text{pw}, \Delta)$ *for some function* f_1, *then there exists a protection strategy such that at most* $f_2(\text{pw}, \Delta) \geq f_1(\text{pw}, \Delta)$ *vertices are burned for some function* f_2.

Proof. First we prove the following claim: Consider a graph of cutwidth cw. If the number of initially burned vertices is bounded by $g_1(\text{cw})$ for some function g_1, then there exists a protection strategy such that at most $g_2(\text{cw}) \geq g_1(\text{cw})$ vertices are burned for some function g_2. We will prove this by induction on cw.

The claim is obviously true when the cutwidth is 0, since the graph cannot contain any edge. Suppose now that the claim is true for any graph of cutwidth at most k, $k > 0$. We show that it also holds for a graph of cutwidth $k + 1$. Let $H = (V, E)$ be such a graph and $F \subseteq V$ be the set of initially burned vertices with $|F| \leq g_1(\text{cw}(H))$. Consider a linear layout $L = (v_1, \ldots, v_n)$ of H such that for every $i = 1, \ldots, n - 1$, there are at most $k + 1$ edges between $\{v_1, \ldots, v_i\}$ and $\{v_{i+1}, \ldots, v_n\}$. For every $s \in F$ and $i \geq 0$, we define inductively the following sets, where $R_0(s) = L_0(s) = \{s\}$

$$R_i(s) = \begin{cases} \{s = v_k, v_{k+1}, \ldots, v_{k'}\} & \text{if } \exists v_{k'} \in N^i(s) : v_{k'} = \text{argmax}_{v \in N^i(s)} \, d_L(s, v) \\ R_{i-1}(s) & \text{otherwise} \end{cases}$$

(1)

$$L_i(s) = \begin{cases} \{s = v_k, v_{k-1}, \ldots, v_{k'}\} & \text{if } \exists v_{k'} \subset N^i(s) : v_{k'} = \text{argmin}_{v \in N^i(s)} \, d_L(s, v) \\ L_{i-1}(s) & \text{otherwise} \end{cases}$$

(2)

We are now in position to define the set $B_i(s)$, called a *bubble*, by $B_i(s) = L_i(s) \cup R_i(s)$ for all $i \geq 0$. Informally speaking, the bubble $B_i(s)$ corresponds to the *effect zone* of s after i steps of propagation, *i.e.*, every vertex that gets burned after i steps (starting at s) must be inside the bubble $B_i(s)$. The idea of the proof is to show that every bubble can be "isolated" from the rest of the graph in a bounded number of steps by surrounding it with firefighters (see Fig. 3). We then show that the inductive hypothesis can be applied on each bubble, which will prove the claim. Let $s_1, s_2 \in F$. We say that two bubbles $B_i(s_1)$ and $B_i(s_2)$ for some $i \geq 0$ *overlap* if $B_i(s_1) \cap B_i(s_2) \neq \emptyset$.

Let us consider an initially burned vertex $s \in F$ and its bubble $B_{2 \cdot \text{cw}(H)}(s)$. Let $B'_{2 \cdot \text{cw}(H)}(s)$ be the union of $B_{2 \cdot \text{cw}(H)}(s)$ with every other bubble $B_{2 \cdot \text{cw}(H)}(s')$, $s' \in F$, that overlap with $B_{2 \cdot \text{cw}(H)}(s)$. By definition, we know that the number of edges with an endpoint in $B'_{2 \cdot \text{cw}(H)}(s)$ and the other one in $V \setminus B'_{2 \cdot \text{cw}(H)}(s)$ is less or equal to $2 \cdot \text{cw}(H)$. Thus, we define the strategy that consists in protecting one vertex $v \in V \setminus B'_{2 \cdot \text{cw}(H)}(s)$ adjacent to a vertex in $B'_{2 \cdot \text{cw}(H)}(s)$ at each step $t = 1, \ldots, 2 \cdot \text{cw}(H)$. Let F' be the set of vertices burned at step $2 \cdot \text{cw}(H)$. Since $\Delta(H) \leq 2 \cdot \text{cw}(H)$, we deduce that $|F'|$ is less or equal

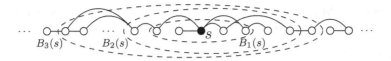

Fig. 3. A linear layout of a graph of cutwidth two. Dashed ellipses represent the bubbles associated to an initially burned vertex s.

to $|F| \cdot \Delta(H)^{2 \cdot \text{cw}(H)} \le g_1(\text{cw}(H)) \cdot (2 \cdot \text{cw}(H))^{2 \cdot \text{cw}(H)}$ hence bounded by a function of $\text{cw}(H)$. Let us consider the subgraph $H' = H[B'_{2 \cdot \text{cw}(H)}(s)]$. Observe that we can safely remove every edge uv from H' for which $u, v \in F'$. Indeed, such edge cannot have any influence during the subsequent steps of propagation. By the definition of a bubble and the overlapping of bubbles, this implies that the cutwidth of H' is decreased by one and thus is now at most k. Therefore, we can apply our inductive hypothesis to H' which tells us that there is a strategy for H' such that at most $g'_2(\text{cw}(H'))$ vertices are burned for some function g'_2. By Lemma 1, this strategy uses at most $g'_2(\text{cw}(H'))$ steps to be applied. It follows that the number of burned vertices in H after applying this strategy is at most the number of burned vertices from step 1 to the step $2 \cdot \text{cw}(H) + g'_2(\text{cw}(H'))$ which is $|F| \cdot \Delta(H)^{2 \cdot \text{cw}(H) + g'_2(\text{cw}(H'))} \le g_1(\text{cw}(H)) \cdot (2 \cdot \text{cw}(H))^{2 \cdot \text{cw}(H) + g'_2(\text{cw}(H'))}$ which is bounded by a function of $\text{cw}(H)$. From now on, one can see that the previous argument can be applied iteratively to each bubble. Since the number of bubbles is bounded by $g_1(\text{cw}(H))$ (there is at most one bubble for each vertex initially on fire), we deduce that the total number of burned vertices is bounded by $g_2(\text{cw}(H))$ some function g_2. This concludes the proof of the claim.

We are now in position to prove the theorem. Let G be a graph. Suppose that the number of initially burned vertices in G is at most $f_1(\text{pw}(G), \Delta(G))$ for some function f_1. We know that $\text{pw}(G) \le \text{cw}(G)$ and $\Delta(G) \le 2 \cdot \text{cw}(G)$ [17]. Thus the number of burned vertices is at most $f'_1(\text{cw}(G))$ for some function f'_1. From the above claim we deduce that there exists a strategy such that at most $f'_2(\text{cw}(G))$ vertices get burned. Since $\text{cw}(G) \le \text{pw}(G) \cdot \Delta(G)$ [6], it follows that the number of burned vertices is bounded by $f_2(\text{pw}(G), \Delta(G))$ for some function f_2. This completes the proof. $\square$

Remark 1. Notice that Theorem 2 is still valid even if the number of firefighters available at each step is not the same (for example if there are b_1 firefighters at time step one, b_2 firefighters during the second time step, etc.).

We are now in position to give the main result of this section.

Theorem 3. *The* FIREFIGHTER *problem is fixed-parameter tractable with respect to the combined parameter "pathwidth" and "maximum degree" of the input graph.*

From the proof of Theorem 2 and the fact that $\text{cw}(G) \le \frac{\text{bw}(G)(\text{bw}(G)+1)}{2}$ [4] for any graph G, we easily deduce the following theorem.

Theorem 4. *The* FIREFIGHTER *problem is fixed-parameter tractable with respect to the parameter "cutwidth" and to the parameter "bandwidth".*

5   Firefighting on Dense Graphs

As trees are the less dense graphs, it seems natural to ask for the tractability of the problem when the graph is essentially made up of cliques. In the following we show that even if the graph can be partitioned into two cliques (also known as a co-bipartite graph), the problem turns out to be NP-complete. Notice that the problem is trivial for cliques.

Theorem 5. *The* FIREFIGHTER *problem is NP-complete and W[1]-hard for the parameter k even on co-bipartite graphs.*

We note that if the budget b is fixed, then one can solve the problem in polynomial time on co-bipartite graphs. To see this, observe that there are at most 3 propagation steps in such a graph. Hence the total number of protected vertices is bounded by a constant, which implies that the problem is polynomial-time solvable [2].

As a final result, we show that the problem is fixed-parameter tractable with respect to the parameter cluster vertex deletion number, that is the minimum number of vertices that have to be deleted to get a disjoint union of complete graphs. We first discuss the motivation for this parameter. Whenever a problem is hard on graphs of bounded treewidth/pathwidth, it is a common research direction to ask for the parameterized complexity of the problem with respect to the larger parameter vertex cover. However, the class of graphs of small vertex cover is rather limited and, hence, looking for the complexity of the problem for parameters that generalize it tends to be a more relevant approach. Among them the cluster vertex deletion number appears to be an interesting intermediate parameterization between vertex cover and cliquewidth [9].

Theorem 6. *For any fixed $b > 0$, the* FIREFIGHTER *problem is fixed-parameter tractable with respect to the parameter "cluster vertex deletion".*

6   Conclusion

The main result of this paper is that the FIREFIGHTER problem is NP-complete even on trees of pathwidth three but fixed-parameter tractable with respect to the combined parameter "pathwidth" and "maximum degree" of the input graph. The combination of these two results with the NP-completeness of the problem on trees of bounded degree [10] indicates that the complexity of the problem depends heavily on the degree and the pathwidth of the graph. We left as an open question whether the problem is polynomial-time solvable on graphs of pathwidth two.

References

1. Anshelevich, E., Chakrabarty, D., Hate, A., Swamy, C.: Approximability of the firefighter problem. Algorithmica **62**(1–2), 520–536 (2012)
2. Bazgan, C., Chopin, M., Cygan, M., Fellows, M.R., Fomin, F.V., van Leeuwen, E.J.: Parameterized complexity of firefighting. J. Comput. Syst. Sci. **80**(7), 1285–1297 (2014)
3. Bazgan, C., Chopin, M., Ries, B.: The firefighter problem with more than one firefighter on trees. Discrete Appl. Math. **161**(7–8), 899–908 (2013)
4. Bodlaender, H.L.: Classes of graphs with bounded tree-width. Bull. EATCS **36**, 116–128 (1988)
5. Cai, L., Verbin, E., Yang, L.: Firefighting on trees: $(1 - 1/e)$–approximation, fixed parameter tractability and a subexponential algorithm. In: Hong, S.-H., Nagamochi, H., Fukunaga, T. (eds.) ISAAC 2008. LNCS, vol. 5369, pp. 258–269. Springer, Heidelberg (2008)
6. Chung, F.R., Seymour, P.D.: Graphs with small bandwidth and cutwidth. In: Graph Theory and combinatorics 1988 Proceedings of the Cambridge Combinatorial Conference in Honour of Paul Erdös, Annals of Discrete Mathematics, vol. 43, pp. 113–119 (1989)
7. Costa, V., Dantas, S., Dourado, M.C., Penso, L., Rautenbach, D.: More fires and more fighters. Discrete Appl. Math. **161**(16–17), 2410–2419 (2013)
8. Develin, M., Hartke, S.G.: Fire containment in grids of dimension three and higher. Discrete Appl. Math. **155**(17), 2257–2268 (2007)
9. Doucha, M., Kratochvíl, J.: Cluster vertex deletion: a parameterization between vertex cover and clique-width. In: Rovan, B., Sassone, V., Widmayer, P. (eds.) MFCS 2012. LNCS, vol. 7464, pp. 348–359. Springer, Heidelberg (2012)
10. Finbow, S., King, A., MacGillivray, G., Rizzi, R.: The firefighter problem for graphs of maximum degree three. Discrete Math. **307**(16), 2094–2105 (2007)
11. Fomin, F.V., Heggernes, P., van Leeuwen, E.J.: Making Life Easier for Firefighters. In: Kranakis, E., Krizanc, D., Luccio, F. (eds.) FUN 2012. LNCS, vol. 7288, pp. 177–188. Springer, Heidelberg (2012)
12. Garey, M.R., Johnson, D.S.: Computers and Intractability: A Guide to the Theory of NP-Completeness. W.H Freeman and Company, New York (1979)
13. Hartnell, B.: Firefighter! an application of domination, Presentation. In: 10th Conference on Numerical Mathematics and Computing, University of Manitoba in Winnipeg, Canada (1995)
14. Hartnell, B., Li, Q.: Firefighting on trees: how bad is the greedy algorithm? Congressus Numerantium **145**, 187–192 (2000)
15. Iwaikawa, Y., Kamiyama, N., Matsui, T.: Improved approximation algorithms for firefighter problem on trees. IEICE Trans. Inf. Syst. **E94.D**(2), 196–199 (2011)
16. King, A., MacGillivray, G.: The firefighter problem for cubic graphs. Discrete Math. **310**(3), 614–621 (2010)
17. Korach, E., Solel, N.: Tree-width, path-width, and cutwidth. Discrete Appl. Math. **43**(1), 97–101 (1993)
18. MacGillivray, G., Wang, P.: On the firefighter problem. J. Comb. Math. Comb. Comput. **47**, 83–96 (2003)
19. Ng, K.L., Raff, P.: A generalization of the firefighter problem on ZxZ. Discrete Appl. Math. **156**(5), 730–745 (2008)

AND-compression of NP-complete Problems: Streamlined Proof and Minor Observations

Holger Dell[(⊠)]

Cluster of Excellence, MMCI, Saarland University, Saarbrucken, Germany
hdell@mmci.uni-saarland.de

Abstract. Drucker [8] proved the following result: Unless the unlikely complexity-theoretic collapse coNP ⊆ NP/poly occurs, there is no AND-compression for SAT. The result has implications for the compressibility and kernelizability of a whole range of NP-complete parameterized problems. We present a streamlined proof of Drucker's theorem.

An AND-compression is a deterministic polynomial-time algorithm that maps a set of SAT-instances $x_1, \ldots, x_t$ to a single SAT-instance y of size $\mathrm{poly}(\max_i |x_i|)$ such that y is satisfiable if and only if all x_i are satisfiable. The "AND" in the name stems from the fact that the predicate "y is satisfiable" can be written as the AND of all predicates "x_i is satisfiable". Drucker's theorem complements the result by Bodlaender et al. [3] and Fortnow and Santhanam [10], who proved the analogous statement for OR-compressions, and Drucker's proof not only subsumes their result but also extends it to *randomized* compression algorithms that are allowed to have a certain probability of failure.

Drucker [8] presented two proofs: The first uses information theory and the minimax theorem from game theory, and the second is an elementary, iterative proof that is not as general. In our proof, we realize the iterative structure as a generalization of the arguments of Ko [12] for P-selective sets, which use the fact that tournaments have dominating sets of logarithmic size. We generalize this fact to hypergraph tournaments. Our proof achieves the full generality of Drucker's theorem, avoids the minimax theorem, and restricts the use of information theory to a single, intuitive lemma about the average noise sensitivity of compressive maps. To prove this lemma, we use the same information-theoretic inequalities as Drucker.

1 Introduction

The influential "OR-conjecture" by Bodlaender et al. [3] asserts that t instances $x_1, \ldots, x_t$ of SAT cannot be mapped in polynomial time to an instance y of size $\mathrm{poly}(\max_i |x_i|)$ so that y is a yes-instance if and only if at least one x_i is a yes-instance. Conditioned on the OR-conjecture, the "composition framework" of Bodlaender et al. [3] has been used to show that many different problems in parameterized complexity do not have polynomial kernels. Fortnow

---

Holger Dell: work done as a postdoc at LIAFA, Université Paris Diderot. Extended abstract version from September 28, 2014.

© Springer International Publishing Switzerland 2014
M. Cygan and P. Heggernes (Eds.): IPEC 2014, LNCS 8894, pp. 184–195, 2014.
DOI: 10.1007/978-3-319-13524-3_16

and Santhanam [10] were able to prove that the OR-conjecture holds unless coNP $\subseteq$ NP/poly, thereby connecting the OR-conjecture with a standard hypothesis in complexity theory.

The results of [3,10] can be used not only to rule out deterministic kernelization algorithms, but also to rule out randomized kernelization algorithms with one-sided error, as long as the success probability is bigger than zero; this is the same as allowing the kernelization algorithm to be a coNP-algorithm. Left open was the question whether the complexity-theoretic hypothesis coNP $\not\subseteq$ NP/poly (or some other hypothesis believed by complexity theorists) suffices to rule out kernelization algorithms that are randomized and have two-sided error. Drucker [8] resolves this question affirmatively; his results can rule out kernelization algorithms that have a constant gap in their error probabilities. This result indicates that randomness does not help to decrease the size of kernels significantly.

With the same proof, Drucker [8] resolves a second important question: whether the "AND-conjecture", which has also been formulated by Bodlaender et al. [3] analogous to the OR-conjecture, can be derived from existing complexity-theoretic assumptions. This is an intriguing question in itself, and it is also relevant for parameterized complexity as, for some parameterized problems, we can rule out polynomial kernels under the AND-conjecture, but we do not know how to do so under the OR-conjecture. Drucker [8] proves that the AND-conjecture is true if coNP $\not\subseteq$ NP/poly holds.

The purpose of this paper is to discuss Drucker's theorem and its proof. To this end, we attempt to present a simpler proof of his theorem. Our proof in Sect. 3 gains in simplicity with a small loss in generality: the bound that we get is worse than Drucker's bound by a factor of two. Using the slightly more complicated approach deferred to the full version of this paper, it is possible to get the same bounds as Drucker. These differences, however, do not matter for the basic version of the main theorem, which we state in Sect. 1.1 and further discuss in Sect. 1.2. For completeness, we briefly discuss a formulation of the composition framework in Sect. 1.3.

1.1 Main Theorem: Ruling Out OR- and AND-compressions

An AND-compression A for a language $L \subseteq \{0,1\}^*$ is a polynomial-time reduction that maps a set $\{x_1, \ldots, x_t\}$ to some instance $y \doteq A(\{x_1, \ldots, x_t\})$ of a language $L' \subseteq \{0,1\}^*$ such that $y \in L'$ holds if and only if $x_1 \in L$ and $x_2 \in L$ and $\ldots$ and $x_t \in L$. By De Morgan's law, the same A is an OR-compression for $\overline{L} \doteq \{0,1\}^* \setminus L$ because $y \in \overline{L'}$ holds if and only if $x_1 \in \overline{L}$ or $x_2 \in \overline{L}$ or $\ldots$ or $x_t \in \overline{L}$. Drucker [8] proved that an OR-compression for L implies that $L \in$ NP/poly $\cap$ coNP/poly, which is a complexity consequence that is closed under complementation, that is, it is equivalent to $\overline{L} \in$ NP/poly $\cap$ coNP/poly. For this reason, and as opposed to earlier work [3,7,10], it is without loss of generality that we restrict our attention to OR-compressions for the remainder of this paper. We now formally state Drucker's theorem.

Theorem 1 (Drucker's theorem). *Let $L, L' \subseteq \{0,1\}^*$ be languages, let e_s, $e_c \in [0,1]$ be error probabilities with $e_s + e_c < 1$, and let $\epsilon > 0$. Assume that there exists a randomized polynomial-time algorithm A that maps any set $x = \{x_1, \ldots, x_t\} \subseteq \{0,1\}^n$ for some n and t to $y = A(x)$ such that:*

- *(Soundness) If all x_i's are no-instances of L, then y is a no-instance of L' with probability $\geq 1 - e_s$.*
- *(Completeness) If exactly one x_i is a yes-instance of L, then y is a yes-instance of L' with probability $\geq 1 - e_c$.*
- *(Size bound) The size of y is bounded by $t^{1-\epsilon} \cdot \mathrm{poly}(n)$.*

Then $L \in \mathsf{NP}/\mathsf{poly} \cap \mathsf{coNP}/\mathsf{poly}$.

The procedure A above does not need to be a "full" OR-compression, which makes the theorem more general. In particular, A is *relaxed* in two ways: it only needs to work, or be analyzed, in the case that all input instances have the same length; this is useful in hardness of kernelization proofs as it allows similar instances to be grouped together. Furthermore, A only needs to work, or be analyzed, in the case that at most one of the input instances is a yes-instance of L; we believe that this property will be useful in future work on hardness of kernelization.

The fact that "relaxed" OR-compressions suffice in Theorem 1 is implicit in the proof of Drucker [8], but not stated explicitly. Before Drucker's work, Fortnow and Santhanam [10] proved the special case of Theorem 1 in which $e_c = 0$, but they only obtain the weaker consequence $L \in \mathsf{coNP}/\mathsf{poly}$, which prevents their result from applying to AND-compressions in a non-trivial way. Moreover, their proof uses the full completeness requirement and does not seem to work for relaxed OR-compressions.

1.2 Comparison and Overview of the Proof

The simplification of our proof stems from two main sources: 1. The "scaffolding" of our proof, its overall structure, is more modular and more similar to arguments used previously by Ko [12], Fortnow and Santhanam [10], and Dell and Van Melkebeek [7] for compression-type procedures and Dell et al. [5] for isolation procedures. 2. While the information-theoretic part of our proof uses the same set of information-theoretic inequalities as Drucker's, the simple version in Sect. 3 applies these inequalities to distributions that have a simpler structure. Moreover, our calculations have a somewhat more mechanical nature.

Both Drucker's proof and ours use the relaxed OR-compression A to design a P/poly-reduction from L to the *statistical distance problem*, which is known to be in the intersection of NP/poly and coNP/poly by previous work (cf. Xiao [15]). Drucker [8] uses the minimax theorem and a game-theoretic sparsification argument to construct the polynomial advice of the reduction. He also presents an alternative proof [9, Sect. 3] in which the advice is constructed without these arguments and also without any explicit invocation of information theory; however, the alternative proof does not achieve the full generality of his theorem, and we feel that avoiding information theory entirely leads to a less intuitive proof structure. In contrast, our proof achieves full generality up to a factor of two in the simplest proof, it avoids game theoretic arguments, and it limits information theory to a single, intuitive lemma about the average noise sensitivity of compressive maps.

Using this information-theoretic lemma as a black box, we design the P/poly-reduction in a purely combinatorial way: We generalize the fact that tournaments have dominating sets of logarithmic size to *hypergraph tournaments*; these are complete t-uniform hypergraphs with the additional property that, for each hyperedge, one of its elements gets "selected". In particular, for each set $e \subseteq \overline{L}$ of t no-instances, we select one element of e based on the fact that A's behavior on e somehow proves that the selected instance is a no-instance of L. The advice of the reduction is going to be a small dominating set of this hypergraph tournament on the set of no-instances of L. The crux is that we can efficiently test, with the help of the statistical distance problem oracle, whether an instance is dominated or not. Since any instance is dominated if and only if it is a no-instance of L, this suffices to solve L.

In the information-theoretic lemma, we generalize the notion of average noise sensitivity of Boolean functions (which can attain two values) to compressive maps (which can attain only relatively few values compared to the input length). We show that compressive maps have small average noise sensitivity. Drucker's "distributional stability" is a closely related notion, which we make implicit use of in our proof. Using the latter notion as the anchor of the overall reduction, however, leads to some additional technicalities in Drucker's proof, which we also run into in the full version of this paper where we obtain the same bounds as Drucker's theorem. In Sect. 3 we instead use the average noise sensitivity as the anchor of the reduction, which avoids these technicalities at the cost of losing a factor of two in the bounds.

1.3 Application: The Composition Framework

We briefly describe a modern variant of the composition framework that is sufficient to rule out kernels of size $O(k^{d-\epsilon})$ using Theorem 1. It is almost identical to Lemma 1 of [6,7] and the notion defined by Hermelin and Wu [11, Definition 2.2]. By applying the framework for unbounded d, we can also use it to rule out polynomial kernels.

Definition 2. Let L be a language, and let Π with parameter k be a parameterized problem. A *d-partite composition of L into Π* is a polynomial-time algorithm A that maps any set $x = \{x_1, \ldots, x_t\} \subseteq \{0, 1\}^n$ for some n and t to $y = A(x)$ such that:

(1) If all x_i's are no-instances of L, then y is a no-instance of Π.
(2) If exactly one x_i is a yes-instance of L, then y is a yes-instance of Π.
(3) The parameter k of y is bounded by $t^{1/d+o(1)} \cdot \mathrm{poly}(n)$.

This notion of composition has one crucial advantage over previous notions of OR-composition: The algorithm A does not need to work, or be analyzed, in the case that two or more of the x_i's are yes-instances.

Definition 3. Let Π be a parameterized problem. We call Π *d-compositional* if there exists an NP-hard or coNP-hard problem L that has a d-partite composition algorithm into Π.

The above definition encompasses both AND-compositions and OR-compositions because an AND-composition of L into Π is the same as an OR-composition of $\overline{L}$ into $\overline{\Pi}$. We have the following corollary of Drucker's theorem.

Corollary 4. *If* coNP $\not\subseteq$ NP/poly, *then no d-compositional problem has kernels of size* $O(k^{d-\epsilon})$. *Moreover, this even holds when the kernelization algorithm is allowed to be a randomized algorithm with at least a constant gap in error probability.*

Proof. Let L be an NP-hard or coNP-hard problem that has a d-partite composition A' into Π. Assume for the sake of contradiction that Π has a kernelization algorithm with soundness error at most e_s and completeness error at most e_c so that $e_s + e_c$ is bounded by a constant smaller than one. The concatenation of A' with the assumed $O(k^{d-\epsilon'})$-kernelization gives rise to an algorithm A that satisfies the conditions of Theorem 1, for example with $\epsilon = \epsilon'/(2d)$. Therefore, we get $L \in (\text{coNP/poly} \cap \text{NP/poly})$ and thus coNP $\subseteq$ NP/poly, a contradiction. ∎

Several variants of the framework provided by this corollary are possible:

1. n order to rule out poly(k)-kernels for a parameterized problem Π, we just need to prove that Π is d-compositional for all $d \in \mathbb{N}$; let's call Π *compositional* in this case. One way to show that Π is compositional is to construct a single *composition* from a hard problem L into Π; this is an algorithm as in Definition 2, except that we replace (3) with the bound $k \leq t^{o(1)} \text{poly}(n)$.
2. ince all x_i's in Definition 2 are promised to have the same length, we can consider a padded version $\tilde{L}$ of the language L in order to filter the input instances of length n of the original L into a polynomial number of equivalence classes. Each input length of $\tilde{L}$ in some interval $[p_1(n), p_2(n)]$ corresponds to one equivalence class of length-n instances of L. So long as $\tilde{L}$ remains NP-hard or coNP-hard, it is sufficient to consider a composition from $\tilde{L}$ into Π. Bodlaender, Jansen, and Kratsch [4, Definition 4] formalize this approach.
3. The composition algorithm can also use randomness, as long as the overall probability gap of the concatenation of composition and kernelization is not negligible.
4. In the case that L is NP-hard, Fortnow and Santhanam [10] and Dell and Van Melkebeek [7] prove that the composition algorithm can also be a coNP-algorithm or even a coNP oracle communication game in order to get the collapse. Interestingly, this does not seem to follow from Drucker's proof nor from the proof presented here, and it seems to require the full completeness condition for the OR-composition. Kratsch [13] and Kratsch, Philip, and Ray [14] exploit these variants of the composition framework to prove kernel lower bounds.

2   Preliminaries

For any set $R \subseteq \{0,1\}^*$ and any $\ell \in \mathbb{N}$, we write $R_\ell \doteq R \cap \{0,1\}^\ell$ for the set of all length-ℓ strings inside of R. For any $t \in \mathbb{N}$, we write $[t] \doteq \{1, \ldots, t\}$. For a set V, we write $\binom{V}{\leq t}$ for the set of all subsets $x \subseteq V$ that have size at most t. We will work over a finite alphabet, usually $\Sigma = \{0,1\}$. For a vector $a \in \Sigma^t$, a number $j \in [t]$, and a value $y \in \Sigma$, we write $a|_{j \leftarrow y}$ for the string that coincides with a except in position j, where it has value y. For background in complexity theory, we defer to the book by Arora and Barak [2]. We assume some familiarity with the complexity classes NP and coNP as well as their non-uniform versions NP/poly and coNP/poly.

2.1   Distributions and Randomized Mappings

A *distribution* on a finite ground set Ω is a function $\mathcal{D}: \Omega \to [0,1]$ with $\sum_{\omega \in \Omega} \mathcal{D}(\omega) = 1$. The *support* of $\mathcal{D}$ is the set $\operatorname{supp} \mathcal{D} = \{\omega \in \Omega \mid \mathcal{D}(\omega) > 0\}$. The *uniform distribution* $\mathcal{U}_\Omega$ on Ω is the distribution with $\mathcal{U}_\Omega(\omega) = \frac{1}{|\Omega|}$ for all $\omega \in \Omega$. We often view distributions as *random variables*, that is, we may write $f(\mathcal{D})$ to denote the distribution $\mathcal{D}'$ that first produces a sample $\omega \sim \mathcal{D}$ and then outputs $f(\omega)$, where $f: \Omega \to \Omega'$. We use any of the following notations: $\mathcal{D}'(\omega') = \Pr(f(\mathcal{D}) = \omega') = \Pr_{\omega \sim \mathcal{D}}(f(\omega) = \omega') = \sum_{\omega \in \Omega} \mathcal{D}(\omega) \cdot \Pr(f(\omega) = \omega')$. The last term $\Pr(f(\omega) = \omega')$ in this equation is either 0 or 1 if f is a deterministic function, but we will also allow f to be a *randomized mapping*, that is, f has access to some "internal" randomness. This is modeled as a function $f: \Omega \times \{0,1\}^r \to \Omega'$ for some $r \in \mathbb{N}$, and we write $f(\mathcal{D})$ as a short-hand for $f(\mathcal{D}, \mathcal{U}_{\{0,1\}^r})$. That is, the internal randomness consists of a sequence of independent and fair coin flips.

2.2   Statistical Distance

The *statistical distance* $d(X, Y)$ between two distributions X and Y on Ω is

$$d(X, Y) = \max_{T \subseteq \Omega} \big| \Pr(X \in T) - \Pr(Y \in T) \big|. \tag{1}$$

The statistical distance between X and Y is a number in $[0,1]$, with $d(X, Y) = 0$ if and only if $X = Y$ and $d(X, Y) = 1$ if and only if the support of X is disjoint from the support of Y. It is an exercise to show the standard equivalence between the statistical distance and the 1-norm:

$$d(X, Y) = \frac{1}{2} \cdot \big\| X - Y \big\|_1 = \frac{1}{2} \sum_{\omega \in \Omega} \big| \Pr(X = \omega) - \Pr(Y = \omega) \big|.$$

2.3   The Statistical Distance Problem

For $\mathcal{U} = \mathcal{U}_{\{0,1\}^n}$ and $0 \leq \delta < \Delta \leq 1$, let $\operatorname{SD}_{\leq \delta}^{\geq \Delta}$ be the following promise problem:

yes-instances: Circuits $C, C': \{0,1\}^n \to \{0,1\}^*$ with $d(C(\mathcal{U}), C'(\mathcal{U})) \geq \Delta$.
no-instances: Circuits $C, C': \{0,1\}^n \to \{0,1\}^*$ with $d(C(\mathcal{U}), C'(\mathcal{U})) \leq \delta$.

The statistical distance problem is not known to be polynomial-time computable, and in fact it is not believed to be. On the other hand, the problem is also not believed to be NP-hard because the problem is computationally easy in the following sense.

Theorem 5 (Xiao [15] + Adleman [1]). *If $\delta < \Delta$ are constants, we have* $\mathrm{SD}^{\geq \Delta}_{\leq \delta} \in \left(\mathsf{NP/poly} \cap \mathsf{coNP/poly} \right).$

Moreover, the same holds when $\delta = \delta(n)$ and $\Delta = \Delta(n)$ are functions of the input length that satisfy $\Delta - \delta \geq \frac{1}{\mathrm{poly}(n)}$.

This is the only fact about the SD-problem that we will use in this paper. Slightly stronger versions of this theorem are known, but we do not list them here due to the 12-page limit imposed by the publisher.

3   Ruling Out OR-compressions

In this section we prove Theorem 1: Any language L that has a relaxed OR-compression is in $\mathsf{coNP/poly} \cap \mathsf{NP/poly}$. We rephrase the theorem in a form that reveals the precise inequality between the error probabilities and the compression ratio needed to get the complexity consequence.

Theorem 6 (ϵt-compressive version of Drucker's theorem). *Let $L, L' \subseteq \{0,1\}^*$ be languages and $e_s, e_c \in [0,1]$ be some constants denoting the error probabilities. Let $t = t(n) > 0$ be a polynomial and $\epsilon > 0$. Let*

$$A \colon \binom{\{0,1\}^n}{\leq t} \to \{0,1\}^{\epsilon t} \tag{2}$$

be a randomized $\mathsf{P/poly}$-algorithm such that, for all $x \in \binom{\{0,1\}^n}{\leq t}$,

- *if $|x \cap L| = 0$, then $A(x) \in \overline{L'}$ holds with probability $\geq 1 - e_s$, and*
- *if $|x \cap L| = 1$, then $A(x) \in L'$ holds with probability $\geq 1 - e_c$.*

If $e_s + e_c < 1 - \sqrt{(2\ln 2)\epsilon}$, then $L \in \mathsf{NP/poly} \cap \mathsf{coNP/poly}$.

This is Theorem 7.1 in Drucker [9]. However, there are two noteworthy differences:

1. Drucker obtains complexity consequences even when $e_s + e_c < 1 - \sqrt{(\ln 2/2)\epsilon}$ holds, which makes his theorem more general. The difference stems from the fact that we optimized the proof in this section for simplicity and not for the optimality of the bound. He also obtains complexity consequences under the (incomparable) bound $e_s + e_c < 2^{-\epsilon-3}$. Using the slightly more complicated setup deferred to the full version of this paper, we would be able to achieve both of these bounds.

2. To get a meaningful result for OR-compression of NP-complete problems, we need the complexity consequence $L \in \mathsf{NP/poly} \cap \mathsf{coNP/poly}$ rather than just $L \in \mathsf{NP/poly}$. To get the stronger consequence, Drucker relies on the fact that the statistical distance problem $\mathrm{SD}_{\leq \delta}^{\geq \Delta}$ has statistical zero knowledge proofs. This is only known to be true when $\Delta^2 > \delta$ holds, which translates to the more restrictive assumption $(e_s + e_c)^2 < 1 - \sqrt{(\ln 2/2)\epsilon}$ in his theorem. We instead use Theorem 5, which does not go through statistical zero knowledge and proves more directly that $\mathrm{SD}_{\leq \delta}^{\geq \Delta}$ is in $\mathsf{NP/poly} \cap \mathsf{coNP/poly}$ whenever $\Delta > \delta$ holds. Doing so in Drucker's paper immediately improves all of his $L \in \mathsf{NP/poly}$ consequences to $L \in \mathsf{NP/poly} \cap \mathsf{coNP/poly}$.

To obtain Theorem 1, the basic version of Drucker's theorem, as a corollary of Theorem 6, none of these differences matter. This is because we could choose $\epsilon > 0$ to be sufficiently smaller in the proof of Theorem 1, which we provide now before we turn to the proof of Theorem 6.

Proof (of Theorem 1). Let A be the algorithm assumed in Theorem 1, and let $C \geq 2$ be large enough so that the output size of A is bounded by $t^{1-1/C} \cdot C \cdot n^C$. We transform A into an algorithm as required for Theorem 6. Let $\epsilon > 0$ be a small enough constant so that $e_s + e_c < 1 - \sqrt{(2 \ln 2)\epsilon}$. Moreover, let $t(n)$ be a large enough polynomial so that $(t(n))^{1-1/C} \cdot C \cdot n^C < \epsilon t(n)$ holds. Then we restrict A to a family of functions $A_n \colon \binom{\{0,1\}^n}{\leq t(n)} \to \{0,1\}^{<\epsilon t(n)}$. Now a minor observation is needed to get an algorithm of the form (2): The set $\{0,1\}^{<\epsilon t}$ can be efficiently encoded in $\{0,1\}^{\epsilon t}$ (which changes the output language from L' to some L''). Thus we constructed a family A_n as required by Theorem 6, which proves the claim. ∎

3.1   ORs are Sensitive to Yes-instances

The semantic property of relaxed OR-compressions is that they are "L-sensitive": They show a dramatically different behavior for all-no input sets vs. input sets that contain a single yes-instance of L. The following simple fact is the only place in the overall proof where we use the soundness and completeness properties of A.

Lemma 7. *For all distributions X on $\binom{\overline{L}}{<t}$ and all $v \in L$, we have*

$$d\Big(A(X), \, A(X \cup \{v\})\Big) \geq \Delta \doteq 1 - (e_s + e_c). \tag{3}$$

Proof. The probability that $A(X)$ outputs an element of L' is at most e_s, and similarly, the probability that $A(X \cup \{v\})$ outputs an element of L' is at least $1 - e_c$. By (1) with $T = L'$, the statistical distance between the two distributions is at least Δ. ∎

Despite the fact that relaxed OR-compressions are sensitive to the presence or absence of a yes-instance, we argue next that their behavior *within* the set of no-instances is actually quite predictable.

3.2   The Average Noise Sensitivity of Compressive Maps is Small

Relaxed OR-compressions are in particular compressive maps. The following lemma says that the average noise sensitivity of any compressive map is low. Here, "average noise sensitivity" refers to the difference in the behavior of a function when the input is subject to random noise; in our case, we change the input in a single random location and notice that the behavior of a compressive map does not change much.

Lemma 8. *Let $t \in \mathbb{N}$, let X be the uniform distribution on $\{0,1\}^t$, and let $\epsilon > 0$. Then, for all randomized mappings $f \colon \{0,1\}^t \to \{0,1\}^{\epsilon t}$, we have*

$$\underset{j \sim \mathcal{U}_{[t]}}{\mathbf{E}} \quad d\Big(f(X|_{j \leftarrow 0}) \,,\, f(X|_{j \leftarrow 1})\Big) \quad \leq \delta \doteq \sqrt{2 \ln 2 \cdot \epsilon}. \tag{4}$$

We defer the purely information-theoretic and mechanical calculation that yields Lemma 8 to the full version of this paper. We remark that, in the special case where $f \colon \{0,1\}^t \to \{0,1\}$ is a Boolean function, the left-hand side of (4) coincides with the usual definition of the average noise sensitivity.

We translate Lemma 8 to our relaxed OR-compression A as follows.

Lemma 9. *Let $A \colon \binom{\{0,1\}^n}{\leq t} \to \{0,1\}^{\epsilon t}$. For all $e \in \binom{\{0,1\}^n}{t}$, there is $v \in e$ with*

$$d\Big(A(\mathcal{U}_{2^e} \setminus \{v\}) \,,\, A(\mathcal{U}_{2^e} \cup \{v\})\Big) \leq \delta. \tag{5}$$

Here $\mathcal{U}_{2^e}$ samples a subset of e uniformly at random. Note that we replaced the expectation over j from (4) with the mere existence of an element v in (5) since this is all we need; the stronger property also holds.

Proof. To prove the claim, let $v_1, \dots, v_t$ be the elements of e in lexicographic order. For $b \in \{0,1\}^t$, let $g(b) \subseteq e$ be such that $v_i \in g$ holds if and only if $b_i = 1$. We define the randomized mapping $f \colon \{0,1\}^t \to \{0,1\}^{\epsilon t}$ as follows:

$$f(b_1, \dots, b_t) \doteq A\Big(g(b)\Big).$$

Then $f(X|_{j \leftarrow 0}) = A(\mathcal{U}_{2^e} \setminus \{v_j\})$ and $f(X|_{j \leftarrow 1}) = A(\mathcal{U}_{2^e} \cup \{v_j\})$. The claim follows from Lemma 8 with $v \doteq v_j$ for some j that minimizes the statistical distance in (4). ∎

This lemma suggest the following tournament idea. We let $V = \overline{L}_n$ be the set of no-instances, and we let them compete in matches consisting of t players each. That is, a match corresponds to a hyperedge $e \in \binom{V}{t}$ of size t and every such hyperedge is present, so we are looking at a complete t-uniform hypergraph. We say that a player $v \in e$ is "selected" in the hyperedge e if the behavior of A on $\mathcal{U}_{2^e} \setminus \{v\}$ is not very different from the behavior of A on $\mathcal{U}_{2^e} \cup \{v\}$, that is, if (5) holds. The point of this construction is that v being selected proves that v must be a no-instance because (3) does not hold. We obtain a "selector" function $S \colon \binom{V}{t} \to V$ that, given e, selects an element $v = S(e) \in e$. We call S a *hypergraph tournament* on V.

3.3 Hypergraph Tournaments have Small Dominating Sets

Tournaments are complete directed graphs, and it is well-known that they have dominating sets of logarithmic size. A straightforward generalization applies to hypergraph tournaments $S\colon \binom{V}{t} \to V$. We say that a set $g \in \binom{V}{t-1}$ *dominates* a vertex v if $v \in g$ or $S(g \cup \{v\}) = v$ holds. A set $\mathcal{D} \subseteq \binom{V}{t-1}$ is a *dominating set* of S if all vertices $v \in V$ are dominated by at least one element in $\mathcal{D}$.

Lemma 10. *Let V be a finite set, and let $S\colon \binom{V}{t} \to V$ be a hypergraph tournament. Then S has a dominating set $\mathcal{D} \subseteq \binom{V}{t-1}$ of size at most $t \log |V|$.*

Proof. We construct the set $\mathcal{D}$ inductively. Initially, it has $k = 0$ elements. After the k-th step of the construction, we will preserve the invariant that $\mathcal{D}$ is of size exactly k and that $|R| \leq (1 - 1/t)^k \cdot |V|$ holds, where R is the set of vertices that are not yet dominated, that is,

$$R = \left\{ v \in V \;\middle|\; \notin g \text{ and } S(g \cup \{v\}) \neq v \text{ holds for all } g \in \mathcal{D} \right\}.$$

If $0 < |R| < t$, we can add an arbitrary edge $g^* \in \binom{V}{t-1}$ with $R \subseteq g^*$ to $\mathcal{D}$ to finish the construction. Otherwise, the following averaging argument, shows that there is an element $g^* \in \binom{R}{t-1}$ that dominates at least a $1/t$-fraction of elements $v \in R$:

$$\frac{1}{t} = \mathop{\mathbf{E}}_{e \in \binom{R}{t}} \Pr_{v \in e}\left(S(e) = v \right) = \mathop{\mathbf{E}}_{g \in \binom{R}{t-1}} \Pr_{v \in R - g}\left(S(g \cup \{v\}) = v \right).$$

Thus, the number of elements of R left undominated by g^* is at most $(1 - 1/t) \cdot |R|$, so the inductive invariant holds. Since $(1 - 1/t)^k \cdot |V| \leq \exp(-k/t) \cdot |V| < 1$ for $k = t \log |V|$, we have $R = \emptyset$ after $k \leq t \log |V|$ steps of the construction, and in particular, $\mathcal{D}$ has at most $t \log |V|$ elements. ∎

3.4 Proof of the Main Theorem: Reduction to Statistical Distance

Proof (of Theorem 6). We describe a deterministic P/poly reduction from L to the statistical distance problem $\mathrm{SD}_{\leq \delta}^{\geq \Delta}$ with $\Delta = 1 - (e_s + e_c)$ and $\delta = \sqrt{(2 \ln 2)\epsilon}$. The reduction outputs the conjunction of polynomially many instances of $\mathrm{SD}_{\leq \delta}^{\geq \Delta}$. Since $\mathrm{SD}_{\leq \delta}^{\geq \Delta}$ is contained in the intersection of NP/poly and coNP/poly by Theorem 5, and since this intersection is closed under taking polynomial conjunctions, we obtain $L \in \mathsf{NP/poly} \cap \mathsf{coNP/poly}$. Thus it remains to find such a reduction. To simplify the discussion, we describe the reduction in terms of an algorithm that solves L and uses $\mathrm{SD}_{\leq \delta}^{\geq \Delta}$ as an oracle. However, the algorithm only makes non-adaptive queries at the end of the computation and accepts if and only if all oracle queries accept; this corresponds to a reduction that maps an instance of L to a conjunction of instances of $\mathrm{SD}_{\leq \delta}^{\geq \Delta}$ as required.

To construct the advice at input length n, we use Lemma 9 with $t = t(n)$ to obtain a hypergraph tournament S on $V = \overline{L}_n$, which in turn gives rise to a small dominating set $\mathcal{D} \subseteq \binom{V}{t-1}$ by Lemma 10. We remark the triviality that if $|V| \leq t = \mathrm{poly}(n)$, then we can use V, the set of all no-instances of L at this input length, as the advice. Otherwise, we define the hypergraph tournament S for all $e \in \binom{V}{t}$ as follows:

$$S(e) \doteq \min \left\{ v \in e \mid d(A(\mathcal{U}_{2^e} \setminus \{v\}) , A(\mathcal{U}_{2^e} \cup \{v\})) \leq \delta \right\}.$$

By Lemma 9, the set over which the minimum is taken is non-empty, and thus S is well-defined. Furthermore, the hypergraph tournament has a dominating set $\mathcal{D}$ of size at most tn by Lemma 10. As advice for input length n, we choose this set $\mathcal{D}$. Now we have $v \in \overline{L}$ if and only if v is dominated by $\mathcal{D}$. The idea of the reduction is to efficiently check the latter property.

The algorithm works as follows: Let $v \in \{0,1\}^n$ be an instance of L given as input. If $v \in g$ holds for some $g \in \mathcal{D}$, the algorithm rejects v and halts. Otherwise, it queries the SD-oracle on the instance $(A(\mathcal{U}_{2^g}), A(\mathcal{U}_{2^g} \cup \{v\}))$ for each $g \in \mathcal{D}$. If the oracle claims that all queries are yes-instances, our algorithm accepts, and otherwise, it rejects.

First note that distributions of the form $A(\mathcal{U}_{2^g})$ and $A(\mathcal{U}_{2^g} \cup \{v\})$ can be be sampled by using polynomial-size circuits, and so they form syntactically correct instances of the SD-problem: The information about A, g, and v is hard-wired into these circuits, the input bits of the circuits are used to produce a sample from $\mathcal{U}_{2^g}$, and they serve as internal randomness of A in case A is a randomized algorithm.

It remains to prove the correctness of the reduction. If $v \in L$, we have for all $g \in \mathcal{D} \subseteq \overline{L}$ that $v \notin g$ and that the statistical distance of the query corresponding to g is at least $\Delta = 1 - (e_s + e_c)$ by Lemma 7. Thus all queries that the reduction makes satisfy the promise of the SD-problem and the oracle answers the queries correctly, leading our reduction to accept. On the other hand, if $v \notin L$, then, since $\mathcal{D}$ is a dominating set of $\overline{L}$ with respect to the hypergraph tournament S, there is at least one $g \in \mathcal{D}$ so that $v \in g$ or $S(g \cup \{v\}) = v$ holds. If $v \in g$, the reduction rejects. The other case implies that the statistical distance between $A(\mathcal{U}_{2^g})$ and $A(\mathcal{U}_{2^g} \cup \{v\})$ is at most δ. The query corresponding to this particular g therefore satisfies the promise of the SD-problem, which means that the oracle answers correctly on this query and our reduction rejects. ∎

Acknowledgments. I would like to thank Andrew Drucker, Martin Grohe, and others for encouraging me to pursue the publication of this manuscript, David Xiao for pointing out Theorem 5 to me, Andrew Drucker, Dániel Marx, and anonymous referees for comments on an earlier version of this paper, and Dieter van Melkebeek for some helpful discussions.

References

1. Adleman, L.M.: Two theorems on random polynomial time. In: Proceedings of the 19th Annual Symposium on Foundations of Computer Science (FOCS), pp. 75–83 (1978)

2. Arora, S., Barak, B.: Computational Complexity - A Modern Approach. Cambridge University Press (2009)
3. Bodlaender, H.L., Downey, R.G., Fellows, M.R., Hermelin, D.: On problems without polynomial kernels. J. Comput. Syst. Sci. **75**(8), 423–434 (2009)
4. Bodlaender, H.L., Jansen, B.M.P., Kratsch, S.: Kernelization lower bounds by cross-composition. SIAM J. Discrete Math. **28**(1), 277–305 (2014)
5. Dell, H., Kabanets, V., van Melkebeek, D., Watanabe, O.: Is Valiant-Vazirani's isolation probability improvable? Comput. Complex. **22**(2), 345–383 (2013)
6. Dell, H., Marx, D.: Kernelization of packing problems. In: Proceedings of the 23rd Annual ACM-SIAM Symposium on Discrete Algorithms (SODA), pp. 68–81 (2012)
7. Dell, H., van Melkebeek, D.: Satisfiability allows no nontrivial sparsification unless the polynomial-time hierarchy collapses. J. ACM **61**(4), 27 (2014)
8. Drucker, A.: New limits to classical and quantum instance compression. In: Proceedings of the 53rd Annual Symposium on Foundations of Computer Science (FOCS), pp. 609–618 (2012)
9. Drucker, A.: New limits to classical and quantum instance compression. Technical report TR12-112 rev. 2, Electronic Colloquium on Computational Complexity (ECCC) (2013). http://eccc.hpi-web.de/report/2012/112/
10. Fortnow, L., Santhanam, R.: Infeasibility of instance compression and succinct PCPs for NP. J. Comput. Syst. Sci. **77**(1), 91–106 (2011)
11. Hermelin, D., Wu, X.: Weak compositions and their applications to polynomial lower bounds for kernelization. In: Proceedings of the 23rd Annual ACM-SIAM Symposium on Discrete Algorithms (SODA), pp. 104–113 (2012)
12. Ko, K.I.: On self-reducibility and weak P-selectivity. J. Comput. Syst. Sci. **26**, 209–211 (1983)
13. Kratsch, S.: Co-nondeterminism in compositions: a kernelization lower bound for a Ramsey-type problem. In: Proceedings of the 23rd Annual ACM-SIAM Symposium on Discrete Algorithms (SODA), pp. 114–122 (2012)
14. Kratsch, S., Philip, G., Ray, S.: Point line cover: the easy kernel is essentially tight. In: Proceedings of the 25th Annual ACM-SIAM Symposium on Discrete Algorithms (SODA), pp. 1596–1606 (2014)
15. Xiao, D.: New perspectives on the complexity of computational learning, and other problems in theoretical computer science. Ph.D. thesis, Princeton University (2009). ftp://ftp.cs.princeton.edu/techreports/2009/866.pdf

Editing to a Graph of Given Degrees

Petr A. Golovach[1,2]($\boxtimes$)

[1] Department of Informatics, University of Bergen, Bergen, Norway
`petr.golovach@ii.uib.no`
[2] Steklov Institute of Mathematics at St. Petersburg, Russian Academy of Sciences,
St. Petersburg, Russia

Abstract. We consider the EDITING TO A GRAPH OF GIVEN DEGREES problem that for a graph G, non-negative integers d, k and a function $\delta \colon V(G) \to \{1, \ldots, d\}$, asks whether it is possible to obtain a graph G' from G such that the degree of v is $\delta(v)$ for any vertex v by at most k vertex or edge deletions or edge additions. We construct an FPT-algorithm for EDITING TO A GRAPH OF GIVEN DEGREES parameterized by $d + k$. We complement this result by showing that the problem has no polynomial kernel unless NP $\subseteq$ coNP/poly.

1   Introduction

The aim of graph editing or modification problems is to change a given graph by applying a bounded number of specified operations in order to satisfy a certain property. Many basic problems like CLIQUE, INDEPENDENT SET or FEEDBACK (EDGE OR VERTEX) SET can be seen as graph editing problems. It is common to allow combinations of vertex deletions, edge deletions and edge additions, but other operations, like edge contractions, are considered as well.

The systematic study of the vertex deletion problems was initiated by Lewis and Yannakakis [16]. They considered hereditary non-trivial properties. A property is hereditary if it holds for any induced subgraph of a graph that satisfy the property, and a property is non-trivial if it is true for infinitely many graphs and false for infinitely many graphs. Lewis and Yannakakis [16] proved that for any non-trivial hereditary property, the corresponding vertex deletion problem is NP-hard, and for trivial properties the problem can be solved in polynomial time. The edge deletion problems were considered by Yannakakis [23], Alon et al. [1]. The case when edge additions and deletions are allowed and the property is the inclusion in some hereditary graph class was considered by Natanzon et al. [20] and Burzyn et al. [5].

As typically graph editing problems are NP-hard, it is natural to use the parameterized complexity framework to analyze them. Cai [6] proved that for

The research leading to these results has received funding from the European Research Council under the European Union's Seventh Framework Programme (FP/2007-2013)/ERC Grant Agreement n. 267959 and the Government of the Russian Federation (grant 14.Z50.31.0030).

M. Cygan and P. Heggernes (Eds.): IPEC 2014, LNCS 8894, pp. 196–207, 2014.
DOI: 10.1007/978-3-319-13524-3_17

any property defined by a finite set of forbidden induced subgraphs, the editing problem is FPT when parameterized by the bound on the number of vertex deletions, edge deletions and edge additions. Building up on this result, Khot and Raman [15] gave a complete characterization of the parameterized complexity for hereditary properties.

As it could be seen from the aforementioned results, the editing problems are well investigated for hereditary properties. For properties of other types, a great deal less is known, and the graph editing problems where the aim is to obtain a graph that satisfies degree constraints belong to this class. Investigation of the parameterized complexity of such problems were initiated by Moser and Thilikos in [18], Cai and Yang [8] and Mathieson and Szeider [17] (see also [9,13] for related results).

In particular, Mathieson and Szeider [17] considered different variants of the following problem:

EDITING TO A GRAPH OF GIVEN DEGREES
    Instance: A graph G, non-negative integers d, k and a function
        $\delta \colon V(G) \to \{1, \ldots, d\}$.
Parameter 1: d.
Parameter 2: k.
    Question: Is it possible to obtain a graph G' from G such that
        $d_{G'}(v) = \delta(v)$ for each $v \in V(G')$ by at most k operations
        from the set S?

They classified the parameterized complexity of the problem for

$$S \subseteq \{\text{vertex deletion}, \text{edge deletion}, \text{edge addition}\}.$$

They showed that EDITING TO A GRAPH OF GIVEN DEGREES is W[1]-hard when parameterized by k and the unparameterized version is NP-complete if vertex deletion is in S. If $S \subseteq \{\text{edge deletion}, \text{edge addition}\}$, then the problem can be solved in polynomial time. For $\{\text{vertex deletion}\} \subseteq S \subseteq \{\text{vertex deletion}, \text{edge deletion}, \text{edge addition}\}$, they proved that EDITING TO A GRAPH OF GIVEN DEGREES is *Fixed Parameter Tractable* (FPT) when parameterized by $d + k$. Moreover, the FPT result holds for a more general version of the problem where vertices and edges have costs and the degree constraints are relaxed: for each $v \in V(G')$, $d_{G'}(v)$ should be in a given set $\delta(v) \subseteq \{1, \ldots, d\}$. The proof given by Mathieson and Szeider [17] uses a logic-based approach that does not provide practically feasible algorithms. They used the observation that EDITING TO A GRAPH OF GIVEN DEGREES can be reduced to the instances with graphs whose degrees are bounded by a function of k and d. By a result of Seese [22], the problem of deciding any property that can be expressed in first-order logic is FPT for graphs of bounded degree when parameterized by the length of the sentence defining the property. In particular, to obtain their FPT-result, Mathieson and Szeider constructed a non-trivial first-order logic formula that expresses the property that a graph with vertices of given degrees can be obtained by at most k editing operations. For the case $S \subseteq \{\text{vertex deletion}, \text{edge deletion}\}$, they improved the aforementioned result by showing that EDITING TO A GRAPH OF GIVEN DEGREES has a polynomial kernel when parameterized by $d + k$.

In Sect. 3 we construct an FPT-algorithm for EDITING TO A GRAPH OF GIVEN DEGREES parameterized by $k + d$ for the case when S includes vertex deletion and edge addition that runs in time $2^{O(kd^2 + k \log k)} \cdot poly(n)$ for n-vertex graphs, i.e., we give the first feasible algorithm for the problem. Our algorithm is based on the random separation techniques introduced by Cai et al. [7]. We complement this result by showing in Sect. 4 that EDITING TO A GRAPH OF GIVEN DEGREES parameterized by $k + d$ has no polynomial kernel unless NP $\subseteq$ coNP/poly if {vertex deletion, edge addition} $\subseteq$ S. This resolves an open problem by Mathieson and Szeider [17]. The proof uses the cross-composition framework introduced by Bodlaender et al. [3,4]. Due to space restrictions, some proofs and technical details are omitted or just sketched in this extended abstract. The full version of the paper is available at [14].

2   Basic Definitions and Preliminaries

Graphs. We consider only finite undirected graphs without loops or multiple edges. The vertex set of a graph G is denoted by $V(G)$ and the edge set is denoted by $E(G)$.

For a set of vertices $U \subseteq V(G)$, $G[U]$ denotes the subgraph of G induced by U, and by $G - U$ we denote the graph obtained form G by the removal of all the vertices of U, i.e., the subgraph of G induced by $V(G) \setminus U$. If $U = \{u\}$, we write $G - u$ instead of $G - \{u\}$. Respectively, for a set of edges $L \subseteq E(G)$, $G[L]$ is a subgraph of G induced by L, i.e., the vertex set of $G[L]$ is the set of vetices of G incident to the edges of L, and L is the set of edges of $G[L]$. For a non-empty set U, $\binom{U}{2}$ is the set of unordered pairs of elements of U. For a set of edges L, by $G - L$ we denote the graph obtained from G by the removal of all the edges of L. Respectively, for $L \subseteq \binom{V(G)}{2}$, $G + L$ is the graph obtained from G by the addition of the edges that are elements of L. If $L = \{a\}$, then for simplicity, we write $G - a$ or $G + a$.

For a vertex v, we denote by $N_G(v)$ its *(open) neighborhood*, that is, the set of vertices which are adjacent to v, and for a set $U \subseteq V(G)$, $N_G(U) = (\cup_{v \in U} N_G(v)) \setminus U$. The *closed neighborhood* $N_G[v] = N_G(v) \cup \{v\}$, and for a positive integer r, $N_G^r[v]$ is the set of vertices at distance at most r from v. For a set $U \subseteq V(G)$ and a positive integer r, $N_G^r[U] = \cup_{v \in U} N_G^r[v]$. The *degree* of a vertex v is denoted by $d_G(v) = |N_G(v)|$.

A *walk* in G is a sequence $P = v_0, e_1, v_1, e_2, \ldots, e_s, v_s$ of vertices and edges of G such that $v_0, \ldots, v_s \in V(G)$, $e_1, \ldots, e_s \in E(G)$, and for $i \in \{1, \ldots, s\}$, $e_i = v_{i-1}v_i$; v_0, v_s are the *end-vertices* of the trail, and $v_1, \ldots, v_{s-1}$ are the *internal* vertices. A walk is *closed* if $v_0 = v_s$. Sometimes we write $P = v_0, \ldots, v_s$ to denote a trail $P = v_0, e_1, \ldots, e_s, v_s$ omitting edges. A walk is a *trail* if $e_a, \ldots, e_s$ are pairwise distinct, and a trail is a *path* if $v_0, \ldots, v_s$ are pairwise distinct except maybe v_0, v_s.

Parameterized Complexity. Parameterized complexity is a two dimensional framework for studying the computational complexity of a problem. One dimension is the input size n and another one is a parameter k. It is said that a problem is *fixed parameter tractable* (or FPT), if it can be solved in time $f(k) \cdot n^{O(1)}$ for some function f. A *kernelization* for a parameterized problem is a polynomial algorithm that maps each instance (x, k) with the input x and the parameter k to an instance (x', k') such that (i) (x, k) is a YES-instance if and only if (x', k') is a YES-instance of the problem, and (ii) the size of x' is bounded by $f(k)$ for a computable function f. The output (x', k') is called a *kernel*. The function f is said to be a *size* of a kernel. Respectively, a kernel is *polynomial* if f is polynomial. We refer to the books of Flum and Grohe [10], and Niedermeier [21] for detailed introductions to parameterized complexity.

Solutions of Editing to a Graph of Given Degrees. Let (G, δ, d, k) be an instance of EDITING TO A GRAPH OF GIVEN DEGREES. Let $U \subset V(G)$, $D \subseteq E(G - U)$ and $A \subseteq \binom{V(G) \setminus U}{2}$. If the vertex deletion, edge deletion or edge addition is not in S, then it is assumed that $U = \emptyset$, $D = \emptyset$ or $A = \emptyset$ respectively. We say that (U, D, A) is a *solution* for (G, δ, d, k), if $|U| + |D| + |A| \leq k$, and for the graph $G' = G - U - D + A$, $d_{G'}(v) = \delta(v)$ for $v \in V(G')$. We also say that G' is obtained by editing with respect to (U, D, A).

3 FPT-Algorithm

Throughout this section we assume that $S = \{$vertex deletion, edge deletion, edge addition$\}$, i.e., the all three editing operations are allowed, unless we explicitly specify the set of allowed operations. We prove the following theorem.

Theorem 1. EDITING TO A GRAPH OF GIVEN DEGREES *can be solved in time* $2^{O(kd^2 + k \log k)} \cdot poly(n)$ *for n-vertex graphs.*

Due to space restriction, we only sketch the proof here. The complete description of the algorithm is given in [14].

Proof. We construct an FPT-algorithm for EDITING TO A GRAPH OF GIVEN DEGREES parameterized by $k + d$. The algorithm is based on the random separation techniques introduced by Cai et al. [7] (see also [2]).

Let (G, δ, d, k) be an instance of EDITING TO A GRAPH OF GIVEN DEGREES, and let $n = |V(G)|$.

Preprocessing. At this stage of the algorithm our main goal is to reduce the original instance of the problem to a bounded number of instances with the property that for any vertex v, the degree of v is at most $\delta(v)$.

We apply the following rule.

Vertex Deletion Rule. If G has a vertex v with $d_G(v) > \delta(v) + k$, then delete v and set $k = k - 1$. If $k < 0$, then stop and return a NO-answer.

We exhaustively apply the rule until we either stop and return a NO-answer or obtain an instance of the problem such that the degree of any vertex v is at most $\delta(v) + k$. In the last case it is sufficient to solve the problem for the obtained instance, and if it has a solution (U, D, A), then the solution for the initial instance can be obtained by adding the deleted vertices to U. From now we assume that we do not stop while applying the rule, and to simplify notations, assume that (G, δ, d, k) is the obtained instance. Notice that for any $v \in V(G)$, $d_G(v) \leq \delta(v) + k \leq d + k$. Suppose that $v \in V(G)$ and $d_G(v) > \delta(v)$. Then if the considered instance has a solution, either v or at least one of its neighbors should be deleted or at least one of incident to v edges have to be deleted. It implies that we can branch as follows.

Branching Rule. If G has a vertex v with $d_G(v) > \delta(v)$, then stop and return a NO-answer if $k = 0$, otherwise branch as follows.

- For each $u \in N_G[v]$, solve the problem for $(G - u, \delta, d, k - 1)$, and if there is a solution (U, D, A), then stop and return $(U \cup \{u\}, D, A)$.
- For each $u \in N_G(v)$, solve the problem for $(G - uv, \delta, d, k - 1)$, and if there is a solution (U, D, A), then stop and return $(U, D \cup \{uv\}, A)$.

If none of the instances have a solution, then return a NO-answer.

It is straightforward to observe that by the exhaustive application of the rule we either solve the problem or obtain at most $(2(k+d)+1)^k$ instances of the problem such that the original instance has a solution if and only if one of the new instances has a solution, and for each of the obtained instances, the degree of any vertex v is upper bounded by $\delta(v)$. Now it is sufficient to explain how to solve EDITING TO A GRAPH OF GIVEN DEGREES for such instances.

To simplify notations, from now we assume that for (G, δ, d, k), $d_G(v) \leq \delta(v)$ for $v \in V(G)$. Let $Z = \{v \in V(G) | d_G(v) < \delta(v)\}$.

We apply the following rule.

Stopping Rule. If $|Z| > 2k$, then stop and return a NO-answer. If $Z = \emptyset$, then stop and return the trivial solution $(\emptyset, \emptyset, \emptyset)$. If $Z \neq \emptyset$ and $k = 0$, then stop and return a NO-answer.

Then we exhaustively apply the following rule.

Isolates Removing Rule. If G has a vertex v with $d_G(v) = \delta(v) = 0$, then delete v.

Finally on this stage, we solve small instances.

Small Instance Rule. If G has at most $3kd^2 - 1$ edges, then solve EDITING TO A GRAPH OF GIVEN DEGREES.

From now we assume that we do not stop at this stage of the algorithm and, as before, denote by (G, δ, d, k) the obtained instance and assume that $n = |V(G)|$. We have that G has at least $3kd^2$ edges, $|Z| \leq 2k$, $Z \neq \emptyset$, $k \geq 1$, and for any isolated vertex v, $\delta(v) \neq 0$, i.e., $v \in Z$. Notice that since $Z \neq \emptyset$, $d \geq 1$.

Random Separation. Now we apply the random separation technique. We start with constructing a randomized algorithm and then explain how it can be derandomized.

We color the vertices of G independently and uniformly at random by two colors. In other words, we partition $V(G)$ into two sets R and B. We say that the vertices of R are *red*, and the vertices of B are *blue*.

Let $P = v_0, \ldots, v_s$ be a walk in G. We say that P is an R-*connecting* walk if either $s \leq 1$ or for any $i \in \{0, \ldots, s - 2\}$, $\{v_i, v_{i+1}, v_{i+2}\} \cap R \neq \emptyset$, i.e., for any three consecutive vertices of P, at least one of them is red. We also say that two vertices x, y are R-*equivalent* if there is an R-connecting walk that joins them. Clearly, R-equivalence is an equivalence relation on R. Therefore, it defines the corresponding partition of R into equivalence classes. Denote by R_0 the set of red vertices that can be joined with some vertex of Z by an R-connecting walk. Notice that R_0 is a union of some equivalence classes. Denote by $R_1, \ldots, R_t$ the remaining classes, i.e., it is a partition of $R \setminus R_0$ such that any two vertices x, y are in the same set if and only if x and y are R-connected; notice that it can happen that $t = 0$. Observe that for any distinct $i, j \in \{0, \ldots, t\}$, $N_G^3[R_i] \cap R_j = \emptyset$ because any two vertices of R at distance at most 3 in G are R-equivalent. For $i \in \{0, \ldots, t\}$, let $r_i = |R_i|$.

Our aim is to find a solution (U, D, A) for (G, δ, d, k) such that

- $U \cap B = \emptyset$,
- $R_0 \subseteq U$,
- for any $i \in \{1, \ldots, t\}$, either $R_i \subseteq U$ or $R_i \cap U = \emptyset$,
- the edges of D are not incident to the vertices of $N_G(U)$;

i.e., U is a union of equivalence classes of R that contains the vertices of R_0. We call such a solution *colorful*.

Let $B_0 = (Z \cap B) \cup N_G(R_0)$, and for $i \in \{1, \ldots, t\}$, let $B_i = N_G(R_i)$. Notice that each $B_i \subseteq B$, and for distinct $i, j \in \{0, \ldots, t\}$, the distance between any $u \in B_i$ and $v \in B_j$ is at least two, i.e., $u \neq v$ and $uv \notin E(G)$. For a vertex $v \in V(G) \setminus R$, denote by $def(v) = \delta(v) - d_{G-R}(v)$. Recall that $d_G(v) \leq \delta(v)$. Therefore, $def(v) \geq 0$. Notice also that $def(v)$ could be positive only for vertices of the sets $B_0, \ldots, B_t$. For each $i \in \{0, \ldots, t\}$, if $v \in B_i$, then either $v \in Z$ or v is adjacent to a vertex of R_i. Hence, $def(v) > 0$ for the vertices of $B_0, \ldots, B_t$. For a set $A \subseteq \binom{V(G)}{2} \setminus E(G)$, denote by $d_{G,A}(v)$ the number of elements of A incident to v for $v \in V(G)$.

We construct a dynamic programming algorithm that consecutively constructs tables T_i for $i = 0, \ldots, t$ such that T_i is either empty, or contains the unique zero element, or contains lists of all the sequences $(d_1, \ldots, d_p)$ of positive integers, $d_1 \leq \ldots \leq d_p$, such that

(i) there is a set $U \subseteq R_0 \cup \ldots \cup R_i$, $R_0 \subseteq U$, and for any $j \in \{1, \ldots, i\}$, either $R_j \subseteq U$ or $R_j \cap U = \emptyset$,
(ii) there is a set $A \subseteq \binom{V(G)}{2} \setminus E(G)$ of pairs of vertices of $B_0 \cup \ldots \cup B_i$,
(iii) $d_1 + \ldots + d_p + |U| + |A| \leq k$,

and the graph $G' = G - U + A$ has the following properties:

(iv) $d_{G'}(v) \leq \delta(v)$ for $v \in V(G')$, and $d_{G'}(v) < \delta(v)$ for exactly p vertices $v = v_1, \ldots, v_p$,

(v) $\delta(v_j) - d_{G'}(v_j) = d_i$ for $j \in \{1, \ldots, p\}$.

For each sequence $(d_1, \ldots, d_p)$, the algorithm also keeps the sets U, A for which (i)–(v) are fulfilled and $|U| + |A|$ is minimum. The table contains the unique zero element if

(vi) there is a set $U \subseteq R_0 \cup \ldots \cup R_i$, $R_0 \subseteq U_i$, and for any $j \in \{1, \ldots, i\}$, either $R_j \subseteq U$ or $R_j \cap U = \emptyset$,

(vii) there is a set $A \subseteq \binom{V(G)}{2} \setminus E(G)$ of pairs of vertices of $B_0 \cup \ldots \cup B_i$,

(viii) $|U| + |A| \le k$, and

(ix) for the graph $G' = G - U + A$, $d_{G'}(v) = \delta(v)$ for $v \in V(G')$.

For the zero element, the table stores the corresponding sets U and A for which (vi)–(ix) are fulfilled.

Now we explain how we construct the tables for $i \in \{0, \ldots, t\}$.

Construction of T_0. Initially we set $T_0 = \emptyset$. If $\sum_{v \in B_0} def(v) > 2(k - |R_0|)$, then we stop, i.e., $T_0 = \emptyset$. Otherwise, we consider the auxiliary graph $H_0 = G[B_0]$. For all sets $A \subseteq \binom{V(H_0)}{2} \setminus E(H_0)$ such that for any $v \in V(H_0)$, $d_{H_0, A}(v) \le def(v)$, and $\sum_{v \in B_0} def(v) - |A| + |R_0| \le k$, we construct the collection of positive integers $Q = \{def(v) - d_{H_0, A}(v) | v \in B_0 \text{ and } def(v) - d_{H_0, A}(v) > 0\}$ (notice that some elements of Q could be the same). If $Q \ne \emptyset$, then we arrange the elements of Q in increasing order and put the obtained sequence $(d_1, \ldots, d_p)$ of positive integers together with $U = R_0$ and A in T_0. If there is A such that $Q = \emptyset$, then we put the zero element in T_0 together with $U = R_0$ and A, delete all other elements of T_0 and then stop, i.e., T_0 contains the unique zero element in this case.

Construction of T_i for $i \ge 1$. We assume that T_{i-1} is already constructed. Initially we set $T_i = T_{i-1}$. If $T_i = \emptyset$ or T_i contains the unique zero element, then we stop. Otherwise, we consecutively consider all sequences $(d_1, \ldots, d_p)$ from T_{i-1} with the corresponding sets U, A. If $\sum_{v \in B_i} def(v) + \sum_{j=1}^p d_j > 2(k - |R_i| - |U| - |A|)$, then we stop considering $(d_1, \ldots, d_p)$. Otherwise, let $G' = G - U + A$, and let $u_1, \ldots, u_p$ be the vertices of G' with $d_j = \delta(u_j) - d_{G'}(u_j)$ for $j \in \{1, \ldots, p\}$. We consider an auxiliary graph H_i obtained from $G[B_i]$ by the addition of p pairwise adjacent vertices $u_1, \ldots, u_p$. We set $def(u_j) = d_j$ for $j \in \{1, \ldots, p\}$. For all sets $A' \subseteq \binom{V(H_i)}{2} \setminus E(H_i)$ such that for any $v \in V(H_i)$, $d_{H_i, A'}(v) \le def(v)$, and $\sum_{v \in V(H_i)} def(v) - |A'| + |R_i| + |A| + |U| \le k$, we construct the collection of positive integers $Q = \{def(v) - d_{H_i, A'}(v) | v \in V(H_i) \text{ and } def(v) - d_{H_i, A'}(v) > 0\}$. If $Q \ne \emptyset$, then we arrange the elements of Q in increasing order and obtain the sequence $(d'_1, \ldots, d'_q)$ of positive integers together with $U'' = U \cup R_i$ and $A'' = A \cup A'$. If $(d'_1, \ldots, d'_q)$ is not in T_i, then we add it in T_i together with U'', A''. If $(d'_1, \ldots, d'_q)$ is already in T_i together with some sets U''', A''', we replace U''' and A''' by U'' and A'' respectively if $|U''| + |A''| < |U'''| + |A'''|$. If there is A' such that $Q = \emptyset$, then we put the zero element in T_i together with $U'' = U \cup U_i$ and $A'' = A \cup A'$, delete all other elements of T_i and then stop, i.e., T_i contains the unique zero element in this case.

We use the final table T_t to find a colorful solution for (G, δ, d, k) if it exists.

- If T_r contains the zero element with U, A, then $(U, \emptyset, A)$ is a colorful solution.
- If T_r contains a sequence $(d_1, \ldots, d_p)$ with U, A such that $3(d_1 + \ldots + d_p)/2 + |U| + |A| \leq k$ and $r = d_1 + \ldots + d_p$ is even, then let $G' = G - U + A$ and find the vertices $u_1, \ldots, u_p$ of G' such that $\delta(u_i) - d_{G'}(u_i) = d_i$ for $i \in \{1, \ldots, p\}$. Then greedily find a matching D in G' with $h = r/2$ edges $x_1 y_1, \ldots, x_h y_h$ such that $x_1, \ldots, x_h$ and $y_1, \ldots, y_h$ are distinct from the vertices of $\{u_1, \ldots, u_p\} \cup N_G(U)$ and not adjacent to $u_1, \ldots, u_p$. Then we construct the set A' as follows. Initially $A' = \emptyset$. Then for each $i \in \{1, \ldots, r\}$, we consecutively select next d_i vertices $w_1, \ldots, w_{d_i} \in \{x_1, \ldots, x_h, y_1, \ldots, y_h\}$ in such a way that each vertex is selected exactly once and add in A' the pairs $u_1 w_1, \ldots, u_i w_{d_i}$. Then we output the solution $(U, D, A \cup A')$.
- In all other cases we have a NO-answer.

The described algorithm finds a colorful solution if it exists. To find a solution, we run the randomized algorithm N times. If we find a solution after some run, we return it and stop. If we do not obtain a solution after N runs, we return a NO-answer. We show that it is sufficient to run the algorithm $N = 2^{O(dk^2)}$ times.

The algorithm can be derandomized by standard techniques (see [2,7]). We replace random colorings by the colorings induced by *universal sets*. Let n and r be positive integers, $r \leq n$. An (n, r)-*universal set* is a collection of binary vectors of length n such that for each index subset of size r, each of the 2^r possible combinations of values appears in some vector of the set. It is known that an (n, r)-universal set can be constructed in FPT-time with the parameter r. The best construction is due to Naor et al. [19]. They obtained an (n, r)-universal set of size $2^r \cdot r^{O(\log r)} \log n$, and proved that the elements of the sets can be listed in time that is linear in the size of the set.

To apply this technique in our case, we construct an (n, r)-universal set $\mathcal{U}$ for $r = \min\{4kd^2, n\}$. Then we let $V(G) = \{v_1, \ldots, v_n\}$ and for each element of $\mathcal{U}$, i.e., a binary vector $x = (x_1, \ldots, x_n)$, we consider the coloring of G induced by x; a vertex v_i is colored red if $x_i = 1$, and v_i is blue otherwise. Then if (G, δ, d, k) has a solution (U, D, A), then for one of these colorings, the vertices of $N_G^2[Z] \cup N_G^3[U]$ are colored correctly with respect to the solution, i.e., the vertices of U are red and all other vertices of the set are blue. In this case the instance has a colorful solution, and our algorithm finds it.   □

We conclude the section by the observation that a simplified variant of our algorithm solves EDITING TO A GRAPH OF GIVEN DEGREES for $S = \{$vertex deletion, edge addition$\}$. We have to modify the branching rule to exclude edge deletions. Also on the preprocessing stage we do not need the small instance rule. On the random separation stage, we simplify the algorithm by the observation that we have a colorful solution if and only if the table T_t has the zero element. It gives us the following corollary.

Corollary 1. EDITING TO A GRAPH OF GIVEN DEGREES *can be solved in time* $2^{O(kd^2 + k \log k)} \cdot poly(n)$ *for* n-*vertex graphs for* $S = \{$*vertex deletion, edge addition*$\}$.

4   Kernelization Lower Bound

In this section we show that it is unlikely that EDITING TO A GRAPH OF GIVEN DEGREES parameterized by $k + d$ has a polynomial kernel if {vertex deletion, edge addition} $\subseteq S$. The proof uses the cross-composition technique introduced by Bodlaender et al. [3,4]. We need the following definitions (see [3,4]).

Let Σ be a finite alphabet. An equivalence relation $\mathcal{R}$ on the set of strings Σ^* is called a *polynomial equivalence relation* if the following two conditions hold:

(i) there is an algorithm that given two strings $x, y \in \Sigma^*$ decides whether x and y belong to the same equivalence class in time polynomial in $|x| + |y|$,
(ii) for any finite set $S \subseteq \Sigma^*$, the equivalence relation $\mathcal{R}$ partitions the elements of S into a number of classes that is polynomially bounded in the size of the largest element of S.

Let $L \subseteq \Sigma^*$ be a language, let $\mathcal{R}$ be a polynomial equivalence relation on Σ^*, and let $Q \subseteq \Sigma^* \times \mathbb{N}$ be a parameterized problem. An *OR-cross-composition of L into Q* (with respect to $\mathcal{R}$) is an algorithm that, given t instances $x_1, x_2, \ldots, x_t \in \Sigma^*$ of L belonging to the same equivalence class of $\mathcal{R}$, takes time polynomial in $\sum_{i=1}^{t} |x_i|$ and outputs an instance $(y, k) \in \Sigma^* \times \mathbb{N}$ such that:

(i) the parameter value k is polynomially bounded in $\max\{|x_1|, \ldots, |x_t|\} + \log t$,
(ii) the instance (y, k) is a YES-instance for Q if and only if at least one instance x_i is a YES-instance for L for $i \in \{1, \ldots, t\}$.

It is said that L *OR-cross-composes into* Q if a cross-composition algorithm exists for a suitable relation $\mathcal{R}$.

In particular, Bodlaender et al. [3,4] proved the following theorem.

Theorem 2. ([3,4]) *If an NP-hard language L OR-cross-composes into the parameterized problem Q, then Q does not admit a polynomial kernelization unless* NP $\subseteq$ coNP/poly.

It is well-known that the CLIQUE problem is NP-complete for regular graphs [12]. We need a special variant of CLIQUE for regular graphs where a required clique is small with respect to the degree.

SMALL CLIQUE IN A REGULAR GRAPH
    Instance: Positive integers d and k, $k \geq 2$, $k^2 < d$, and a d-regular
           graph G.
    Question: Is there a clique with k vertices in G?

Lemma 1. SMALL CLIQUE IN A REGULAR GRAPH *is* NP-*complete.*

Now we are ready to prove the main result of the section.

Theorem 3. EDITING TO A GRAPH OF GIVEN DEGREES *parameterized by* $k+d$ *has no polynomial kernel unless* NP $\subseteq$ coNP/poly *if vertex deletions and edge additions are allowed.*

Proof. First, we consider the case when the all three editing operations are allowed, i.e., $S = \{$vertex deletion, edge deletion, edge addition$\}$.

We show that SMALL CLIQUE IN A REGULAR GRAPH OR-cross-composes into EDITING TO A GRAPH OF GIVEN DEGREES.

We say that that two instances (G_1, d_1, k_1) and (G_2, d_2, k_2) of SMALL CLIQUE IN A REGULAR GRAPH are *equivalent* if $|V(G_1)| = |V(G_2)|$, $d_1 = d_2$ and $k_1 = k_2$. Notice that this is a polynomial equivalence relation.

Let $(G_1, d, k), \ldots, (G_t, d, k)$ be equivalent instances of SMALL CLIQUE IN A REGULAR GRAPH, $n = |V(G_i)|$ for $i \in \{1, \ldots, t\}$. We construct the instance (G', δ, d', k') of EDITING TO A GRAPH OF GIVEN DEGREES as follows.

- Construct copies of $G_1, \ldots, G_t$.
- Construct $p = k(d - k + 1)$ pairwise adjacent vertices $u_1, \ldots, u_p$.
- Construct $k + 1$ pairwise adjacent vertices $w_0, \ldots, w_k$ and join each w_j with each u_h by an edge.
- Set $\delta(v) = d$ for $v \in V(G_1) \cup \ldots \cup V(G_t)$, $\delta(u_i) = p + k + 1$ for $i \in \{1, \ldots, p\}$, and $\delta(w_j) = p + k$ for $j \in \{0, \ldots, k\}$.
- Set $d' = p + k + 1$ and $k' = k(d - k + 2)$.

Denote the obtained graph G'.

Clearly, $k' + d' = O(n^2)$, i.e., the parameter value is polynomially bounded in n. We show that (G', δ, d', k') is a YES-instance of EDITING TO A GRAPH OF GIVEN DEGREES if and only if (G_i, k, d) is a YES-instance of SMALL CLIQUE IN A REGULAR GRAPH for some $i \in \{1, \ldots, t\}$.

Suppose that (G_i, k, d) is a YES-instance of SMALL CLIQUE IN A REGULAR GRAPH for some $i \in \{1, \ldots, t\}$. Then G_i has a clique K of size k. Let $\{v_1, \ldots, v_q\} = N_{G_i}(K)$. For $j \in \{1, \ldots, q\}$, let $d_j = |N_{G_i}(v_j) \cap K|$. Because G_i is a d-regular graph, $d_1 + \ldots + d_q = k(d - k + 1) = p$. We construct the solution (U, D, A) for (G', δ, d', k') as follows. We set $U = K$ in the copy of G_i, and let $D = \emptyset$. Observe that to satisfy the degree conditions, we have to add d_j edges incident to each v_j in the copy of G_i and add one edge incident to each u_h. To construct A, we consecutively consider the vertices v_j in the copy of G_i for $j = 1, \ldots, q$. For each v_j, we greedily select d_j vertices $x_1, \ldots, x_{d_j}$ in $\{u_1, \ldots, u_p\}$ that were not selected before and add $v_j x_1, \ldots, v_j x_{d_j}$ to A. It is straightforward to verify that (U, D, A) is a solution and $|U| + |D| + |A| = k + p = k'$.

Assume now that (U, D, A) is a solution for (G', δ, d', k').

We show that $U \cap (\{u_1, \ldots, u_p\} \cup \{w_0, \ldots, w_k\}) = \emptyset$. To obtain a contradiction, assume that $|U \cap (\{u_1, \ldots, u_p\} \cup \{w_0, \ldots, w_k\})| = h > 0$. Let $X = (\{u_1, \ldots, u_p\} \cup \{w_0, \ldots, w_k\}) \setminus U$. Because $\{u_1, \ldots, u_p\} \cup \{w_0, \ldots, w_k\}$ has $k(d - k + 1) + k + 1 = k' + 1$ vertices, X has $k' + 1 - h > 0$ vertices. Let $G'' = G' - U$. Observe that for $v \in X$, $\delta(v) - d_{G''}(v) \geq h$. Because the vertices of X are pairwise adjacent, the set A has at least $|X|h = (k' + 1 - h)h$ elements. But $|A| \leq k' - |U| \leq k' - h$. Because $(k' - h + 1)h > k' - h$, we obtain a contradiction.

Next, we claim that $|U| = k$ and $D = \emptyset$. Because $U \cap (\{u_1, \ldots, u_p\} \cup \{w_0, \ldots, w_k\}) = \emptyset$, $\sum_{j=1}^p (\delta(u_j) - d_{G'}(u_j)) = p$ and the vertices $u_1, \ldots, u_p$ are pairwise adjacent, A contains at least p elements. Moreover, A has at least p edges with one end-vertex in $\{u_1, \ldots, u_p\}$ and another in $V(G_1) \cup \ldots \cup V(G_t)$ for the copies of $G_1, \ldots, G_t$ in (G', δ, d', k'). Hence, $|U| + |D| \leq k' - |A| \leq k' - p = k$. Suppose that $|U| = s < k$ and $|D| = h$. Let also $D' = D \cap (E(G_1) \cup \ldots \cup E(G_t))$ and $h' = |D'|$. Let $G'' = G' - U - D'$. Because $G_1, \ldots, G_t$ are d-regular, $\sum_{v \in V(G'')}(\delta(v) - d_{G''}(v)) \leq sd + 2h' \leq sd + 2h \leq sd + 2(k - s)$. Therefore, A contains at most $sd + 2(k - s)$ edges with one end-vertex in $V(G_1) \cup \ldots \cup V(G_t)$. Notice that $sd + 2(k-s) \leq (k-1)d+2$ because $d > k^2 \geq 4$. But $p - (k-1)d - 2 = k(d - k + 1) - (k-1)d - 2 = d - k^2 + k - 2 > 0$ as $d > k^2$, and we have no p edges with one end-vertex in $\{u_1, \ldots, u_p\}$ and another in $V(G_1) \cup \ldots \cup V(G_t)$; a contradiction. Hence, $|U| = k$ and $D = \emptyset$.

Now we show that U is a clique. Suppose that U has at least two non-adjacent vertices. Let $G'' = G' - U$. Because $G_1, \ldots, G_t$ are d-regular, $\sum_{v \in V(G'')}(\delta(v) - d_{G''}(v)) \geq k(d - k + 1) + 2 = p + 2$. Recall that A has at least p edges with one end-vertex in $\{u_1, \ldots, u_p\}$ and another in $V(G_1) \cup \ldots \cup V(G_t)$. Because $|U| = k$ and $k' = p + k$, A consists of p edges with one end-vertex in $\{u_1, \ldots, u_p\}$ and another in $V(G_1) \cup \ldots \cup V(G_t)$. But to satisfy the degree restrictions for the vertices of $V(G_1) \cup \ldots \cup V(G_t)$, we need at least $p+2$ such edges; a contradiction.

We have that $U \subseteq V(G_1) \cup \ldots \cup V(G_t)$ is a clique of size k. Because the copies of $G_1, \ldots, G_t$ in (G', δ, d', k') are disjoint, U is a clique in some G_i.

It remains to apply Theorem 2. Because SMALL CLIQUE IN A REGULAR GRAPH is NP-complete by Lemma 1, EDITING TO A GRAPH OF GIVEN DEGREES parameterized by $k + d$ has no polynomial kernel unless NP $\subseteq$ coNP/poly.

To prove the theorem for $S = \{$vertex deletion, edge addition$\}$, it is sufficient to observe that for the constructed instance (G', δ, d', k') of EDITING TO A GRAPH OF GIVEN DEGREES, any solution (U, D, A) has $D = \emptyset$, i.e., edge deletions are not used. Hence, the same arguments prove the claim.   ⊔

5   Conclusion

We proved that EDITING TO A GRAPH OF GIVEN DEGREES is FPT when parameterized by $k + d$ for $\{$vertex deletion, edge addition$\} \subseteq S \subseteq \{$vertex deletion, edge deletion, edge addition$\}$, but does not admit a polynomial kernel. Our algorithm runs in time $2^{O(kd^2 + k \log k)} \cdot poly(n)$ for n-vertex graph. Hence, it is natural to ask whether this running time could be improved. Another open question is whether the same random separation approach could be applied for more general variants of the problem. Recall that Mathieson and Szeider [17] proved that the problem is FPT for the case when vertices and edges have costs and the degree constraints are relaxed: for each $v \in V(G')$, $d_{G'}(v)$ should be in a given set $\delta(v) \subseteq \{1, \ldots, d\}$. It would be interesting to construct a feasible algorithm for this case. Some interesting results in this direction were recently obtained by Froese et al. [11].

References

1. Alon, N., Shapira, A., Sudakov, B.: Additive approximation for edge-deletion problems. In: FOCS, pp. 419–428. IEEE Computer Society (2005)
2. Alon, N., Yuster, R., Zwick, U.: Color-coding. J. ACM **42**(4), 844–856 (1995)
3. Bodlaender, H.L., Jansen, B.M.P., Kratsch, S.: Cross-composition: a new technique for kernelization lower bounds. In: STACS. LIPIcs, vol. 9, pp. 165–176. Schloss Dagstuhl - Leibniz-Zentrum fuer Informatik (2011)
4. Bodlaender, H.L., Jansen, B.M.P., Kratsch, S.: Kernelization lower bounds by cross-composition. CoRR abs/1206.5941 (2012)
5. Burzyn, P., Bonomo, F., Durán, G.: NP-completeness results for edge modification problems. Discrete Appl. Math. **154**(13), 1824–1844 (2006)
6. Cai, L.: Fixed-parameter tractability of graph modification problems for hereditary properties. Inf. Process. Lett. **58**(4), 171–176 (1996)
7. Cai, L., Chan, S.M., Chan, S.O.: Random separation: a new method for solving fixed-cardinality optimization problems. In: Bodlaender, H.L., Langston, M.A. (eds.) IWPEC 2006. LNCS, vol. 4169, pp. 239–250. Springer, Heidelberg (2006)
8. Cai, L., Yang, B.: Parameterized complexity of even/odd subgraph problems. J. Discrete Algorithms **9**(3), 231–240 (2011)
9. Cygan, M., Marx, D., Pilipczuk, M., Pilipczuk, M., Schlotter, I.: Parameterized complexity of Eulerian deletion problems. In: Kolman, P., Kratochvíl, J. (eds.) WG 2011. LNCS, vol. 6986, pp. 131–142. Springer, Heidelberg (2011)
10. Flum, J., Grohe, M.: Parameterized Complexity Theory. Texts in Theoretical Computer Science. An EATCS Series. Springer, Berlin (2006)
11. Froese, V., Nichterlein, A., Niedermeier, R.: Win-win kernelization for degree sequence completion problems. CoRR abs/1404.5432 (2014)
12. Garey, M.R., Johnson, D.S.: Computers and Intractability: A Guide to the Theory of NP-Completeness. W. H. Freeman, New York (1979)
13. Golovach, P.A.: Editing to a connected graph of given degrees. CoRR abs/1308.1802 (2013)
14. Golovach, P.A.: Editing to a graph of given degrees. CoRR abs/1311.4768 (2013)
15. Khot, S., Raman, V.: Parameterized complexity of finding subgraphs with hereditary properties. Theor. Comput. Sci. **289**(2), 997–1008 (2002)
16. Lewis, J.M., Yannakakis, M.: The node-deletion problem for hereditary properties is np-complete. J. Comput. Syst. Sci. **20**(2), 219–230 (1980)
17. Mathieson, L., Szeider, S.: Editing graphs to satisfy degree constraints: A parameterized approach. J. Comput. Syst. Sci. **78**(1), 179–191 (2012)
18. Moser, H., Thilikos, D.M.: Parameterized complexity of finding regular induced subgraphs. J. Discrete Algorithms **7**(2), 181–190 (2009)
19. Naor, M., Schulman, L., Srinivasan, A.: Splitters and near-optimal derandomization. In: 36th Annual Symposium on Foundations of Computer Science (FOCS 1995), pp. 182–191. IEEE (1995)
20. Natanzon, A., Shamir, R., Sharan, R.: Complexity classification of some edge modification problems. Discrete Appl. Math. **113**(1), 109–128 (2001)
21. Niedermeier, R.: Invitation to Fixed-Parameter Algorithms, Oxford Lecture Series in Mathematics and Its Applications, vol. 31. Oxford University Press, Oxford (2006)
22. Seese, D.: Linear time computable problems and first-order descriptions. Math. Struct. Comput. Sci. **6**(6), 505–526 (1996)
23. Yannakakis, M.: Node- and edge-deletion NP-complete problems. In: Lipton, R.J., Burkhard, W.A., Savitch, W.J., Friedman, E.P., Aho, A.V. (eds.) STOC, pp. 253–264. ACM (1978)

Polynomial Kernels and User Reductions for the Workflow Satisfiability Problem

Gregory Gutin[1], Stefan Kratsch[2], and Magnus Wahlström[1]([⊠])

[1] Royal Holloway, University of London, London, UK
{G.Gutin,Magnus.Wahlstrom}@rhul.ac.uk
[2] TU Berlin, 10587 Berlin, Germany
stefan.kratsch@tu-berlin.de

Abstract. The *workflow satisfiability problem (WSP)* is a problem of practical interest that arises whenever tasks need to be performed by authorized users, subject to constraints defined by business rules. We are required to decide whether there exists a *plan* – an assignment of tasks to authorized users – such that all constraints are satisfied.

The WSP is, in fact, the conservative Constraint Satisfaction Problem (i.e., for each variable, here called *task*, we have a unary authorization constraint) and is, thus, NP-complete. It was observed by Wang and Li (2010) that the number k of tasks is often quite small and so can be used as a parameter, and several subsequent works have studied the parameterized complexity of WSP regarding parameter k.

We take a more detailed look at the kernelization complexity of $WSP(\Gamma)$ when Γ denotes a finite or infinite set of allowed constraints. Our main result is a dichotomy for the case that all constraints in Γ are regular: (1) We are able to reduce the number n of users to $n' \leq k$. This entails a kernelization to size poly(k) for finite Γ, and, under mild technical conditions, to size poly($k+m$) for infinite Γ, where m denotes the number of constraints. (2) Already $WSP(R)$ for some $R \in \Gamma$ allows no polynomial kernelization in $k + m$ unless the polynomial hierarchy collapses.

1   Introduction

A business process is a collection of interrelated tasks that are performed by users in order to achieve some objective. In many situations, a task can be performed only by certain *authorized* users. Additionally, either because of the particular requirements of the business logic or security requirements, we may require that certain sets of tasks cannot be performed by some sets of users [7]. Such constraints include *separation-of-duty*, which may be used to prevent sensitive combinations of tasks being performed by a single user, and *binding-of-duty*, which requires that a particular combination of tasks is performed by the same user. The use of constraints in workflow management systems to enforce security policies has been studied extensively in the last fifteen years; see, e.g., [3,7,17].

It is possible that the combination of constraints and authorization lists is "unsatisfiable", in the sense that there does not exist an assignment of users to

© Springer International Publishing Switzerland 2014
M. Cygan and P. Heggernes (Eds.): IPEC 2014, LNCS 8894, pp. 208–220, 2014.
DOI: 10.1007/978-3-319-13524-3_18

tasks (called a *plan*) such that all constraints are satisfied and every task is performed by an authorized user. A plan that satisfies all constraints and allocates an authorized user to each task is called *valid*. The workflow satisfiability problem (WSP) takes a workflow specification as input and returns a valid plan if one exists and NO otherwise. It is important to determine whether a business process is satisfiable or not, since an unsatisfiable one can never be completed without violating the security policy encoded by the constraints and authorization lists.

Wang and Li [17] were the first to observe that the number k of tasks is often quite small and so can be considered as a parameter. As a result, WSP can be studied as a parameterized problem. Wang and Li [17] proved that, in general, WSP is W[1]-hard, but WSP is fixed-parameter tractable[1] (FPT) if we consider some special types of practical constraints which include separation-of-duty and binding-of-duty constraints. Crampton et al. [9] found a faster fixed-parameter algorithm to solve the special cases of WSP studied in [17] and showed that the algorithm can be used for a wide family of constraints called regular (in fact, regular constraints include all constraints studied in [17]). Subsequent research has demonstrated the existence of fixed-parameter algorithms for WSP in the presence of other constraint types [5,8]. In particular, Cohen et al. [5] showed that WSP with only so-called user-independent constraints is FPT. Crampton et al. [9] also launched the study of polynomial and partially polynomial kernels (in the latter only the number of users is required to be bounded by a polynomial in k), but obtained results only for concrete families of constraints.

In this work, we explore the kernelization properties of WSP in more detail. We study both the possibility of polynomial kernels and of simplifying WSP instances by reducing the set of users;[2] reductions of the latter type have been called partial kernels previously. Our goal is to determine for which types of constraints such user-limiting reductions are possible, i.e., for which sets Γ does the problem WSP(Γ) of WSP restricted to using constraint types (i.e., relations) from Γ admit a reduction to poly(k) users? We study this question for both finite and infinite sets Γ of regular constraints, and show a strong separation: Essentially, either every instance with k tasks can be reduced to at most k users, or there is no polynomial-time reduction to poly(k) users unless the polynomial hierarchy collapses. (However, some technical issues arise for the infinite case.)

Our results. Our main result is a dichotomy for the WSP(Γ) problem when Γ contains only regular relations. We show two results. On the one hand, if every relation $R \in \Gamma$ is *intersection-closed* (see Sect. 4), then we give a polynomial-time reduction which reduces the number of users in an instance to $n' \leq k$, without increasing the number of tasks k or constraints m. This applies even if Γ is infinite, given a natural assumption on computable properties of the relations. On the other hand, we show that given even a single relation R which is regular but not intersection-closed, the problem WSP(R) restricted to using only the relation R admits no polynomial kernel, and hence no reduction to

---

[1] For an introduction to fixed-parameter algorithms and complexity, see, e.g., [12].

[2] Such reductions are of interest by themselves as some practical WSP algorithms iterate over users in search for a valid plan [6].

poly(k) users, unless the polynomial hierarchy collapses. For finite sets Γ, this gives a dichotomy in a straight-forward manner: For every finite set Γ of regular relations, WSP(Γ) admits a polynomial kernel if every $R \in \Gamma$ is intersection-closed, and otherwise not unless the polynomial hierarchy collapses.

However, for infinite sets Γ things get slightly more technical, for two reasons: (1) An instance with k tasks and few users could still be exponentially large due to the number of constraints, analogously to the result that HITTING SET admits no polynomial kernel parameterized by the size of the ground set [11] (cf. [13]). (2) More degenerately, without any restriction on Γ, an instance could be exponentially large simply due to the encoding size of a single constraint (e.g., one could interpret a complete WSP instance on k tasks as a single constraint on these k tasks). Both these points represent circumstances that are unlikely to be relevant for practical WSP instances. We make two restrictions to cope with this: (1) We allow the number m of constraints as an extra parameter, since it could be argued that $m \leq \text{poly}(k)$ in practice. (2) We require that each constraint of arity $r \leq k$ can be expressed by poly(r) bits. E.g., this allows unbounded arity forms of all standard constraints. Using this, we obtain a more general dichotomy: For any (possibly infinite) set Γ of regular relations, WSP(Γ) admits a kernel of size poly($k+m$) if every $R \in \Gamma$ is intersection-closed, otherwise not, unless the polynomial hierarchy collapses.

Note that prior to our work there was no conjecture on how a polynomial kernel dichotomy for all regular constraints may look like (we cannot offer such a conjecture for the more general case of user-independent constraints). The positive part follows by generalizing ideas of Crampton et al. [9]; the negative part is more challenging, and requires more involved arguments, especially to show the completeness of the dichotomy (see Sect. 4.2).

Organization. We define WSP formally and introduce a number of different constraint types, including regular constraints, in Sect. 2. In Sect. 3 we give several lower bounds for the kernelization of WSP(Γ). In Sect. 4 we prove our main result, namely the dichotomy for regular constraints. We conclude in Sect. 5. Full proofs of all nontrivial assertions are deferred to the full version.

2   Preliminaries

We define a *workflow schema* to be a tuple (S, U, A, C), where S is the set of tasks in the workflow, U is the set of users, $A\colon S \to 2^U$ assigns each task $s \in S$ an *authorization list* $A(s) \subseteq U$, and C is a set of workflow constraints. A *workflow constraint* is a pair $c = (L, \Theta)$, where $L \subseteq S$ is the *scope* of the constraint and Θ is a set of functions from L to U that specifies those assignments of elements of U to elements of L that *satisfy* the constraint c. Given $T \subseteq S$ and $X \subseteq U$, a *plan* is a function $\pi\colon T \to X$; a plan $\pi\colon S \to U$ is called a *complete plan*. Given a workflow constraint (L, Θ), $T \subseteq S$, and $X \subseteq U$, a plan $\pi\colon T \to X$ *satisfies* (L, Θ) if either $L \setminus T \neq \emptyset$, or π restricted to L is contained in Θ. A plan $\pi\colon T \to X$ is *eligible* if π satisfies every constraint in C. A plan $\pi\colon T \to X$ is *authorized* if $\pi(s) \in A(s)$ for all $s \in T$. A plan is *valid* if it is both authorized and eligible. For an algorithm that runs on an instance (S, U, A, C) of WSP, we will measure the running time in terms of $n = |U|, k = |S|$, and $m = |C|$.

2.1 WSP Constraints and Further Notation

Let us first recall some concrete constraints that are of interest for this work:

$(=, T, T'), (\neq, T, T')$: These generalize the binary *binding-of-duty* and *separation-of-duty* constraints and were previously studied in [9,17]. They demand that there exist $s \in T$ and $s' \in T'$ which are assigned to the same (resp. different) users. We shorthand $(s = s')$ and $(s \neq s')$ if $T = \{s\}$ and $T' = \{s'\}$.

(t_ℓ, t_r, T): A plan π satisfies (t_ℓ, t_r, T), also called a *tasks-per-user counting constraint*, if a user performs either no tasks in T or between t_ℓ and t_r tasks. Tasks-per-user counting constraints generalize the cardinality constraints which have been widely adopted by the WSP community [1,2,14,16].

$(\leq t, T), (\geq t, T)$: These demand that the tasks in T are assigned to at most t (resp. at least t) different users. They generalize *binding-of-duty* and *separation-of-duty*, respectively, and enforce security and diversity [6].

All these constraints share the property that satisfying them depends only on the partition of tasks that is induced by the plan. Formally, a constraint (L, Θ) is *user-independent* if for any $\theta \in \Theta$ and permutation $\psi: U \to U$, $\psi \circ \theta \in \Theta$.

Regular and user-independent constraints. For $T \subseteq S$ and $u \in U$ let $\pi: T \to u$ denote the plan that assigns every task of T to u. We call a constraint $c = (L, \Theta)$ *regular* if it satisfies the following condition: For any partition $L_1, \ldots, L_p$ of L such that for every $i \in [p] = \{1, \ldots, p\}$ there exists an eligible plan $\pi: L \to U$ and user u such that $\pi^{-1}(u) = L_i$, then the plan $\bigcup_{i=1}^{p}(L_i \to u_i)$, where all u_i's are distinct, is eligible. Regular constraints are a special class of user-independent constraints, but not every user-independent constraint is regular. Crampton et al. [9] show that constraints of the type $(\neq, T, T')$; $(=, T, T')$, where at least one of the sets T, T' is a singleton, and tasks-per-user counting constraints of the form (t_ℓ, t_r, T), where $t_\ell = 1$, are regular. In general, $(=, T, T')$ is not regular [9].

Since regular constraints are of central importance to this paper, we introduce some further notation and terminology. Below, we generally follow Crampton et al. [9]. Let $W = (S, U, A, C)$ be a workflow schema, and π an eligible (complete) plan for W. Then $\sim_\pi$ is the equivalence relation on S defined by π, where $s \sim_\pi s'$ if and only if $\pi(s) = \pi(s')$. We let S/π be the set of equivalence classes of $\sim_\pi$, and for a task $s \in S$ we let $[s]_\pi$ denote the equivalence class containing s.

For a constraint $c = (L, \Theta)$, a set $T \subseteq L$ of tasks is *c-eligible* if there is a plan $\pi: L \to U$ that satisfies c, such that $T \in L/\pi$. It is evident from the definition that c is regular if and only if the following holds: For every plan $\pi: L \to U$, π satisfies c if and only if every equivalence class $T \in L/\pi$ is c-eligible. In this sense, a regular constraint c is entirely defined by the set of c-eligible sets of tasks. It is clear that regular constraints are closed under conjunction, i.e., if every constraint $c \in C$ is regular, then the constraint defined by the conjunction of the constraints in C is regular.

In a similar sense, if $c = (L, \Theta)$ is user-independent but not necessarily regular, then c can be characterized on the level of partitions of L: Let $\pi, \pi' \colon L \to U$ be two plans such that $L/\pi = L/\pi'$. Then either both π and π' are eligible for c, or neither is. Overloading the above terminology, if c is a user-independent constraint, then we say that a partition L/π is c-eligible if a plan π generating the partition would satisfy the constraint. We may thus refer to the partition L/π itself as either eligible or ineligible. As with regular constraints, user-independent constraints are closed under conjunction.

Describing constraints via relations. We will frequently describe constraint types in terms of relations. In the following, we restrict ourselves to user-independent constraints. Let $R \subseteq \mathbb{N}^r$ be an r-ary relation, and $(s_1, \ldots, s_r) \in S'$ a tuple of tasks, with repetitions allowed (i.e., we may have $s_i = s_j$ for some $i \neq j$, $i, j \in [r]$). An *application* $R(s_1, \ldots, s_r)$ (of R) is a constraint (L, Θ) where $L = \{s_i : i \in [r]\}$ and $\Theta = \{\pi \colon L \to \mathbb{N} \mid (\pi(s_1), \ldots, \pi(s_r)) \in R\}$. Here, we identify users $U = \{u_1, \ldots, u_n\}$ with integers $[n] = \{1, \ldots, n\}$. We say that R is *user-independent (regular)* if every constraint $R(s_1, \ldots, s_n)$ resulting from an application of R is user-independent (regular). In particular, a user-independent relation R can be defined on the level of partitions, in terms of whether each partition L/π of its arguments is eligible or not, and a regular relation can be defined in terms of eligible sets, as above.

Given a (possibly infinite) set Γ of relations as above, a workflow schema *over* Γ is one where every constraint is an application of a relation $R \in \Gamma$, and WSP(Γ) denotes the WSP problem restricted to workflow schemata over Γ. To cover cases of constraints of unbounded arity, we allow Γ to be infinite.

Well-behaved constraint sets. To avoid several degenerate cases associated with infinite sets Γ we make some standard assumptions on our constraints. We say that a set Γ of user-independent relations is *well-behaved* if the following hold: (1) Every relation $R \in \Gamma$ can be encoded using $poly(r)$ bits, where r is the arity of R; note that this does not include the space needed to specify the scope of an application of R. (2) For every application $c = (L, \Theta)$ of a relation $R \in \Gamma$, we can test in polynomial time whether a partition of L is c-eligible; we can also test in polynomial time whether a set $S \subseteq L$ is c-eligible, and if not, then we can (if possible) find a c-eligible set S' with $S \subset S' \subseteq L$. All relations corresponding to the concrete constraints mentioned above, are well-behaved.

2.2 Kernelization

A *parameterized problem* $\mathcal{Q}$ is a subset of $\Sigma^* \times \mathbb{N}$ for some finite alphabet Σ. A *kernelization* of $\mathcal{Q}$ is a polynomial-time computable function $K \colon (x, k) \mapsto (x', k')$ such that $(x, k) \in \mathcal{Q}$ if and only if $(x', k') \in \mathcal{Q}$, and such that $|x'|, k' \leq h(k)$ for some $h(k)$. Here, (x, k) is an *instance* of $\mathcal{Q}$, and $h(k)$ is the *size* of the kernel. We say that K is a *polynomial kernelization* if $h(k) = k^{O(1)}$.

Our main tool for studying existence of polynomial kernels is kernelization-preserving reductions. Given two parameterized problems Q_1 and Q_2, a *polynomial parametric transformation (PPT)* from Q_1 to Q_2 is a polynomial time computable function $\Psi\colon (x,k) \mapsto (x',k')$ such that for every input (x,k) of Q_1 we have $(x',k') \in Q_2$ if and only if $(x,k) \in Q_1$, and such that $k' \le p(k)$ for some $p(k) = k^{O(1)}$. Note that if Q_2 has a polynomial kernel and if there is a PPT from Q_1 to Q_2, then Q_1 has a *polynomial compression*, i.e., a kernel-like reduction to an instance of a different problem with total output size $k^{O(1)}$. Furthermore, for many natural problems (including all considered in this paper), we are able to complete these reductions using NP-completeness to produce a polynomial kernel for Q_1. Conversely, by giving PPTs from problems that are already known not to admit polynomial compressions (under some assumption) we rule out polynomial kernels for the target problems. For more background on kernelization we refer the reader to the recent survey by Lokshtanov et al. [15].

2.3   Implementations and Implications

Let $W = (S, U, A, C)$ be a workflow schema and $T \subseteq S$ a set of tasks. The *projection of W onto T* is a constraint $c = (T, \Theta)$ where $\pi\colon T \to U$ is contained in Θ if and only if there is a valid and complete plan π' for W that extends π. Further, let $R \subseteq \mathbb{N}^r$ be a user-independent relation, $Q = \{q_1, \ldots, q_r\}$ a set of r distinct tasks, and Γ a set of relations. We say that Γ *implements* R if, for any r-tuple $\mathcal{A} = (A(q_1), \ldots, A(q_r))$ of authorization lists, there is a workflow schema $W = (S, U, A, C)$ over Γ that can be computed in polynomial time, such that the projection of W onto T for some $T \subseteq S$ is equivalent to $R(q_1, \ldots, q_r)$ for every plan $\pi\colon \{q_1, \ldots, q_r\} \to U$ authorized with respect to $\mathcal{A}$, where furthermore $|S| + |C|$ does not depend on $\mathcal{A}$ and U equals $\bigcup_{i \in [r]} A(q_i)$ plus a constant number of local users, i.e., new users who will not be authorized to perform any task outside of $S \setminus T$.

Lemma 1. *Let Γ and Γ' be finite workflow constraint languages such that Γ' implements R for every $R \in \Gamma$. Then there is a PPT from $WSP(\Gamma)$ to $WSP(\Gamma')$, both with respect to parameter k and $k + m$.*

3   Lower Bounds for Kernelization

In this section we begin our investigation of the preprocessing properties of the WORKFLOW SATISFIABILITY PROBLEM. We establish lower bounds against polynomial kernels for WSP for several widely-used constraint types. Like for many other problems, e.g., HITTING SET(n) or CNF SAT(n), there is little hope to get polynomial kernels for WSP when we allow an unbounded number of constraints of arbitrary arity, cf. [10,11,13]. As an example, we give Lemma 2, whose proof uses a PPT from CNF SAT(n) to WSP(≥ 2) with only two users.

Lemma 2. *Let WSP($\geq$ 2) be the WSP problem with constraints ($\geq$ 2, L) for task sets L of arbitrary arity. Then WSP($\geq$ 2) admits no polynomial kernelization with respect to the number k of tasks unless the polynomial hierarchy collapses, even if the number of users is restricted to $n = 2$.*

In our further considerations we will avoid such cases, by either taking m as an additional parameter or by restricting Γ to be finite, which implies bounded arity (namely the maximum arity over the finitely many $R \in \Gamma$). We also assume that all constraints are well-behaved (cf. Sect. 2.1). We then have the following, showing that bounding the number of users implies a polynomial kernel.

Proposition 1. *Let Γ be a set of relations. If Γ is finite, then WSP(Γ) has a polynomial kernel under parameter $(k+n)$; if Γ is infinite but Γ is well-behaved, then WSP(Γ) has a polynomial kernel under parameter $(k + m + n)$.*

The following lemma addresses a special case of ternary constraint $R(a, b, c)$ and proves that WSP(R) already admits no polynomial kernelization in terms of $k + m$. This lemma will be a cornerstone of the dichotomy in the following section. We also get immediate corollaries for constraints $(=, S, S')$ and $(\leq t, S)$ since $(=, \{a\}, \{b, c\})$ and $(\leq 2, \{a, b, c\})$ fulfill the requirement of the lemma.

Lemma 3. *Let $R(a, b, c)$ be a ternary user-independent constraint which is satisfied by plans with induced partition $\{\{a, b\}, \{c\}\}$ or $\{\{a, c\}, \{b\}\}$, but not by plans with partition $\{\{a\}, \{b\}, \{c\}\}$. Then WSP(R) does not admit a polynomial kernel with respect to parameter $k + m$ unless the polynomial hierarchy collapses.*

Corollary 1. *WSP($(=, S, S')$) and WSP($(\leq t, S)$) do not admit a kernelization to size polynomial in $k + m$ unless the polynomial hierarchy collapses.*

4   A Dichotomy for Regular Constraints

In this section, we present a dichotomy for the kernelization properties of WSP(Γ) when Γ is a well-behaved set of regular relations.

Let us describe the dichotomy condition. Let $c = (L, \Theta)$ be a regular constraint, and $E_R \subseteq 2^L$ the set of c-eligible subsets of L; for ease of notation, we let $\emptyset \in E_R$. Note that by regularity, E_R defines c. We say that c is intersection-closed if for any $T_1, T_2 \in E_R$ it holds that $T_1 \cap T_2 \in E_R$. Similarly, we say that a regular relation $R \in \Gamma$ is intersection-closed if every application $R(s_1, \dots, s_r)$ of R is. Note (1) that this holds if and only if an application $R(s_1, \dots, s_r)$ of R with r distinct tasks s_i is intersection-closed, and (2) that the conjunction of intersection-closed constraints again defines an intersection-closed constraint. Finally, a set Γ of relations is intersection-closed if every relation $R \in \Gamma$ is. Our dichotomy results will essentially say that WSP(Γ) admits a polynomial kernel if and only if Γ is intersection-closed; see Theorem 1 below.

The rest of the section is laid out as follows. In Sect. 4.1 we show that if Γ is regular, intersection-closed, and well-behaved, then WSP(Γ) admits a reduction to $n' \leq k$ users; by Proposition 1, this implies a polynomial kernel under parameter $(k + m)$, and under parameter (k) if Γ is finite. In Sect. 4.2 we show that for any single relation R that is not intersection-closed, the problem WSP(R) admits no polynomial kernel, by application of Lemma 3. In Sect. 4.3 we consider the implications of these results for the existence of efficient user-reductions.

In summary, we will show the following result for kernelization. Again, a discussion of the consequences for user-reductions is deferred until Sect. 4.3.

Theorem 1. *Let Γ be a possibly infinite set of well-behaved regular relations. If every relation in Γ is intersection-closed, then WSP(Γ) admits a polynomial-time many-one reduction down to $n' \leq k$ users, implying a polynomial kernel under parameter $k + m$ (and a polynomial kernel under parameter k if Γ is finite). Otherwise, WSP(Γ) admits no kernel of size poly($k + m$) unless the polynomial hierarchy collapses (even if Γ consists of a single such relation R).*

4.1 A User Reduction for Intersection-Closed Constraints

We now give a procedure that reduces a WSP instance $W = (S, U, A, C)$ with n users, k tasks and m constraints to one with $k' \leq k$ tasks, $n' \leq k'$ users and $m' \leq m$ constraints, under the assumption that every constraint $c \in C$ occurring in the instance is intersection-closed and that our language is well-behaved. (This has been called a *partial kernel* in other work [4].) The approach is as in Crampton et al. [9], e.g., Theorem 6.5 of [9], but becomes more involved due to having to work in full generality; we also use a more refined marking step that allows us to decrease the number of users from k^2 to k, a significant improvement. As noted (Proposition 1), under the appropriate further assumption on the constraints, this gives a polynomial kernel under parameter $k + m$ or k.

We begin by noting a consequence of sets closed under intersection.

Lemma 4. *Let $c = (L, \Theta)$ be an intersection-closed constraint, and let $T \subseteq L$ be c-ineligible. If there is a superset T' of T which is c-eligible, then there is a task $s \in L \setminus T$ such that every c-eligible superset T' of T contains s.*

We refer to the task s guaranteed by the lemma as a *required addition* to T by c. Note that assuming well-behavedness, we can make this lemma constructive, i.e., in polynomial time we can test whether a set T is eligible for a constraint, whether it has an eligible superset, and find all required additions if it does. This can be done by first asking for an eligible superset T' of T, then greedily finding a minimal set $T \subset T'' \subseteq T'$. Then every $s \in T'' \setminus T$ is a required addition.

Our reduction proceeds in three phases. First, we detect all binary equalities implied by the constraints i.e., all explicit or implicit constraints ($s = s'$), and handle them separately by merging tasks, intersecting their authorization lists. The output of this phase is an instance where any plan which assigns to every task a unique user is eligible (though such a plan may not be authorized); in particular, since our constraints are regular, we have that all singleton sets of tasks are c-eligible for every constraint c of the instance.

The second phase of the kernel is a user-marking process, similar to the kernels in [9] but with a stronger bound on the number of users. This procedure is based around attempting to produce a system of *distinct representatives* for $\{A(s) : s \in S\}$, i.e., to find a plan $\pi \colon S \to U$ such that π is authorized and $\pi(s) \neq \pi(s')$ for every $s \neq s'$. Via Hall's theorem, this procedure either succeeds, or produces a set T of tasks such that fewer than $|T|$ users are authorized to perform any task in T. In the latter case, we mark all these users, discard the tasks T, and repeat the procedure. Eventually, we end up with a (possibly empty) set of tasks S' which allows for a set of distinct representatives, and mark these representatives as well. Refer to a task s as *easy* if it was appointed a representative in this procedure, and *hard* if it was not (i.e., if it was a member of a set T of discarded tasks). We discard every non-marked user, resulting in a partially polynomial kernel with $k' \leq k$ tasks and $n' \leq k' \leq k$ users.

Finally, to establish the correctness of the kernelization, we give a procedure that, given a partial plan for the set of hard tasks, either extends the plan to a valid complete plan or derives that no such extended plan exists.

Lemma 5. *Let (S, U, A, C) be a workflow schema with k tasks, n users, and m constraints, with at least one equality constraint $(s = s')$, $s \neq s'$. In polynomial time, we can produce an equivalent instance (S', U', A', C') with at most $k - 1$ tasks, n users, and m constraints. Furthermore, if the constraints in C were given as applications $R(\ldots)$ of some relations R, $R \in \Gamma$, then the constraints in C' can be given the same way.*

We now show the detection of equalities.

Lemma 6. *Let (S, U, A, C) be a workflow schema where every constraint is regular and intersection-closed. Then we can in polynomial time reduce the instance to the case where every singleton $\{s\}$, $s \in S$, is eligible.*

Next, we describe the user-marking procedure in detail. We assume that Lemma 6 has been applied, i.e., that all singleton sets are eligible.

1. Let $M = \emptyset$, let S be the set of all tasks, and U the set of all users.
2. While $\{A(s) \cap U : s \in S\}$ does not admit a system of distinct representatives:
   Let $T \subseteq S$ such that $|\bigcup_{s \in T} A(s)| < |T|$. Let $U_T = \bigcup_{s \in T} A(s)$. Add U_T to M, remove U_T from U, and remove from S every task s such that $A(s) \subseteq M$.
3. Add to M the distinct representatives of the remaining tasks S, if any.
4. Discard all users not occurring in M from the instance.

We refer to the set M of users produced above as the *marked* users, and let $S_{\text{hard}} \subseteq [k]$ be the set of hard tasks, i.e., the set of tasks removed in Step 2 of the procedure. Finally, we show the correctness of the above procedure.

Lemma 7. *Let (S, U, A, C) be a workflow schema where all constraints are regular and intersection-closed, and where all singleton sets are eligible. There is a valid complete plan for the instance if and only if there is a valid complete plan only using marked users.*

Putting the above pieces together yields the following theorem.

Theorem 2. *The WSP, restricted to well-behaved constraint languages where every constraint is regular and intersection-closed admits a kernel with $m' \leq m$ constraints, $k' \leq k$ tasks, and $n' \leq k'$ users.*

4.2 Kernel Lower Bounds for Non-intersection-closed Constraints

We now give the other side of the dichotomy by showing that within the setting of regular constraints, even a single relation R which is not intersection-closed can be used to construct a kernelization lower bound, following one of the constructions in Sect. 3. First, we need an auxiliary lemma.

Lemma 8. *Let R be a (satisfiable) regular relation R which is not intersection-closed, let $=$ be the binary equality relation, and let $\Gamma = \{R, =\}$. Then either Γ implements a relation matching that of Lemma 3, or Γ implements the binary disequality relation $\neq$.*

Proof (sketch). Let $c = (L, \Theta)$ be an application of R. We first attempt to find an eligible plan π with two sets $T, T' \in L/\pi$ such that $T \cup T'$ is not c-eligible; in this case, we can use T and T' to implement binary disequality. Otherwise, for any eligible partition L/π of L, merging sets in L/π yields a new eligible partition; in particular, eligible sets are closed under complementation. We now have two remaining cases: either there are two (possibly overlapping) c-eligible sets T, T' such that $T \cup T'$ is ineligible, or the c-eligible sets are closed under union. In the former case, we can use $\overline{T}$, $\overline{T'}$, and $\overline{T \cup T'}$ (through some case analysis) to construct a relation compatible with Lemma 3; in the latter case, the c-eligible sets are closed under both union and complementation, implying that they are closed under intersection, contrary to our assumptions. □

Using $\neq$, we can more easily construct a relation $R(a, b, c)$ as in Lemma 3.

Lemma 9. *Let R be a regular relation which is not intersection-closed, and let $=$ and $\neq$ denote the binary equality and disequality relations. Then $\Gamma = \{R, =, \neq\}$ implements a relation $R'(a, b, c)$ as in Lemma 3.*

Proof (sketch). Let $c = (L, \Theta)$ be an application of R. We say that a pair of sets $P, Q \subset L$ is a *counterexample witness* if P and Q are both c-eligible while $P \cap Q$ is not. We consider a counterexample that is *minimal* under the following priorities: (1) Merging tasks, by adding a constraint $(s = s')$; (2) Picking minimal-cardinality sets P, Q as a counterexample witness; (3) Adding as many disequality constraints $(s \neq s')$ as possible. It can be verified that if W is a minimal workflow schema under these conditions which admits some counterexample witness P, Q, then projecting down to tasks $a \in P \cap Q$, $b \in P - a$, $c \in Q - a$ must define a relation $R'(a, b, c)$ compatible with the conditions of Lemma 3. □

Theorem 3. *Let R be a regular relation which is not intersection-closed. Then $WSP(R)$ admits no kernel of size $poly(k + m)$ unless PH collapses.*

This finishes the proof of Theorem 1.

Table 1. Overview of results for typical user-independent constraints. We recall that the WSP problem is FPT with respect to k when all constraints are user-independent.

	Regular	∩-cl.	poly(k) user reduction		Bounded arity resp. finite Γ		Well-behaved infinite Γ	
$(\neq, T, T')$ $(\geq 2, T)$	Yes	Yes	Yes	[9]	$PK(k)$	[9]	$PK(k+m)$	Corollary 2
$(1, t_u, T)$	Yes	Yes	Yes	[9]	$PK(k)$	[9]	$PK(k+m)$	Corollary 2
(t_l, t_u, T)		No	No	Corollary 3	No $PK(k+m)$	Theorem 1	No $PK(k+m)$	
$(=, s, T')$	Yes	No	No	Corollary 3	No $PK(k+m)$	Theorem 2	No $PK(k+m)$	Corollary 2
$(=, T, T')$	No	N.A						
$(\geq t, T)$	No	N.A.	Yes	[9]	$PK(k)$	[9]	$PK(k+m)$	Proposition 1
$(\leq t, T)$	No	N.A	No	Corollary 3	No $PK(k+m)$	Corollary 1	No $PK(k+m)$	Corollary 1
reg.+∩-cl.	Yes	Yes	Yes	Theorem 2	$PK(k)$	Theorem 1	$PK(k+m)$	Corollary 2
Regular	Yes	No	No	Corollary 3	No $PK(k+m)$	Theorem 1	No $PK(k+m)$	Corollary 2

Corollary 2. *Let Γ be a set of regular relations. If Γ is well-behaved, then $WSP(\Gamma)$ admits a polynomial kernel in parameter $k + m$ if Γ is intersection-closed, otherwise not, unless the polynomial hierarchy collapses. If Γ is finite, then the same dichotomy holds for parameter k instead of $k + m$.*

4.3   On User Bounds for WSP

In this section we return to the question of preprocessing WSP down to a number of users that is polynomial in the number k of tasks. As seen above, the positive side of our kernel dichotomy relies directly on a procedure that reduces the number of users in an instance, while the lower bounds refer entirely to the total size of the instance. Could there be a loophole here, allowing the number of users to be bounded without directly resulting in a polynomial kernel? Alas, it seems that while such a result cannot be excluded, it might not be very useful.

Corollary 3. *Let Γ be a set of user-independent relations containing at least one relation which is regular but not intersection-closed. Unless the polynomial hierarchy collapses, any polynomial-time procedure that reduces the number of users in an instance down to poly(k) must in some cases increase either the number k of tasks, the number m of constraints, or the coding length of individual constraints superpolynomially in $k + m$.*

5   Conclusion

In this paper, we have considered kernelization properties of the workflow satisfiability problem $WSP(\Gamma)$ restricted to use only certain types $R \in \Gamma$ of constraints. We have focused on the case that all relations $R \in \Gamma$ are regular. For this case, we showed that $WSP(\Gamma)$ admits a reduction down to $n' \leq k$ users if every $R \in \Gamma$ is *intersection-closed* (and obeys some natural assumptions on efficiently computable properties), otherwise (under natural restrictions) no such reduction is possible unless the polynomial hierarchy collapses. In particular, this implies a dichotomy on the kernelizability of WSP under the parameters k for finite Γ, and $k + m$ for infinite languages Γ (subject to the aforementioned computability assumptions). This extends kernelization results of Crampton et al. [9], and represents the first kernelization lower bounds for regular languages. Some results are summarized in Table 1.

An interesting open problem is to extend this result beyond regular constraints, e.g., to general user-independent constraints.

References

1. American National Standards Institute. ANSI INCITS 359–2004 for role based access control (2004)
2. Bertino, E., Bonatti, P.A., Ferrari, E.: TRBAC: a temporal role-based access control model. ACM Trans. Inf. Syst. Secur. **4**(3), 191–233 (2001)
3. Bertino, E., Ferrari, E., Atluri, V.: The specification and enforcement of authorization constraints in workflow management systems. ACM Trans. Inf. Syst. Secur. **2**(1), 65–104 (1999)
4. Betzler, N., Bredereck, R., Niedermeier, R.: Partial kernelization for rank aggregation: theory and experiments. In: Raman, V., Saurabh, S. (eds.) IPEC 2010. LNCS, vol. 6478, pp. 26–37. Springer, Heidelberg (2010)
5. Cohen, D., Crampton, J., Gagarin, A., Gutin, G., Jones, M.: Iterative plan construction for the workflow satisfiability problem. CoRR, abs/1306.3649v2 (2013)
6. Cohen, D., Crampton, J., Gagarin, A., Gutin, G., Jones, M.: Engineering algorithms for workflow satisfiability problem with user-independent constraints. In: Chen, J., Hopcroft, J.E., Wang, J. (eds.) FAW 2014. LNCS, vol. 8497, pp. 48–59. Springer, Heidelberg (2014)
7. Crampton, J.: A reference monitor for workflow systems with constrained task execution. In: SACMAT, pp. 38-47. ACM (2005)
8. Crampton, J., Crowston, R., Gutin, G., Jones, M., Ramanujan, M.S.: Fixed-parameter tractability of workflow satisfiability in the presence of seniority constraints. In: Fellows, M., Tan, X., Zhu, B. (eds.) FAW-AAIM 2013. LNCS, vol. 7924, pp. 198–209. Springer, Heidelberg (2013)
9. Crampton, J., Gutin, G., Yeo, A.: On the parameterized complexity and kernelization of the workflow satisfiability problem. ACM Trans. Inf. Syst. Secur. **16**(1), 4 (2013)
10. Dell, H., van Melkebeek, D.: Satisfiability allows no nontrivial sparsification unless the polynomial-time hierarchy collapses. In: Schulman, L.J. (ed.) STOC, pp. 251-260. ACM (2010)
11. Dom, M., Lokshtanov, D., Saurabh, S.: Incompressibility through colors and IDs. In: Albers, S., Marchetti-Spaccamela, A., Matias, Y., Nikoletseas, S., Thomas, W. (eds.) ICALP 2009, Part I. LNCS, vol. 5555, pp. 378–389. Springer, Heidelberg (2009)
12. Downey, R.G., Fellows, M.R.: Fundamentals of Parameterized Complexity. Texts in Computer Science. Springer, (2013)
13. Hermelin, D., Kratsch, S., Sołtys, K., Wahlström, M., Wu, X.: A completeness theory for polynomial (Turing) kernelization. In: Gutin, G., Szeider, S. (eds.) IPEC 2013. LNCS, vol. 8246, pp. 202–215. Springer, Heidelberg (2013)
14. Joshi, J., Bertino, E., Latif, U., Ghafoor, A.: A generalized temporal role-based access control model. IEEE Trans. Knowl. Data Eng. **17**(1), 4–23 (2005)

15. Lokshtanov, D., Misra, N., Saurabh, S.: Kernelization – preprocessing with a guarantee. In: Bodlaender, H.L., Downey, R., Fomin, F.V., Marx, D. (eds.) Fellows Festschrift 2012. LNCS, vol. 7370, pp. 129–161. Springer, Heidelberg (2012)
16. Sandhu, R.S., Coyne, E.J., Feinstein, H.L., Youman, C.E.: Role-based access control models. IEEE Comput. **29**(2), 38–47 (1996)
17. Wang, Q., Li, N.: Satisfiability and resiliency in workflow authorization systems. ACM Trans. Inf. Syst. Secur. **13**(4), 40 (2010)

Finding Shortest Paths Between Graph Colourings

Matthew Johnson[1], Dieter Kratsch[2(✉)], Stefan Kratsch[3],
Viresh Patel[4], and Daniël Paulusma[1]

[1] School of Engineering and Computing Sciences, Durham University, Durham, UK
{matthew.johnson2,daniel.paulusma}@durham.ac.uk
[2] LITA, Université de Lorraine, 57045 Metz Cedex 01, France
dieter.kratsch@univ-lorraine.fr
[3] Technische Universität Berlin, Berlin, Germany
stefan.kratsch@tu-berlin.de
[4] Queen Mary, University of London, London, UK
viresh.patel@qmul.ac.uk

Abstract. The k-colouring reconfiguration problem asks whether, for
a given graph G, two proper k-colourings α and β of G, and a positive
integer ℓ, there exists a sequence of at most ℓ proper k-colourings of G
which starts with α and ends with β and where successive colourings
in the sequence differ on exactly one vertex of G. We give a complete
picture of the parameterized complexity of the k-colouring reconfiguration problem for each fixed k when parameterized by ℓ. First we show
that the k-colouring reconfiguration problem is polynomial-time solvable
for $k = 3$, settling an open problem of Cereceda, van den Heuvel and
Johnson. Then, for all $k \geq 4$, we show that the k-colouring reconfiguration problem, when parameterized by ℓ, is fixed-parameter tractable
(addressing a question of Mouawad, Nishimura, Raman, Simjour and
Suzuki) but that it has no polynomial kernel unless the polynomial
hierarchy collapses.

1 Introduction

Graph colouring has its origin in a nineteenth century map colouring problem
and has now been an active area of research for more than 150 years, finding many
applications within and beyond Computer Science and Mathematics. Given a
graph $G = (V, E)$ and a positive integer k, a k-colouring of G is a map $c\colon V \to
\{1, \ldots, k\}$; it is proper if $c(u) \neq c(v)$ for all u, v with $uv \in E$. The problem of
deciding whether a graph has a proper k-colouring for fixed $k \geq 3$ was an early
example of an NP-complete problem. If, however, one knows that a graph has a
proper k-colouring, or several of them, one may wish to know more about them
such as how many there are or what structural properties they have.

Supported by EPSRC (EP/G043434/1), by a Scheme 7 grant from the London
Mathematical Society, and by the German Research Foundation (KR 4286/1).

© Springer International Publishing Switzerland 2014
M. Cygan and P. Heggernes (Eds.): IPEC 2014, LNCS 8894, pp. 221–233, 2014.
DOI: 10.1007/978-3-319-13524-3_19

One way to study these questions is to consider the k-colouring reconfiguration graph: given a graph G, the k-colouring reconfiguration graph $R_k(G)$ of G is a graph whose vertices are the proper k-colourings of G and where an edge is present between two k-colourings if and only if the two k-colourings differ on only a single vertex of G.

There are several algorithmic questions one can ask about the graph $R_k(G)$ such as whether $R_k(G)$ is connected, whether there exists a path between two given vertices of $R_k(G)$, or how long is the shortest path between two given vertices of $R_k(G)$. (Note that in general $R_k(G)$ has size exponential in the size of G, making these questions highly non-trivial.) It is the latter question, stated formally below, that we address in this paper.

k-COLOURING RECONFIGURATION

Instance: An n-vertex graph $G = (V, E)$, two proper k-colourings α and β and a positive integer ℓ.

Question: Is there a path in the reconfiguration graph of G between α and β of length at most ℓ?

General Motivation. Reconfiguration graphs can be defined for any search problem: the vertices correspond to all solutions to the problem and the edges are defined by a symmetric adjacency relation normally chosen to represent a smallest possible change between solutions. They arise naturally when one wishes to understand the solution space for a search problem.

There has been much research over the last 10 years on the structure and algorithmic aspects of reconfiguration graphs, not only for k-COLOURING [1,2, 5,8–10] but also for many other problems, such as SATISFIABILITY [11], INDE-PENDENT SET [7,17], LIST EDGE COLOURING [13,15], $L(2,1)$-LABELING [14], SHORTEST PATH [3,4,18], and SUBSET SUM [16]. From these studies, the follow-ing subtle phenomenon has been observed, which one would like to better under-stand: it is often (but not always) the case that NP-complete search problems give rise to PSPACE-complete reconfiguration problems, whereas polynomial-time solvable search problems often give rise to polynomial-time solvable recon-figuration problems. For further background we refer the reader to the recent survey of van den Heuvel [12].

Reconfiguration graphs are also important for constructing and analyzing algorithms that sample or count solutions to a search problem. Indeed, under-standing connectivity properties of the k-colouring reconfiguration graph is fun-damental in analyzing certain randomized algorithms for sampling and counting k-colourings of a graph and in analyzing certain cases of the Glauber dynamics in statistical physics (see Sect. 5 of [12]).

Our Results. Our first result, which we prove in Sect. 2, shows that k-COLOURING RECONFIGURATION can be solved in polynomial time when $k = 3$, which settles a problem raised by Cereceda et al. [10]. Note that the cases $k = 1, 2$ are easily seen to be polynomial-time solvable.

In [10], Cereceda et al. were mainly concerned with determining whether, given a graph G and two proper 3-colourings α and β, there exists *any* path between α and β in $R_k(G)$. They found a polynomial-time algorithm to solve this problem and further showed that, for certain instances, their algorithm in fact finds a shortest path between α and β (a precise statement is given in Sect. 2). Here we complete their result by giving an algorithm for all instances.

Theorem 1. 3-COLOURING RECONFIGURATION *can be solved in time* $O(n^2)$.

For $k \geq 4$, we cannot expect a polynomial-time algorithm for k-COLOURING RECONFIGURATION: Bonsma and Cereceda [5] showed that, for each $k \geq 4$, the problem of determining if there is *any* path between two given proper k-colourings of a given graph is PSPACE-complete. On the other hand, our second result (proven in Sect. 3) is that for each $k \geq 4$, k-COLOURING RECONFIGURATION is fixed-parameter tractable when parameterized by the path length ℓ.

Recall that, informally, a parameterized problem is a decision problem (in our case k-COLOURING RECONFIGURATION) in which every problem instance I has an associated integer parameter p (in our case the path length ℓ). A parameterized problem is *fixed-parameter tractable* (FPT) if every instance I can be solved in time $f(p)|I|^c$ where f is a computable function that only depends on p and c is a constant independent of p.

Theorem 2. *For each fixed* $k \geq 4$, k-COLOURING RECONFIGURATION *can be solved in time* $O((k \cdot \ell)^{\ell^2 + \ell} \cdot \ell n^2)$. *In particular, for each fixed* $k \geq 4$, k-COLOURING RECONFIGURATION *is* FPT *when parameterized by* ℓ.

Once a problem is shown to be FPT (and it is unlikely that the problem is polynomial-time solvable), one can go further and ask whether it has a *polynomial kernel*. It is well known that a problem is FPT with respect to a parameter p if and only if it can be *kernelized*, i.e., if and only if, for any instance (I, p) of the given parameterized problem, it is possible to compute in polynomial time an *equivalent instance* (I', p') such that $|I'|, p' \leq g(p)$ for some computable function g (two problem instances are equivalent if and only if they are both yes-instances or both no-instances). If $g(p)$ is a polynomial, then the given parameterized problem is said to have a *polynomial kernel*. We prove the following theorem in Sect. 4.

Theorem 3. *For each fixed* $k \geq 4$, k-COLOURING RECONFIGURATION *parameterized by* ℓ *does not admit a polynomial kernel unless* NP $\subseteq$ coNP/poly.

In fact Theorem 3 holds even when we restrict attention to inputs where the two proper k-colourings of the input graph differ in only two vertices (note that the problem becomes trivial if the two given k-colourings differ in only one vertex).

Our three results give a *complete* picture of the parameterized complexity of k-COLOURING RECONFIGURATION for each fixed k when parameterized by ℓ.

Related work. Fixed-parameter tractability of k-COLOURING RECONFIGURATION was proved independently in recent work of Bonsma and Mouawad [6]. They also prove various hardness results for other parameterizations of k-COLOURING RECONFIGURATION. In particular, they proved that if k is part of the input then

k-COLOURING RECONFIGURATION is W[1]-hard when parameterized only by ℓ (note that the problem, when parameterized only by k, is para-PSPACE-complete due to the aforementioned result of Bonsma and Cereceda [5]).

Mouawad et al. [20] were the first to consider reconfiguration problems in the context of parameterized complexity. For various NP-complete search problems, they showed that determining whether there exists a path of length at most ℓ in the reconfiguration graph between two given vertices is W[1]-hard (when ℓ is the parameter); they asked if there exists an NP-complete problem for which the corresponding reconfiguration problem, parameterized by ℓ, is FPT. Theorem 2 and [6] give the second positive answer to this question, the first being an FPT algorithm for a reconfiguration problem related to VERTEX COVER [19]. However, perhaps surprisingly, Theorem 1 shows that there even exists an NP-complete problem for which the corresponding shortest path problem in the reconfiguration graph is polynomial-time solvable, and thus trivially FPT when parameterized by ℓ.

As mentioned earlier, deciding whether there exists *any* path in $R_k(G)$ between two k-colourings α and β of an input graph G is polynomial-time solvable for $k \leq 3$ [10] and PSPACE-complete for $k \geq 4$ [5]. The problem remains PSPACE-complete for bipartite graphs when $k \geq 4$, for planar graphs when $4 \leq k \leq 6$ and for planar bipartite graphs for $k = 4$ [5].

The algorithmic question of whether $R_k(G)$ is connected for a given G is addressed in [8,9], where it is shown that the problem is coNP-complete for $k = 3$ and bipartite G, but polynomial-time solvable for planar bipartite G.

Finally, the study of the diameter of $R_k(G)$ raises interesting questions. In [10] it is shown that every component of $R_3(G)$ has diameter polynomial (in fact quadratic) in the size of G. On the other hand, for $k \geq 4$, explicit constructions [5] are given of graphs G for which $R_k(G)$ has at least one component with diameter exponential in the size of G. It is known that if G is a $(k-2)$-degenerate graph then $R_k(G)$ is connected and it is conjectured that in this case $R_k(G)$ has diameter polynomial in the size of G [8]; for graphs of treewidth $k-2$ the conjecture has been proved in the affirmative [1].

2   A Polynomial-Time Algorithm for $k = 3$

In this section we consider 3-COLOURING RECONFIGURATION and prove Theorem 1. Some proofs are omitted for reasons of space.

First some definitions needed throughout the paper. Let $G = (V, E)$ be a graph on n vertices, and let α and β be two proper k-colourings of G. For *any* two colourings c and d, we say that c and d *agree* on a vertex u if $c(u) = d(u)$ and that otherwise they *disagree* on u. An $(\alpha \rightarrow \beta)$-*recolouring* R of *length* $\ell = |R|$ is a sequence of proper colourings $c_0, \ldots, c_\ell$ where $c_0 = \alpha$ and $c_\ell = \beta$, and, for $1 \leq q \leq \ell$, c_q and c_{q-1} disagree on at most one vertex. So possibly $c_q = c_{q-1}$ though in this case c_q could be deleted and the sequence that remained would also be an $(\alpha \rightarrow \beta)$-recolouring. The set $\{c_{q-1}c_q : c_{q-1} \neq c_q\}$ is a set of edges in the reconfiguration graph corresponding to a walk from α to β.

In this section, α and β are 3-colourings. The three colours are 1, 2 and 3, and we think of them cyclically: so when, for example, we refer to a colour one greater than a we mean $a + 1 \mod 3$. A cycle in G is *fixed* with respect to a 3-colouring if the two neighbours of each vertex on the cycle are not coloured alike (one can see that this implies that the cycle is coloured in this way in every other colouring in the same component of $R_3(G)$ since one cannot change the colour of just one vertex and obtain another proper 3-colouring).

Cereceda et al. [10] provided a partial solution to 3-COLOURING RECONFIG-URATION. They were interested in recognizing whether or not α and β belong to the same component of the reconfiguration graph. They introduced a polynomial-time algorithm that we will call FINDPATH(G, α, β) that

- correctly determines when α and β belong to different components of $R_3(G)$;
- finds an $(\alpha \rightarrow \beta)$-recolouring of G, of length $O(n^2)$, when α and β belong to the same component of $R_3(G)$;
- moreover, if G contains a fixed cycle with respect to α, the $(\alpha \rightarrow \beta)$-recolouring found is the shortest possible.

We also note that it is possible to recognize in time $O(n^2)$ whether or not there is a fixed cycle (this is described in [10], but is an easy exercise). We need to show how to find a shortest possible $(\alpha \rightarrow \beta)$-recolouring of G in the case where α and β are known to belong to the same component of G, and G contains no fixed cycle with respect to α. We assume now that these conditions hold.

We require a further notion related to colourings called a *height* function (that extends a concept introduced in [10]). Let $S = c_0, c_1, \ldots$ be a sequence of colourings where c_i and c_{i-1} disagree on exactly one vertex and $c_0 = \alpha$. The height function is denoted h^S and has domain $S \times V$ and its range is the set of integers. For each $v \in V$, $h^S(c_0, v) = 0$. For $i > 0$, for each $v \in V$:

$$h^S(c_i, v) = \begin{cases} h^S(c_{i-1}, v), & \text{if } c_i(v) = c_{i-1}(v); \\ h^S(c_{i-1}, v) + 2, & \text{if } c_i(v) \equiv c_{i-1}(v) + 1 \mod 3; \\ h^S(c_{i-1}, v) - 2, & \text{if } c_i(v) \equiv c_{i-1}(v) - 1 \mod 3. \end{cases}$$

So each vertex has height 0 initially and is raised or lowered by 2 when its colour is increased or decreased as we move along the sequence of colourings. For any $(\alpha \rightarrow \beta)$-recolouring R, let the *total height* of R be $H(R) = \sum_{v \in V} |h^R(\beta, v)|$.

Lemma 1. *Let R be a $(\alpha \rightarrow \beta)$-recolouring of length ℓ. Then $\ell \geq \frac{1}{2} H(R)$.*

Proof. For each colouring in R, the height of only one vertex differs from the previous colouring in R and the height difference is 2. Thus, for each vertex v, at least $|h^R(\beta, v)|/2$ distinct colourings in R are needed and the lemma follows.   □

Lemma 2. *For any colouring c, for any sequence of colourings S from α to c, for each vertex v in V,*

$$2(c(v) - \alpha(v)) \equiv h^S(c, v) \mod 6 \tag{1}$$

Proof. We use induction on the length of S. If S contains only one colouring, then this is α, and both sides of (1) are zero with $c = \alpha$.

Suppose that S is longer and that c' is its penultimate colouring. We must show that if (1) is true for c', then it is also true for c. If c and c' agree on v, then we are done. If c and c' disagree on v, then we need only to notice that

$$2(c(v) - c'(v)) \equiv h^S(c, v) - h^S(c', v) \bmod 6$$

and each side of (1) changes by the same amount if we replace c' by c. □

Some more terminology. If an edge is oriented, then we can define its *weight* with respect to a colouring c. The *weight* of an edge oriented from u to v is a value $w(c, \overrightarrow{uv}) \in \{-1, 1\}$ such that $w(c, \overrightarrow{uv}) \equiv c(v) - c(u) \bmod 3$. To orient a path is to orient each edge so that a directed path is obtained. The weight of an oriented path $w(c, \overrightarrow{P})$ is the sum of the weight of its edges.

Lemma 3. *For any colouring c, for any sequence of colourings S from α to c, for each pair of vertices u, v in V, for each oriented path $\overrightarrow{P}$ from u to v,*

$$h^S(c, u) = h^S(c, v) + w(c, \overrightarrow{P}) - w(\alpha, \overrightarrow{P}). \tag{2}$$

Proof. We use induction on the length of S. If S contains only one colouring, then this is α, and both sides of (2) are zero with $c = \alpha$.

Suppose that S is longer and that c' is the penultimate colouring in the sequence. We must show that if (2) is true for c', then it is also true for c. Let x be the vertex on which c' and c disagree.

Suppose that $x \notin \{u, v\}$. If $\overrightarrow{P}$ does not contain x, then clearly the weight of the path is the same for c' and c. If $\overrightarrow{P}$ does contain x, then let $\overrightarrow{yx}$ and $\overrightarrow{xz}$ be the edges of $\overrightarrow{P}$ that x belongs to. As c and c' are proper and $c(x) \neq c'(x)$, we must have $c(y) = c'(y) = c'(z) = c(z)$. Thus

$$w(c, \overrightarrow{yx}) + w(c, \overrightarrow{xz}) = c(x) - c(y) + c(z) - c(x) = 0,$$
$$w(c', \overrightarrow{yx}) + w(c', \overrightarrow{xz}) = c'(x) - c'(y) + c'(z) - c'(x) = 0.$$

So $w(c, \overrightarrow{P}) = w(c', \overrightarrow{P})$ and both sides of (2) are unchanged when c' replaces c.

Suppose that $x = u$. Let y be the vertex adjacent to x on $\overrightarrow{P}$. Suppose that $h^S(c, x) = h^S(c', x) + 2$; that is, the colour of x is increased (as c replaces c'). Then $c(x) \equiv c'(x) + 1 \bmod 3$ and so $c(y) \equiv c'(x) - 1 \bmod 3$. Thus $w(c', \overrightarrow{xy}) = -1$, and, as $c(y) \equiv c(x) + 1 \bmod 3$, $w(c, \overrightarrow{xy}) = 1$, which gives $w(c, \overrightarrow{P}) = w(c', \overrightarrow{P}) + 2$ and (2) remains satisfied. If the height of x is instead lowered, a similar argument can be used. The case $x = v$ can also be proved in this way. □

If β is obtained from α by an $(\alpha \rightarrow \beta)$-recolouring, then the vertices can be ordered by their heights. Lemma 3 tells us that this ordering is the same for all $(\alpha \rightarrow \beta)$-recolourings and can be found by considering only α, β and paths in G. Let y be the vertex that is a median vertex in this ordering (if $|V|$ is even, arbitrarily choose one of the two vertices in the middle of the ordering). Let g be a function defined on V such that for all $v \in V$

$$g(v) = w(\beta, \overrightarrow{P_{vy}}) - w(\alpha, \overrightarrow{P_{vy}}).$$

Considering Lemma 3, we see that $g(v)$ is the height of v relative to y with respect to β, and that ordering the vertices by g is equivalent to ordering them by height so y is also a median of this ordering.

For any integer k congruent to $2(\beta(y) - \alpha(y))$ mod 6, let

$$J(k) = \sum_{v \in V} |k + g(v)|.$$

We observe that if k is the height of y, then $J(k)$ is the sum of the vertices' heights. Let (k_1, k_2) be the unique pair in the set $\{(0,0), (2,-4), (4,-2)\}$ such that $k_1 \equiv k_2 \equiv 2(\beta(y) - \alpha(y))$ mod 6. (Notice that, by Lemma 2, k_1 and k_2 are two possible values for the height of y when β is obtained by a recolouring sequence.)

Lemma 4. *Let $k \equiv 2(\beta(y) - \alpha(y))$ mod 6 be an integer. Then $J(k)$ is at least $\min\{J(k_1), J(k_2)\}$, and for any $(\alpha \rightarrow \beta)$-recolouring R, $|R| \geq \frac{1}{2} \min\{J(k_1), J(k_2)\}$.*

Lemma 5. *Let $k \equiv 2(\beta(y) - \alpha(y))$ mod 6 be an integer. If S is a recolouring sequence from α to c such that, for all $v \in V$, $h^S(c, v) = k + g(v)$, then $c = \beta$.*

Lemma 6. *Let $k \equiv 2(\beta(y) - \alpha(y))$ mod 6 be an integer. Then there exists an $(\alpha \rightarrow \beta)$-recolouring R of length ℓ such that $\ell = \frac{1}{2} J(k)$.*

Proof. We will define R by describing how to recolour from α to a colouring c such that $h^R(c, v) = k + g(v)$. By Lemma 5, $c = \beta$. Let $h(v)$ denote $k + g(v)$.

As we go from one colouring to the next we change the height of one vertex v by 2. If this change is always such that the difference between the current height of v and $k + g(v)$ is reduced by 2, then we will have $\ell = \frac{1}{2} J(k)$.

More definitions: for a vertex u in G and colouring c, a *maximal rising path* from u is a path on vertices $u = v_0, v_1, \dots v_t$ such that, for $1 \leq i \leq t$, $c(v_i) \equiv c(v_{i-1}) + 1$ mod 3, and v_t has no neighbours coloured $c(v_t) + 1$ mod 3. A maximal rising path can easily be found: we just repeatedly look for the next vertex along and if none with the required colour can be found we are done; we never return to a vertex that we have already met as this would mean we had found a fixed cycle. A *maximal falling path* from u is the same except that the colours decrease rather than increase moving along the path from u, and one can be found in an analogous way. (That is, the colours along a rising path are, for example, $231231231231 \cdots$, and along a falling path are, for example, $321321321321 \cdots$)

We need to describe how, at each step, to choose a vertex v to recolour and say what its "new" colour should be. Let c denote the current colouring and S the sequence of colourings found so far (so $h^S(c, x)$ is the current height of a vertex x).

1. Find a vertex x for which $|h(x) - h^S(c, x)|$ is maximum.
2. If $h(x) - h^S(c, x) > 0$, find a maximal rising path from x. Else find a maximal falling path from x. In either case, let v be the end-vertex of the path.
3. Change the colour of v so that $|h(v) - h^S(c, v)|$ is reduced by 2.

We must show that $h(v) \neq h^S(c, v)$ and that the new colouring is proper. We will treat the case that $h(x) - h^S(c, x) > 0$ (the other case is identical in form).

Let p be the number of edges in the maximal rising path P from x to v. Let $\overrightarrow{P}$ be the orientation from x to v. Applying Lemma 3 twice to x and v and then subtracting, we find that

$$h^S(c, x) = h^S(c, v) + w(c, \overrightarrow{P}) - w(\alpha, \overrightarrow{P}),$$
$$h(x) = h(v) + w(\beta, \overrightarrow{P}) - w(\alpha, \overrightarrow{P})$$
$$h(x) - h^S(c, x) = h(v) - h^S(c, v) + w(\beta, \overrightarrow{P}) - w(c, \overrightarrow{P}).$$

Note that $w(c, \overrightarrow{P}) = p$ and that $w(\beta, \overrightarrow{P}) \leq p$ since the weight of a path cannot be more than the number of edges. Thus $0 < h(x) - h^S(c, x) \leq h(v) - h^S(c, v)$ and so $h(v) > h^S(c, v)$. As reducing $|h(v) - h^S(c, v)|$ requires increasing the colour at v by 1, and it is at the end of a maximal rising path, the new colouring is proper.   □

Proof of Theorem 1. The algorithm FINDPATH(G, α, β) can be used to determine whether there is a path from α to β of length at most ℓ except when α and β are in the same component of $R_3(G)$ and G contains no fixed cycles with respect to α. In this case, a path of length ℓ can be found if and only if $\ell \leq \frac{1}{2} \min\{J(k_1), J(k_2)\}$. This follows from Lemmas 4 and 6.

Though the running time of FINDPATH is not analyzed in detail in [10], it is easy to prove that it is $O(n^2)$. We omit the details, but it is also straightforward to show that $J(k_1)$ and $J(k_2)$ can be found in time $O(n^2)$. Moreover, if one wishes to find the path from α to β this can be done by using the algorithm in the proof of Lemma 6 which can also be adapted to run in time $O(n^2)$.   □

3   An FPT Algorithm for k-Colouring Reconfiguration

In this section we will present our FPT algorithm for k-COLOURING RECONFIG-URATION when parameterized by ℓ. Let $G = (V, E)$ be a graph on n vertices, and let α, β be two proper k-colourings of G. First we prove three lemmas concerning the vertices that might be recoloured if a path between α and β of length at most ℓ does exist. That is, we assume that (G, α, β, ℓ) is a yes-instance of k-COLOURING RECONFIGURATION. This means that there exists an $(\alpha \rightarrow \beta)$-recolouring $R = c_0, \ldots, c_\ell$. We assume that R has *minimum length*.

We say that R *recolours* a vertex u if $c_q(u) \neq \alpha(u)$ for some q. Notice that if for each recoloured vertex u we find the least q such that $c_q(u) \neq \alpha(u)$, these values must be distinct (else c_q and c_{q-1} disagree on more than one vertex). Thus the number of distinct vertices recoloured by R is at most ℓ. We will prove something stronger. For $0 \leq q \leq \ell$, let W_q be the set of vertices on which c_0 and c_q disagree, that is, $W_q = \{u \in V : c_0(u) \neq c_q(u)\}$.

Lemma 7. *For all q with $1 \leq q \leq \ell$, the set W_q has size $|W_q| \leq q$.*

Proof. Suppose this is false and let r be the smallest value such that $|W_r| > r$. So $|W_{r-1}| \leq r-1$ (clearly $r-1 \geq 0$ as W_0 is the empty set). Then there are (at least) two vertices v_1, v_2 in $W_r \setminus W_{r-1}$, and so, for $i \in \{1,2\}$, $c_{r-1}(v_i) = c_0(v_i) \neq c_r(v_i)$, and c_r and c_{r-1} disagree on more than one vertex; a contradiction. □

For any $u \in V$, let $N(u)$ be the set of neighbours of u. For any $v \in N(u)$, let $N(u,v) = \{w \in N(u) : \alpha(w) = \alpha(v)\}$; that is, the set of neighbours of u with the same colour as v in α. Let $A_0 = \{v \in V : \alpha(v) \neq \beta(v)\}$ be the set of vertices on which α and β disagree. For $i \geq 1$, let $A_i = \bigcup_{u \in A_{i-1}} \{v \in N(u) : |N(u,v)| \leq \ell\}$. That is, to find A_i consider each vertex u in A_{i-1} and partition $N(u)$ into colour classes (according to the colouring α). Vertices in $N(u)$ that belong to colour classes of size at most ℓ belong to A_i. Note that two sets A_h and A_i need not be disjoint. Our first goal is to show that each vertex recoloured by R must be in $A^* = \bigcup_{h=0}^{\ell-1} A_h$. We will then show that the size of A^* is bounded by a function of $k + \ell$. This will then enable us to use brute-force to find R or some other $(\alpha \to \beta)$-recolouring of G (if it exists).

Lemma 8. *Each vertex recoloured by R belongs to A^*.*

Proof. For $i \geq 0$, let $L_i = A_i \setminus (\bigcup_{h<i} A_j)$ be the set of vertices that are in A_i but not in any A_h with $h < i$. Let z be the greatest value such that R recolours a vertex in L_z; denote this vertex by v_z. By definition, every vertex in A_0 is recoloured by R. Let $v_0 \in A_0$. We claim that also for $1 \leq i \leq z-1$, there is a vertex $v_i \in L_i$ that is recoloured by R. Then, as $v_0, \ldots, v_z$ are distinct vertices and R has length ℓ, we have $z \leq \ell - 1$ proving the lemma. For contradiction, assume there is a set L_i ($1 \leq i \leq z-1$) that contains no vertex recoloured by R.

From R we construct a new recolouring sequence R' by ignoring every recolouring step done to a vertex in $V \setminus \bigcup_{h<i} L_h$. For $0 \leq q \leq \ell$, let d_q be a colouring of G such that

- if $u \in \bigcup_{h<i} L_h$, $d_q(u) = c_q(u)$;
- if $u \notin \bigcup_{h<i} L_h$, $d_q(u) = \alpha(u)$.

Let R' be the sequence $d_0, \ldots, d_\ell$. Note that $d_0 = \alpha$, as $d_0(u)$ is either $c_0(u)$ or $\alpha(u)$, and $c_0 = \alpha$. Moreover, if $u \in \bigcup_{h<i} L_h = \bigcup_{h<i} A_i$ then $d_\ell(u) = c_\ell(u) = \beta(u)$, and if $u \notin \bigcup_{h<i} L_h$ then $d_\ell(u) = \alpha(u) = \beta(u)$ (since α and β only disagree on vertices in A_0); thus $d_\ell = \beta$. This means that if we can show that $d_1, \ldots, d_{\ell-1}$ are proper colourings, then R' is an $(\alpha \to \beta)$-recolouring. We will prove this first.

Assume to the contrary that R' contains a colouring d_q that is not proper. Then there is an edge uv with $d_q(u) = d_q(v)$. If u and v both belong to $\bigcup_{h<i} L_h$ then $c_q(u) = c_q(v)$, and if neither belong to $\bigcup_{h<i} L_h$ then $\alpha(u) = \alpha(v)$. Both cases are not possible, as c_q and α are proper colourings. Hence we may assume, without loss of generality, that $u \in \bigcup_{h<i} L_h$ and $v \notin \bigcup_{h<i} L_h$. Then $c_q(u) = d_q(u) = d_q(v) = \alpha(v)$ by the definition of d_q.

As $v \in N(u)$, the set $N(u,v)$ exists. First suppose $|N(u,v)| \leq \ell$. Then $v \in A_i$ by the definition of A_i. Hence $v \in L_h$ for some $h \leq i$. As $v \notin \bigcup_{h<i} L_h$, we obtain

$v \in L_i$. By assumption, no vertex of L_i is recoloured by R. Hence $c_q(v) = \alpha(v)$ and thus $c_q(u) = c_q(v)$ contradicting the fact that c_q is a proper k-colouring.

Now suppose $|N(u,v)| > \ell$. Because $c_q(u) = \alpha(v)$ and c_q is proper, we find that $c_q(w) \neq c_q(u) = \alpha(v) = \alpha(w)$ for all $w \in N(u,v)$. Thus $W_q \supseteq N(u,v)$ and so $|W_q| \geq |N(u,v)| > \ell \geq q$ contradicting the fact that $|W(q)| \leq q$ by Lemma 7. So, d_q must be proper. We conclude that R' is an $(\alpha \rightarrow \beta)$-recolouring of length ℓ.

We now proceed as follows. Recall that $v_z \in L_z$. Then there is a pair of colourings c_q and c_{q+1} that differ only on v_z. Because $v_z \in L_z$, $v_z \notin \bigcup_{h<i} L_h$. Hence, d_q and d_{q+1} are identical colourings. We remove d_q from R' to obtain another $(\alpha \rightarrow \beta)$-recolouring, which has length shorter than ℓ, contradicting that R has minimum length. This completes the proof.   □

Lemma 9 gives a bound on $|A^*|$ depending only on k and ℓ (proof omitted).

Lemma 9. *The set A^* has size $|A^*| \leq \ell \cdot (k\ell)^{\ell}$.*

We are now ready to present our FPT algorithm and prove Theorem 2.

Proof of Theorem 2. Let $k \geq 1$, and let (G, α, β, ℓ) be an instance of k-COLOURING RECONFIGURATION, where G is a graph on n vertices, and α, β are two proper k-colourings of G. Our algorithm does as follows. First compute the set A^* in $O(n^2)$ time. By Lemma 9, we find that $|A^*| \leq \ell \cdot (k\ell)^{\ell}$. By Lemma 8, we only have to search for a path of length at most ℓ in $R_k(G)$ among the vertices of A^*. By allowing consecutive recolourings to be equal we may restrict our search to $(\alpha \rightarrow \beta)$-recolourings of length exactly ℓ. Use brute force to enumerate all possible sequences of pairs (v_i, c_i), such that for all $0 \leq i \leq \ell - 1$, v_i is a vertex in A^* and c_i is a colour in $\{1, \ldots, k\}$. For each such sequence do as follows. Starting from α, recolour v_i with colour c_i for $i = 0, \ldots, \ell - 1$. As soon as this results in a k-colouring that is not proper, stop considering the sequence. If not, check whether the resulting colouring is equal to β. If this happens, then there is a path of length ℓ in $R_k(G)$. Hence, return **yes**. Otherwise, that is, if no sequence has this property, return **no**. Processing one sequence takes time $O(\ell n^2)$. By using Lemma 9, the number of sequences is at most $(|A^*| \cdot k)^{\ell} \leq ((\ell \cdot (k \cdot \ell)^{\ell}) \cdot k)^{\ell} \leq (k \cdot \ell)^{\ell^2 + \ell}$, leading to a total running time of $O((k \cdot \ell)^{\ell^2 + \ell} \cdot \ell n^2)$. This completes the proof.   □

4   A Lower Bound for Kernelization for $k \geq 4$

In this section we sketch the proof of Theorem 3, which states that k-COLOURING RECONFIGURATION parameterized by the maximum path length ℓ does not admit a polynomial kernelization for $k \geq 4$ unless NP $\subseteq$ coNP/poly. Theorem 3 is proved by a polynomial parameter transformation from the HITTING SET problem parameterized by the number m of sets in the input. It is known that this rules out polynomial kernels for the target problem, unless NP $\subseteq$ coNP/poly.

The main idea for the reduction is to create a 4-coloured tree that serves as a selection gadget for each set, which requires a recolouring at its root. This in turn requires a chain of earlier recolourings starting in one of the leaves; the selection of possible leaves encodes the elements of the set. Finally, recolouring any leaf requires a recolouring in a set of vertices corresponding to the ground set; this encodes the selection of a hitting set. Crucially, the height of the tree construction, which factors into the number ℓ of needed recolourings, can be bounded polynomially in the input parameter m.

The selection trees are composed of claws on four vertices a, b, c, d each, where c is the center vertex. For each of these vertices, α and β colour are the same, but we may (through adjacent gadgets) require a recolouring of d. The latter will be only possible by first recolouring a or b. To ensure this, several colours will be forbidden for a, b, c, d by adjacency to a global k-clique:

1. For a we have $\alpha(a) = \beta(a) = 2$, and, using adjacency to the k-clique, only colours 2 and 4 allow proper k-colourings.
2. For b we have $\alpha(b) = \beta(b) = 3$, and only colours 3 and 4 are possible.
3. For c we have $\alpha(c) = \beta(c) = 1$, and only colours 1, 2, and 3 are possible.
4. For d we have $\alpha(d) = \beta(d) = 4$, and only colours 1 and 4 are possible.

If we need to recolour d then it can only change to colour 1. This requires to first recolour c to either 2 or 3. This in turn, depending on choice of colour 2 or 3, necessitates a recolouring of a to 4 or b to 4. Thus, locally, we make a choice out of two options using constant number of recolourings. By building a tree structure from such claws, always making d-vertices of new claws adjacent to the a- or b-vertex of the current claw, we can make a one out of n choice at cost of $O(\log n)$ recolourings.

By standard arguments when reducing from a HITTING SET instance with m sets (recall that m is the parameter) we have a ground set size of $n \leq 2^m$. Thus, the choice of element to hit in each set costs only $O(\log n) = O(m)$ recolourings per set. To relate the different choices we make a set of n vertices that are adjacent to the corresponding leaves in each selection gadget. If we end up with a recolouring in a leaf of a selection gadget then this requires a recolouring of the corresponding one among these n vertices. By correct choice of number of recolourings and detailed analysis, we can enforce that at most p out of n vertices can be recoloured. Note that this involves also recolouring almost all vertices back to their initial colour since α and β will agree on almost all vertices (which is necessary to make the graph exponentially large in the parameter value). The whole recolouring from α to β is then possible within the chosen number of steps if and only if the given set family has a hitting set of size at most p.

5 Conclusions

We showed that k-COLOURING RECONFIGURATION is fixed-parameter tractable for any fixed $k \geq 1$, when parameterized by the number of recolourings ℓ. It is a natural question to ask whether a single-exponential FPT algorithm can be achieved for this problem. We also proved that the k-COLOURING RECONFIG-URATION problem is polynomial-time solvable for $k = 3$, which solves the open problem of Cereceda et al. [10], and that it has no polynomial kernel for all $k \geq 4$, when parameterized by ℓ (up to the standard assumption that $\mathsf{NP} \not\subseteq \mathsf{coNP/poly}$).

Acknowledgements. We are grateful to several reviewers for insightful comments that greatly improved our presentation.

References

1. Bonamy, M., Bousquet, N.: Recoloring bounded treewidth graphs. Electron. Notes Discrete Math. **44**, 257–262 (2013)
2. Bonamy, M., Johnson, M., Lignos, I.M., Patel, V., Paulusma, D.: Reconfiguration graphs for vertex colourings of chordal and chordal bipartite graphs. J. Comb. Optim. **27**, 132–143 (2014)
3. Bonsma, P.: The Complexity of rerouting shortest paths. In: Rovan, B., Sassone, V., Widmayer, P. (eds.) MFCS 2012. LNCS, vol. 7464, pp. 222–233. Springer, Heidelberg (2012)
4. Bonsma, P.: Rerouting shortest paths in planar graphs. In: D'Souza, D., Kavitha, T., Radhakrishnan, J. (eds.) IARCS Annual Conference on Foundations of Software Technology and Theoretical Computer Science (FSTTCS 2012), LIPIcs, pp. 337–349. Schloss Dagstuhl-Leibniz-Zentrum für Informatik, Wadern (2012)
5. Bonsma, P., Cereceda, L.: Finding paths between graph colourings: PSPACE-completeness and superpolynomial distances. Theor. Comput. Sci. **410**, 5215–5226 (2009)
6. Bonsma, P., Mouawad, A.E.: The complexity of bounded length graph recolouring, Manuscript (2014). arXiv:1404.0337
7. Bonsma, P., Kamiński, M., Wrochna, M.: Reconfiguring independent sets in claw-free graphs. In: Ravi, R., Gørtz, I.L. (eds.) SWAT 2014. LNCS, vol. 8503, pp. 86–97. Springer, Heidelberg (2014)
8. Cereceda, L., van den Heuvel, J., Johnson, M.: Connectedness of the graph of vertex-colourings. Discrete Math. **308**, 913–919 (2008)
9. Cereceda, L., van den Heuvel, J., Johnson, M.: Mixing 3-colourings in bipartite graphs. Eur. J. Comb. **30**, 1593–1606 (2009)
10. Cereceda, L., van den Heuvel, J., Johnson, M.: Finding paths between 3-colourings. J. Graph Theory **67**, 69–82 (2010)
11. Gopalan, P., Kolaitis, P.G., Maneva, E.N., Papadimitriou, C.H.: The connectivity of boolean satisfiability: computational and structural dichotomies. SIAM J. Comput. **38**, 2330–2355 (2009)
12. van den Heuvel, J.: The complexity of change. In: Blackburn, S.R., Gerke, S., Wildon, M. (eds.) Surveys in Combinatorics 2013. London Mathematical Society Lecture Note Series, pp. 127–160. Cambridge University Press, Cambridge (2013)
13. Ito, T., Kamiński, M., Demaine, E.D.: Reconfiguration of list edge-colorings in a graph. In: Dehne, F., Gavrilova, M., Sack, J.-R., Tóth, C.D. (eds.) WADS 2009. LNCS, vol. 5664, pp. 375–386. Springer, Heidelberg (2009)
14. Ito, T., Kawamura, K., Ono, H., Zhou, X.: Reconfiguration of list $L(2,1)$-labelings in a graph. In: Chao, K.-M., Hsu, T., Lee, D.-T. (eds.) ISAAC 2012. LNCS, vol. 7676, pp. 34–43. Springer, Heidelberg (2012)
15. Ito, T., Kawamura, K., Zhou, X.: An improved sufficient condition for reconfiguration of list edge-colorings in a tree. In: Ogihara, M., Tarui, J. (eds.) TAMC 2011. LNCS, vol. 6648, pp. 94–105. Springer, Heidelberg (2011)

16. Ito, T., Demaine, E.D.: Approximability of the subset sum reconfiguration problem. In: Ogihara, M., Tarui, J. (eds.) TAMC 2011. LNCS, vol. 6648, pp. 58–69. Springer, Heidelberg (2011)
17. Kamiński, M., Medvedev, P., Milanič, M.: Complexity of independent set reconfigurability problems. Theor. Comput. Sci. **439**, 9–15 (2012)
18. Kamiński, M., Medvedev, P., Milanič, M.: Shortest paths between shortest paths. Theor. Comput. Sci. **412**, 5205–5210 (2011)
19. Mouawad A.E., Nishimura, N., Raman, V.: Vertex cover reconfiguration and beyond, Manuscript (2014) arXiv:1402.4926
20. Mouawad, A.E., Nishimura, N., Raman, V., Simjour, N., Suzuki, A.: On the parameterized complexity of reconfiguration problems. In: Gutin, G., Szeider, S. (eds.) IPEC 2013. LNCS, vol. 8246, pp. 281–294. Springer, Heidelberg (2013)

Shortest Paths in Nearly Conservative Digraphs

Zoltán Király$^{(\boxtimes)}$

Department of Computer Science and Egerváry Research Group (MTA-ELTE),
Eötvös University, Pázmány Péter Sétány 1/C, Budapest, Hungary
kiraly@cs.elte.hu

Abstract. We introduce the following notion: a digraph $D = (V, A)$
with arc weights $c : A \to \mathbb{R}$ is called nearly conservative if every negative
cycle consists of two arcs. Computing shortest paths in nearly conservative digraphs is NP-hard, and even deciding whether a digraph is nearly
conservative is coNP-complete.

We show that the "All Pairs Shortest Path" problem is fixed parameter
tractable with various parameters for nearly conservative digraphs. The
results also apply for the special case of conservative mixed graphs.

Keywords: Conservative weights · All Pairs Shortest Paths · FPT algorithm · Mixed graph

1 Introduction

We are given a digraph $D = (V, A)$, a weight (or a length) function $c : A \to \mathbb{R}$
is called conservative (on D) if no directed cycle with negative total weight
("negative cycle" for short) exists, and c is called λ-**nearly conservative** if
every negative cycle consists of at most λ arcs.

The APSP (All Pairs Shortest Paths) problem we are going to solve has two
parts, first we must decide whether c is λ-nearly conservative, next, if the answer
for the previous question is YES, then for all (ordered) pairs $s \neq t$ of vertices the
task is to determine the length of the shortest (directed and simple) path from
s to t.

In this paper we concentrate on the case $\lambda = 2$, a 2-nearly conservative
weight function c is simply called *nearly conservative* in this paper. A mixed
graph $G = (V, E, A)$ on vertex set V has the set E of undirected edges and the
set A of directed edges (i.e., arcs). A weight function $c : E \cup A \to \mathbb{R}$ is called
conservative if no cycle with negative total weight exists. For a mixed graph we
can associate a digraph by replacing each undirected edge e having endvertices
u and v by two arcs uv and vu with weights $c(uv) = c(vu) = c(e)$. It is an easy
observation that the resulting c is nearly conservative on the resulting digraph if
and only if the original weight function was conservative on the original mixed
graph, and in this case the solution of the APSP problem remains the same.

Zoltán Király—Research was supported by grants (no. CNK 77780 and no. K
109240) from the National Development Agency of Hungary, based on a source from
the Research and Technology Innovation Fund.

M. Cygan and P. Heggernes (Eds.): IPEC 2014, LNCS 8894, pp. 234–245, 2014.
DOI: 10.1007/978-3-319-13524-3_20

Arkin and Papadimitriou proved in [1] that the problems of detecting negative cycles and finding the shortest path in the absence of negative cycles are both NP-hard in mixed graphs. Consequently, checking whether c is nearly conservative on D is coNP-complete, and solving the APSP problem in the case c is nearly conservative on D is NP-hard. In this paper we give FPT algorithms for this problem related to various parameters.

Though it was a surprise to the author, he could not find any algorithm for dealing with these problems (despite the fact that many paper are written about the Chinese Postman problem on mixed graphs). We only found two more papers that are somehow related to this topic. In [4] for the special case of skew-symmetric graphs shortest "regular" paths are found in polynomial time if no negative "regular" cycle exist. In [2] for the similar special case of bidirected graphs minimum mean edge-simple cycles are found in polynomial time, this is essentially the same as finding minimum mean "regular" cycles in skew-symmetric graphs. The class of nearly conservative graphs seems to be not studied (and defined) in the literature, as well as we could not find any FPT result about APSP.

For defining the parameters we are going to use, we first define the notion of negative trees. Given D and c, we associate an undirected graph $F = (V, E)$ as follows. Edge-set E consists of pairs $u \neq v$ of vertices for which both uv and vu are arcs in A, and $c(uv) + c(vu) < 0$. We can construct F in time $O(|A|)$, and can also check whether it is a forest. We claim that if F is not a forest, then c is not nearly conservative on D, so our algorithm can stop with this decision. If F contains a cycle, then it corresponds to two oppositely directed cycles of D, and the sum of the total weights of these two cycles are negative, proving that c is not nearly conservative.

From now on we will suppose that F is a forest, and we call its nontrivial components (that have at least one edge) the **negative trees**.

Our first parameter k_0 is the number of negative trees, and we give an $O(2^{k_0} \cdot n^4)$ algorithm for the APSP problem (where $n = |V|$). Later we refine this algorithm for parameter k_1, which is the maximum number of negative trees in any strongly connected component of D, and finally for parameter k_2, which is the maximum number of negative trees in any weakly 2-connected block of any strongly connected component of D (for the definitions see the next section). Our final algorithm also runs in time $O(2^{k_2} \cdot n^4)$. Consequently, if there is a constant γ such that every weakly 2-connected block of any strongly connected component of D has at most γ negative trees, then we have a polynomial algorithm.

The preliminary version of this paper appeared in [6] for the special case of mixed graphs. In that paper we also gave a strongly polynomial algorithm for finding shortest exact walk (a walk with given number of edges) in any non-conservative mixed graph.

2   Definitions

For our input digraph D we may assume it is simple. An arc from u to v is called a *loose arc* if there is another arc from u to v with a smaller weight. In a shortest path between s and t (if $s \neq t$) neither loops nor loose arcs can appear. Consequently, as a preprocessing, we can safely delete these (and also keep only one copy from multiple arcs having the same weight).

However for our purposes multiple arcs will be useful, so we will use them for describing the algorithm. We use the convention that the notation uv always refers to the shortest arc from u to v.

We call an arc uv of D **special** if vu is also an arc, and moreover $c(uv) + c(vu) < 0$. Other arcs are called *ordinary*. For a special arc uv the special arc vu is called its *opposite*. As a part of the preprocessing, we add some loose arcs to D. For every special arc uv we add an arc a from v to u with weight $c(a) = -c(uv)$. By the definition of special arcs, these are really loose arcs, as $-c(uv) > c(vu)$. We call these arcs *added ordinary arcs*, or shortly *loose arcs*. We call the improved digraph also D, and its arc set is called A. Arc set A is decomposed into $A = A_s \cup A_o$, where A_s is the set of special arcs, and A_o is the set of ordinary (original or added) arcs. (The main purpose of this procedure is the following. We will sometimes work in the ordinary subdigraph $D_o = (V, A_o)$, and we need to maintain the same reachability: if there is a path from s to t in D, then there is also a path from s to t in D_o.) Our main property remained true: if c is nearly conservative on D, then every negative cycle consists of two oppositely directed special arcs. Remark: special arcs may have positive length, so loose arcs may have negative length. We call a path *ordinary* if all its arcs are ordinary. Note that by the assumptions $|A| \leq 2n^2$, where $n = |V|$.

Given D and c, we associate an undirected graph $F = (V, E)$ as follows. Edge-set E consists of unordered pairs $u \neq v$ of vertices for which uv is a special arc in A_s. As we detailed in the Introduction, if F is not a forest, then c is not nearly conservative on D. We consider this process as the last phase of the preprocessing: we determine F, and if it is not a forest, then we stop with the answer "NOT NEARLY CONSERVATIVE".

From now on we suppose that F is a forest, and we call its nontrivial components (that have at least one edge) the **negative trees**. If T is a negative tree, then $\mathbf{V(T)}$ denotes its vertex set, and $\mathbf{A(T)}$ denotes the set of special arcs that correspond to its edges. If $s, t \in V(T)$ are two vertices of T, then $\mathbf{d^T(s, t)}$ denotes the length of unique path from s to t in $A(T)$.

A walk from v_0 to v_ℓ (or a $v_0 v_\ell$-walk) is a sequence

$$W = v_0, a_1, v_1, a_2, v_2, \ldots, v_{\ell-1}, a_\ell, v_\ell$$

where $v_i \in V$ for all i, and a_j is an arc from v_{j-1} to v_j for all j. A walk is closed if $v_0 = v_\ell$. A closed walk is also called here a $v_0 v_0$-walk. A number ℓ of arcs used by a walk W is denoted by $|W|$. The length (or weight) $c(W)$ of a walk W is defined as $\sum_{j=1}^{\ell} c(a_j)$. If W_1 is a $s_1 v$-walk and W_2 is a vt_2-walk, then their concatenation is denoted by $W_1 + W_2$. For a walk W we use the notation $W[v_i, v_j]$ for the corresponding part $v_i, a_{i+1}, \ldots, a_j, v_j$ if $i < j$.

A walk W is **special-simple** if no special arc is contained twice in it, moreover, if W contains special arc uv, then it does not contain its opposite vu. A walk is a *path* if all the vertices $v_0, \ldots, v_\ell$ are distinct. A closed walk is a *cycle* if all the vertices $v_0, \ldots, v_\ell$ are distinct, with the exception of $v_0 = v_\ell$. If $|W| = \ell = 0$, then we call the walk also an empty path (its length is 0), and in this paper unconventionally the empty path will also be considered as an empty cycle. The distance $d_D(s,t) = d(s,t)$ of t from s is the length of the shortest path from s to t (where $s, t \in V$).

The relation: there is a path in D from s to t and also from t to s, is obviously an equivalence relation, its classes are called the strongly connected components of D. (Notice that a negative tree always resides in one strongly connected component.) A weakly 2-connected block of a digraph is a 2-connected block of the underlying undirected graph (where arcs are replaced with undirected edges).

An algorithm is FPT for a problem with input size n and parameter k if there is an absolute constant γ, and a function f such that the running time is $f(k) \cdot O(n^\gamma)$. (Originally FPT stands for "fixed parameter tractable", and it is an attribute of the problem, however in the literature usually the corresponding algorithms are also called FPT.) In this paper we give FPT algorithms for the APSP problem for nearly conservative digraphs.

In the simplest version we assume that there is just one negative tree and it is spanning V. Next we give an algorithm for the case where we still have only one negative tree, but it is not spanning V. These algorithms are polynomial and simple.

Then we use various parameters: k_0 is the number of negative trees in D, k_1 is the maximum number of negative trees in any strongly connected component of D, and k_2 is the maximum number of negative trees in any weakly 2-connected block of any strongly connected component of D. (Clearly $k_0 \geq k_1 \geq k_2$.) The main goal of this paper to give an $O(2^{k_2} \cdot n^4)$ algorithm for the APSP problem for the case $\lambda = 2$, i.e., for deciding whether c is nearly conservative on D, and if it is, then for calculating the distances $d_D(s,t)$ for each (ordered) pair of vertices $s, t \in V$.

In the next section we show some lemmas. In Sect. 4 we give some polynomial algorithms for the case of one negative tree. In Sect. 5 we give an FPT algorithm where the parameter k_0 is the total number of negative trees in D. Next, in Sect. 6 we extend it to the case where k_2 only bounds the number of negative trees in any weakly 2-connected block of any strongly connected component.

Our main goal is only giving the length of the shortest paths, in Sect. 7 we detail how the actual shortest paths themselves can be found.

Finally in Sect. 8 we conclude the results, show their consequences to mixed graphs, and pose some open problems.

3 Lemmas

In this section we formulate some lemmas. Though each of them can be easily proved using the newly introduced notions and the statements of the preceding lemmas, we could not find these statements in the literature (neither in an implicit form).

We premise some unusual aspects of nearly conservative weight functions. Usually shortest path algorithms use the following two facts about conservative weight functions. If P is a shortest sx-path and Q is a shortest xt-path, then $P+Q$ contains an st-path not longer than $c(P)+c(Q)$. If P is a shortest st-path containing vertices u and v (in this order), then $P[u,v]$ is a shortest uv-path. These two statements are NOT true for nearly conservative weight functions.

Remember that $D = (V, A_s \cup A_o)$ is the improved digraph with loose arcs, and the associated graph F is a forest.

Lemma 1. *Weight function c is nearly conservative on D if and only if there is no negative special-simple closed walk.*

Proof. If C is a negative cycle consisting of at least three arcs, then it is also a negative special-simple closed walk. On the other hand, suppose that C is a negative special-simple closed walk with a minimum number of arcs, and assume that C is not a cycle, that is there are $0 < i < j \le \ell$ such that $v_i = v_j$. Now C decomposes into two special-simple closed walks with less arcs, clearly at least one of them has negative length, a contradiction.   □

Lemma 2. *If c is nearly conservative on D, and $s, t \in V$, and Q is a special-simple st-walk, then we also have an st-path P with $c(P) \le c(Q)$, and P contains only arcs of Q.*

Proof. Let Q be a shortest special-simple st-walk (which exists by the previous lemma and as c is nearly conservative) having the minimum number of arcs.

By the previous lemma, if $s = t$, then the empty path serves well as P. So we may assume that $s \ne t$ and Q is not a path, i.e., there are $0 \le i < j \le \ell$ such that $v_i = v_j$. Now Q decomposes to a special-simple sv_i-walk Q_1, a special-simple closed walk C through v_i and an special-simple $v_j t$-walk Q_2. By the previous lemma C is nonnegative, so $c(Q_1 + Q_2) \le c(Q)$, consequently $Q_1 + Q_2$ is a not longer special-simple st-walk with less number of arcs, a contradiction.   □

Suppose T is a negative tree, $u, v \in V(T)$, and P is a uv-path in $D' = D - A(T)$. If $c(P) < -d^T(v, u)$, then c is not nearly conservative on D because otherwise $P + P^T_{vu}$ would be a negative special-simple closed walk, where P^T_{vu} is the vu-path in $A(T)$. Otherwise, if $c(P) \ge -d^T(v, u)$, then we have a uv-path P' in D' consisting of loose arcs such that $c(P') \le c(P)$. Using this train of thought we get the following lemmas that play the central role in our algorithms.

Lemma 3. *Let T be a negative tree, and assume that c is nearly conservative on D. If P is a shortest st-path using some vertex of $V(T)$, then let u be the first vertex of P in $V(T)$, and let v be the last vertex of P in $V(T)$. Then $P[u,v]$ uses only special arcs from $A(T)$. Consequently, if $s, t \in V(T)$, then $d(s,t) = d^T(s,t)$.*

Proof. Remember that a uv-path in $A(T)$ may have positive length. Fortunately, by the definition of u and v, there are no vertices of P preceding u or following v inside $V(T)$, and this fact can be used successfully.

Suppose P is a shortest st-path. By the observation made before the lemma, for any $u', v' \in V(T)$, any subpath of form $P[u', v']$ that uses no arcs from $A(T)$ can be replaced by loose arcs without increasing the length. After we made all these replacements, we replaced $P[u, v]$ by a special-simple uv-walk Q' such that Q' contains only arcs in $A(T)$ and loose arcs, and $c(Q') \leq c(P[u, v])$. By Lemma 2, Q' contains a uv-path P' with $c(P') \leq c(Q')$. We got P' by eliminating cycles, if any cycle had positive length, then we get $c(P') < c(Q')$. Suppose now that all eliminated cycles had zero length, meaning that each one had the form x, a, y, yx, x, where a is the loose arc from x to y and yx is the special arc from y to x. If after deleting all these cycles P' still has a loose arc a from x to y then it can be replaced safely with the special arc xy yielding again a path strictly shorter than Q'. Thus the only possibility where we can only get a P' with the same length (as Q') is that the special-simple uv-walk Q' consisted of the uv-path P_{uv}^T inside $A(T)$ and additionally some zero length cycle described above, and moreover $P' = P_{uv}^T$. Now we claim that in this case the path $P[u, v]$ used only arcs from $A(T)$, i.e., it was also P_{uv}^T. Otherwise there are vertices $x, y \in V(T)$ such that x is on P_{uv}^T, y is not on it, and Q' contains one loose arc and one special arc between x and y. However in this case vertex x had to be included twice in path P, a contradiction.

To finish the proof observe that $P[s, u] + P' + P[v, t]$ is an st-path, and in the case $P[u, v] \neq P_{uv}^T$ it would be shorter than the shortest path P.   □

Lemma 4. *Let T be a negative tree, and assume that c is nearly conservative on digraph $D' = D - A(T)$ defining distance function d'. Then c is nearly conservative on D if and only if for any pair of vertices $u, v \in V(T)$ we have $d'(u, v) \geq -d^T(v, u)$.*

Proof. We showed that the condition is necessary. Suppose that C is a negative cycle in D having at least three arcs. If it has at most one vertex in $V(T)$, then it is also a negative cycle in D'. We claim that we can construct a special-simple negative closed walk C' which uses only loose arcs and arcs in $A(T)$. To achieve this goal, repeatedly take any subpath $C[u, v]$, where $u, v \in V(T)$, but inner vertices of $C[u, v]$ are in $V - V(T)$. By the condition $c(C[u, v]) \geq -d^T(v, u)$, which means that changing $C[u, v]$ to the uv-path consisting of loose arcs does not increase the length of C. We arrived at a contradiction, as the special-simple closed walk C' contains a negative cycle which is impossible by the definition of loose arcs.   □

4 Polynomial Algorithms for the Case $k_0 = 1$

First we give an $O(n^2)$ algorithm for the very restricted case, where we have only one negative tree T, and moreover it spans V. We claim first that c is nearly conservative on D if and only if for each ordinary arc uv we have $c(uv) \geq -d^T(v, u)$. If $c(uv) < -d^T(v, u)$, then we have a negative special-simple closed walk, so c is not nearly conservative by Lemma 1. Suppose now that $c(uv) \geq -d^T(v, u)$ holds for each ordinary arc uv, and C is a negative cycle in D with at least three arcs. As in the proof of Lemma 4, replace each ordinary arc uv of C by a uv-path consisting of loose arcs, this does not increase the length. We arrive at special-simple closed walk using only special and loose arcs that is negative. However this contradicts to the definition of loose arcs. We also got that in this case for any pair $s, t \in V$ the length of the shortest path is $d^T(s, t)$ by Lemma 3. Consequently it is enough to give an $O(n^2)$ algorithm for calculating distances $d^T(s, t)$. We suppose that $V = \{1, \ldots, n\}$ and initialize a length-n all-zero array D_u for each vertex u. Then we fill up these arrays in a top-down fashion starting from the root vertex 1. Let $\mathcal{P}$ denote the subset of vertices already processed, initially it is $\{1\}$. If a parent u of an unprocessed vertex v is already processed, we process v: for each processed vertex x we set $D_v(x) = c(vu) + D_u(x)$, and set $D_x(v) = D_x(u) + c(uv)$, and put v into $\mathcal{P}$.

Next we give an $O(n^4)$ algorithm for the case where we have only one negative tree T, but we do not assume it to span V.

In digraph $D' = D - A(T) = D_o = (V, A_o)$ using the Floyd-Warshall algorithm (see in any lecture notes, e.g., in [3]), it is easy to check whether c is conservative on D' in time $O(n^3)$. If it is not conservative, then we return with output "NOT NEARLY CONSERVATIVE" (as in this case c clearly cannot be nearly conservative on D), and if it is conservative, then this algorithm also calculates the length $d'(s, t)$ of all shortest paths in D' (for $s, t \in V$). If vertex t is not reachable from vertex s, then it gives $d'(s, t) = +\infty$ (remember that reachability is the same in D' as in D).

Then we calculate the distances $d^T(u, v)$ in time $O(n^2)$ as in the previous section. By Lemma 4, c is nearly conservative on D if and only if for all pairs $u, v \in V_T$ we have $d'(u, v) \geq -d^T(u, v)$, this can be checked in time $O(n^2)$. It remains to calculate the pairwise distances. If P is a shortest st-path, then it is either a ordinary path (having length $d'(s, t)$), or it has a first arc $uu' \in A(T)$ and a last arc $v'v \in A(T)$. The part $P[u, v]$ must reside inside $A(T)$ by Lemma 3.

Lemma 5. *If c is nearly conservative on D, and T is the only negative tree, then the distance $d(s, t)$ is*

$$d(s, t) = \min\left(d'(s, t), \min_{u, v \in V_T}[d'(s, u) + d^T(u, v) + d'(v, t)]\right).$$

Proof. This is a consequence of Lemma 3. The trick used here is that a shortest su-path and a shortest vt-path in D' need not be arc-disjoint, this is the main purpose for which we introduced the notion of special-simple, so for the relation LHS $\leq$ RHS we have to use Lemma 2. $\qquad\square$

These values can be easily calculated for all pairs in total time $O(n^4)$, so we are done.

5   FPT Algorithm for Parameter k_0

In this section we suppose that there are at most k_0 negative trees in D. Let $T_1, \ldots, T_{k_0}$ be the negative trees, remember that we defined $A(T_i)$ as the set of special arcs that correspond to the edges of T_i. We denote by V_T the vertex set $\bigcup_i V(T_i)$.

First we compute distances d^{T_i} for all $1 \leq i \leq k_0$ in total time $\sum O(|V(T_i)|^2) = O(n^2)$. Next we compute distances d' in digraph $D' = D - \bigcup_i A(T_i) = D_o$ in time $O(n^3)$, or stop if c is not nearly conservative on D'.

We use dynamic programming for the calculation remained. For all $J \subseteq \{1, \ldots, k_0\}$ we define the J-subproblem as follows. Solve the APSP problem in digraph $D_J = D - \bigcup_{i \in \{1, \ldots, k_0\} - J} A(T_i)$, and let d_J denote the corresponding distance function if c is nearly conservative on D_J (otherwise, if c is not nearly conservative on D_J for any J, we stop). We already solved the $\emptyset$-subproblem, $d_\emptyset \equiv d'$.

Lemma 6. *Suppose we solved the $(J - i)$-subproblem for every $i \in J$ and found that c is nearly conservative on D_{J-i}. By Lemma 4, we can check whether c is conservative on D_J using only distance functions d^{T_i} and d_{J-i} for one element $i \in J$. If yes, then we have*

$$d_J(s,t) = \min\left(d_\emptyset(s,t), \min_{i \in J}[\min_{u,v \in V(T_i)} (d_\emptyset(s,u) + d^{T_i}(u,v) + d_{J-i}(v,t))] \right)$$

Proof. First we show that LHS $\geq$ RHS. Let P be a shortest path in D_J. Either P is disjoint from $\bigcup_{j \in J} V(T_j)$, in this case its length is $d_\emptyset(s,t)$ in graph D_J. The other possibility is that P has some first vertex u in $\bigcup_{j \in J} V(T_j)$, say $u \in V(T_i)$. Let v denote the last vertex of P in $V(T_i)$. That is, $P[s,u]$ goes inside $D_\emptyset$ and $P[v,t]$ goes inside D_{J-i}, and, by Lemma 3, $P[u,v]$ goes inside $A(T_i)$.

To show that LHS $\leq$ RHS we only need to observe that if P_1 is an su-path in $D_\emptyset$, P_2 is a uv-path in $A(T_i)$, and P_3 is a vt-path in D_{J-i}, then $P_1 + P_2 + P_3$ is a special-simple st-walk.

Remember that, 'solving the APSP problem' is defined in this paper as first checking nearly conservativeness, and if c is nearly conservative, then calculate all shortest paths. As solving one subproblem needs $O(n^4)$ steps, we proved the following.

Theorem 1. *If D has k_0 negative trees, then the dynamic programming algorithm given in this section correctly solves the APSP problem in time $O(2^{k_0} \cdot n^4)$.*

The weak blocks of a digraph refer to the 2-connected blocks of the underlying undirected graph. It is well known that the block-tree of an undirected graph can be determined in time $O(n^2)$ by DFS. If we have this decomposition and we also calculated APSP inside every weak block, then we can also calculate APSP for the whole digraph in additional time $O(n^3)$. Consequently we have

Corollary 1. *If every weak block of D contains at most k_0' negative trees, then we can solve the APSP problem in time $O(2^{k_0'} \cdot n^4)$.*

6 General FPT Algorithm for Parameters k_1 and k_2

Suppose every strongly connected component of D contains at most k_1 negative trees. By the previous section we can solve the APSP problem inside each
strongly connected component in total time $O(2^{k_1} \cdot n^4)$. If for any of them we
found that c is not nearly conservative, then we stop and report the fact that
c is not nearly conservative on D. Henceforth in this section we assume that
for every strongly connected component K of D, c is nearly conservative on K.
(In this situation clearly c is nearly conservative on D.) The distance function
restricted to component K is denoted by d_K. If $s, t \in V(K)$, then every st-path
goes inside K, thus $d(s,t) = d_K(s,t)$. It remains to calculate APSP in D for
pairs s, t, that are in different strongly connected components.

We construct a new acyclic digraph D^* by first substituting every strongly
connected component K by acyclic digraph D_K^* as follows. Suppose $V(K) =
\{x_1^K, x_2^K, \ldots, x_r^K\}$, the vertex set of D_K^* will consist of $2r$ vertices, $\{a_1^K,
a_2^K, \ldots, a_r^K, b_1^K, b_2^K, \ldots, b_r^K\}$. For each $1 \le i, j \le r$ the digraph D_K^* contains
arc $a_i^K b_j^K$ with length $d_K(x_i^K, x_j^K)$.

In order to finish the construction of D^*, for every arc $x_i^K x_j^L$ of D connecting
two different strongly connected components $K \ne L$, digraph D^* contains the
arc $b_i^K a_j^L$ with length $c(x_i^K x_j^L)$. It is easy to see that D^* is truly acyclic and
has $2n$ vertices. As D^* is a simple digraph, paths can be given by only listing
the sequence of its vertices. We can calculate APSP in D^* in time $O(n^3)$ by the
method of Morávek [7] (see also in [3]) if we run this famous algorithm from all
possible sources s. It gives distance function d_{D^*} (where if t is not reachable from
s, then we write $d_{D^*}(s,t) = +\infty$). The total running time is still $O(2^{k_1} \cdot n^4)$.
We remark that if every strongly connected component has a spanning negative
tree, then the running time is $O(n^3)$.

Theorem 2. *Suppose $s = x_{i_0}^{K_0} \in V(K_0)$ and $t = x_{j_r}^{K_r} \in V(K_r)$ where $K_0 \ne K_r$
are different strongly connected components of D. Then the shortest st-path in
D has length exactly $d_{D^*}(a_{i_0}^{K_0}, b_{j_r}^{K_r})$.*

Proof. Vertex t is not reachable from s in D if and only if $b_{j_r}^{K_r}$ is not reachable
from $a_{i_0}^{K_0}$ in D^*. Otherwise, suppose that $a_{i_0}^{K_0}, b_{j_0}^{K_0}, a_{i_1}^{K_1}, b_{j_1}^{K_1}, \ldots, a_{i_r}^{K_r}, b_{j_r}^{K_r}$ is a
shortest path P in D^*. For $0 \le \ell \le r$ let path P_ℓ be a shortest path in D from
$x_{i_\ell}^{K_\ell}$ to $x_{j_\ell}^{K_\ell}$, this path obviously goes inside K_ℓ. We can construct an st-path Q
in D with the same length as P has in D^*: $Q = P_0 + x_{j_0}^{K_0} x_{i_1}^{K_1} + P_1 + x_{j_1}^{K_1} x_{i_2}^{K_2} +
P_2 + \cdots + P_{r-1} + x_{j_{r-1}}^{K_{r-1}} x_{i_r}^{K_r} + P_r$.

For the other direction, suppose that there are strongly connected components $K_0, K_1, \ldots, K_r$, such that the shortest st-path Q in D meets these components in this order, and for all ℓ the path Q arrives into K_ℓ at vertex $x_{i_\ell}^{K_\ell}$ and
leaves K_ℓ at vertex $x_{j_\ell}^{K_\ell}$. As Q is a shortest path it clearly contains a path of
length $d_{K_\ell}(x_{i_\ell}{}^{K_\ell}, x_{j_\ell}^{K_\ell})$ inside K_ℓ for each ℓ, consequently the following path has
the same length in D^*: $P = a_{i_0}^{K_0}, b_{j_0}^{K_0}, a_{i_1}^{K_1}, b_{j_1}^{K_1}, \ldots, a_{i_r}^{K_r}, b_{j_r}^{K_r}$. $\square$

Using Corollary 1 we easily get the following more general statements.

Corollary 2. *If every weak block of any strongly connected component of D contains at most k_2 negative trees, then we can solve the APSP problem in time $O(2^{k_2} \cdot n^4)$.*

Corollary 3. *If there is an absolute constant γ, such that in any weak block of any strongly connected component of D there are at most γ negative trees, then there is a polynomial time algorithm for the APSP problem that runs in time $O_\gamma(n^4)$.*

7 Finding the Paths

In this section we assume that c is nearly conservative on D.

We usually are not only interested in the lengths of the shortest paths, but also some (implicit) representation of the paths themselves. The requirement for this representation is that for any given s and t, one shortest st-path P must be computable from it in time $O(\ell)$ if ℓ is the number of arcs in P.

It is well known (see e.g., in [3]) that both the algorithm of Floyd and Warshall and the algorithm of Morávek can compute predecessor matrices Π (by increasing the running time by a constant factor only), with the property that for each $s \neq t$ the entry $\Pi(s,t)$ points to the last-but-one vertex of a shortest st-path. This representation clearly satisfies the requirement described in the previous paragraph.

For a digraph H let Π_H denote the predecessor matrix of this type, and suppose that for each strongly connected component K we computed Π_K, and we also computed Π_{D^*}. Then Π_D is easily computable as follows. Suppose that $s = x_{i_0}^{K_0}$ and $t = x_{j_r}^{K_r}$, and $\Pi_{D^*}(a_{i_0}^{K_0}, b_{j_r}^{K_r}) = a_{i_r}^{K_r}$. If $i_r \neq j_r$, then define $\Pi_D(s,t) = \Pi_{K_r}(x_{i_r}^{K_r}, x_{j_r}^{K_r})$, otherwise let $b_{j_{r-1}}^{K_{r-1}} = \Pi_{D^*}(a_{i_0}^{K_0}, a_{i_r}^{K_r})$ and define $\Pi_D(s,t) = x_{j_{r-1}}^{K_{r-1}}$.

It remained to compute the predecessor matrices Π_K in the case where K is a strongly connected component of D. In accordance with Sect. 5 from now on we call K as D (and forget the other vertices of the digraph), and the matrix we are going to determine is simply Π.

If s and t are vertices of the same negative tree T_i, then the method given in the first paragraph of Sect. 4 easily calculates $\Pi(s,t) = \Pi_{A(T_i)}(s,t)$. Next we call the Floyd-Warshall algorithm on D', and it can give $\Pi_{D'}$, then during the dynamic programming algorithm we determine matrices Π_{D_J} for all J.

Given s and t, by Lemma 6 if the minimum is $d_\emptyset(s,t)$, then $\Pi_{D_J}(s,t) = \Pi_{D_\emptyset}(s,t)$, otherwise we find i, u, v giving the minimum value. If $v \neq t$, then $\Pi_{D_J}(s,t) = \Pi_{D_{J-i}}(s,t)$, otherwise if $v = t$ but $u \neq v$, then $\Pi_{D_J}(s,t) = \Pi_{A(T_i)}(s,t)$, and finally if $v = t = u \neq s$ then $\Pi_{D_J}(s,t) = \Pi_{D_\emptyset}(s,t)$.

Extending this setup for weak blocks is obvious.

8   Conclusion and Open Problems

We gave FPT algorithms for the NP-hard APSP problem in nearly conservative graphs regarding with various parameters.

For mixed graphs we have the following consequence. As nonnegative undirected edges can be replaced by two opposite arcs, we may assume that every undirected edge has negative length. Here the negative trees are the nontrivial components made up by undirected edges, and APSP problem is to check whether c is conservative on a mixed graph G, and if YES, then calculate the pairwise distances.

Remember, that for mixed graphs the APSP problem contains checking conservativeness, and if c is conservative on the mixed graph, then all shortest paths should be calculated.

Corollary 4. *If every weak block of any strongly connected component of a mixed graph contains at most k_2 negative trees, then we can solve the APSP problem in time $O(2^{k_2} \cdot n^4)$.*

Finally we pose three open problems. A weight function is even-nearly conservative if every negative cycle consist of an even number of arcs.

Question 1. Is there an FPT algorithm for shortest paths if c is 3-nearly conservative? (The parameter should not contain the number of negative triangles.)

Question 2. Is there a polynomial or FPT algorithm for recognizing even-nearly conservative weights? This would be interesting even if we restrict the digraph to be symmetric (i.e., every arc has its opposite).

Question 3. Is there an FPT algorithm for shortest paths if c is λ-nearly conservative, using some parameter k of "inconvenient components" (should be defined accordingly) and also λ?

Acknowledgment. The author is thankful to András Frank who asked a special case of this problem, and also to Dániel Marx who proposed the generalization to nearly conservative digraphs.

References

1. Arkin, E.M., Papadimitriou, C.H.: On negative cycles in mixed graphs. Oper. Res. Lett. **4**(3), 113–116 (1985)
2. Babenko, M.A., Karzanov, A.V.: Minimum mean cycle problem in bidirected and skew-symmetric graphs. Discrete Optim. **6**, 92–97 (2009)
3. Cormen, T.H., Leiserson, C.E., Rivest, R.L., Stein, C.: Introduction to Algorithms, 3rd edn. MIT Press, Cambridge (2009)
4. Goldberg, A.V., Karzanov, A.V.: Path problems in skew-symmetric graphs. Combinatorica **16**(3), 353–382 (1996)

5. Karp, R.M.: A characterization of the minimum cycle mean in a digraph. Discrete Math. **23**, 309–311 (1978)
6. Király, Z.: Shortest paths in mixed graphs. Egres Technical Report TR-2012-20. http://www.cs.elte.hu/egres/
7. Morávek, J.: A note upon minimal path problem. J. Math. Anal. Appl. **30**, 702–717 (1970)

Reconfiguration over Tree Decompositions

Amer E. Mouawad[1]([✉]), Naomi Nishimura[1],
Venkatesh Raman[2], and Marcin Wrochna[3]

[1] David R. Cheriton School of Computer Science,
University of Waterloo, Ontario, Canada
{aabdomou,nishi}@uwaterloo.ca
[2] Institute of Mathematical Sciences, Chennai, India
vraman@imsc.res.in
[3] Institute of Computer Science, Uniwersytet Warszawski, Warsaw, Poland
mw290715@students.mimuw.edu.pl

Abstract. A vertex-subset graph problem Q defines which subsets of the vertices of an input graph are feasible solutions. The reconfiguration version of a vertex-subset problem Q asks whether it is possible to transform one feasible solution for Q into another in at most ℓ steps, where each step is a vertex addition or deletion, and each intermediate set is also a feasible solution for Q of size bounded by k. Motivated by recent results establishing W[1]-hardness of the reconfiguration versions of most vertex-subset problems parameterized by ℓ, we investigate the complexity of such problems restricted to graphs of bounded treewidth. We show that the reconfiguration versions of most vertex-subset problems remain PSPACE-complete on graphs of treewidth at most t but are fixed-parameter tractable parameterized by $\ell + t$ for all vertex-subset problems definable in monadic second-order logic (MSOL). To prove the latter result, we introduce a technique which allows us to circumvent cardinality constraints and define reconfiguration problems in MSOL.

1 Introduction

Reconfiguration problems allow the study of structural and algorithmic questions related to the solution space of computational problems, represented as a *reconfiguration graph* where feasible solutions are represented by nodes and adjacency by edges [6,16,18]; a path is equivalent to the step-by-step transformation of one solution into another as a *reconfiguration sequence* of *reconfiguration steps*.

Reconfiguration problems have so far been studied mainly under classical complexity assumptions, with most work devoted to deciding whether it is possible to find a path between two solutions. For several problems, this question has been shown to be PSPACE-complete [4,18,19], using reductions that construct examples where the length ℓ of reconfiguration sequences can be exponential in the size of the input graph. It is therefore natural to ask whether we can

A.E. Mouawad and N. Nishimura—Research supported by the Natural Science and Engineering Research Council of Canada.

© Springer International Publishing Switzerland 2014
M. Cygan and P. Heggernes (Eds.): IPEC 2014, LNCS 8894, pp. 246–257, 2014.
DOI: 10.1007/978-3-319-13524-3_21

achieve tractability if we allow the running time to depend on ℓ or on other properties of the problem, such as a bound k on the size of feasible solutions. These results motivated Mouawad et al. [22] to study reconfiguration under the *parameterized complexity* framework [12], showing the W[1]-hardness of VERTEX COVER RECONFIGURATION (VC-R), FEEDBACK VERTEX SET RECONFIGURATION (FVS-R), and ODD CYCLE TRANSVERSAL RECONFIGURATION (OCT-R) parameterized by ℓ, and of INDEPENDENT SET RECONFIGURATION (IS-R), INDUCED FOREST RECONFIGURATION (IF-R), and INDUCED BIPARTITE SUBGRAPH RECONFIGURATION (IBS-R) parameterized by $k + \ell$ [22].

Here we focus on reconfiguration problems restricted to $\mathscr{C}_t$, the class of graphs of treewidth at most t. In Sect. 3, we show that a large number of reconfiguration problems, including the six aforementioned problems, remain PSPACE-complete on $\mathscr{C}_t$, answering a question left open by Bonsma [5]. The result is in fact stronger in that it applies to graphs of bounded bandwidth and even to the question of finding a reconfiguration sequence of *any* length.

In Sect. 4, using an adaptation of Courcelle's cornerstone result [8], we present a meta-theorem proving that the reconfiguration versions of all vertex-subset problems definable in monadic second-order logic become tractable on $\mathscr{C}_t$ when parameterized by $\ell + t$. Since the running times implied by our meta-theorem are far from practical, we consider the reconfiguration versions of problems defined in terms of hereditary graph properties in Sect. 5. In particular, we first introduce signatures to succinctly represent reconfiguration sequences and define "generic" procedures on signatures which can be used to exploit the structure of nice tree decompositions. We use these procedures in Sect. 5.2 to design algorithms solving VC-R and IS-R in $\mathcal{O}^\star(4^\ell(t+1)^\ell)$ time (the $\mathcal{O}^\star$ notation suppresses factors polynomial in n, ℓ, and t). In Sect. 5.4, we extend the algorithms to solve OCT-R and IBS-R in $\mathcal{O}^\star(2^{\ell t}4^\ell(t+1)^\ell)$ time, as well as FVS-R and IF-R in $\mathcal{O}^\star(t^{\ell t}4^\ell(t+1)^\ell)$ time. We further demonstrate in Sect. 5.3 that VC-R and IS-R parameterized by ℓ can be solved in $\mathcal{O}^\star(4^\ell(3\ell+1)^\ell)$ time on planar graphs by an adaptation of Baker's shifting technique [1].

2   Preliminaries

For general graph theoretic definitions, we refer the reader to the book of Diestel [11]. We assume that each input graph G is a simple undirected graph with vertex set $V(G)$ and edge set $E(G)$, where $|V(G)| = n$ and $|E(G)| = m$. The *open neighborhood* of a vertex v is denoted by $N_G(v) = \{u \mid uv \in E(G)\}$ and the *closed neighborhood* by $N_G[v] = N_G(v) \cup \{v\}$. For a set of vertices $S \subseteq V(G)$, we define $N_G(S) = \{v \notin S \mid uv \in E(G), u \in S\}$ and $N_G[S] = N_G(S) \cup S$. We drop the subscript G when clear from context. The subgraph of G induced by S is denoted by $G[S]$, where $G[S]$ has vertex set S and edge set $\{uv \in E(G) \mid u, v \in S\}$. Given two sets $S_1, S_2 \subseteq V(G)$, we let $S_1 \Delta S_2 = \{S_1 \setminus S_2\} \cup \{S_2 \setminus S_1\}$ denote the symmetric difference of S_1 and S_2.

We say a graph problem Q is a *vertex-subset* problem whenever feasible solutions for Q on input G correspond to subsets of $V(G)$. Q is a *vertex-subset minimization (maximization)* problem whenever feasible solutions for Q correspond to subsets of $V(G)$ of size at most (at least) k, for some integer k. The *reconfiguration graph* of a vertex-subset minimization (maximization) problem Q, $R_{\mathrm{MIN}}(G, k)$ ($R_{\mathrm{MAX}}(G, k)$), has a node for each $S \subseteq V(G)$ such that $|S| \leq k$ ($|S| \geq k$) and S is a feasible solution for Q. We say k is the *maximum (minimum) allowed capacity* for $R_{\mathrm{MIN}}(G, k)$ ($R_{\mathrm{MAX}}(G, k)$). Nodes in a reconfiguration graph are adjacent if they differ by the addition or deletion of a single vertex.

Definition 1. *For any vertex-subset problem Q, graph G, positive integers k and ℓ, $S_s \subseteq V(G)$, and $S_t \subseteq V(G)$, we define four decision problems:*

- Q-MIN(G, k): *Is there $S \subseteq V(G)$ such that $|S| \leq k$ and S is a feasible solution for Q?*
- Q-MAX(G, k): *Is there $S \subseteq V(G)$ such that $|S| \geq k$ and S is a feasible solution for Q?*
- Q-MIN-R(G, S_s, S_t, k, ℓ): *For $S_s, S_t \in V(R_{\mathrm{MIN}}(G, k))$, is there a path of length at most ℓ between the nodes for S_s and S_t in $R_{\mathrm{MIN}}(G, k)$?*
- Q-MAX-R(G, S_s, S_t, k, ℓ): *For $S_s, S_t \in V(R_{\mathrm{MAX}}(G, k))$, is there a path of length at most ℓ between the nodes for S_s and S_t in $R_{\mathrm{MAX}}(G, k)$?*

For ease of description, we present our positive results for paths of length exactly ℓ, as all our algorithmic techniques can be generalized to shorter paths. Throughout, we implicitly consider reconfiguration problems as parameterized problems with ℓ as the parameter. The reader is referred to the books of Downey and Fellows [12], Flum and Grohe [15], and Niedermeier [23] for more on parameterized complexity.

In Sect. 5, we consider problems that can be defined using graph properties, where a *graph property Π* is a collection of graphs closed under isomorphism, and is *non-trivial* if it is non-empty and does not contain all graphs. A graph property is *polynomially decidable* if for any graph G, it can be decided in polynomial time whether G is in Π. The property Π is *hereditary* if for any $G \in \Pi$, any induced subgraph of G is also in Π. For a graph property Π, $R_{\mathrm{MAX}}(G, k)$ has a node for each $S \subseteq V(G)$ such that $|S| \geq k$ and $G[S]$ has property Π, and $R_{\mathrm{MIN}}(G, k)$ has a node for each $S \subseteq V(G)$ such that $|S| \leq k$ and $G[V(G) \setminus S]$ has property Π. We use Π-MIN-R and Π-MAX-R instead of Q-MIN-R and Q-MAX-R, respectively, to denote *reconfiguration problems for Π*; examples include VC-R, FVS-R, and OCT-R for the former and IS-R, IF-R, and IBS-R for the latter, for Π defined as the collection of all edgeless graphs, forests, and bipartite graphs, respectively.

Due to space limitations, most proofs have been omitted from the current version of the paper. The affected propositions, lemmas, and theorems have been marked with a star.

Proposition 1. *Given Π and a collection of graphs $\mathscr{C}$, if Π-MIN-R parameterized by ℓ is fixed-parameter tractable on $\mathscr{C}$ then so is Π-MAX-R.*

Proof. Given an instance (G, S_s, S_t, k, ℓ) of Π-MAX-R, where $G \in \mathscr{C}$, we solve the Π-MIN-R instance $(G, V(G) \setminus S_s, V(G) \setminus S_t, n - k, \ell)$. Note that the parameter ℓ remains unchanged.

It is not hard to see that there exists a path between the nodes corresponding to S_s and S_t in $R_{\mathrm{MAX}}(G, k)$ if and only if there exists a path of the same
length between the nodes corresponding to $V(G) \setminus S_s$ and $V(G) \setminus S_t$ in R_{MIN}
$(G, n - k)$.   □

We obtain our results by solving Π-MIN-R, which by Proposition 1 implies
results for Π-MAX-R. We always assume Π to be non-trivial, polynomially
decidable, and hereditary.

Our algorithms rely on dynamic programming over graphs of bounded tree-
width. A *tree decomposition* of a graph G is a pair $\mathcal{T} = (T, \chi)$, where T is a
tree and χ is a mapping that assigns to each node $i \in V(T)$ a vertex subset X_i
(called a *bag*) such that: (1) $\bigcup_{i \in V(T)} X_i = V(G)$, (2) for every edge $uv \in E(G)$,
there exists a node $i \in V(T)$ such that the bag $\chi(i) = X_i$ contains both u and
v, and (3) for every $v \in V(G)$, the set $\{i \in V(T) \mid v \in X_i\}$ forms a connected
subgraph (subtree) of T. The *width* of any tree decomposition $\mathcal{T}$ is equal to
$\max_{i \in V(T)} |X_i| - 1$. The *treewidth* of a graph G, $tw(G)$, is the minimum width
of a tree decomposition of G.

For any graph of treewidth t, we can compute a tree decomposition of width
t and transform it into a nice tree decomposition of the same width in linear
time [21], where a rooted tree decomposition $\mathcal{T} = (T, \chi)$ with root *root* of a
graph G is a *nice tree decomposition* if each of its nodes is either (1) a leaf node
(a node i with $|X_i| = 1$ and no children), (2) an introduce node (a node i with
exactly one child j such that $X_i = X_j \cup \{v\}$ for some vertex $v \notin X_j$; v is said
to be *introduced* in i), (3) a forget node (a node i with exactly one child j such
that $X_i = X_j \setminus \{v\}$ for some vertex $v \in X_j$; v is said to be *forgotten* in i), or
(4) a join node (a node i with two children p and q such that $X_i = X_p = X_q$).
For node $i \in V(T)$, we use T_i to denote the subtree of T rooted at i and V_i to
denote the set of vertices of G contained in the bags of T_i. Thus $G[V_{root}] = G$.

3   PSPACE-Completeness

We define a simple intermediary problem that highlights the essential elements of
a PSPACE-hard reconfiguration problem. Given a pair $H = (\Sigma, E)$, where Σ is
an alphabet and $E \subseteq \Sigma^2$ a binary relation between symbols, we say that a word
over Σ is an H-*word* if every two consecutive symbols are in the relation. If one
looks at H as a digraph (possibly with loops), a word is an H-word if and only
if it is a walk in H. The H-WORD RECONFIGURATION problem asks whether
two given H-words of equal length can be transformed into one another (in any
number of steps) by changing one symbol at a time so that all intermediary steps
are also H-words.

A *Thue system* is a pair (Σ, R), where Σ is a finite alphabet and $R \subseteq \Sigma^* \times \Sigma^*$
is a set of rules. A rule can be applied to a word by replacing one subword by
the other, that is, for two words $s, t \in \Sigma^*$, we write $s \leftrightarrow_R t$ if there is a rule
$\{\alpha, \beta\} \in R$ and words $u, v \in \Sigma^*$ such that $s = u\alpha v$ and $t = u\beta v$. The reflexive
transitive closure of this relation defines an equivalence relation $\leftrightarrow_R^*$, where words

s, t are equivalent if and only if one can be reached from the other by repeated application of rules. The *word problem* of R is the problem of deciding, given two words $s, t \in \Sigma^*$, whether $s \leftrightarrow_R^* t$. A Thue system is called *c-balanced* if for each $\{\alpha, \beta\} \in R$ we have $|\alpha| = |\beta| = c$. The following fact is a folklore variant [2] of the classic proof of undecidability for general Thue systems [24].

Lemma 1 (*). *There exists a 2-balanced Thue system whose word problem is PSPACE-complete.*

A simple but technical reduction from Lemma 1 allows us to show the PSPACE-completeness of H-WORD RECONFIGURATION. The simplicity of the problem statement allows for easy reductions to various reconfiguration problems, as exemplified in Theorem 1. Similar reductions apply to the reconfiguration versions of, e.g., k-COLORING [7] and SHORTEST PATH [20] – a comprehensive discussion is available in an online manuscript by the fourth author [25].

Lemma 2 (*). *There exists a digraph H for which H-WORD RECONFIGURATION is PSPACE-complete.*

Theorem 1. *There exists an integer b such that VC-R, FVS-R, OCT-R, IS-R, IF-R, and IBS-R are PSPACE-complete even when restricted to graphs of treewidth at most b.*

Proof. Let $H = (\Sigma, R)$ be the digraph obtained from Lemma 2. We show a reduction from H-WORD RECONFIGURATION to VC-R.

For an integer n, we define G_n as follows. The vertex set contains vertices v_i^a for all $i \in \{1, \ldots, n\}$ and $a \in \Sigma$. Let $V_i = \{v_i^a \mid a \in \Sigma\}$ for $i \in \{1, \ldots, n\}$. The edge set of G_n contains an edge between every two vertices of V_i for $i \in \{1, \ldots, n\}$ and an edge $v_i^a v_{i+1}^b$ for all $(a, b) \notin R$ and $i \in \{1, \ldots, n-1\}$. The sets $V_i \cup V_{i+1}$ give a tree decomposition of width $b = 2|\Sigma|$.

Let $k = n \cdot (|\Sigma| - 1)$ and consider a vertex cover S of G_n of size k. For all i, since $G_n[V_i]$ is a clique, S contains all vertices of V_i except at most one. Since $|S| = \sum_i (|V_i| - 1)$, S contains all vertices except exactly one from each set V_i, say $v_i^{s_i}$ for some $s_i \in \Sigma$. Now $s_1 \ldots s_n$ is an H-word ($s_i s_{i+1} \in R$, as otherwise $v_i^{s_i} v_{i+1}^{s_{i+1}}$ would be an uncovered edge) and any H-word can be obtained in a similar way, giving a bijection between vertex covers of G_n of size k and H-words of length n.

Consider an instance $s, t \in \Sigma^*$ of H-WORD RECONFIGURATION. We construct the instance $(G_n, S_s, S_t, k + 1, \ell)$ of VC-R, where $n = |s| = |t|$, $\ell = 2^{n|\Sigma|}$ (that is, we ask for a reconfiguration sequence of any length) and S_s and S_t are the vertex covers of size k that correspond to s and t, respectively. Any reconfiguration sequence between such vertex covers starts by adding a vertex (since G_n has no vertex cover of size $k - 1$) and then removing another (since vertex covers larger than $k + 1$ are not allowed), which corresponds to changing one symbol of an H-word. This gives a one-to-one correspondence between reconfiguration sequences of H-words and reconfiguration sequences (of exactly twice the length) between vertex covers of size k. The instances are thus equivalent.

This proof can be adapted to FVS-R and OCT-R by replacing edges with cycles, e.g. triangles [22]. For IS-R, IF-R, and IBS-R, we simply need to consider set complements of solutions for VC-R, FVS-R, and OCT-R, respectively. □

4   A Meta-Theorem

In contrast to Theorem 1, in this section we show that a host of reconfiguration problems definable in monadic second-order logic (MSOL) become fixed-parameter tractable when parameterized by $\ell + t$. First, we briefly review the syntax and semantics of MSOL over graphs. The reader is referred to the excellent survey by Martin Grohe [17] for more details.

We have an infinite set of *individual variables*, denoted by lowercase letters x, y, and z, and an infinite set of *set variables*, denoted by uppercase letters X, Y, and Z. A *monadic second-order formula (MSOL-formula)* ϕ over a graph G is constructed from *atomic formulas* $\mathcal{E}(x,y)$, $x \in X$, and $x = y$ using the usual Boolean connectives as well as existential and universal quantification over individual and set variables. We write $\phi(x_1,\ldots,x_r,X_1,\ldots,X_s)$ to indicate that ϕ is a formula with free variables $x_1,\ldots,x_r$ and $X_1,\ldots,X_s$, where free variables are variables not bound by quantifiers.

For a formula $\phi(x_1,\ldots,x_r,X_1,\ldots,X_s)$, a graph G, vertices $v_1,\ldots,v_r$, and sets $V_1,\ldots,V_r$, we write $G \models \phi(v_1,\ldots,v_r,V_1,\ldots,V_r)$ if ϕ is satisfied in G when $\mathcal{E}$ is interpreted by the adjacency relation $E(G)$, the variables x_i are interpreted by v_i, and variables X_i are interpreted by V_i. We say that a vertex-subset problem Q is *definable in monadic second-order logic* if there exists an MSOL-formula $\phi(X)$ with one free set variable such that $S \subseteq V(G)$ is a feasible solution of problem Q for instance G if and only if $G \models \phi(S)$. For example, an independent set is definable by the formula $\phi_{\text{IS}}(X) = \forall_x \forall_y (x \in X \land y \in X) \rightarrow \neg \mathcal{E}(x,y)$.

Theorem 2 (Courcelle [8]). *There is an algorithm that given a MSOL-formula $\phi(x_1,\ldots,x_r,X_1,\ldots,X_s)$, a graph G, vertices $v_1,\ldots,v_r \in V(G)$, and sets $V_1,\ldots,V_s \subseteq V(G)$ decides whether $G \models \phi(v_1,\ldots,v_r,V_1,\ldots,V_s)$ in $\mathcal{O}(f(tw(G),|\phi|) \cdot n)$ time, for some computable function f.*

Theorem 3. *If a vertex-subset problem Q is definable in monadic second-order logic by a formula $\phi(X)$, then Q-MIN-R and Q-MAX-R parameterized by $\ell + tw(G) + |\phi|$ are fixed-parameter tractable.*

Proof. We provide a proof for Q-MIN-R as the proof for Q-MAX-R is analogous. Given an instance (G, S_s, S_t, k, ℓ) of Q-MIN-R, we build an MSOL-formula $\omega(X_0, X_\ell)$ such that $G \models \omega(S_s, S_t)$ if and only if the corresponding instance is a yes-instance. Since the size of ω will be bounded by a function of $\ell + |\phi|$, the statement will follow from Theorem 2.

As MSOL does not allow cardinality constraints, we overcome this limitation using the following technique. We let $L \subseteq \{-1, +1\}^\ell$ be the set of all sequences of length ℓ over $\{-1, +1\}$ which do not violate the maximum allowed capacity. In other words, given S_s and k, a sequence σ is in L if and only if for all $\ell' \leq \ell$

it satisfies $|S_s| + \sum_{i=1}^{\ell'} \sigma[i] \le k$, where $\sigma[i]$ is the ith element in sequence σ. We let $\omega = \bigvee_{\sigma \in L} \omega_\sigma$ and

$$\omega_\sigma(X_0, X_\ell) = \exists_{X_1,\dots,X_{\ell-1}} \bigwedge_{0 \le i \le \ell} \phi(X_i) \wedge \bigwedge_{1 \le i \le \ell} \psi_{\sigma[i]}(X_{i-1}, X_i)$$

where $\psi_{-1}(X_{i-1}, X_i)$ means X_i is obtained from X_{i-1} by removing one element and $\psi_{+1}(X_{i-1}, X_i)$ means it is obtained by adding one element. Formally, we have:

$$\psi_{-1}(X_{i-1}, X_i) = \exists_x \, x \in X_{i-1} \wedge x \notin X_i \wedge \forall y \, (y \in X_i \leftrightarrow (y \in X_{i-1} \wedge y \ne x))$$

$$\psi_{+1}(X_{i-1}, X_i) = \exists_x \, x \notin X_{i-1} \wedge x \in X_i \wedge \forall y \, (y \in X_i \leftrightarrow (y \in X_{i-1} \vee y = x))$$

It is easy to see that $G \models \omega_\sigma(S_s, S_t)$ if and only if there is a reconfiguration sequence from S_s to S_t (corresponding to $X_0, X_1, \dots, X_\ell$) such that the ith step removes a vertex if $\sigma[i] = -1$ and adds a vertex if $\sigma[i] = +1$. Since $|L| \le 2^\ell$, the size of the MSOL-formula ω is bounded by an (exponential) function of $\ell + |\phi|$. $\qquad\square$

5   Dynamic Programming Algorithms

Throughout this section we will consider one fixed instance (G, S_s, S_t, k, ℓ) of Π-MIN-R and a nice tree decomposition $\mathcal{T} = (T, \chi)$ of G. Moreover, similarly to the previous section, we will ask, for a fixed sequence $\sigma \in \{-1, +1\}^\ell$, whether $G \models \omega_\sigma(S_s, S_t)$ holds. That is, we ask whether there is a reconfiguration sequence which at the ith step removes a vertex when $\sigma[i] = -1$ and adds a vertex when $\sigma[i] = +1$. The final algorithm then asks such a question for every sequence σ which does not violate the maximum allowed capacity: $|S_s| + \sum_{i=1}^{\ell'} \sigma[i] \le k$ for all $\ell' \le \ell$. This will add a factor of at most 2^ℓ to the running time.

5.1   Signatures as Equivalence Classes

A reconfiguration sequence can be described as a sequence of steps, each step specifying which vertex is being removed or added. To obtain a more succinct representation, we observe that in order to propagate information up from the leaves to the root of a nice tree decomposition, we can ignore vertices outside of the currently considered bag (X_i) and only indicate whether a step has been used by a vertex in any previously processed bags, i.e. a vertex in $V_i \setminus X_i$.

Definition 2. *A* signature τ *over a set* $X \subseteq V(G)$ *is a sequence of steps* $\tau[1], \dots, \tau[\ell] \in X \cup \{\text{used}, \text{unused}\}$. *Steps from* X *are called* vertex steps.

The total number of signatures over a bag X of at most t vertices is $(t + 2)^\ell$. Our dynamic programming algorithms start by considering a signature with only unused steps in each leaf node, specify when a vertex may be added/removed in introduce nodes by replacing unused steps with vertex steps ($\tau[i] = \text{unused}$

becomes $\tau[i] = v$ for the introduced vertex v), merge signatures in join nodes, and replace vertex steps with **used** steps in forget nodes.

For a set $S \subseteq V(G)$ and a bag X, we let $\tau(i, S) \subseteq S \cup X$ denote the set of vertices obtained after executing the first i steps of τ: the ith step adds $\tau[i]$ if $\tau[i] \in X$ and $\sigma[i] = +1$, removes it if $\tau[i] \in X$ and $\sigma[i] = -1$, and does nothing if $\tau[i] \in \{\text{used}, \text{unused}\}$.

A valid signature must ensure that no step deletes a vertex that is absent or adds a vertex that is already present, and that the set of vertices obtained after applying reconfiguration steps to $S_s \cap X$ is the set $S_t \cap X$. Additionally, because Π is hereditary, we can check whether this property is at least locally satisfied (in $G[X]$) after each step of the sequence. More formally, we have the following definition.

Definition 3. *A signature τ over X is valid if*

(1) $\tau[i] \in \tau(i - 1, S_s \cap X)$ whenever $\tau[i] \in X$ and $\sigma[i] = -1$,
(2) $\tau[i] \notin \tau(i - 1, S_s \cap X)$ whenever $\tau[i] \in X$ and $\sigma[i] = +1$,
(3) $\tau(\ell, S_s \cap X) = S_t \cap X$, and
(4) $G[X \setminus \tau(i, S_s \cap X)] \in \Pi$ for all $i \leq \ell$.

It is not hard to see that a signature τ over X is valid if and only if $\tau(0, S_s \cap X), \ldots, \tau(\ell, S_s \cap X)$ is a well-defined path between $S_s \cap X$ and $S_t \cap X$ in $R_{\text{MIN}}(G[X], n)$. We will consider only valid signatures. The dynamic programming algorithms will enumerate exactly the signatures that can be extended to valid signatures over V_i in the following sense:

Definition 4. *A signature π over V_i extends a signature π over X_i if it is obtained by replacing some **used** steps with vertex steps from $V_i \setminus X_i$*

However, for many problems, the fact that S is a solution for $G[X]$ for each bag X does not imply that S is a solution for G, and checking this 'local' notion of validity will not be enough – the algorithm will have to maintain additional information. One such example is the OCT-R problem, which we discuss in Sect. 5.4.

5.2   An Algorithm for VC-R

To process nodes of the tree decomposition, we now define ways of generating signatures from other signatures. The introduce operation determines all ways that an introduced vertex can be represented in a signature, replacing **unused** steps in the signature of its child.

Definition 5. *Given a signature τ over X and a vertex $v \notin X$, the introduce operation, $introduce(\tau, v)$ returns the following set of signatures over $X \cup \{v\}$: for every subset I of indices i for which $\tau[i] = $ **unused**, consider a copy τ' of τ where for all $i \in I$ we set $\tau'[i] = v$, check if it is valid, and if so, add it to the set.*

In particular $\tau \in introduce(\tau, v)$ and $|introduce(\tau, v)| \leq 2^{\ell}$. All signatures obtained through the introduce operation are valid, because of the explicit check.

Definition 6. *Given a signature τ over X and a vertex $v \in X$, the* forget *operation, returns a new signature $\tau' = forget(\tau, v)$ over $X \setminus \{v\}$ such that for all $i \leq \ell$, we have $\tau'[i] = $ used if $\tau[i] = v$ and $\tau'[i] = \tau[i]$ otherwise.*

Since $\tau'(i, S_s \cap X \setminus \{v\}) = \tau(i, S_s \cap X) \setminus \{v\}$, it is easy to check that the forget operation preserves validity.

Definition 7. *Given two signatures τ_1 and τ_2 over $X \subseteq V(G)$, we say τ_1 and τ_2 are* compatible *if for all $i \leq \ell$:*

(1) $\tau_1[i] = \tau_2[i] = $ unused,
(2) $\tau_1[i] = \tau_2[i] = v$ for some $v \in X$, or
(3) either $\tau_1[i]$ or $\tau_2[i]$ is equal to used *and the other is equal to* unused.

For two compatible signatures τ_1 and τ_2, the join *operation returns a new signature $\tau' = join(\tau_1, \tau_2)$ over X such that for all $i \leq \ell$ we have, respectively:*

(1) $\tau'[i] = $ unused,
(2) $\tau'[i] = v$, and
(3) $\tau'[i] = $ used.

Since $\tau' = join(\tau_1, \tau_2)$ is a signature over the same set as τ_1 and differs from τ_1 only by replacing some unused steps with used steps, the join operation preserves validity, that is, if two compatible signatures τ_1 and τ_2 are valid then so is $\tau' = join(\tau_1, \tau_2)$.

Let us now describe the algorithm. For each $i \in V(T)$ we assign an initially empty table A_i. All tables corresponding to internal nodes of T will be updated by simple applications of the introduce, forget, and join operations.

Leaf nodes. Let i be a leaf node, that is $X_i = \{v\}$ for some vertex v. We let $A_i = introduce(\tau, v)$, where τ is the signature with only unused steps.

Introduce nodes. Let j be the child of an introduce node i, that is $X_i = X_j \cup \{v\}$ for some $v \notin X_j$. We let $A_i = \bigcup_{\tau \in A_j} introduce(\tau, v)$.

Forget nodes. Let j be the child of a forget node i, that is $X_i = X_j \setminus \{v\}$ for some $v \in X_j$. We let $A_i = \{forget(\tau, v) \mid \tau \in A_j\}$.

Join nodes. Let j and h be the children of a join node i, that is $X_i = X_j = X_h$. We let $A_i = \{join(\tau_j, \tau_h) \mid \tau_j \in A_j, \tau_h \in A_h, \text{ and } \tau_j \text{ is compatible with } \tau_h\}$.

The operations were defined so that the following lemma holds by induction. The theorem then follows by making the algorithm accept when A_{root} contains a signature τ such that no step of τ is unused.

Lemma 3. *For $i \in V(T)$ and a signature τ over X_i, $\tau \in A_i$ if and only if τ can be extended to a signature over V_i that is valid.*

Theorem 4 (*). *VC-R and IS-R can be solved in $\mathcal{O}^{\star}(4^{\ell}(t+1)^{\ell})$ time on graphs of treewidth t.*

5.3   VC-R in Planar Graphs

Using an adaptation of Baker's approach for decomposing planar graphs [1], also known as the *shifting technique* [3,10,13], we show a similar result for VC-R and IS-R on planar graphs. The idea is that at most ℓ elements of a solution will be changed, and thus if we divide the graph into $\ell + 1$ parts, one of these parts will be unchanged throughout the reconfiguration sequence. The shifting technique allows the definition of the $\ell+1$ parts so that removing one (and replacing it with simple gadgets to preserve all needed information) yields a graph of treewidth at most $3\ell - 1$.

Theorem 5 (*). VC-R *and* IS-R *are fixed-parameter tractable on planar graphs when parameterized by* ℓ. *Moreover, there exists an algorithm which solves both problems in* $\mathcal{O}^\star(4^\ell(3\ell + 1)^\ell)$ *time.*

We note that, by a simple application of the result of Demaine et al. [9], Theorem 5 generalizes to H-minor-free graphs and only the constants of the overall running time of the algorithm are affected.

5.4   An Algorithm for OCT-R

In this section we show how known dynamic programming algorithms for problems on graphs of bounded treewidth can be adapted to reconfiguration. The general idea is to maintain a view of the reconfiguration sequence just as we did for VC-R and in addition check if every reconfiguration step gives a solution, which can be accomplished by maintaining (independently for each step) any information that the original algorithm would maintain. We present the details for OCT-R (where Π is the collection of all bipartite graphs) as an example.

In a dynamic programming algorithm for VC on graphs of bounded treewidth, it is enough to maintain information about what the solution's intersection with the bag can be. This is not the case for OCT. One algorithm for OCT works in time $\mathcal{O}^\star(3^t)$ by additionally maintaining a bipartition of the bag (with the solution deleted) [14,15]. That is, at every bag X_i, we would maintain a list of assignments $X \to \{\texttt{used}, \texttt{left}, \texttt{right}\}$ with the property that there exists a subset S of V_i and a bipartition L, R of $G[V_i \setminus S]$ such that $X_i \cap S, X_i \cap L$, and $X_i \cap R$ are the $\texttt{used}$, $\texttt{left}$, and $\texttt{right}$ vertices, respectively. A signature for OCT-R will hence additionally store a bipartition for each step (except for the first and last sets S_s and S_t, as we already assume them to be solutions).

Definition 8. *An* OCT-signature τ *over a set* $X \subseteq V(G)$ *is a sequence of steps* $\tau[1], \ldots, \tau[\ell] \in X \cup \{\texttt{used}, \texttt{unused}\}$ *together with an entry* $\tau[i, v] \in \{\texttt{left}, \texttt{right}\}$ *for every* $1 \leq i \leq \ell - 1$ *and* $v \in X \setminus \tau(i, S_s \cap X)$.

There are at most $(t + 2)^\ell 2^{t(\ell-1)}$ different OCT-signatures. In the definition of validity, we replace the last condition with the following, stronger one:

(4) For all $1 \leq i \leq \ell - 1$, the sets $\{v \mid \tau[i, v] = \texttt{left}\}$ and $\{v \mid \tau[i, v] = \texttt{right}\}$ give a bipartition of $G[X \setminus \tau(p, S_s \cap X)]$.

In the definition of the join operation, we additionally require two signatures to have equal $\tau[i, v]$ entries (whenever defined) to be considered compatible; the operation copies them to the new signature. In the definition of the forget operation, we delete any $\tau[i, v]$ entries, where v is the vertex being forgotten. In the introduce operation, we consider (and check the validity of) a different copy for each way of replacing **unused** steps with v steps and each way of assigning $\{\texttt{left}, \texttt{right}\}$ values to new $\tau[i, v]$ entries, where v is the vertex being introduced. As before, to each node we assign an initially empty table of OCT-signatures and fill them bottom-up using these operations. Lemma 3, with the new definitions, can then be proved again by induction.

Theorem 6 (*). OCT-R and IBS-R can be solved in $\mathcal{O}^\star(2^{t\ell}4^\ell(t+1)^\ell)$ time on graphs of treewidth t.

Similarly, using the classical $\mathcal{O}^\star(2^{\mathcal{O}(t \log t)})$ algorithm for FVS and IF (which maintains what partition of X_i the connected components of V_i can produce), we get the following running times for reconfiguration variants of these problems.

Theorem 7 (*). FVS-R and IF-R can be solved in $\mathcal{O}^\star(t^{\ell t}4^\ell(t+1)^\ell)$ time on graphs of treewidth t.

6 Conclusion

We have seen in Sect. 5.4 that, with only minor modifications, known dynamic programming algorithms for problems on graphs of bounded treewidth can be adapted to reconfiguration. It is therefore natural to ask whether the obtained running times can be improved via more sophisticated algorithms which exploit properties of the underlying problem or whether these running times are optimal under some complexity assumptions. Moreover, it would be interesting to investigate whether the techniques presented for planar graphs can be extended to other problems or more general classes of sparse graphs. In particular, the parameterized complexity of "non-local" reconfiguration problems such as FVS-R and OCT-R remains open even for planar graphs.

References

1. Baker, B.S.: Approximation algorithms for NP-complete problems on planar graphs. J. ACM **41**(1), 153–180 (1994)
2. Bauer, G., Otto, F.: Finite complete rewriting systems and the complexity of the word problem. Acta Inf. **21**(5), 521–540 (1984)
3. Bodlaender, H.L., Koster, A.M.C.A.: Combinatorial optimization on graphs of bounded treewidth. Comput. J. **51**(3), 255–269 (2008)
4. Bonsma, P.: The complexity of rerouting shortest paths. In: Rovan, B., Sassone, V., Widmayer, P. (eds.) MFCS 2012. LNCS, vol. 7464, pp. 222–233. Springer, Heidelberg (2012)
5. Bonsma, P.: Rerouting shortest paths in planar graphs. In: Leibniz International Proceedings in Informatics (LIPIcs), FSTTCS 2012, vol. 18, pp. 337–349 (2012)

6. Cereceda, L., van den Heuvel, J., Johnson, M.: Connectedness of the graph of vertex-colourings. Discrete Math. **308**(56), 913–919 (2008)
7. Cereceda, L., van den Heuvel, J., Johnson, M.: Finding paths between 3-colorings. J. Graph Theory **67**(1), 69–82 (2011)
8. Courcelle, B.: The monadic second-order logic of graphs. I. Recognizable sets of finite graphs. Inf. Comput. **85**(1), 12–75 (1990)
9. Demaine, E., Hajiaghayi, M., Kawarabayashi, K.: Algorithmic graph minor theory: secomposition, approximation, and coloring. In: Proceedings of the 46th Annual IEEE Symposium on Foundations of Computer Science, pp. 637–646, October 2005
10. Demaine, E.D., Hajiaghayi, M.: The bidimensionality theory and its algorithmic applications. Comput. J. **51**(3), 292–302 (2008)
11. Diestel, R.: Graph Theory. Springer, Heidelberg (2005). (Electronic Edition)
12. Downey, R.G., Fellows, M.R.: Parameterized Complexity. Springer, New York (1997)
13. Eppstein, D.: Diameter and treewidth in minor-closed graph families. Algorithmica **27**(3), 275–291 (2000)
14. Fiorini, S., Hardy, N., Reed, B., Vetta, A.: Planar graph bipartization in linear time. Discrete Appl. Math. **156**(7), 1175–1180 (2008)
15. Flum, J., Grohe, M.: Parameterized Complexity Theory. Springer, Heidelberg (2006)
16. Gopalan, P., Kolaitis, P.G., Maneva, E.N., Papadimitriou, C.H.: The connectivity of boolean satisfiability: computational and structural dichotomies. SIAM J. Comput. **38**(6), 2330–2355 (2009)
17. Grohe, M.: Logic, graphs, and algorithms. Electron. Colloquium Comput. Complex. (ECCC) **14**(091), 3 (2007)
18. Ito, T., Demaine, E.D., Harvey, N.J.A., Papadimitriou, C.H., Sideri, M., Uehara, R., Uno, Y.: On the complexity of reconfiguration problems. Theor. Comput. Sci. **412**(12–14), 1054–1065 (2011)
19. Ito, T., Kamiński, M., Demaine, E.D.: Reconfiguration of list edge-colorings in a graph. Discrete Appl. Math. **160**(15), 2199–2207 (2012)
20. Kamiński, M., Medvedev, P., Milanič, M.: Shortest paths between shortest paths. Theor. Comput. Sci. **412**(39), 5205–5210 (2011)
21. Kloks, T. (ed.): Treewidth Computations and Approximations. LNCS, vol. 842. Springer, Heidelberg (1994)
22. Mouawad, A.E., Nishimura, N., Raman, V., Simjour, N., Suzuki, A.: On the parameterized complexity of reconfiguration problems. In: Gutin, G., Szeider, S. (eds.) IPEC 2013. LNCS, vol. 8246, pp. 281–294. Springer, Heidelberg (2013)
23. Niedermeier, R.: Invitation to Fixed-Parameter Algorithms. Oxford University Press, Oxford (2006)
24. Post, E.L.: Recursive unsolvability of a problem of Thue. J. Symbol. Logic **12**(1), 1–11 (1947)
25. Wrochna, M.: Reconfiguration in bounded bandwidth and treedepth (2014). arXiv:1405.0847

Finite Integer Index of Pathwidth and Treewidth

Jakub Gajarský[1], Jan Obdržálek[1], Sebastian Ordyniak[1]([⊠]), Felix Reidl[2],
Peter Rossmanith[2], Fernando Sánchez Villaamil[2], and Somnath Sikdar[2]

[1] Faculty of Informatics, Masaryk University, Brno, Czech Republic
{gajarsky,obdrzalek,ordyniak}@fi.muni.cz
[2] Theoretical Computer Science, Department of Computer Science,
RWTH Aachen University, Aachen, Germany
{reidl,rossmani,fernando.sanchez,sikdar}@cs.rwth-aachen.de

Abstract. We show that the optimization versions of the PATHWIDTH and TREEWIDTH problems have a property called *finite integer index* when the inputs are restricted to graphs of bounded pathwidth and bounded treewidth, respectively. They do not have this property in general graph classes. This has interesting consequences for kernelization of both these (optimization) problems on certain sparse graph classes. In the process we uncover an interesting property of path and tree decompositions, which might be of independent interest.

1 Introduction

One way of efficiently solving decision and optimization problems on graphs are so-called *reduction algorithms*. Each such algorithm is characterized by a set of reduction rules (which locally modify a graph) and a finite set of graphs. The problem is then solved by repeatedly applying the reduction rules (until none can be applied) and checking whether the resulting graph is in the given finite set. If so, then the answer is true, otherwise it is false. This approach has been successfully applied to many different problems. An often cited example is the result of Arnborg et al. [1]. They show that for all "finite state" decision problems (which include all MSO definable problems) on graphs of bounded treewidth there is a well-defined set of reduction rules such that the resulting algorithm works in linear time.

The results of [1] have been later restated by Bodlaender and de Fluiter [4] in a different, more direct way, which avoided the original algebraic setting. (We give only a sketch here—see [4] and Sect. 2 for formal definitions). Let a *t-boundaried graph* be a graph where t vertices (the terminals) are identified,

Research funded by DFG-Project RO 927/12-1 "Theoretical and Practical Aspects of Kernelization", by the Czech Science Foundation, project GA14-03501S, and by Employment of Newly Graduated Doctors of Science for Scientific Excellence (CZ.1.07/2.3.00/30.0009).

© Springer International Publishing Switzerland 2014
M. Cygan and P. Heggernes (Eds.): IPEC 2014, LNCS 8894, pp. 258–269, 2014.
DOI: 10.1007/978-3-319-13524-3_22

and for two t-boundaried graphs G and H let $G \oplus H$ be the graph obtained by "gluing" the corresponding terminals of G and H. Let P be a decision problem (graph property) and let us define a relation $\sim_{P,t}$ as:

$$G_1 \sim_{P,t} G_2 \equiv \quad \text{for all} \quad t - \text{boundaried graphs } H : P(G_1 \oplus H) \quad \text{iff} \quad P(G_2 \oplus H)$$

Then we say the property P has *finite index* if and only if for all $t \geq 0$ the relation $\sim_{P,t}$ has finitely many equivalence classes. As one can see, if a property has finite index, we can use a reduction rule which replaces each t-boundaried subgraph by the smallest graph in its equivalence class. Since by Courcelle [5] all MSO-definable properties have finite index, the result of [1] follows.

More importantly, in [4] Bodlaender and de Fluiter also generalize these ideas further to obtain reduction systems to *optimization problems*, by defining a property called *finite integer index* (*FII*) (see Definition 4.5 in [4] and Sect. 2.3). This property is similar to finite index, but it additionally incorporates the solution size, making it applicable to optimization problems. Both finite index and finite integer index extend the notion of *finite state* introduced by Langston and Fellows in [7].

Recently there has been a flurry of work on algorithmic meta-theorems that rely crucially on finite integer index [2,8–10]. Together the main results behind these meta-theorems can be summarized as follows: Graph problems that have finite integer index admit linear kernels under certain conditions on a sparse graph class.

However, proving that a problem has finite integer index is generally not easy. Bodlaender and de Fluiter showed that the optimization versions of several well-known problems are of finite integer index (see Theorems 4.3 and 4.4 in [4]). These include MAX INDUCED d-DEGREE SUBGRAPH, MAX INDEPENDENT SET, MIN VERTEX COVER, MIN p-DOMINATING SET for all $p \geq 1$. They also showed that problems such as MAX CUT, LONGEST PATH, and LONGEST CYCLE do not have finite integer index. Bodlaender et al. [2] in their influential (*Meta*) *Kernelization* paper give a sufficiency condition, which they call *strong monotonicity*, for a problem to have FII. We also refer to their work for an extensive compendium of problems that have finite integer index.

Our contribution. We show that the problems PATHWIDTH and TREEWIDTH have finite integer index, if we restrict the general equivalence relation to graphs classes of bounded pathwidth and bounded treewidth, respectively. It is rather easy to give counterexamples that show that both these problems *in general* do not have FII, and we provide such examples later in Theorem 3.

Interestingly, having FII even in this restricted setting still allows us to apply the meta kernelization framework of [9,10] to PATHWIDTH and TREEWIDTH. This is because the framework does not require FII in its full generality; the reduction rule used there, only replaces subgraphs that have constant treewidth to begin with. Our result therefore has the following consequences with respect to kernelization with structural parameters as introduced in [9,10]. (For definitions of the graph classes mentioned below see e.g. [11].)

1. With the size of a modulator to bounded treedepth[1] as parameter, the optimization versions of PATHWIDTH and TREEWIDTH admit
   (a) a linear kernel on hereditary graph classes of bounded expansion;
   (b) a quadratic vertex kernel on hereditary graph classes of locally bounded expansion;
   (c) a polynomial kernel on nowhere-dense graphs.
2. When parameterized by the size of a modulator to constant pathwidth (respectively, treewidth), the optimization version of PATHWIDTH (respectively, TREEWIDTH) has a linear kernel on graph classes excluding a fixed graph as a topological minor.

Along the way, we prove Corollaries 1 and 2, which provide some nice insight into path and tree decompositions.

The rest of the paper is organized as follows. We introduce notation and important definitions in Sect. 2. We then prove our results for pathwidth in Sect. 3 and extend it to treewidth in Sect. 4. In Sect. 5 we show that neither treewidth nor pathwidth have FII on arbitrary graphs. After stating the consequences of our results for kernelization in Sect. 6, we conclude in Sect. 7.

2   Preliminaries

We use standard notation from graph theory as can be found, e.g., in [6]. All graphs considered in this paper are finite, undirected and simple. Let G be a graph. We denote the vertex set of G by $V(G)$ and the edge set of G by $E(G)$. Let $X \subseteq V(G)$ be a set of vertices of G. The *subgraph of G induced by X*, denoted $G[X]$, is the graph with vertex set X and edges $E(G) \cap \{\{u, v\} : u, v \in X\}$. By $G \setminus X$ we denote the subgraph of G induced by $V(G) \setminus X$.

Let G be a graph and A and B two sets of vertices of G. A set S of vertices of G is called an (A, B)-*separator* (or a separator between A and B) if the graph $G \setminus S$ contains no path between a vertex in A and a vertex in B. We say S is a *minimal (A, B)-separator* if there is no set $S' \subseteq V(G)$ with $|S'| < |S|$ such that S' is an (A, B)-separator.

2.1   Boundaried Graphs

A *t-boundaried graph* $\widetilde{G}$ is a pair $(G, \partial(G))$, where $G = (V, E)$ is a graph and $\partial(G) \subseteq V$ is a set of t vertices with distinct labels from the set $\{1, \ldots, t\}$. The graph G is called the *underlying unlabeled graph* and $\partial(G)$ is called the *boundary*. In the sequel we use $\widetilde{G}$, $\widetilde{H}$ etc. to denote t-boundaried graphs.

For t-boundaried graphs $\widetilde{G}_1 = (G_1, \partial(G_1))$ and $\widetilde{G}_2 = (G_2, \partial(G_2))$, we let $\widetilde{G}_1 \oplus \widetilde{G}_2$ denote the graph obtained by taking the disjoint union of G_1 and G_2 and identifying each vertex in $\partial(G_1)$ with the vertex in $\partial(G_2)$ with the same

---

[1] A *modulator to treedepth d* of graph G is a set $X \subseteq V(G)$ s.t. the treedepth of $G - X$ is at most $d - 1$. Modulators to bounded treewidth and pathwidth are defined similarly.

label, and then making the graph simple, if necessary. Note that the operation ⊕ "destroys" the boundaries of two t-boundaried graphs and creates an ordinary graph.

2.2 Treewidth and Pathwidth

A *tree decomposition* $\mathcal{T}$ of an (undirected) graph $G = (V, E)$ is a pair (T, χ), where T is a tree and χ is a function that assigns each tree node t a set $\chi(t) \subseteq V$ of vertices such that the following conditions hold:

(P1) For every vertex $u \in V$, there is a tree node t such that $u \in \chi(t)$.
(P2) For every edge $\{u, v\} \in E(G)$ there is a tree node t such that $u, v \in \chi(t)$.
(P3) For every vertex $v \in V(G)$, the set of tree nodes t with $v \in \chi(t)$ forms a subtree of T.

The sets $\chi(t)$ are called *bags* of the decomposition $\mathcal{T}$ and $\chi(t)$ is the bag associated with the tree node t. The *width* of a tree decomposition (T, χ) is the size of a largest bag minus 1. A tree decomposition of minimum width is called *optimal*. The *treewidth* of a graph G, denoted by $\mathrm{tw}(G)$, is the width of an optimal tree decomposition of G. For a subtree T' of T we denote by $G[T']$ the subgraph of G induced by $\bigcup_{t \in V(T')} \chi(t)$.

Let $\mathcal{T} = (T, \chi)$ be a tree decomposition of a graph G and let G' be an induced subgraph of G. The *projection* of $\mathcal{T}$ onto G', denoted by $\mathcal{T}|G'$, is the pair (T, χ') where $\chi'(t) = \chi(t) \cap V(G')$ for every $t \in V(T)$. It is well-known that $\mathcal{T}|G'$ is a tree decomposition of G'.

A *path decomposition* of a graph G is a tree decomposition (T, χ) such that T is a path instead of a tree. All notions and definitions introduced for tree decompositions above apply in the same way for path decompositions. The *pathwidth* of G, denoted by $\mathrm{pw}(G)$, is the width of an optimal path decomposition of G.

We consider the following problems:

TREEWIDTH (tw)
Input: Graph G and integer k.
Question: Is $\mathrm{tw}(G) \le k$?

PATHWIDTH (pw)
Input: Graph G and integer k.
Question: Is $\mathrm{pw}(G) \le k$?

It is well-known that every bag of a path or tree decomposition is a separator in the underlying graph. We will use the following formulation of this fact.

Proposition 1. *Let $\mathcal{T} = (T, \chi)$ be a tree decomposition (path decomposition) of a graph G, $t \in V(T)$, and let T_1 and T_2 be two sets of nodes of $T \setminus \{t\}$ such that $\{t\}$ separates T_1 from T_2 in T. Then, $\chi(t)$ separates $G[T_1]$ from $G[T_2]$ in G.*

2.3   Finite Integer Index

A *graph problem* Π is a set of pairs (G, ξ), where G is a graph and $\xi \in \mathbf{N}_0$, such that for all graphs G_1, G_2 and all $\xi \in \mathbf{N}_0$, if G_1 is isomorphic to G_2, then $(G_1, \xi) \in \Pi$ if and only if $(G_2, \xi) \in \Pi$. For instance, the PATHWIDTH and the TREEWIDTH problem defined in the previous subsection are graph problems.

Let Π be a graph problem and let $\widetilde{G}_1 = (G_1, \partial(G_1))$, $\widetilde{G}_2 = (G_2, \partial(G_2))$ be two t-boundaried graphs. We say that $\widetilde{G}_1 \equiv_{\Pi, t} \widetilde{G}_2$ if there exists an integer constant $\Delta_{\Pi, t}(\widetilde{G}_1, \widetilde{G}_2)$ such that for all t-boundaried graphs $\widetilde{H} = (H, \partial(H))$ and for all $\xi \in \mathbf{N}$: $(\widetilde{G}_1 \oplus \widetilde{H}, \xi) \in \Pi$ iff $(\widetilde{G}_2 \oplus \widetilde{H}, \xi + \Delta_{\Pi, t}(\widetilde{G}_1, \widetilde{G}_2)) \in \Pi$. We say that Π has *finite integer index in the class of graphs* $\mathcal{F}$ if, for every $t \in \mathbf{N}$, the number of equivalence classes of $\equiv_{\Pi, t}$ which have a non-empty intersection with $\mathcal{F}$ is finite.

In this paper we focus on two concrete equivalence relations: $\equiv_{\mathrm{pw}, t}$ and $\equiv_{\mathrm{tw}, t}$. Two t-boundaried graphs $\widetilde{G}_1, \widetilde{G}_2$ are equivalent $\widetilde{G}_1 \equiv_{\mathrm{pw}, t} \widetilde{G}_2$ if there exists an integer constant $\Delta_{\mathrm{pw}, t}(\widetilde{G}_1, \widetilde{G}_2)$ such that for all t-boundaried graphs $\widetilde{H}$ and for all $\xi \in \mathbf{N}$ it holds that $\mathrm{pw}(\widetilde{G}_1 \oplus \widetilde{H}) \leq \xi$ iff $\mathrm{pw}(\widetilde{G}_2 \oplus \widetilde{H}) \leq \xi + \Delta_{\mathrm{pw}, t}(\widetilde{G}_1, \widetilde{G}_2)$. The relation $\equiv_{\mathrm{tw}, t}$ is defined analogously for treewidth.

2.4   Characteristics of Path and Tree Decompositions

One of the tools needed in the following are characteristics of path decompositions and tree decompositions, which have been introduced in [3]. Because the definition of these characteristics is quite technical and the properties we require have already been shown in [3], we will not provide a formal definition. Instead, we will only state the required properties and refer the reader to [3] for details and proofs.

The concept of a *characteristic* of a partial path decomposition of a graph—or equivalently the characteristic of a path decomposition of a boundaried graph— was introduced by Bodlaender and Kloks in [3, Definition 4.4]. Informally, the characteristic of a path decomposition $\mathcal{P}$ of $\widetilde{G}$ compactly represents all the information required to compute, for any $\widetilde{H}$, the ways $\mathcal{P}$ can be extended into a path decomposition of the graph $\widetilde{G} \oplus \widetilde{H}$. This information can then be used to compute the pathwidth of the graph $\widetilde{G} \oplus \widetilde{H}$. Importantly, the number of characteristics of path decompositions of width at most w of any t-boundaried graph only depends on t and w, but not on the the graph itself.

Proposition 2 ([3, Lemma 4.1]).   *Let $\widetilde{G}$ be a t-boundaried graph and w an integer. Then the number of characteristics of path decompositions of width at most w of $\widetilde{G}$ is bounded by a function of t and w.*

For integer w, the *full set of (path decomposition) characteristics of $\widetilde{G}$ of width at most w* (as defined in [3, Definition 4.6]), denoted by $\mathrm{FSCP}_w(\widetilde{G})$, is the set of all characteristics of path decompositions of $\widetilde{G}$ of width at most w. We denote by $\mathrm{FSCP}(\widetilde{G})$ the (possible infinite) set $\bigcup_{w \in \mathbf{N}} \mathrm{FSCP}_w(\widetilde{G})$.

Proposition 3 ([3, Sect. 4.3]). *Let $\widetilde{H}$, $\widetilde{G}_1$ and $\widetilde{G}_2$ be t-boundaried graphs, and let $\mathcal{P}$ be a path decomposition of $\widetilde{G}_1 \oplus \widetilde{H}$. If the (unique) characteristic of $\mathcal{P}|G_1$ is in $FSCP(\widetilde{G}_2)$, then there is a path decomposition of $\widetilde{G}_2 \oplus \widetilde{H}$ that has the same width as $\mathcal{P}$.*

Proof (Sketch). For $i \in \{1, 2\}$, let $\mathcal{P}_i$ be any path decomposition of $\widetilde{G}_i$ such that the content of the last bag of $\mathcal{P}_i$ is $\partial(G_i)$ and let $\mathcal{P}_3$ be any path decomposition of $\widetilde{H}$ such that the content of the first bag of $\mathcal{P}_3$ is $\partial(H)$. Furthermore, for $i \in \{1, 2\}$, let $\mathcal{P}_{i,3}$ be the path decomposition of $\widetilde{G}_i \oplus \widetilde{H}$ obtained from $\mathcal{P}_i$ and $\mathcal{P}_3$ by appending the first bag of $\mathcal{P}_3$ to the last bag of $\mathcal{P}_i$, let $p_{i,3}$ be the bag of $\mathcal{P}_{i,3}$ that corresponds to the last bag of $\mathcal{P}_i$, and let $l_{i,3}$ be the last bag of $\mathcal{P}_{i,3}$.

Now assume that we run the algorithm described in [3, Sect. 4.3] on the path decomposition $\mathcal{P}_{i,3}$ and let $F(p_{i,3})$ and $F(l_{i,3})$ be the full set of characteristics of partial path decompositions computed at the node $p_{i,3}$ and the node $l_{i,3}$, respectively, of width at most the width of $\mathcal{P}$. Then, by the definition of a full set of characteristics, we obtain that $F(p_{1,3})$ contains the characteristic of $\mathcal{P}|G_1$ and that $F(l_{1,3})$ contains the characteristic of $\mathcal{P}$. Moreover, the characteristic of $\mathcal{P}$ in $F(l_{1,3})$ is generated by the algorithm from the characteristic of $\mathcal{P}|G_1$ in $F(p_{1,3})$. By the assumptions of the Proposition, we have that the characteristic of $\mathcal{P}|G_1$ is contained in $FSCP(\widetilde{G}_2)$ and hence also in $F(p_{2,3})$. Hence, because the path decompositions $\mathcal{P}_{1,3}$ and $\mathcal{P}_{2,3}$ are identical with respect to everything behind the nodes $p_{1,3}$ and $p_{2,3}$, respectively, we obtain that the characteristic of $\mathcal{P}$ is also contained in $F(l_{2,3})$, witnessing that $\widetilde{G}_2 \oplus \widetilde{H}$ has a path decomposition with the same width as $\mathcal{P}$.

The above Proposition illuminates the usefulness of characteristics to show FII for the PATHWIDTH problem. In particular, it follows that if $FSCP(\widetilde{G}_1) = FSCP(\widetilde{G}_2)$, then $\widetilde{G}_1 \equiv_{\mathrm{pw},t} \widetilde{G}_2$, for all t-boundaried graphs $\widetilde{G}_1$ and $\widetilde{G}_2$. Hence, the full set of characteristics of a boundaried graph fully describes its equivalence class with respect to $\equiv_{\mathrm{pw},t}$. However, as mentioned above the full set of characteristics of a boundaried graph can be infinite. We will show in the next section that if we consider FII with respect to a class $\mathcal{C}$ of graphs of bounded pathwidth, then it is sufficient to consider the set $FSCP_{(\mathrm{pw}(\widetilde{G})+t)}(\widetilde{G})$ instead of $FSCP(\widetilde{G})$ for every t-boundary graph $\widetilde{G} = (G, \partial(G))$ with $G \in \mathcal{C}$. Because $\mathrm{pw}(\widetilde{G})$ is bounded by a constant the set $FSCP_{(\mathrm{pw}(\widetilde{G})+t)}$ is finite due to Proposition 2.

In the following we introduce characteristics for tree decompositions of boundaried graphs. All the explanations for characteristics of path decompositions transfer to characteristics of tree decompositions and we will not repeat them here. In [3, Definition 5.9] the authors define the *characteristic* of a tree decomposition of a boundaried graph. They show the following:

Proposition 4 ([3, Remark below Lemma 5.3]). *Let $\widetilde{G}$ be a t-boundaried graph and w an integer. Then the number of characteristics of tree decompositions of width at most w of $\widetilde{G}$ is bounded a function of t and w.*

For an integer w, the *full set of (tree decomposition) characteristics* of $\widetilde{G}$ of width at most w (as defined in [3, Definition 5.11]), denoted by $\text{FSCT}_w(\widetilde{G})$, is the set of all characteristics of tree decompositions of $\widetilde{G}$ of width at most w. We denote by $\text{FSCT}(\widetilde{G})$ the (possible infinite) set $\bigcup_{w \in \mathbf{N}} \text{FSCT}_w(\widetilde{G})$.

Proposition 5 ([3, Sect. 5.3]). *Let $\widetilde{H}$, $\widetilde{G}_1$ and $\widetilde{G}_2$ be t-boundaried graphs, and let T be a tree decomposition of $\widetilde{G}_1 \oplus \widetilde{H}$. If the (unique) characteristic of $\mathcal{P}|G_1$ is in $\text{FSCT}(\widetilde{G}_2)$, then there is a tree decomposition of $\widetilde{G}_2 \oplus \widetilde{H}$ that has the same width as T.*

3   Pathwidth has FII on Graphs of Small Pathwidth

As stated in the previous section, we will make use of characteristics of path decompositions of boundaried graphs to show FII for the PATHWIDTH problem in a class of graphs of bounded pathwidth. In particular, we will show that the equivalence relation $\equiv$ defined by $\widetilde{G}_1 \equiv \widetilde{G}_2$ if and only $\text{FSCP}_{(\text{pw}(G_1)+t)}(\widetilde{G}_1) = \text{FSCP}_{(\text{pw}(G_2)+t)}(\widetilde{G}_2)$ is a refinement of the equivalence relation $\equiv_{\text{pw},t}$. Central to our proof is the following lemma, which we believe to be interesting in its own right.

Lemma 1. *Let $\widetilde{G}_1, \widetilde{G}_2$ be two t-boundaried graphs, $G = \widetilde{G}_1 \oplus \widetilde{G}_2$, and $\mathcal{P} = (P, \chi)$ be a path decomposition of G. Then there is a path decomposition $\mathcal{P}' = (P', \chi')$ of G of the same width as $\mathcal{P}$ such that $\mathcal{P}'|G_1$ has width at most $\text{pw}(G_1) + t$.*

Proof. If $\mathcal{P}|G_1$ has width at most $\text{pw}(G_1) + t$, then $\mathcal{P}' := \mathcal{P}$ is the required path decomposition of G. Otherwise, there is a bag $p \in V(P)$ such that $|\chi(p) \cap V(G_1)| > \text{pw}(G_1) + t + 1$. Call such a bag p a *bad bag* of $\mathcal{P}$. The next claim shows that we can eliminate the bad bags of $\mathcal{P}$ one by one without introducing new bad bags. Hence, we obtain the desired path decomposition $\mathcal{P}'$ from $\mathcal{P}$ by a repeated application of the following claim:

Claim 1. *There is a path decomposition $\mathcal{P}'' = (P'', \chi'')$ of G of the same width as $\mathcal{P}$ such that the set of bad bags of $\mathcal{P}''$ is a proper subset of the set of bad bags of $\mathcal{P}$. Moreover, the bag p is no longer a bad bag of $\mathcal{P}''$.*

Let $\chi_{G_1}(p)$ be the set of vertices $\chi(p) \cap V(G_1)$ and let S be a minimal separator between $\chi_{G_1}(p)$ and $\partial(G_1)$ in the graph G. Since $\partial(G_1)$ separates $\chi_{G_1}(p)$ from $\partial(G_1)$ and is of cardinality at most t, we obtain that $|S| \leq t$. Let W be the set of all vertices reachable from $\chi_{G_1}(p)$ in $G \setminus S$, and let $\mathcal{P}_W = (P_W, \chi_W)$ be an optimal path decomposition of $G[W]$. Then, because $W \subseteq V(G_1)$, it follows that the width $\mathcal{P}_W$ is at most the pathwidth of G_1.

To obtain the desired path decomposition $\mathcal{P}''$, where p is not a bad bag anymore, we delete all vertices of W from the bags of $\mathcal{P}$ and, instead, insert the path decomposition $\mathcal{P}_W$ between p and an arbitrary neighbor of p in P. To ensure Property P3 of a path decomposition for the vertices in $\chi(p) \setminus V(G_1)$, we add $\chi(p) \setminus V(G_1)$ to every bag of $\mathcal{P}_W$ in $\mathcal{P}''$. Furthermore, to cover the edges between S and W in G we also need to add S to p and every bag of $\mathcal{P}_W$. Because $\chi(p)$ does not necessarily contain all vertices of S, this could potentially violate the Property P3 of a path decomposition. To get around this we will add a vertex $s \in S$ to every bag $p' \in V(P)$ in between p and any bag containing s, i.e., we complete $\mathcal{P}''$ into a valid path decomposition in a minimal way. This completes the construction of $\mathcal{P}''$ and it remains to argue that adding these vertices from S does not increase the width of any bag in $\mathcal{P}$. Suppose it does, and let p_2 be a bag where we add more vertices than we remove. It follows that there is a bag $p_1 \in V(P)$ such that p_2 lies on the path from p_1 to p in P and $|R| < |S'|$, where $R = \chi(p_2) \cap W$ and $S' = (\chi(p_1) \setminus \chi(p_2)) \cap S$. Note that in $\mathcal{P}|G[W \cup S']$ we have $\chi_{G_1[W \cup S']}(p_2) = R$. Because of Proposition 1 applied to $\mathcal{P}|G_1[W \cup S']$, R separates $\chi_{G_1[W \cup S']}(p)$ from S' in $G_1[W \cup S']$.

We claim that $S'' = (S \setminus S') \cup R$ is a separator between $\chi_{G_1}(p)$ and $\partial(G_1)$. Since $|S''| < |S|$, this would contradict the minimality of S. Let Π be a path between $\chi_{G_1}(p)$ and $\partial(G_1)$. Since $\chi_{G_1}(p) \subseteq W \cup S$, Π has to intersect S in order to reach $\partial(G_1)$. Let s be the first vertex of Π which intersects S (note that the subpath from $\chi_{G_1}(p)$ to s of Π lies entirely in W). Either $s \in S \setminus S'$ and therefore $s \in S''$, or $s \in S'$ and the subpath from $\chi_{G_1}(p)$ to s of Π lies entirely in $W \cup S'$, and therefore Π has to intersect $R \subseteq S''$ in order to reach s. It follows that S'' is indeed a separator between $\chi_{G_1}(p)$ and $\partial(G_1)$, completing the proof. $\qquad \square$

We note here that the bound for the pathwidth given in the above lemma is essentially tight. To see this consider the complete bipartite graph G that has t vertices on one side (side A) and $t+1$ vertices on the other side (side B). Let $\widetilde{G}_1$ be the graph $G[A]$ with boundary A, let $\widetilde{G}_2$ be the graph G with boundary A, and let $\mathcal{P}$ be an optimal path decomposition of $\widetilde{G}_1 \oplus \widetilde{G}_2 = G$. Then, because G is a complete bipartite graph, whose smaller side is A, it holds that $\mathcal{P}$ contains a bag that contains A. Consequently, $\mathrm{pw}(\mathcal{P}'|G_1) = t-1$ (for every path decomposition $\mathcal{P}'$ of G that has the same width as $\mathcal{P}$), however, $\mathrm{pw}(G_1) = 0$.

Corollary 1. *Let $\widetilde{G}_1$ and $\widetilde{G}_2$ be two t-boundaried graphs and $G = \widetilde{G}_1 \oplus \widetilde{G}_2$. Then there is an optimal path decomposition $\mathcal{P}$ of G such that $\mathcal{P}|G_1$ has width at most $\mathrm{pw}(G_1) + t$.*

The following lemma shows that $\equiv$ is a refinement of $\equiv_{\mathrm{pw},t}$.

Lemma 2. *Let $\widetilde{G}_1$ and $\widetilde{G}_2$ be two t-boundaried graphs. If $FSCP_{(\mathrm{pw}(G_1)+t)}(\widetilde{G}_1) = FSCP_{(\mathrm{pw}(G_2)+t)}(\widetilde{G}_2)$, then $\widetilde{G}_1 \equiv_{\mathrm{pw},t} \widetilde{G}_2$.*

Proof. Let $\widetilde{G}_1$ and $\widetilde{G}_2$ be two t-boundaried graphs such that $FSCP_{(\mathrm{pw}(G_1)+t)}(\widetilde{G}_1) = FSCP_{(\mathrm{pw}(G_2)+t)}(\widetilde{G}_2)$. We show that $\mathrm{pw}(\widetilde{G}_1 \oplus \widetilde{H}) \leq \xi$ if and only if $\mathrm{pw}(\widetilde{G}_2 \oplus \widetilde{H}) \leq \xi$ for any t-boundaried graph $\widetilde{H}$ and any $\xi \in \mathbf{N}$. This implies $\widetilde{G}_1 \equiv_{\mathrm{pw},t} \widetilde{G}_2$ with $\varDelta_{\mathrm{pw},t}(\widetilde{G}_1, \widetilde{G}_2) = 0$.

Let $\widetilde{H}$ and ξ be such that $\mathrm{pw}(\widetilde{G}_1 \oplus \widetilde{H}) \leq \xi$. It follows from Corollary 1 that there is a path decomposition $\mathcal{P} = (P, \chi)$ of $\widetilde{G}_1 \oplus \widetilde{H}$ of width at most ξ such that $\mathcal{P}|G_1$ is a path decomposition of G_1 of width at most $\mathrm{pw}(G_1) + t$.

Hence, there is a characteristic in $\mathrm{FSCP}_{(\mathrm{pw}(G_1)+t)}(\widetilde{G}_1)$ corresponding to $\mathcal{P}|G_1$. Because $\mathrm{FSCP}_{(\mathrm{pw}(G_1)+t)}(\widetilde{G}_1) = \mathrm{FSCP}_{(\mathrm{pw}(G_2)+t)}(\widetilde{G}_2)$, we have that $\widetilde{G}_2$ has the same characteristic. It now follows from Proposition 3 that there is a path decomposition of $\widetilde{G}_2 \oplus \widetilde{H}$ that has the same width as $\mathcal{P}$ and hence $\mathrm{pw}(\widetilde{G}_2 \oplus \widetilde{H}) \leq \xi$, as required. Because the reverse direction is analogous, this concludes the proof of the lemma. $\square$

We are now ready to show the main result of this section, i.e., that the PATH-WIDTH problem has FII on graphs of bounded pathwidth.

Theorem 1. *For $w \in \mathbf{N}$, let $\mathcal{G}_w$ be a class of graphs that have pathwidth at most w. Then, the problem PATHWIDTH has FII in $\mathcal{G}_w$.*

Proof. We say that two t-boundaried graphs $\widetilde{G}_1$ and $\widetilde{G}_2$ with $G_1, G_2 \in \mathcal{G}_w$ are *equivalent*, denoted by $\equiv_{\mathrm{FSCP}}$ whenever $\mathrm{FSCP}_{(\mathrm{pw}(G_1)+t)}(\widetilde{G}_1) = \mathrm{FSCP}_{(\mathrm{pw}(G_2)+t)}(\widetilde{G}_2)$. Because $\mathrm{pw}(G_1) \leq w$ and $\mathrm{pw}(G_2) \leq w$, it follows from Proposition 2 that the number of equivalence classes of $\equiv_{\mathrm{FSCP}}$ is finite for every $t \in \mathbf{N}$. Furthermore, because of Lemma 2 it holds that $\equiv_{\mathrm{FSCP}}$ is a refinement of $\equiv_{\mathrm{pw},t}$, which concludes the proof of the theorem. $\square$

4   Treewidth has FII on Graphs of Small Treewidth

As the main ideas of the proof for treewidth are the same as for pathwidth (see the previous section) we will not repeat them here but instead only present the steps of the proof that differ significantly.

Lemma 3. *Let $\widetilde{G}_1$ and $\widetilde{G}_2$ be two t-boundaried graphs, $G = \widetilde{G}_1 \oplus \widetilde{G}_2$, and $\mathcal{T} = (T, \chi)$ be a tree decomposition of G. Then there is a tree decomposition $\mathcal{T}' = (T', \chi')$ of G with the same width as $\mathcal{T}$ such that $\mathcal{T}'|G_1$ has width at most $\mathrm{tw}(G_1) + t$.*

Proof. If $\mathcal{T}|G_1$ has width at most $\mathrm{tw}(G_1) + t$, then $\mathcal{T}' := \mathcal{T}$ is the required tree decomposition of G. Hence, there is a bag $p \in V(T)$ such that $|\chi(p) \cap V(G_1)| > \mathrm{tw}(G_1) + t + 1$. We call such a bag p a *bad* bag of $\mathcal{T}$. The next claim shows that we can eliminate the bad bags of $\mathcal{T}$ one by one without introducing new bad bags. Hence, we obtain the desired tree decomposition $\mathcal{T}'$ from $\mathcal{T}$ by a repeated application of the following claim.

Claim 2. *There is a tree decomposition $\mathcal{T}'' = (T'', \chi'')$ of G of the same width as $\mathcal{T}$ such that the set of bad bags of $\mathcal{T}''$ is a proper subset of the set of bad bags of $\mathcal{T}$. Moreover, the bag p is no longer a bad bag of $\mathcal{T}''$.*

Let $\chi_{G_1}(p)$ be the set of vertices in $\chi(p) \cap V(G_1)$ and let S be a minimal separator between $\chi_{G_1}(p)$ and $\partial(G_1)$ in the graph G. Then, because $\partial(G_1)$ is a separator between $\chi_{G_1}(p)$ and $\partial(G_1)$ of cardinality at most t, we obtain that $|S| \leq t$. Let W be the set of all vertices reachable from $\chi_{G_1}(p)$ in $G \setminus S$, and let $\mathcal{T}_W = (T_W, \chi_W)$ be an optimal tree decomposition of $G[W]$. Then, because $W \subseteq V(G_1)$, it follows that the width $\mathcal{T}_W$ is at most the treewidth of G_1.

To obtain the desired tree decomposition T'', where p is not a bad bag anymore, we delete all vertices of W from the bags of T and, instead, insert the tree decomposition T_W by connecting any node of T_W via an edge to p in T. However, to cover the edges between S and W in G we also need to add S to p and every bag of T_W. Because $\chi(p)$ does not necessarily contain all vertices of S, this could potentially violate the property P3 of a tree decomposition. To get around this we will add a vertex $s \in S$ to every bag $p' \in V(T)$ that is on a path between p and any bag containing s in T, i.e., we complete T'' into a valid tree decomposition in a minimal way. This completes the construction of T'' and it remains to argue that adding these vertices from S does not increase the width of any bag in T. Suppose it does, and let p_2 be a bag where we add more vertices than we remove. Let $S' \subseteq S$ be the set of added vertices and $R = \chi(p_2) \cap W$ the set of removed vertices. It follows that $|R| < |S'|$ and the bag p_2 separates in T the set of bags containing a vertex from S' from the bag p. Note that in $T|G[W \cup S']$ we have $\chi_{G_1[W \cup S']}(p_2) = R$. Because of Proposition 1 applied to $T|G_1[W \cup S']$, R separates $\chi_{G_1[W \cup S']}(p)$ from S' in $G_1[W \cup S']$.

We claim that $S'' = (S \setminus S') \cup R$ is a separator between $\chi_{G_1}(p)$ and $\partial(G_1)$. Since $|S''| < |S|$, this would contradict the minimality of S. Let Π be a path between $\chi_{G_1}(p)$ and $\partial(G_1)$. Since $\chi_{G_1}(p) \subseteq W \cup S$, Π has to intersect S in order to reach $\partial(G_1)$. Let s be the first vertex of Π which intersects S (note that the subpath from $\chi_{G_1}(p)$ to s of Π lies entirely in W). Either $s \in S \setminus S'$ and therefore $s \in S''$, or $s \in S'$ and the subpath from $\chi_{G_1}(p)$ to s of Π lies entirely in $W \cup S'$, and therefore Π has to intersect $R \subseteq S''$ in order to reach s. It follows that S'' is indeed a separator between $\chi_{G_1}(p)$ and $\partial(G_1)$, completing the proof. □

Corollary 2. *Let $\widetilde{G}_1$ and $\widetilde{G}_2$ be two t-boundaried graphs and $G = \widetilde{G}_1 \oplus \widetilde{G}_2$. Then there is an optimal tree decomposition T of G such that $T|G_1$ has width at most $tw(G_1) + t$.*

Employing a technical lemma similar to Lemma 2, we obtain our main result of this section.

Theorem 2. *For $w \in \mathbf{N}$, let $\mathcal{G}_w$ be a class of graphs that have treewidth at most w. Then, the problem TREEWIDTH has FII in $\mathcal{G}_w$.*

Proof. The proof is analogous to the proof of Theorem 1.   □

5 FII of Pathwidth and Treewidth on Arbitrary Graphs

In the previous section we have seen that PATHWIDTH and TREEWIDTH have FII on the class of graphs of bounded pathwidth or treewidth, respectively. It is hence natural to ask whether the same holds true for PATHWIDTH and TREEWIDTH on the class of all graphs. The following theorem establishes that this is not the case.

Theorem 3. *The problems PATHWIDTH and TREEWIDTH do not have FII.*

Proof. For $w, t \in \mathbf{N}$, let $\widetilde{G}_w = (G_w, \partial(G_w))$ be the t-boundaried complete graph with $w + t$ vertices. We claim that $G_w \not\equiv_{\mathrm{pw},t} G_{w+1}$ and $G_w \not\equiv_{\mathrm{tw},t} G_{w+1}$ for every $w \in \mathbf{N}$ with $w > t$. This shows that neither $\equiv_{\mathrm{pw},t}$ nor $\equiv_{\mathrm{tw},t}$ is finite and concludes the proof of the theorem.

Let $\widetilde{H}_1 = \widetilde{G}_w$ and $\widetilde{H}_2 = \widetilde{G}_{w+1}$. Then, $\mathrm{pw}(\widetilde{G}_w \oplus \widetilde{H}_1) = \mathrm{tw}(\widetilde{G}_w \oplus \widetilde{H}_1) = t + w$ and $\mathrm{pw}(\widetilde{G}_{w+1} \oplus \widetilde{H}_1) = \mathrm{tw}(\widetilde{G}_{w+1} \oplus \widetilde{H}_1) = t + w + 1$ but $\mathrm{pw}(\widetilde{G}_w \oplus \widetilde{H}_2) = \mathrm{tw}(\widetilde{G}_w \oplus \widetilde{H}_2) = t + w + 1$ and $\mathrm{pw}(\widetilde{G}_{w+1} \oplus \widetilde{H}_2) = \mathrm{tw}(\widetilde{G}_{w+1} \oplus \widetilde{H}_2) = t + w + 1$, as required.   □

6   Application to Kernelization in Sparse Graph Classes

Using the results presented in [9], we immediately obtain the following:

Corollary 3. *Let $\mathcal{G}$ be a hereditary graph class, t a constant, $G \in \mathcal{G}$, and let k be the size of a treedepth-t-modulator of G. Then* TREEWIDTH *and* PATHWIDTH *admit*

1. *a linear kernel if $\mathcal{G}$ has bounded expansion;*
2. *a quadratic vertex kernel if $\mathcal{G}$ has locally bounded expansion;*
3. *a polynomial kernel if $\mathcal{G}$ is nowhere-dense*

when parameterized by k.

Using the result presented in [10], we obtain the following:

Corollary 4. *Let H be a graph, t a constant and let G be a graph excluding H as a topological minor. Then* TREEWIDTH *admits a linear kernel when parameterized by the size of a treewidth-t-modulator of G and* PATHWIDTH *admits a linear kernel when parameterized by the size of a pathwidth-t-modulator of G.*

Furthermore, both kernelization algorithms only take time linear in $|V(G)|$.

7   Conclusion

We have shown that the problems TREEWIDTH and PATHWIDTH have finite integer index if restricted to (boundaried) graphs of constant treewidth or pathwidth, respectively. This result directly implies that certain kernelization for these two problems exist in sparse graph classes, if parameterized by suitable structural parameters. We see this as an encouragement to revisit problems that might not have FII in general, but do so if restricted to graphs with small width measures.

In particular, it would be useful to develop sufficient conditions akin to the ones presented in [2] to quickly check whether a problem has FII in that restricted setting and thus is amenable to the reduction rule.

Acknowledgement. We thank Hans L. Bodlaender for valuable discussions about the properties of characteristics of path and tree decompositions.

References

1. Arnborg, S., Courcelle, B., Proskurowski, A., Seese, D.: An algebraic theory of graph reduction. J. ACM **40**(5), 1134–1164 (1993)
2. Bodlaender, H.L., Fomin, F.V., Lokshtanov, D., Penninkx, E., Saurabh, S., Thilikos, D.M.: (Meta) kernelization. In: 50th Annual IEEE Symposium on Foundations of Computer Science, FOCS 2009, pp. 629–638. IEEE Computer Society, Atlanta, Georgia, USA, 25–27 Oct 2009
3. Bodlaender, H.L., Kloks, T.: Efficient and constructive algorithms for the pathwidth and treewidth of graphs. J. Algorithms **21**(2), 358–402 (1996)
4. Bodlaender, H.L., van Antwerpen-de Fluiter, B.: Reduction algorithms for graphs of small treewidth. Inf. Comput. **167**(2), 86–119 (2001)
5. Courcelle, B.: The Monadic second-order theory of graphs. I. Recognizable sets of finite graphs. Inf. Comput. **85**, 12–75 (1990)
6. Diestel, R.: Graph Theory, 4th edn. Springer, Heidelberg (2010)
7. Fellows, M.R., Langston, M.A.: An analogue of the Myhill-Nerode Theorem and its use in computing finite-basis characterizations. In: Proceedings of the 30th Annual Symposium on Foundations of Computer Science, pp. 520–525. IEEE Computer Society (1989)
8. Fomin, F.V., Lokshtanov, D., Saurabh, S., Thilikos, D.M.: Bidimensionality and kernels. In: Proceedings of the Twenty-first Annual ACM-SIAM Symposium on Discrete Algorithms, SODA '10, pp. 503–510. Society for Industrial and Applied Mathematics (2010)
9. Gajarský, J., Hliněný, P., Obdržálek, J., Ordyniak, S., Reidl, F., Rossmanith, P., Sánchez Villaamil, F., Sikdar, S.: Kernelization using structural parameters on sparse graph classes. In: Bodlaender, H.L., Italiano, G.F. (eds.) ESA 2013. LNCS, vol. 8125, pp. 529–540. Springer, Heidelberg (2013)
10. Kim, E.J., Langer, A., Paul, C., Reidl, F., Rossmanith, P., Sau, I., Sikdar, S.: Linear kernels and single-exponential algorithms via protrusion decompositions. In: Fomin, F.V., Freivalds, R., Kwiatkowska, M., Peleg, D. (eds.) ICALP 2013, Part I. LNCS, vol. 7965, pp. 613–624. Springer, Heidelberg (2013)
11. Nešetřil, J., Ossona de Mendez, P.: Sparsity (graphs, structures, and algorithms). Algorithms and Combinatorics, vol. 28, 465 pp. Springer, Berlin (2012)

A Parameterized Study of Maximum Generalized Pattern Matching Problems

Sebastian Ordyniak$^{(\boxtimes)}$ and Alexandru Popa

Faculty of Informatics, Masaryk University, Brno, Czech Republic
sordyniak@gmail.com, popa@fi.muni.cz

Abstract. The generalized function matching (GFM) problem has been intensively studied starting with [7]. Given a pattern p and a text t, the goal is to find a mapping from the letters of p to non-empty substrings of t, such that applying the mapping to p results in t. Very recently, the problem has been investigated within the framework of parameterized complexity [9].

In this paper we study the parameterized complexity of the optimization variant of GFM (called Max-GFM), which has been introduced in [1]. Here, one is allowed to replace some of the pattern letters with some special symbols "?", termed wildcards or don't cares, which can be mapped to an arbitrary substring of the text. The goal is to minimize the number of wildcards used.

We give a complete classification of the parameterized complexity of Max-GFM and its variants under a wide range of parameterizations, such as, the number of occurrences of a letter in the text, the size of the text alphabet, the number of occurrences of a letter in the pattern, the size of the pattern alphabet, the maximum length of a string matched to any pattern letter, the number of wildcards and the maximum size of a string that a wildcard can be mapped to.

1 Introduction

In the generalized function matching problem one is given a text t and a pattern p and the goal is to decide whether there is a match between p and t, where a single letter of the pattern is allowed to match multiple letters of the text (we say that p GF-matches t). For example, if the text is $t = xyyx$ and the pattern is $p = aba$, then a generalized function match (on short, GF-match) is $a \to x, b \to yy$, but if $t = xyyz$ and $p = aba$, then there is no GF-match. If, moreover, the matching is required to be injective, then we term the problem generalized parameterzied matching (GPM). In [1], Amir and Nor describe applications of GFM in various areas such as software engineering, image searching, DNA analysis, poetry and music analysis, or author validation. GFM is also related to areas such as (un-)avoidable patterns [12], word equations [13] and the ambiguity of morphisms [11].

Sebastian Ordyniak — Research funded by Employment of Newly Graduated Doctors of Science for Scientific Excellence (CZ.1.07/2.3.00/30.0009).

M. Cygan and P. Heggernes (Eds.): IPEC 2014, LNCS 8894, pp. 270–281, 2014.
DOI: 10.1007/978-3-319-13524-3_23

GFM has a long history starting from 1979. Ehrenfeucht and Rozenberg [7] show that GFM is NP-complete. Independently, Angluin [2,3] studies a more general variant of GFM where the pattern may contain also letters of the text alphabet. Angluin's paper received a lot of attention, especially in the learning theory community [17,18,20] (see [14] for a survey) but also in many other areas.

Recently, a systematic study of the classical complexity of a number of variants of GFM and GPM under various restrictions has been carried out [8,19]. It was shown that GFM and GPM remain **NP**-complete for many natural restrictions. Moreover, the study of GFM and its variants within the framework of parameterized complexity has recently been initiated [9].

In this paper we study the parameterized complexity of the optimization variant of GFM (called Max-GFM) and its variants, where one is allowed to replace some of the pattern letters with some special symbols "?", termed wildcards or don't cares, which can be mapped to an arbitrary substring of the text. The goal is to minimize the number of wildcards used. The problem was first introduced to the pattern matching community by Amir and Nor [1]. They show that if the pattern alphabet has constant size, then a polynomial algorithm can be found, but that the problem is **NP**-complete otherwise. Then, in [4], it is shown the **NP**-hardness of the GFM (without wildcards) and the **NP**-hardness of the GFM when the function f is required to be an injection (named GPM). More specifically, GFM is **NP**-hard even if the text alphabet is binary and each letter of the pattern is allowed to map to at most two letters of the text [4]. In the same paper it is given a $\sqrt{OPT}$ approximation algorithm for the optimization variant of GFM where the goal is to search for a pattern p' that GF-matches t and has the smallest Hamming distance to p. In [5] the optimization versions of GFM and GPM are proved to be **APX**-hard.

OUR RESULTS. Before we discuss our results, we give formal definitions of the problems. In the following let t be a text over an alphabet Σ_t and let $p = p_1 \ldots p_m$ be a pattern over an alphabet Σ_p. We say that p *GF-matches* t if there is a function $f : \Sigma_p \to \Sigma_t^+$ such that $f(p_1) \ldots f(p_m) = t$. To improve the presentation we will sometimes abuse notation by writing $f(p)$ instead of $f(p_1) \ldots f(p_m)$. Let k be a natural number. We say that a pattern p *k-GF-matches* t if there is a text p' over alphabet $\Sigma_p \cup \{?_1, \ldots, ?_k\}$ of Hamming distance at most k from p such that p' GF-matches t.

Problem 1. (MAXIMUM GENERALIZED FUNCTION MATCHING). Given a text t, a pattern p, and an integer k, decide whether p k-GF-matches t.

The Max-GFM can be seen as the optimization variant of GFM in which we want to replace some of the pattern letters with special wildcard symbols, i.e., the symbols $?_1, \ldots, ?_k$, which can be mapped to any non-empty substring of the text.

We also study the Max-GPM problem. The only difference between Max-GPM and Max-GFM is that for Max-GPM the function f is required to be injective. The notions of GP-matching and k-GP-matching are defined in the natural way, e.g., we say a pattern p *GP-matches* a text t if p *GF-matches* t using an injective function.

In this paper we study the parameterized complexity of the two problems using a wide range of parameters: maximum number of occurrences of a letter in the text $\#\Sigma_t$, maximum number of occurrences of a letter in the pattern $\#\Sigma_p$, size of the text alphabet $|\Sigma_t|$, size of the pattern alphabet $|\Sigma_p|$, the maximum length of a substring of the text that a letter of the pattern alphabet can be mapped to (i.e., $\max_i |f(p_i)|$), the number of wildcard letters $\#?$, and the maximum length of a substring of the text that a wildcard can be mapped to, denoted by $\max |f(?)|$.

Our results are summarized in Table 1. We verified the completeness of our results using a simple computer program. In particular, the program checks for every of the 128 possible combinations of parameters $\mathcal{C}$ that the table contains either: (i) a superset of $\mathcal{C}$ under which Max-GFM/GPM is hard (and thus, Max-GFM/GPM is hard if parameterized by $\mathcal{C}$); or (ii) a subset of $\mathcal{C}$ for which Max-GFM/GPM is fpt (and then we have an fpt result for the set of parameters $\mathcal{C}$). Since some of our results do not hold for both Max-GFM and Max-GPM, we carried out two separate checks, one for Max-GFM and one for Max-GPM.

Table 1. Parameterized Complexity of Max-GFM and Max-GPM.

$\#\Sigma_t$	$|\Sigma_t|$	$\#\Sigma_p$	$|\Sigma_p|$	$\max_i |f(p_i)|$	$\#?$	$\max |f(?)|$	Complexity
par	par	–	–	–	–	–	**FPT** (Cor. 3)
–	par	–	par	par	–	–	**FPT** (Th. 1)
–	par	–	–	par	–	–	**FPT** only GPM (Cor. 1)
–	–	par	par	par	–	par	**FPT** (Cor. 2)
–	–	–	par	par	par	par	**FPT** (Th. 2)
par	–	par	par	par	par	–	**W**[1]-h (Th. 4)
par	–	par	–		par	par	**W**[1]-h (Th. 7)
par	–	par		par	par	par	**W**[1]-h (Th. 5)
–	par	par	par	–	par	par	**W**[1]- h ([9, Th.2.])
–	par	par	par	par	par	–	**W**[1]- h (Th. 6)
–	–	–	par	par	–	par	**W**[1]- h (Th. 3)
–	par	par	–	par	par	par	**para-NP**-h ([1, Cor.1]),
–	par	par	–	par	–	–	**para-NP**-h only GFM [8]
–	–	par	–	par	–	–	**para-NP**-h only GPM [8]

The paper is organized as follows. In Sect. 2 we give preliminaries, in Sect. 3 we present our fixed-parameter algorithms and in Sect. 4 we show our hardness results.

2   Preliminaries

We define the basic notions of Parameterized Complexity and refer to other sources [6,10] for an in-depth treatment. A *parameterized problem* is a set of pairs $\langle \mathbb{I}, k \rangle$, the *instances*, where $\mathbb{I}$ is the main part and k the *parameter*. The parameter is usually a non-negative integer. A parameterized problem is *fixed-parameter tractable (fpt)* if there exists an algorithm that solves any instance $\langle \mathbb{I}, k \rangle$ of size n in time $f(k)n^c$ where f is an arbitrary computable function and c is a constant independent of both n and k. **FPT** is the class of all fixed-parameter tractable decision problems. Because we focus on fixed-parameter tractability of a problem we will sometimes use the notation O^* to suppress exact polynomial dependencies, i.e., a problem with input size n and parameter k can be solved in time $O^*(f(k))$ if it can be solved in time $O(f(k)n^c)$ for some constant c.

Parameterized complexity offers a completeness theory, similar to the theory of NP-completeness, that allows the accumulation of strong theoretical evidence that some parameterized problems are not fixed-parameter tractable. This theory is based on a hierarchy of complexity classes $\textbf{FPT} \subseteq \textbf{W}[1] \subseteq \textbf{W}[2] \subseteq \textbf{W}[3] \subseteq \cdots$ where all inclusions are believed to be strict. An *fpt-reduction* from a parameterized problem P to a parameterized problem Q is a mapping R from instances of P to instances of Q such that (i) $\langle \mathbb{I}, k \rangle$ is a YES-instance of P if and only if $\langle \mathbb{I}', k' \rangle = R(\mathbb{I}, k)$ is a YES-instance of Q, (ii) there is a computable function g such that $k' \leq g(k)$, and (iii) there is a computable function f and a constant c such that R can be computed in time $O(f(k) \cdot n^c)$, where n denotes the size of $\langle \mathbb{I}, k \rangle$.

For our hardness results we will often reduce from the following problem, which is well-known to be **W**[1]-complete [16].

MULTICOLORED CLIQUE
Instance: A k-partite graph $G = \langle V, E \rangle$ with a partition $V_1, \ldots, V_k$ of V.
Parameter: The integer k.
Question: Are there nodes $v_1, \ldots, v_k$ such that $v_i \in V_i$ and $\{v_i, v_j\} \in E$ for all i and j with $1 \leq i < j \leq k$ (i.e. the subgraph of G induced by $\{v_1, \ldots, v_k\}$ is a clique of size k)?

For our hardness proofs we will often make the additional assumptions that (1) $|V_i| = |V_j|$ for every i and j with $1 \leq i < j \leq k$ and (2) $|E_{i,j}| = |E_{r,s}|$ for every i, j, r, and s with $1 \leq i < j \leq k$ and $1 \leq r < s \leq k$, where $E_{i,j} = \{ \{u,v\} \in E \mid u \in V_i \text{ and } v \in V_j \}$ for every i and j as before. To see that MULTICOLORED CLIQUE remains **W**[1]-hard under these additional restrictions we can reduce from MULTICOLORED CLIQUE to its more restricted version using a simple padding construction as follows. Given an instance $\langle G, k \rangle$ of MULTICOLORED CLIQUE we construct an instance of its more restricted version by adding edges (whose endpoints are new vertices) between parts (i.e. $V_1, \ldots, V_k$) that do not already have the maximum number of edges between them and then adding isolated vertices to parts that do not already have the maximum number of vertices.

Even stronger evidence that a parameterized problem is not fixed-parameter tractable can be obtained by showing that the problem remains **NP**-complete even if the parameter is a constant. The class of these problems is called **para-NP**.

A *square* is a string consisting of two copies of the same (non-empty) string. We say that a string is *square-free* if it does not contain a square as a substring.

3   Fixed-Parameter Tractable Variants

In this section we show our fixed-parameter tractability results for Max-GFM and Max-GPM. In particular, we show that Max-GFM and Max-GPM are fixed-parameter tractable parameterized by $|\Sigma_t|$, $|\Sigma_p|$, and $\max_i |f(p_i)|$, and also parameterized by $\#?$, $\max |f(?)|$, $|\Sigma_p|$, and $\max_i |f(p_i)|$. We start by showing fixed-parameter tractability for the parameters $|\Sigma_t|$, $|\Sigma_p|$, and $\max_i |f(p_i)|$. We need the following lemma.

Lemma 1. *Given a pattern $p = p_1 \ldots p_m$ over an alphabet Σ_p, a text $t = t_1 \ldots t_n$ over an alphabet Σ_t, a natural number q, and a function $f : \Sigma_p \to \Sigma_t^+$, then there is a polynomial time algorithm deciding whether p q-GF/GP-matches t using the function f.*

Proof. If we are asked whether p q-GP-matches t and f is not injective, then we obviously provide a negative answer. Otherwise, we use a dynamic programming algorithm that is similar in spirit to an algorithm in [4]. Let $\Sigma_p = \{a_1, \ldots a_k\}$. For every $0 \leq i \leq j \leq n$, we define the function $g(i,j)$ to be the Hamming GFM/GPM-similarity (i.e., m minus the minimum number of wildcards needed) between $t_1 t_2 \ldots t_j$ and $p_1 p_2 \ldots p_i$. Then, we obtain the Hamming GFM/GPM-similarity between p and t as $g(m,n)$. Consequently, if $m - g(m,n) > q$, we return No, otherwise we return Yes.

We now show how to recursively compute $g(i,j)$. If $i = 0$, we set $g(i,j) = 0$ and if $i \leq j$, we set:

$$g(i,j) = \max_{1 \leq k \leq j} \{g(i-1, j-k) + I(t_{j-k+1} \ldots t_j, f(p_i))\}$$

where $I(s_1, s_2)$ is 1 if the strings s_1, and s_2 are the same, and 0 otherwise.

We must first show that the dynamic programming procedure computes the right function and then that it runs in polynomial time. We can see immediately that $g(0,i) = 0$ for all i because in this case the pattern is empty. The recursion step of $g(i,j)$ has two cases: If $t_{j-|f(p_i)|+1} \ldots t_j = f(p_i)$, then it is possible to map p_i to $f(p_i)$, and we can increase the number of mapped letters by one. Otherwise, we cannot increase the Hamming GFM/GPM-similarity. However, we know that p_i has to be set to a wildcard and therefore we find the maximum of the previous results for different length substrings that the wildcard maps to.

It is straightforward to check that $g(i,j)$ can be computed in cubic time. $\square$

Theorem 1. *Max-GFM and Max-GPM parameterized by $|\Sigma_t|$, $|\Sigma_p|$, and $\max_i |f(p_i)|$ are fixed-parameter tractable.*

Proof. Let p, t, and q be an instance of Max-GFM or Max-GPM, respectively. The pattern p q-GF/GP-matches t if and only if there is a function $f : \Sigma_p \to \Sigma_t^+$ such that p q-GF/GP-matches t using f. Hence, to solve Max-GFM/Max-GPM, it is sufficient to apply the algorithm from Lemma 1 to every function $f : \Sigma_p \to \Sigma_t^+$ that could possible constitute to a q-GF/GP-matching from p to t. Because there are at most $(|\Sigma_t|)^{\max_i |f(p_i)| |\Sigma_p|}$ such functions f and the algorithm from Lemma 1 runs in polynomial time, the running time of this algorithm is $O^*((|\Sigma_t|)^{\max_i |f(p_i)| |\Sigma_p|})$, and hence fixed-parameter tractable in $|\Sigma_t|$, $|\Sigma_p|$, and $\max_i |f(p_i)|$. □

Because in the case of Max-GPM it holds that if $|\Sigma_t|$ and $\max_i |f(p_i)|$ is bounded then also Σ_p is bounded by $|\Sigma_t|^{\max_i |f(p_i)|}$, we obtain the following corollary.

Corollary 1. *Max-GPM parameterized by $|\Sigma_t|$ and $\max_i |f(p_i)|$ is fixed-parameter tractable.*

We continue by showing our second tractability result for the parameters $|\Sigma_p|$, $\max_i |f(p_i)|$, $\#?$, and $\max |f(?)|$.

Theorem 2. *Max-GFM and Max-GPM parameterized by $|\Sigma_p|$, $\max_i |f(p_i)|$, $\#?$, $\max |f(?)|$, are fixed-parameter tractable.*

Proof. Let p, t, and q be an instance of Max-GFM or Max-GPM, respectively.

Observe that if we could go over all possible functions $f : \Sigma_p \to \Sigma_t^+$ that could possible constitute to a q-GF/GP-matching from p to t, then we could again apply Lemma 1 as we did in the proof of Theorem 1. Unfortunately, because $|\Sigma_t|$ is not a parameter, the number of these functions cannot be bounded as easily any more. However, as we will show next it is still possible to bound the number of possible functions solely in terms of the parameters. In particular, we will show that the number of possible substrings of t that any letter of the pattern alphabet can be mapped to is bounded by a function of the parameters. Because also $|\Sigma_p|$ is a parameter this immediately implies a bound (only in terms of the given parameters) on the total number of these functions.

Let $c \in \Sigma_p$ and consider any q-GF/GP-matching from p to t, i.e., a text $p' = p'_1 \ldots p'_m$ of Hamming distance at most q to p and a function $f : \Sigma_p \cup \{?_1, \ldots, ?_q\} \to \Sigma_t^+$ such that $f(p'_1) \ldots f(p'_m) = t$. Then either c does not occur in p' or c occurs in p'. In the first case we can assign to c any non-empty substring over the alphabet Σ_t (in the case of Max-GPM one additionally has to ensure that the non-empty substrings over Σ_t that one chooses for distinct letters in Σ_p are distinct). In the second case let p'_i for some i with $1 \leq i \leq m$ be the first occurrence of c in p', let $\overline{p}'_{i-1} = p'_1 \ldots p'_{i-1}$, and let $\overline{p}_{i-1} = p_1 \ldots p_{i-1}$. Furthermore, for every $b \in \Sigma_p \cup \{?_1, \ldots, ?_q\}$ and $w \in (\Sigma_p \cup \{?_1, \ldots, ?_q\})^*$, we denote by $\#(b, w)$ the number of times b occurs in w. Then $f(c) = t_{c_s+1} \ldots t_{c_s + |f(c)|}$ where $c_s = \sum_{j=1}^{i-1} |f(p'_j)|$, which implies that the value of $f(c)$ is fully determined by c_s and $|f(c)|$. Because the number of possible values for $|f(c)|$ is trivially bounded by the parameters (it is bounded by $\max_i |f(p_i)|$), it remains to show that also c_s is bounded by the given parameters.

Because $c_s = \sum_{j=1}^{i-1} |f(p'_j)| = (\sum_{b \in \Sigma_p \cup \{?_1,\ldots,?_q\}} \#(b, \overline{p}'_{i-1})|f(b)|)$, we obtain that the value of c_s is fully determined by the values of $\#(b, \overline{p}'_{i-1})$ and $|f(b)|$ for every $b \in \Sigma_p \cup \{?_1,\ldots,?_q\}$. For every $? \in \{?_1,\ldots,?_q\}$ there are at most 2 possible values for $\#(?, \overline{p}'_{i-1})$ (namely 0 and 1) and there are at most $\max |f(?)|$ possible values for $|f(?)|$. Similarly, for every $b \in \Sigma_p$ there are at most $q+1$ possible values for $\#(b, \overline{p}'_{i-1})$ (the values $\#(b, \overline{p}_{i-1}) - q, \ldots, \#(b, \overline{p}_{i-1}))$ and there are at most $\max_i |f(p_i)|$ possible values for $|f(b)|$. Hence, the number of possible values for c_s is bounded in terms of the parameters, as required. $\square$

Since $|\Sigma_p|$ and $\#\Sigma_p$ together bound $\#?$, we obtain the following corollary.

Corollary 2. *Max-GFM and Max-GPM parameterized by $\#\Sigma_p$, $|\Sigma_p|$, $\max_i |f(p_i)|$, and $\max |f(?)|$ are fixed-parameter tractable.*

Furthermore, because all considered parameters can be bounded in terms of the parameters $\#\Sigma_t$ and $|\Sigma_t|$, we obtain the following corollary as a consequence of any of our above fpt-results.

Corollary 3. *Max-GFM and Max-GPM parameterized by $\#\Sigma_t$ and $|\Sigma_t|$ are fixed-parameter tractable.*

4   Hardness Results

In this subsection we give our hardness results for Max-GFM and Max-GPM. The proofs of the theorems marked with an asterisk ($*$) can be found in the full version of this paper, which is available on arxiv [15].

Theorem 3. ($*$) *Max-GFM and Max-GPM are $W[1]$-hard parameterized by $|\Sigma_p|$, $\max_i |f(p_i)|$, and $\max |f(?)|$ (even if $\max_i |f(p_i)| = 1$ and $\max |f(?)| = 2$).*

Theorem 4. ($*$) *Max-GFM and Max-GPM are $W[1]$-hard parameterized by $\#\Sigma_t$, $\#\Sigma_p$, $|\Sigma_p|$, $\max_i |f(p_i)|$, and $\#?$.*

Theorem 5. ($*$) *(Max-)GFM and (Max-)GPM are $W[1]$-hard parameterized by $\#\Sigma_t$, $\#\Sigma_p$, $\max_i |f(p_i)|$, $\#?$, and $\max |f(?)|$.*

Theorem 6. *Max-GFM and Max-GPM are $W[1]$-hard parameterized by $\#\Sigma_p$, $|\Sigma_p|$, $\max_i |f(p_i)|$, and $\#?$ (even if $\max_i |f(p_i)| = 1$).*

We will show the above theorem by a parameterized reduction from MULTICOL-ORED CLIQUE. Let $G = (V, E)$ be a k-partite graph with partition $V_1, \ldots, V_k$ of V. Let $E_{i,j} = \{\{u, v\} \in E \mid u \in V_i \text{ and } v \in V_j\}$ for every i and j with $1 \leq i < j \leq k$. As we stated in the preliminaries we can assume that $|V_i| = n$ and $|E_{i,j}| = m$ for every i and j with $1 \leq i < j \leq k$.

Let $V_i = \{v_1^i, \ldots, v_n^i\}$, $E_{i,j} = \{e_1^{i,j}, \ldots, e_m^{i,j}\}$, and $k' = 2\binom{k}{2}$. We construct a text t over alphabet Σ_t and a pattern p over alphabet Σ_p from G and k in polynomial time such that:

(C1) the parameters $\#\Sigma_p$, $|\Sigma_p|$, and $\#?$ can be bounded as a function of k.

(C2) p k'-GF/GP-matches t using a function f with $\max_{p \in \Sigma_p} |f(p)| = 1$ if and only if G has a k-clique.

We set $\Sigma_t = \{;, -, \#, +\} \cup \{l_{i,j}, r_{i,j} \mid 1 \le i < j \le k\} \cup \{v_i^j \mid 1 \le i \le n \text{ and } 1 \le j \le k\}$ and $\Sigma_p = \{;, -, \#, D\} \cup \{V_i \mid 1 \le i \le k\}$.

For an edge $e \in E$ between v_l^i and v_s^j where $1 \le i < j \le k$ and $1 \le l, s \le n$, we write $\mathbf{vt}(e)$ to denote the text $v_l^i - v_s^j$. For $l \in \Sigma_p \cup \Sigma_t$ and $i \in \mathbb{N}$ we write $\mathbf{rp}(l, i)$ to denote the text consisting of repeating the letter l exactly i times. We first define a preliminary text t' as follows.

$$\#l_{1,2}; \mathbf{vt}(e_1^{1,2}); \cdots; \mathbf{vt}(e_m^{1,2}); r_{1,2}\# \cdots \#l_{1,k}; \mathbf{vt}(e_1^{1,k}); \cdots; \mathbf{vt}(e_m^{1,k}); r_{1,k}$$
$$\#l_{2,3}; \mathbf{vt}(e_1^{2,3}); \cdots; \mathbf{vt}(e_m^{2,3}); r_{2,3}\# \cdots \#l_{2,k}; \mathbf{vt}(e_1^{2,k}); \cdots; \mathbf{vt}(e_m^{2,k}); r_{2,k}$$
$$\cdots$$
$$\#l_{k-1,k}; \mathbf{vt}(e_1^{k-1,k}); \cdots; \mathbf{vt}(e_m^{k-1,k}); r_{k-1,k}\#$$

We also define a preliminary pattern p' as follows.

$$\#D; V_1 - V_2; D\# \ldots \#D; V_1 - V_k; D$$
$$\#D; V_2 - V_3; D\# \ldots \#D; V_2 - V_k; D$$
$$\cdots$$
$$\#D; V_{k-1} - V_k; D\#$$

Let $r = 2(k' + 1)$. Then t is obtained by preceding t' with the text t'' defined as follows.

$$\#; -\mathbf{rp}(+, r)$$

Similarly, p is obtained by preceding p' with the text p'' defined as follows.

$$\#; -\mathbf{rp}(D, r)$$

This completes the construction of t and p. Clearly, t and p can be constructed from G and k in fpt-time (even polynomial time). Furthermore, because $\#\Sigma_p = r + k' = 2(k' + 1) + k' = 3k' + 1$, $|\Sigma_p| = k + 4$, and $\#? = k'$, condition (C1) above is satisfied. To show the remaining condition (C2), we need the following intermediate lemmas.

Lemma 2. *If G has a k-clique, then p k'-GF/GP-matches to t using a function f with $\max_{p \in \Sigma_p} |f(p)| = 1$.*

Proof. Let $\{v_{h_1}^1, \ldots, v_{h_k}^k\}$ be the vertices and $\{e_{h_{i,j}}^{i,j} \mid 1 \le i < j \le k\}$ be the edges of a k-clique of G with $1 \le h_j \le n$ and $1 \le h_{i,j} \le m$ for every i and j with $1 \le i < j \le k$.

We put k' wildcards on the last k' occurrences of D in p. Informally, these wildcards are mapped in such a way that for every $1 \le i < j \le k$ the substring $; V_i - V_j;$ of the pattern p is mapped to the substring $; \mathbf{vt}(e_{h_{i,j}}^{i,j});$ of the text t. More formally, for i and j with $1 \le i < j \le k$ let $q = (\sum_{o=1}^{o<i}(k-o)) + j$. We map the wildcard on the $2(q-1)$-th occurrence of the letter D in p' with the text

$l_{i,j}; \mathbf{vt}(e_1^{i,j}); \cdots; \mathbf{vt}(e_{h_{i,j}-1}^{i,j})$ and similarly we map the wildcard on the $(2(q-1)+1)$-th occurrence of the letter D in p' with the text $\mathbf{vt}(e_{h_{i,j}+1}^{i,j}); \cdots; \mathbf{vt}(e_m^{i,j}); r_{i,j}$. Note that in this way every wildcard is mapped to a non-empty substring of t and no two wildcards are mapped to the same substring of t, as required.

We then define the k'-GF/GP-matching function f as follows: $f(;) =;$, $f(-) = -$, $f(\#) = \#$, $f(V_i) = v_{h_i}^i$, $f(D) = +$, for every i and h_i with $1 \leq i \leq k$ and $1 \leq h_i \leq n$. It is straightforward to check that f together with the mapping for the wildcards maps the pattern p to the text t. □

Lemma 3. *Let f be a function that k'-GF/GP-matches p to t with $\max_{p \in \Sigma_p} |f(p)| = 1$, then: $f(;) =;$, $f(-) = -$, $f(\#) = \#$, and $f(D) = +$. Moreover, all wildcards have to be placed on all the k' occurrences of D in p'.*

Proof. We first show that $f(D) = +$. Observe that the string t' is square-free (recall the definition of square-free from Sect. 2). It follows that every two consecutive occurrences of pattern letters in p'' have to be mapped to a substring of t''. Because there are $2(k'+1)$ occurrences of D in p'' it follows that at least two consecutive occurrences of D in p'' are not replaced with wildcards and hence D has to be mapped to a substring of t''. Furthermore, since all occurrences of D are at the end of p'', we obtain that D has to be mapped to $+$, as required. Because all occurrences of D in p' have to be mapped to substrings of t' and t' does not contain the letter $+$, it follows that all the k' occurrences of D in p' have to be replaced by wildcards. Since we are only allowed to use at most k' wildcards, this shows the second statement of the lemma. Since no wildcards are used to replace letters in p'' it now easily follows that $f(;) =;$, $f(-) = -$ and $f(\#) = \#$. □

Lemma 4. *If p k'-GF/GP-matches to t using a function f with $\max_{p \in \Sigma_p} |f(p)| = 1$, then G has a k-clique.*

Proof. Let f be a function that k'-GF/GP-matches p to t such that $\max_{p \in \Sigma_p} |f(p)| = 1$. We claim that the set $\{ f(V_i) \mid 1 \leq i \leq k \}$ is a k-clique of G. Because of Lemma 3, we know that $f(\#) = \#$ and that no occurrence of $\#$ in p is replaced by a wildcard. Since the number of occurrences of $\#$ in t is equal to the number of occurrences of $\#$ in p, we obtain that the i-th occurrence of $\#$ in p is mapped to the i-th occurrence of $\#$ in t. Consequently, for every i and j with $1 \leq i < j \leq k$, we obtain that the substring $; V_i - V_j;$ is mapped to a substring of the string $l_{i,j}; \mathbf{vt}(e_1^{i,j}); \cdots; \mathbf{vt}(e_m^{i,j}); r_{i,j}$ in t. Again, using Lemma 3 and the fact that $\max_{p \in \Sigma_p} |f(p)| = 1$, we obtain that both V_i and V_j are mapped to some letter v_l^i and v_s^j for some l and s with $1 \leq l, s \leq n$ such that $\{v_l^i, v_s^j\} \in E$. Hence, $\{ f(V_i) \mid 1 \leq i \leq k \}$ is a k-clique of G. □

Because Condition (C2) is implied by Lemmas 2 and 4, this concludes the proof of Theorem 6.

Theorem 7. *(Max-)GFM and (Max-)GPM are $\mathbf{W[1]}$-hard parameterized by $\#\Sigma_t$, $\#\Sigma_p$, $|\Sigma_p|$, $\#?$, and $\max |f(?)|$.*

We will show the theorem by a parameterized reduction from MULTICOLORED CLIQUE. Let $G = (V, E)$ be a k-partite graph with partition $V_1, \ldots, V_k$ of V. Let $E_{i,j} = \{\, \{u, v\} \in E \mid u \in V_i \text{ and } v \in V_j \,\}$ for every i and j with $1 \le i < j \le k$. Again, as we stated in the preliminaries we can assume that $|V_i| = n$ and $|E_{i,j}| = m$ for every i and j with $1 \le i < j \le k$.

Let $V_i = \{v_1^i, \ldots, v_n^i\}$ and $E_{i,j} = \{e_1^{i,j}, \ldots, e_m^{i,j}\}$. We construct a text t and a pattern p from G and k such that p GF/GP-matches t if and only if G has a k-clique. The alphabet Σ_t consists of:

- the letter $\#$ (used as a separator);
- one letter a_e for every $e \in E$ (representing the edges of G);
- one letter $\#_i$ for every i with $1 \le i \le n$ (used as special separators that group edges from the same vertex);
- the letters $l_{i,j}, r_{i,j}, l_i, r_i$ for every i and j with $1 \le i < j \le k$ (used as dummy letters to ensure injectivity for GPM);
- the letter d_e^v and d^v for every $e \in E$ and $v \in V(G)$ with $v \in e$ (used as dummy letters to ensure injectivity for GPM).

We set $\Sigma_p = \{\#\} \cup \{\, E_{i,j}, L_{i,j}, R_{i,j}, L_i, R_i, A_i \mid 1 \le i < j \le k \,\} \cup \{\, D_{i,j} \mid 1 \le i \le k \text{ and } 1 \le j \le k+1 \,\}$.

For a vertex $v \in V$ and j with $1 \le j \le k$ we denote by $E_j(v)$ the set of edges of G that are incident to v and whose other endpoint is in V_j. Furthermore, for a vertex $v \in V(G)$, we write $\mathbf{e}(v)$ to denote the text $\mathbf{el}(v, E_1(v)) \cdots \mathbf{el}(v, E_k(v)) d^v$, where $\mathbf{el}(v, E')$, for vertex v and a set E' of edges with $E' = \{e_1, \ldots, e_l\}$, is the text $d_{e_1}^v a_{e_1} d_{e_2}^v a_{e_2} \cdots d_{e_l}^v a_{e_l}$.

We first define the following preliminary text and pattern strings. Let t_1 be the text:

$$\#l_{1,2}a_{e_1^{1,2}} \cdots a_{e_m^{1,2}}r_{1,2}\# \cdots \#l_{1,k}a_{e_1^{1,k}} \cdots a_{e_m^{1,k}}r_{1,k}$$
$$\#l_{2,3}a_{e_1^{2,3}} \cdots a_{e_m^{2,3}}r_{2,3}\# \cdots \#l_{2,k}a_{e_1^{2,k}} \cdots a_{e_m^{2,k}}r_{2,k}$$
$$\cdots$$
$$\#l_{k-1,k}a_{e_1^{k-1,k}} \cdots a_{e_m^{k-1,k}}r_{k-1,k}$$

Let t_2 be the text:

$$\#l_1\#_1\mathbf{e}(v_1^1)\#_1 \cdots \#_n\mathbf{e}(v_n^1)\#_n r_1$$
$$\cdots$$
$$\#l_k\#_1\mathbf{e}(v_1^k)\#_1 \cdots \#_n\mathbf{e}(v_n^k)\#_n r_k\#$$

Let p_1 be the pattern:

$$\#L_{1,2}E_{1,2}R_{1,2}\# \ldots \#L_{1,k}E_{1,k}R_{1,k}$$
$$\#L_{2,3}E_{2,3}R_{2,3}\# \ldots \#L_{2,k}E_{2,k}R_{2,k}$$
$$\cdots$$
$$\#L_{k-1,k}E_{k-1,k}R_{k-1,k}$$

For i, j with $1 \le i, j \le k$, let $I(i, j)$ be the letter $E_{i,j}$ if $i < j$, the letter $E_{j,i}$ if $i > j$ and the empty string if $i = j$. We define $\mathbf{p}(1)$ to be the pattern:

$$A_1 D_{1,2} I(1, 2) D_{1,3} I(1, 3) \cdots\cdots D_{1,k} I(1, k) D_{1,k+1} A_1$$

we define $\mathbf{p}(k)$ to be the pattern:

$$A_k D_{k,1} I(k,1) D_{k,2} I(k,2) \cdots\cdots D_{k,k-1} I(k,k-1) D_{k,k+1} A_k$$

and for every i with $1 < i < k$, we define $\mathbf{p}(i)$ to be the pattern:

$$A_i D_{i,1} I(i,1) D_{i,2} I(i,2) \cdots D_{i,i-1} I(i,i-1)$$
$$D_{i,i+1} I(i,i+1) \cdots D_{i,k} I(i,k) D_{i,k+1} A_i$$

Then p_2 is the pattern:

$$\#L_1 \mathbf{p}(1) R_1 \# \cdots \#L_k \mathbf{p}(k) R_k \#$$

We also define t_0 to be the text $\#\#$ and p_0 to be the pattern $\#\#$. Then, t is the concatenation of t_0, t_1 and t_2 and p is a concatenation of p_0, p_1 and p_2.

This completes the construction of t and p. Clearly, t and p can be constructed from G and k in fpt-time (even polynomial time). Furthermore, $\#\Sigma_t = \binom{k}{2} + k + 3$, $\#\Sigma_p = \binom{k}{2} + k + 3$, $|\Sigma_p| = k(k+1) + 3\binom{k}{2} + 3k + 1$ and hence bounded by k, as required. It remains to show that G has a k-clique if and only if p GF/GP-matches t. The proof of this statement can be found in the full version of this paper [15].

References

1. Amir, A., Nor, I.: Generalized function matching. J. Discrete Algorithms **5**(3), 514–523 (2007)
2. Angluin, D.: Finding patterns common to a set of strings (extended abstract). In: Proceedings of the 11h Annual ACM Symposium on Theory of Computing, 30 April–2 May, Atlanta, Georgia, USA, pp. 130–141 (1979)
3. Angluin, D.: Finding patterns common to a set of strings. J. Comput. Syst. Sci. **21**(1), 46–62 (1980)
4. Clifford, R., Harrow, A.W., Popa, A., Sach, B.: Generalised matching. In: Karlgren, J., Tarhio, J., Hyyrö, H. (eds.) SPIRE 2009. LNCS, vol. 5721, pp. 295–301. Springer, Heidelberg (2009)
5. Clifford, R., Popa, A.: (In)approximability results for pattern matching problems. In: Holub, J., Zdárek, J. (eds.) Proceedings of the Prague Stringology Conference 2010, Prague, Czech Republic, 30 August–1 September, pp. 52–62. Prague Stringology Club, Department of Theoretical Computer Science, Faculty of Information Technology, Czech Technical University in Prague (2010)
6. Downey, R.G., Fellows, M.R.: Parameterized Complexity. Monographs in Computer Science. Springer, New York (1999)
7. Ehrenfreucht, A., Rozenberg, G.: Finding a homomorphism between two words in np-complete. Inf. Process. Lett. **9**(2), 86–88 (1979)
8. Fernau, H., Schmid, M.L.: Pattern matching with variables: a multivariate complexity analysis. In: Fischer, J., Sanders, P. (eds.) CPM 2013. LNCS, vol. 7922, pp. 83–94. Springer, Heidelberg (2013)
9. Fernau, H., Schmid, M.L., Villanger, Y.: On the parameterised complexity of string morphism problems. In: Seth, A., Vishnoi, N.K. (eds). IARCS Annual Conference on Foundations of Software Technology and Theoretical Computer Science, FSTTCS 2013. LIPIcs, 12–14 December 2013, Guwahati, India, vol. 24, pp. 55–66. Schloss Dagstuhl - Leibniz-Zentrum fuer Informatik (2013)

10. Flum, J., Grohe, M.: Parameterized Complexity Theory. Texts in Theoretical Computer Science. An EATCS Series, vol. XIV. Springer, Berlin (2006)
11. Freydenberger, D.D., Reidenbach, D., Schneider, J.C.: Unambiguous morphic images of strings. In: De Felice, C., Restivo, A. (eds.) DLT 2005. LNCS, vol. 3572, pp. 248–259. Springer, Heidelberg (2005)
12. Jiang, T., Kinber, E., Salomaa, A., Salomaa, K., Yu, S.: Pattern languages with and without erasing. Int. J. Comput. Math. **50**(3–4), 147–163 (1994)
13. Mateescu, A., Salomaa, A.: Finite degrees of ambiguity in pattern languages. Inform. Théorique et Appl. **28**(3–4), 233–253 (1994)
14. Ng, Y.K., Shinohara, T.: Developments from enquiries into the learnability of the pattern languages from positive data. Theor. Comput. Sci. **397**(13), 150–165 (2008). Forty Years of Inductive Inference: Dedicated to the 60th Birthday of Rolf Wiehagen
15. Ordyniak, S., Popa, A.: A parameterized study of maximum generalized pattern matching problems. CoRR, abs/1402.6109 (2014)
16. Pietrzak, K.: On the parameterized complexity of the fixed alphabet shortest common supersequence and longest common subsequence problems. J. Comput. Syst. Sci. **67**(4), 757–771 (2003)
17. Reidenbach, D.: A non-learnable class of e-pattern languages. Theor. Comput. Sci. **350**(1), 91–102 (2006)
18. Reidenbach, D.: Discontinuities in pattern inference. Theor. Comput. Sci. **397**(13), 166–193 (2008). Forty Years of Inductive Inference: Dedicated to the 60th Birthday of Rolf Wiehagen
19. Schmid, M.L.: A note on the complexity of matching patterns with variables. Inf. Process. Lett. **113**(19–21), 729–733 (2013)
20. Takeshi, S.: Polynomial time inference of extended regular pattern languages. In: Goto, E., Furukawa, K., Nakajima, R., Nakata, I., Yonezawa, A. (eds.) RIMS 1982. LNCS, vol. 147, pp. 115–127. Springer, Heidelberg (1983)

Improved FPT Algorithms for Weighted Independent Set in Bull-Free Graphs

Henri Perret du Cray and Ignasi Sau$^{(\boxtimes)}$

AlGCo Project-team, CNRS, LIRMM, Montpellier, France
henri.perretducray@gmail.com; ignasi.sau@lirmm.fr

Abstract. Very recently, Thomassé, Trotignon and Vušković [WG 2014] have given an FPT algorithm for WEIGHTED INDEPENDENT SET in bull-free graphs parameterized by the weight of the solution, running in time $2^{O(k^5)} \cdot n^9$. In this article we improve this running time to $2^{O(k^2)} \cdot n^7$. As a byproduct, we also improve the previous Turing-kernel for this problem from $O(k^5)$ to $O(k^2)$. Furthermore, for the subclass of bull-free graphs without holes of length at most $2p-1$ for $p \geq 3$, we speed up the running time to $2^{O(k \cdot k^{\frac{1}{p-1}})} \cdot n^7$. As p grows, this running time is asymptotically tight in terms of k, since we prove that for each integer $p \geq 3$, WEIGHTED INDEPENDENT SET cannot be solved in time $2^{o(k)} \cdot n^{O(1)}$ in the class of $\{bull, C_4, \ldots, C_{2p-1}\}$-free graphs unless the ETH fails.

Keywords: Parameterized complexity · FPT algorithm · Bull-free graphs · Independent set · Turing-kernel

1 Introduction

Motivation. Parameterized complexity deals with problems whose instances I come equipped with an additional integer parameter k, and the objective is to obtain algorithms whose running time is of the form $f(k) \cdot \text{poly}(|I|)$, where f is some computable function (see [7,9,17] for an introduction to the field). Such algorithms are called *Fixed-Parameter Tractable* (FPT). A fundamental notion in parameterized complexity is that of *kernelization*, which asks for the existence of polynomial-time preprocessing algorithms that produce equivalent instances whose size depends exclusively (preferably polynomially) on k. We will be only concerned with problems defined on graphs.

In order to obtain efficient FPT algorithms, a usual strategy is to focus on a graph class whose members have a well-defined *structure*, which can then be exploited to design algorithms. This paradigm has been exhaustively used in the last decades to obtain efficient FPT algorithms for graphs that exclude a fixed graph as a *minor*, relying on the structural characterization of this graph class given by Robertson and Seymour in their seminal work [20]. Nevertheless, the

Research supported by the Languedoc-Roussillon Project "Chercheur d'avenir" KERNEL and by the grant EGOS ANR-12-JS02-002-01.

M. Cygan and P. Heggernes (Eds.): IPEC 2014, LNCS 8894, pp. 282–293, 2014.
DOI: 10.1007/978-3-319-13524-3_24

situation is quite different in graphs that exclude a fixed graph as an *induced subgraph*, for which the design of FPT algorithms is still in an incipient stage. Quite recently, the structural description of *claw-free* graphs given by Chudnovsky and Seymour [3] has triggered the design of FPT algorithms in this graph class [4,11,12]. Even more recently, a structural characterization of *bull-free* graphs has been given by Chudnovsky [1,2]. In this article we focus on this latter graph class.

The *bull* is the graph defined by the set of vertices $\{x_1, x_2, x_3, y, z\}$ and the set of edges $\{x_1x_2, x_2x_3, x_3x_1, x_1y, x_2z\}$ (see Fig. 1 for an illustration). For a graph F, a graph G is said to be *F-free* if G does not contain an induced subgraph isomorphic to F. Note that the class of bull-free graphs contains the classes of P_4-free and triangle-free graphs, so in particular it contains all bipartite graphs.

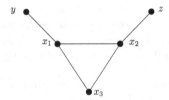

Fig. 1. The bull.

An *independent set* in a graph is a set of pairwise non-adjacent vertices. In a vertex-weighted graph, the *weight* of an independent set is the sum of the weights of its vertices. We are interested in the following parameterized problem.

WEIGHTED INDEPENDENT SET
Input: An graph $G = (V, E)$ with $|V| = n$, a weight function $w : V \to \mathbb{N}$, and a positive integer k.
Parameter: The integer k.
Question: Does G contain an independent set of weight at least k?

The above problem is well-known to be $W[1]$-hard in general graphs [7], and therefore an FPT algorithm is unlikely to exist (see [7,9,17] for the missing definitions). Thus, it is relevant to find graph classes for which the problem admits an FPT algorithm, and for which the non-parameterized version still remains NP-hard. In this direction, Dabrowski et al. [5] gave an FPT algorithm for WEIGHTED INDEPENDENT SET in $\{bull, \overline{P_5}\}$-free graphs, where $\overline{P_5}$ is the complement of a path on 5 vertices. Note that the problem is NP-hard in $\{bull, \overline{P_5}\}$-free graphs, as it is NP-hard in the subclass of triangle-free graphs [19]. Recently, Thomassé et al. [21] generalized this result by giving an FPT algorithm for WEIGHTED INDEPENDENT SET in the class of bull-free graphs, by exploiting the structural results of Chudnovsky [1,2]. This article is the starting point of our work, and its main result is the following.

Theorem 1. (Thomassé et al. [21]). WEIGHTED INDEPENDENT SET *in the class of bull-free graphs can be solved in time $2^{O(k^5)} \cdot n^9$.*

Our results. Our main contribution is to improve the running time of the FPT algorithm of Thomassé et al. [21] stated in Theorem 1, specially in terms of the parameter k.

Theorem 2. WEIGHTED INDEPENDENT SET *in the class of bull-free graphs can be solved in time* $2^{O(k^2)} \cdot n^7$.

We would like to point out that we strongly follow the algorithm of [21], and that our faster algorithm is obtained by improving locally some of the procedures and analyses given in [21]. In particular, one of our main improvements relies on a closer look at the structure of the so-called *basic* bull-free graphs as described by Chudnovsky in her series of papers [1,2].

It is shown in [21, Theorem 7.2] that the FPT algorithm of Theorem 1 actually provides a Turing-kernel[1] of size $O(k^5)$ for WEIGHTED INDEPENDENT SET in bull-free graphs, and that a polynomial kernel is not possible under reasonable complexity hypothesis. Therefore, as our algorithm follows closely that of Theorem 1, from Theorem 2 we immediately obtain the following corollary.

Corollary 1. *There exists a Turing-kernel of size* $O(k^2)$ *for* WEIGHTED INDEPENDENT SET *in the class of bull-free graphs.*

It is natural to ask whether the algorithm of Theorem 2 can be improved for subclasses of bull-free graphs. We prove that it is the case when, in addition to the bull, we exclude the holes[2] of length at most $2p - 1$ for some integer $p \geq 3$ as induced subgraphs. Note that for each $p \geq 3$, the WEIGHTED INDEPENDENT SET problem is NP-hard in the class of $\{bull, C_4, \ldots, C_{2p-1}\}$-free graphs, as for each integer $g \geq 3$, its unweighted version is NP-hard in the class of graphs of girth greater than g [16], that is in $\{C_3, C_4, \ldots, C_g\}$-free graphs, which is a subclass of $\{bull, C_4, \ldots, C_g\}$-free graphs for $g \geq 4$. More precisely, we prove the following theorem.

Theorem 3. *For each integer* $p \geq 3$, WEIGHTED INDEPENDENT SET *in the class of* $\{bull, C_4, \ldots, C_{2p-1}\}$-*free graphs can be solved in time* $2^{O(k \cdot k^{\frac{1}{p-1}})} \cdot n^7$.

In the same way as Corollary 1 follows from Theorem 2, from Theorem 3 we obtain the following corollary. It is worth noting that the multipartite construction given in [21, Theorem 7.1] for ruling out the existence of polynomial kernels actually preserves the property of being $\{bull, C_4, \ldots, C_{2p-1}\}$-free for $p \geq 3$.

Corollary 2. *For each integer* $p \geq 3$, *there exists a Turing-kernel of size* $O(k \cdot k^{\frac{1}{p-1}})$ *for* WEIGHTED INDEPENDENT SET *in the class of* $\{bull, C_4, \ldots, C_{2p-1}\}$-*free graphs.*

---

[1] For a function $g : \mathbb{N} \to \mathbb{N}$, a parameterized problem Π is said to have a *Turing-kernel of size* $g(k)$ if there is an algorithm which, given an input (I, k) together with an oracle for Π that decides whether $(I, k) \in \Pi$ in constant time whenever $|I| \leq g(k)$, decides whether $(I, k) \in \Pi$ in time polynomial in $|I|$ and k.

[2] A *hole* in a graph is an induced cycle of length at least 4.

Finally, we provide lower bounds on the running time on any FPT algorithm that solves WEIGHTED INDEPENDENT SET in the class of $\{bull, C_4, \ldots, C_{2p-1}\}$-free graphs, for $p \geq 3$. These lower bounds rely on the Exponential Time Hypothesis (ETH), which states that there exists a positive real number s such that 3-CNF-SAT with n variables and m clauses cannot be solved in time $2^{sn} \cdot (n+m)^{O(1)}$ (see [15] for more details).

Theorem 4. *For each integer $p \geq 3$, WEIGHTED INDEPENDENT SET cannot be solved in time $2^{o(k)} \cdot n^{O(1)}$ in the class of $\{bull, C_4, \ldots, C_{2p-1}\}$-free graphs unless the ETH fails.*

Note that as p grows, the running time of the algorithm of Theorem 3 tends to $2^{O(k)} \cdot n^7$. As the lower bound given by Theorem 4 holds for *any* fixed integer $p \geq 3$, it follows that, as p grows, the running time of the algorithm of Theorem 3 is asymptotically *tight* with respect to the parameter k.

Organization of the paper. In Sect. 2 we state some definitions and results from [21] that we need in the remaining sections. Section 3 is devoted to the proofs of our main results. Finally, we conclude with some directions for further research in Sect. 4. Due to space limitations, the proofs of the results marked with '[⋆]' can be found in the full version of this article [18].

2   Preliminaries

All the definitions in this section are taken from [21]. We use standard graph-theoretic notation (see [6] for any undefined terminology).

Trigraphs. We need to work with *trigraphs* (see [2]), which are a generalization of graphs in which some edges are left "undecided". Formally, a trigraph consists of a finite set $V(T)$ of vertices and an adjacency function $\theta : \binom{V(T)}{2} \rightarrow \{-1, 0, 1\}$. Two vertices $u, v \in V(T)$ are *strongly adjacent* (resp. *strongly antiadjacent*, resp. *semiadjacent*) if $\theta(uv) = 1$ (resp. $\theta(uv) = -1$, $\theta(uv) = 0$), and in that case u and v constitute a *strong edge* (resp. *strong antiedge*, *switchable pair*). Two vertices $u, v \in V(T)$ are *adjacent* (resp. *antiadjacent*) if $\theta(uv) \in \{0, 1\}$ (resp. $\theta(uv) \in \{-1, 0\}$), and in that case we say that there is an *edge* (resp. *antiedge*) between u and v. Let $\eta(T)$ (resp. $\nu(T)$, $\sigma(T)$) be the set of strongly adjacent (resp. strongly antiadjacent, semiadjacent) pairs of T. That is, a trigraph T is a graph if and only if $\sigma(T) = \emptyset$. For a vertex $v \in V(T)$, $N(v)$ (resp. $\eta(T)$, $\nu(T)$, $\sigma(T)$) denotes the set of vertices in $V(T) \setminus \{v\}$ that are adjacent (resp. strongly adjacent, strongly antiadjacent, semiadjacent) to v. The complement $\overline{T}$ of a trigraph T is the trigraph with $V(\overline{T}) = V(T)$ and $\theta(\overline{T}) = -\theta(T)$. A trigraph is *monogamous* if every vertex belongs to at most one switchable pair. Most trigraphs considered in this paper will be monogamous.

For two disjoint non-empty subsets of vertices A, B of $V(T)$, we say that A is *strongly complete* (resp. *strongly anticomplete*) to B if every vertex in A is strongly adjacent (resp. strongly antiadjacent) to every vertex in B. A *clique* (resp. *strong clique*, *independent set*, *strong independent set*) in T is a set of

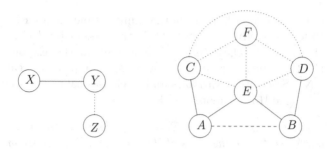

Fig. 2. A homogeneous set X and a homogeneous pair (A,B).

vertices that are pairwise adjacent (resp. strongly adjacent, antiadjacent, strongly antiadjacent). When we speak about the WEIGHTED INDEPENDENT SET problem in a trigraph T, we are interested in finding an independent set in T. We denote by $\alpha(T)$ the maximum weight of an independent set in T (see [21] for the precise restrictions of the weight functions defined in trigraphs).

A *realization* of a trigraph T is any trigraph T' such that $\eta(T) \subseteq \eta(T')$, $\nu(T) \subseteq \nu(T')$, and $\sigma(T') = \emptyset$ (hence T' is a graph). Seen as a trigraph, the *bull* is defined as in Fig. 1, where the corresponding vertices are adjacent or antiadjacent (that is, switchable pairs are allowed). A trigraph is *bull-free* if no induced subtrigraph of it is a bull.

Decomposition of bull-free trigraphs. The algorithm of [21], hence ours as well, is based on a decomposition theorem of bull-free trigraphs that is a simplified version of the one given by Chudnovsky [1,2], and that we proceed to state. We first need two more definitions that will play a fundamental role.

A set $X \subseteq V(T)$ is a *homogeneous set* if $1 < |X| < |V(T)|$ and every vertex in $V(T) \setminus X$ is either strongly complete or strongly anticomplete to X. Thus, $V(T) \setminus X$ can be partitioned into two (possibly empty) sets Y and Z such that X is strongly complete to Y and strongly anticomplete to Z; see Fig. 2 for an illustration, where a solid line means that there are all edges, no line means that there are no edges, and a dashed line means that there is no restriction.

A *homogeneous pair* in T is a pair (A,B) of disjoint non-empty subsets of $V(T)$ such that there exist disjoint (possibly empty) subsets C, D, E, F of $V(T)$ such that the following hold:

- $\{A, B, C, D, E, F\}$ is a partition of $V(T)$;
- $|A \cup B| \geq 3$;
- $|C \cup D \cup E \cup F| \geq 3$;
- A is strongly complete to $C \cup E$ and strongly anticomplete to $D \cup F$;
- B is strongly complete to $D \cup E$ and strongly anticomplete to $C \cup F$; and
- A is not strongly complete nor strongly anticomplete to B.

See again Fig. 2 for an illustration. A homogeneous pair is *small* if $|A \cup B| \leq 6$, and it is *proper* if $C \neq \emptyset$ and $D \neq \emptyset$.

We now define some classes of so-called *basic* trigraphs which will also play an important role in the algorithms. Let T_0 be the class of monogamous trigraphs on at most 8 vertices. Let T_1 be the class of monogamous trigraphs T whose vertex set can be partitioned into (possibly empty) sets $X, K_1, \ldots, K_t$ such that $G[X]$ is triangle free, and $K_1, \ldots, K_t$ are strong cliques that are pairwise anticomplete. According to Chudnovsky's work [1,2], the trigraphs in T_1 satisfy some additional conditions that we will detail in Sect. 3. This closer look at the class T_1 allows us to significantly improve the dependency on k of the algorithm. Finally, let $\overline{T}_1 = \{\overline{T} : T \in T_1\}$. A trigraph is *basic* if it belongs to $T_0 \cup T_1 \cup \overline{T}_1$. We are ready to state the decomposition theorem.

Theorem 5. (Chudnovsky [1,2]). *If T is a bull-free monogamous trigraph, then one of the following holds:*

- T *is basic;*
- T *has a homogeneous set;*
- T *has a small homogeneous pair; or*
- T *has a proper homogeneous pair.*

We say that (X, Y) is a *decomposition* of a trigraph T if (X, Y) is a partition of $V(T)$ and either X is a homogeneous cut of T or $X = A \cup B$ where (A, B) is a small or proper homogeneous pair of T. A decomposition (X, Y) defines two *blocks* T_X and T_Y, whose definition is omitted here, and can be found in [21]. A decomposition (X, Y) is a *homogeneous cut* if X is a homogeneous set or $X = A \cup B$ where (A, B) is a proper homogeneous pair. A homogeneous cut (X, Y) is *minimally-sided* if there is no homogeneous cut (X', Y') with $X' \subsetneq X$.

3   Improved FPT Algorithms in Bull-Free Graphs

In this section we give a proof of Theorem 2. We start by providing a high-level description of the FPT algorithm of [21] in Algorithm 1 below (without giving all the details), which will help us to point out the steps for which we provide an improvement.

As the size of the trigraph T_Y strictly decreases in each recursive step, the overall complexity of Algorithm 1 is easily seen to be upper-bounded by $2^{O(k^5)} \cdot n^9$. (In fact, the algorithm of [21] starts by trying to find a decomposition of T, and if it fails we know by Theorem 5 that T is basic. We reversed the steps in this sketch for the sake of presentation.) Our improvements are the following:

(i) **Improvement in terms of the graph size.** We show that in Step 2, an extreme decomposition (X, Y) of T can be found in time $O(n^6)$.

(ii) **Improvement in terms of the parameter.** We show that in Step 1, the problem can be solved in basic trigraphs in time $O(n^4 m) + 2^{O(k^2)}$.

The two improvements above yield the running time given in Theorem 2. We now proceed to explain these improvements in detail.

Input: A bull-free trigraph T with $|V(T)| = n$ and the parameter k.

Output: 'YES' if $\alpha(T) \geq k$, and an independent set of weight $\alpha(T)$ otherwise.

1. If T is basic, then the problem can be solved in time $O(n^4 m) + 2^{O(k^5)}$, where m is the number of strong edges in T.

2. Otherwise, by Theorem 5, T admits a decomposition. Furthermore, it is shown that T admits a so-called *extreme* decomposition, which is a decomposition (X, Y) such that the block T_X is basic and both T_X and T_Y are bull-free trigraphs. This extreme decomposition can be found in time $O(n^8)$.

   2.1. First, Step 1 is run on the basic bull-free trigraph T_X. If $\alpha(T_X) \geq k$, we answer 'YES' and we stop the algorithm. Otherwise, we use the performed computations to build the weighted trigraph T_Y.

   2.2. The whole algorithm is run recursively on the bull-free trigraph T_Y.

Algorithm 1. Sketch of the FPT algorithm of [21].

Improvement in terms of the graph size. Our first ingredient is the following polynomial-time algorithm running in time $O(n^6)$, which should be compared to the algorithm given by [21, Theorem 4.3] that runs in time $O(n^8)$.

Theorem 6. *There is an algorithm running in time $O(n^6)$ whose input is a trigraph T. The output is a small homogeneous pair of T if some exists. Otherwise, if G has a homogeneous cut, then the output is a minimally-sided homogeneous cut. Otherwise, the output is: "T has no small homogeneous pair, no proper homogenous pair, and no homogenous set".*

The proof of [21, Theorem 4.3] starts by enumerating all sets of vertices of size at most 6 and then it checks whether they define a small homogeneous pair. This procedure takes time $O(n^8)$. Our first improvement is a simple algorithm that finds small homogeneous pairs (A, B) in time $O(n^6)$, if there exists one. Without loss of generality, we can assume that $|A| \geq |B|$. The main idea is to fix the vertices of A and then try to find a suitable B verifying $|A \cup B| \leq 6$. While we have not found a small homogeneous pair, we execute Algorithm 2 below for all possible pairs of positive integers (i, j) such that $3 \leq i + j \leq 6$ and $j \leq i$ (note that there are at most 8 such pairs), in lexicographic order for $i \in \{2, \ldots, 5\}$ and $j \in \{1, \ldots, \min\{1, 6 - i\}\}$.

Lemma 1. *Algorithm 2 is correct and runs in time $O(n^6)$. That is, a small homogeneous pair in a trigraph T can be found in time $O(n^6)$, if it exists.*

Proof. Suppose that T contains a small homogeneous pair (A, B) such that $|A| = i$ and $|B| = j$, and that T does not contain a small homogeneous pair (A', B') with $|A'| = i$ and $|B'| < j$ (such a pair would have been found in previous iterations). We claim that there exists a vertex $v \in R$ that is neither strongly complete nor strongly anticomplete to A, or neither strongly complete nor strongly anticomplete to B. Indeed, otherwise $(A, B \setminus \{v\})$ would be a small homogeneous pair, contradicting the conditions of the algorithm.

Let $B' = B \setminus \{v\}$. At some point, the algorithm will consider the pair (A, B'), and then it will find the corresponding v and check that the found pair is indeed homogeneous. Since $|A| + |B| \leq 6$, these two operations can be done in linear time. Since $i + j - 1$ vertices are guessed, the complexity of the algorithm is $O(n^{i+j}) = O(n^6)$, as $i + j \leq 6$. $\qquad\square$

Input: A trigraph T on n vertices, two positive integers i and j such that
$\quad\quad 3 \leq i + j \leq 6$ and $i \geq j$, and such that T does not contain a small
$\quad\quad$ homogeneous pair (A', B') with $|A'| = i$ and $|B'| < j$.
Output: A small homogeneous pair (A, B) with $|A| = i$ and $|B| = j$, if it exists.
begin
$\quad$ **forall the** *subsets $A \subseteq V$ of size i* **do**
$\quad\quad$ **forall the** *subsets $B' \subseteq V \setminus A$ of size $j - 1$* **do**
$\quad\quad\quad$ $B = B', R = V \setminus (A \cup B')$.
$\quad\quad\quad$ **while** $|B| \neq j$ *and* $R \neq \emptyset$ **do**
$\quad\quad\quad\quad$ pick a new vertex $v \in R$ and remove it from R.
$\quad\quad\quad\quad$ **if** *v is neither strongly complete nor strongly anticomplete to A,*
$\quad\quad\quad\quad$ *or neither strongly complete nor strongly anticomplete to B* **then**
$\quad\quad\quad\quad\quad$ $\llcorner$ add v to B.
$\quad\quad\quad$ **if** $|B| = j$ *and all vertices of $V \setminus (A \cup B)$ are either strongly complete*
$\quad\quad\quad$ *or strongly anticomplete to A and either strongly complete or*
$\quad\quad\quad$ *strongly anticomplete to B* **then**
$\quad\quad\quad\quad$ $\llcorner$ return (A, B).

Algorithm 2. Algorithm for finding a small homogeneous pair of size $i + j$.

The second bottleneck in the proof of [21, Theorem 4.3] is a subroutine that finds a minimally-sided proper homogeneous pair, if it exists, in time $O(n^7)$. We prove the following lemma.

Lemma 2. *[⋆] There exists an algorithm running in time $O(n^6)$ that finds a minimally-sided homogeneous cut in a trigraph T, provided that T has some homogeneous cut.*

Lemmas 1 and 2 together clearly imply Theorem 6.

Improvement in terms of the parameter. We now focus on the improvement in Step 1 of Algorithm 1. It is shown in the proof [21, Lemma 6.1] that WEIGHTED INDEPENDENT SET restricted to the class $\mathcal{T}_1$ admits a kernel of size $O(k^5)$, and this is what gives the function $2^{O(k^5)}$ in the algorithm of Theorem 1, as well as the Turing-Kernel of Corollary 1. In the following we will show that the kernel in the class $\mathcal{T}_1$ can be improved to $f(k) = O(k^2)$, concluding the proof of Theorem 2 and of Corollary 1. This improvement is detailed in the following lemma, which should be compared to [21, Lemma 6.1]. More precisely, in [21, Lemma 6.1] the function f is defined as $f(x) = g(x) + (x - 1)(\binom{g(x)}{2} + 2g(x) + 1)$, where

$g(x) = \binom{x+1}{2} - 1$. We redefine f as $f(x) = 5g(x)$, yielding the desired upper bound.

Lemma 3. *There is an $O(n^4 m)$-time algorithm with the following specifications.*

> **Input:** *A weighted monogamous basic trigraph T on n vertices and m strong edges, in which all vertices have weight at least 1 and all switchable pairs have weight at least 2, with no homogeneous set, and a positive integer k.*
> **Output:** *One of the following true statements:*
> 1. *$n \leq f(k)$;*
> 2. *the number of maximal independent sets in T is at most n^3; or*
> 3. *$\alpha(T) \geq k$.*

Proof. The proof follows closely that of [21, Lemma 6.1]. Let G be the realization of T where all switchable pairs are set to "strong antiedge". We first check whether $n \leq f(k)$ in constant time. If this is not the case, we apply [21, Theorem 5.4] to G, and check whether Output 2 is true. If not, it just remains to prove that Output 3 is a true statement. The running time of the algorithm is $O(n^4 m)$.

Since T is basic, there are three cases to consider. Assume first that $T \in \mathcal{T}_0$. If $k \geq 2$, then $f(k) > 8 \geq n$, so the algorithm should have given Output 1, a contradiction. Thus, $k \leq 1$, and Output 3 is true. If $T \in \overline{\mathcal{T}}_1$, then by [21, Lemma 5.9] T has at most n^3 maximal independent sets, so the algorithm should have given Output 2, a contradiction.

Thus, necessarily $T \in \mathcal{T}_1$. Suppose for contradiction that $\alpha(T) < k$. We consider the decomposition of T into a triangle-free trigraph X and a disjoint union of t strong cliques $K_1, \ldots, K_t$. In contrast to the proof of [21, Lemma 6.1], we will use the following two properties of the class $\mathcal{T}_1$, as described by Chudnovsky [1,2]:

(i) Each vertex of X has neighbors in at most two distinct cliques.
(ii) For each clique $K \in \{K_1, \ldots, K_t\}$, with $K = \{v_1, \ldots, v_r\}$, the neighborhood of K in T is a bipartite trigraph, with bipartition (A, B), such that for all $i \in \{1, \ldots, r\}$, $\mathcal{A}_{i+1} \subseteq \mathcal{A}_i$ and $\mathcal{B}_i \subseteq \mathcal{B}_{i+1}$, where $\mathcal{A}_i = A \cap N(v_i)$ and $\mathcal{B}_i = B \cap N(v_i)$ (see Fig. 3).

We can suppose that $|X| \leq g(k)$, otherwise as $T[X]$ is triangle-free, by Ramsey Theorem it follows that $\alpha(G) \geq k$, so we would have that $\alpha(T) \geq \alpha(G) \geq k$.

For $1 \leq i \leq t$, let us denote by $N(K_i)$ the subset of vertices of X that are adjacent to at least one vertex of K_i. By Property (i) above, it holds that

$$\sum_{i=1}^{t} |N(K_i)| \leq 2|X|. \tag{1}$$

Claim 1. *For each clique $K \in \{K_1, \ldots, K_t\}$, it holds that $|K| \leq 2|N(K)|$.*

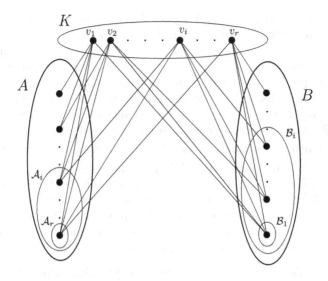

Fig. 3. Adjacency between a clique K and the set X in the proof of Lemma 3.

Proof. Consider an arbitrary $K \in \{K_1, \ldots, K_t\}$, and let $K = \{v_1, \ldots, v_r\}$. Consider the set $N(K)$ as described by Property (ii) above. Let us consider $K' = \{v_{i_1}, \ldots, v_{i_{r'}}\}$, for $1 \leq i_1 < i_2 < \cdots < i_{r'} \leq r$, the set of vertices in K that do not belong to any switchable pair. Since T is monogamous, we have that $r - r' \leq |N(K)|$.

Let us note $\mathcal{V}_j = \overline{\mathcal{A}}_{i_j} \cup \mathcal{B}_{i_j}$, where $\overline{\mathcal{A}}_i = A \cap \overline{N(v_i)}$. Note that any two vertices in K' must have a distinct neighborhood, otherwise they form a homogeneous set, a contradiction. Together with Property (ii), this implies that for all $j \in \{1, \ldots, r' - 1\}$, $\mathcal{V}_j \subsetneq \mathcal{V}_{j+1}$.

Since $\mathcal{B}_i \neq \emptyset$ for all $i \in \{1, \ldots, r\}$, we have that $|\mathcal{V}_{r'}| \geq r'$. And since $\mathcal{V}_{r'} \subseteq N(K')$, we have that $|N(K')| \geq |\mathcal{V}_{r'}| \geq r' = |K'|$.

Therefore, $|K| = r = (r - r') + r' \leq |N(K)| + |N(K')| \leq 2|N(K)|$, and the claim follows. $\qquad\square$

Equation (1) and Claim 1 imply that $\sum\limits_{i=1}^{t} |K_i| \leq 4|X|$, and therefore

$$|V(T)| = |X| + \sum_{i=1}^{t} |K_i| \leq |X| + 4|X| = 5|X|, \qquad (2)$$

that is, $n \leq 5|X|$, and since $|X| \leq g(k)$, the algorithm should have given Output 1, a contradiction. $\qquad\square$

3.1 Independent Set in Bull-Free Graphs Without Small Holes

In this subsection we deal with bull-free graphs without small holes. Namely, we discuss below the main ideas of the faster FPT algorithm of Theorem 3. The proof of the lower bound given by Theorem 4 can be found in [18]. The reduction is from the SPARSE-3-SAT problem, which cannot be solved in time $2^{o(n)}$ unless the ETH fails (see for instance [13]). Our reduction consists of a modification of the classical reduction to show the NP-hardness of INDEPENDENT SET [10].

In order to prove Theorem 3, we use the same algorithm described above for general bull-free graphs, and the improvement in the time bound for $\{bull, C_4, \ldots, C_{2p-1}\}$-free graphs consists in a more careful analysis of the kernel size for the basic class $\mathcal{T}_1$. More precisely, we will prove that the function g such that $|X| \leq g(k)$ can be redefined as $g_p(x) = x(x^{\frac{1}{p-1}} + 2)$. Plugging this function in Equation (2) yields a kernel of size $O(k \cdot k^{\frac{1}{p-1}})$ for the class $\mathcal{T}_1$. Indeed, in the proof of Lemma 3, if T is a $\{bull, C_4, \ldots, C_{2p-1}\}$-free trigraph that belongs to the basic class $\mathcal{T}_1$, the following lemma implies that in this case it holds that $|X| \leq g_p(k)$, hence proving Theorem 3. The proof is inspired from classical arguments in Ramsey theory [6] (see also [14] for recent results on the independence number of triangle-free graphs in terms of several parameters).

Lemma 4. [⋆] *Let $p, k \geq 2$ be two integers and let G be a graph of girth $g(G) \geq 2p$. If $|V(G)| \geq k(k^{\frac{1}{p-1}} + 2)$, then $\alpha(G) \geq k$.*

4   Conclusions and Further Research

We showed in Theorem 2 that WEIGHTED INDEPENDENT SET in bull-free graphs can be solved in time $2^{O(k^2)} \cdot n^7$, and the lower bound of Theorem 4 states that the problem cannot be solved in time $2^{o(k)} \cdot n^{O(1)}$ in bull-free graphs unless the ETH fails. Closing this complexity gap (in terms of k) is an interesting avenue for further research.

It is tempting to try to apply similar techniques for obtaining FPT algorithms for other (NP-hard) problems in bull-free graphs. The INDEPENDENT FEEDBACK VERTEX SET problem may be a natural candidate.

Feghali, Abu-Khzam and Müller [8] have recently shown that the problem of deciding whether the vertices of a graph can be partitioned into a triangle-free subgraph and a disjoint union of cliques is NP-complete in planar and perfect graphs. Note that this problem is closely related to deciding whether a given graph belongs to the class $\mathcal{T}_1$ of basic bull-free graphs. Is this problem NP-complete when restricted to bull-free graphs? The recognition of the class $\mathcal{T}_1$ has also been left as an open question in [21].

References

1. Chudnovsky, M.: The structure of bull-free graphs I - Three-edge-paths with centers and anticenters. J. Comb. Theor. B **102**(1), 233–251 (2012)
2. Chudnovsky, M.: The structure of bull-free graphs II and III - A summary. J. Comb. Theor. B **102**(1), 252–282 (2012)
3. Chudnovsky, M., Seymour P.D.: The structure of claw-free graphs. In: Surveys in Combinatorics, London Mathematical Society Lecture Note Series, vol. 327, pp. 153–171. Cambridge University Press, Cambridge (2005)

4. Cygan, M., Philip, G., Pilipczuk, M., Pilipczuk, M., Wojtaszczyk, J.O.: Dominating set is fixed parameter tractable in claw-free graphs. Theoret. Comput. Sci. **412**(50), 6982–7000 (2011)
5. Dabrowski, K., Lozin, V.V., Müller, H., Rautenbach, D.: Parameterized complexity of the weighted independent set problem beyond graphs of bounded clique number. J. Discrete Algorithms **14**, 207–213 (2012)
6. Diestel, R.: Graph Theory, vol. 173. Springer, Heidelberg (2005)
7. Downey, R.G., Fellows, M.R.: Parameterized Complexity. Springer, Heidelberg (1999)
8. Feghali, C., Abu-Khzam, F.N., Müller, H.: NP-hardness results for partitioning graphs into disjoint cliques and a triangle-free subgraph. CoRR, arxiv:1403.5248 (2014)
9. Flum, J., Grohe, M.: Parameterized Complexity Theory. Springer, Heidelberg (2006)
10. Garey, M., Johnson, D.: Computers and Intractability: A Guide to the Theory of NP-completeness. Freeman, San Francisco (1979)
11. Hermelin, Danny, Mnich, Matthias, van Leeuwen, Erik Jan: Parameterized complexity of induced H-matching on claw-free graphs. In: Epstein, Leah, Ferragina, Paolo (eds.) ESA 2012. LNCS, vol. 7501, pp. 624–635. Springer, Heidelberg (2012)
12. Hermelin, Danny, Mnich, Matthias, van Leeuwen, Erik Jan, Woeginger, Gerhard J.: Domination when the stars are out. In: Aceto, Luca, Henzinger, Monika, Sgall, Jiří (eds.) ICALP 2011, Part I. LNCS, vol. 6755, pp. 462–473. Springer, Heidelberg (2011)
13. Kanj, I.A., Szeider, S.: On the subexponential time complexity of CSP. In: Proceedings of the 27th AAAI Conference on Artificial Intelligence (2013)
14. Lichiardopol, N.: New lower bounds on independence number in triangle-free graphs in terms of order, maximum degree and girth. Discrete Math. **332**, 55–59 (2014)
15. Lokshtanov, D., Marx, D., Saurabh, S.: Lower bounds based on the exponential time hypothesis. Bull. EATCS **105**, 41–72 (2011)
16. Murphy, O.J.: Computing independent sets in graphs with large girth. Discrete Appl. Math. **35**(2), 167–170 (1992)
17. Niedermeier, R.: Invitation to Fixed Parameter Algorithms. Oxford Lecture Series in Mathematics and Its Applications, vol. 31. Oxford University Press, Oxford (2006)
18. Perret du Cray, H., Sau, I.: Improved FPT algorithms for weighted independent set in bull-free graphs. CoRR, arXiv:1407.1706 (2014)
19. Poljak, S.: A note on the stable sets and coloring of graphs. Commentationes Mathematicae Universitatis Carolinae **15**, 307–309 (1974)
20. Robertson, N., Seymour, P.D.: Graph minors. XVI. Excluding a non-planar graph. J. Comb. Theor. B **89**(1), 43–76 (2003)
21. Thomassé, S., Trotignon, N., Vušković, K.: Parameterized algorithm for weighted independent set problem in bull-free graphs. CoRR, arXiv:1310.6205. Short version to appear in the Proceedings of the 40th International Workshop on Graph-Theoretic Concepts in Computer Science (WG), June 2014 (2013)

Improved Parameterized Algorithms for Network Query Problems

Ron Y. Pinter, Hadas Shachnai, and Meirav Zehavi[(✉)]

Department of Computer Science, Technion, 32000 Haifa, Israel
{pinter,hadas,meizeh}@cs.technion.ac.il

Abstract. In the PARTIAL INFORMATION NETWORK QUERY (PINQ) problem, we are given a host graph H, and a pattern $\mathcal{P}$ whose topology is *partially* known. We seek a subgraph of H that *resembles* $\mathcal{P}$. PINQ is a generalization of SUBGRAPH ISOMORPHISM, where the topology of $\mathcal{P}$ is known, and GRAPH MOTIF, where the topology of $\mathcal{P}$ is unknown. This generalization has important applications to bioinformatics, since it addresses the major challenge of analyzing biological networks in the absence of certain topological data. In this paper, we use a non-standard part-algebraic/part-combinatorial hybridization strategy to develop an exact parameterized algorithm as well as an FPT-approximation scheme for PINQ, allowing *near resemblance* between H and $\mathcal{P}$. We thus unify and significantly improve previous results related to network queries.

1 Introduction

With the increasing amount of data on biological networks available, the discovery of conserved patterns has become of major importance. Such patterns can be identified through the use of network queries, which compare the *graph* modeling the network with a given *pattern*. Indeed, the ALIGNMENT NETWORK QUERY (ANQ) and GRAPH MOTIF (GM) problems play a pivotal role in the analysis of biological networks [16,31]. Due to their general nature, they can also be used in analyzing other types of networks, such as social and technical networks [15].

Given a pattern $\mathcal{P}$ and a graph H, GM and ANQ seek a subgraph of H that *resembles* $\mathcal{P}$. GM requires only the connectivity of the solution, while ANQ, a variant of the classic SUBGRAPH ISOMORPHISM (SI) problem, requires resemblance between the *topology* of $\mathcal{P}$ and the solution. The PARTIAL INFORMATION NETWORK QUERY (PINQ) problem, introduced in [27], fits for the common scenario where we have only partial information on the topology of $\mathcal{P}$.

Since network query problems are often NP-hard, there is a growing body of literature studying their parameterized complexity. A problem is *fixed-parameter tractable (FPT)* with respect to a parameter k if it can be solved in time $O^*(f(k))$ for some function f.[1] In this paper, we introduce a non-standard part-algebraic/part-combinatorial hybridization strategy, which we use to develop an

---

[1] The notation O^* hides factors polynomial in the input size.

© Springer International Publishing Switzerland 2014
M. Cygan and P. Heggernes (Eds.): IPEC 2014, LNCS 8894, pp. 294–306, 2014.
DOI: 10.1007/978-3-319-13524-3_25

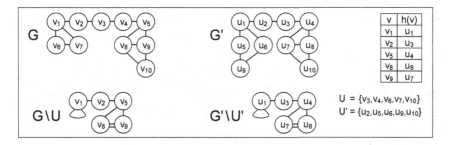

Fig. 1. An example of a homeomorphism h from G to G'.

exact as well as approximation FPT algorithms for a variant of PINQ that allows *insertions and deletions (indels)* of nodes.

1.1 Problem Statement

Given a graph H and a set of graphs $\mathcal{P}$, in PINQ_I we seek a disjoint collection of subgraphs of H, each resembling a different graph in $\mathcal{P}$, whose union is a connected graph. Each of these subgraphs is mapped to the graph it resembles in $\mathcal{P}$, by using a variant of isomorphism allowing to delete degree-2 nodes, called *homeomorphism* (defined below). For biological motivation, see, e.g., [26].

Homeomorphism: Given a graph $G = (V, E)$ and a set U of degree-2 nodes in V, generate the multigraph $G \setminus U$ as follows (see Fig. 1). Delete from G the nodes in U and their incident edges. For every pair $v, u \in V \setminus U$ and simple path connecting them, in which all other nodes belong to U, add an edge $\{v, u\}$. For every $v \in V \setminus U$ and simple cycle in G consisting only of v and nodes in U, add a self-loop to v.

A homeomorphism from $G = (V, E)$ to $G' = (V', E')$ is defined as an isomorphism from $G \setminus U$ to $G' \setminus U'$, where U and U' are subsets of degree-2 nodes in V and V', respectively. To simplify the presentation, we use the term homeomorphism also when referring to a function whose domain is empty.

Definition of PINQ_I: The input for PINQ_I consists of a set of graphs $\mathcal{P} = \{P_1, ..., P_t\}$, where $P_i = (V_i, E_i)$, and a graph $H = (V, E)$ having real numbers as edge-weights, along with a *similarity score* table Δ. The table Δ contains an entry $\Delta(p, h) \in \mathbb{R} \cup \{-\infty\}$ for any pair of nodes p, h, where $p \in V_i$, $1 \leq i \leq t$ and $h \in V$ (an entry $\Delta(p, h) = -\infty$ indicates that p and h cannot be matched). The input contains also the nonnegative integers I_F, I_A and D. Let $k = \sum_{i=1}^{t} |V_i|$ denote the total number of nodes in $\mathcal{P}$ (see Fig. 2(A)).

We now give the definition of a *solution* for PINQ_I (see Fig. 2(B)). Let $\mathcal{S} = (S, V_S^1, ..., V_S^{t+1}, h_1, ..., h_t)$, where $S = (V_S, E_S)$ is a connected subgraph of H, $\{V_S^1, ..., V_S^{t+1}\}$ is a partition of V_S, and h_i is a homeomorphism from P_i to the subgraph of S induced by V_S^i, for all $1 \leq i \leq t$. Let $\text{dom}(f)$ and $\text{ima}(f)$ denote the domain and image of a function f, respectively; denote by $w(e)$ the weight of an edge e. The number of indels and score of $\mathcal{S}$ are defined as follows.

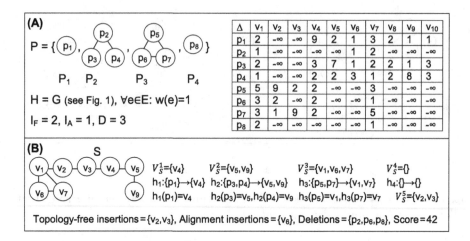

Δ	v_1	v_2	v_3	v_4	v_5	v_6	v_7	v_8	v_9	v_{10}
p_1	2	$-\infty$	$-\infty$	9	2	1	3	2	1	1
p_2	1	$-\infty$	$-\infty$	$-\infty$	$-\infty$	1	2	$-\infty$	$-\infty$	$-\infty$
p_3	2	$-\infty$	$-\infty$	3	7	1	2	2	1	3
p_4	1	$-\infty$	$-\infty$	2	2	3	1	2	8	3
p_5	5	9	2	2	$-\infty$	$-\infty$	3	$-\infty$	$-\infty$	$-\infty$
p_6	3	2	$-\infty$	2	$-\infty$	$-\infty$	1	$-\infty$	$-\infty$	$-\infty$
p_7	3	1	9	2	$-\infty$	$-\infty$	5	$-\infty$	$-\infty$	$-\infty$
p_8	2	$-\infty$	$-\infty$	$-\infty$	$-\infty$	$-\infty$	1	$-\infty$	$-\infty$	$-\infty$

Fig. 2. (A) An input for $\mathrm{PINQ_I}$, where $k = 8$. (B) A solution for the input.

- The number of *free insertions* is $|V_S^{t+1}|$. Informally, this is the number of nodes connecting the subgraphs of S that are mapped to graphs in $\mathcal{P}$.
- The number of *alignment insertions* is the number of unmapped nodes in $\bigcup_{i=1}^{t} V_S^i$, i.e., $\sum_{i=1}^{t} |V_S^i \setminus \mathrm{ima}(h_i)|$. Informally, this is the number of nodes that are not mapped to nodes of graphs in $\mathcal{P}$, and yet belong to the subgraphs of S that are mapped to $\mathcal{P}$.
- The number of *deletions* is the number of unmapped nodes in $\mathcal{P}$, i.e., $\sum_{i=1}^{t} |V_i \setminus \mathrm{dom}(h_i)|$.
- The *score* is the sum of the similarity scores between the matched nodes, and the weights of the edges in E_S, i.e., $\sum_{i=1}^{t} \sum_{p \in \mathrm{dom}(h_i)} \Delta(p, h_i(p)) + \sum_{e \in E_S} w(e)$.

We say that S is a solution if it includes exactly I_F free insertions, I_A alignment insertions and D deletions, and any cycle in S is completely contained in the subgraph induced by V_S^i, for some $1 \leq i \leq t$. The cycle requirement allows us to avoid generalizing the CLIQUE problem, which is W[1]-hard [14].[2] Assuming that there is no solution having less than I_F free insertions,[3] the objective of $\mathrm{PINQ_I}$ is to find the maximum score OPT of a solution.

Relation of $\mathrm{PINQ_I}$ to Known Network Queries: Clearly, PINQ is the special case where $I_F = I_A = D = 0$. Also, ANQ WITH INDELS ($\mathrm{ANQ_I}$) [13] is the special case where $t = 1$. Finally, GM WITH INDELS ($\mathrm{GM_I}$) [10] is the special case where $t = k$, and $\Delta(p, h) \in \{-\infty, 0\}$ for any $p \in V_i, 1 \leq i \leq t$ and $h \in V$.

1.2 Related Work and Our Contribution

ANQ is NP-hard even if the single graph in $\mathcal{P}$ is a path, since this case generalizes the Hamiltonian path problem [18]. GM is NP-hard even if H is a tree [24].

---

[2] Indeed, *without* the cycle requirement, CLIQUE is the special case where $t = k$, $I_F = I_A = D = 0$, $\Delta(p, h) = 0$ for all $p \in V_i, 1 \leq i \leq t$ and $h \in V$, and $w(e) = 1$ for all $e \in E$.

[3] If such solution exists, we can simply reject the input.

Table 1. Parameterized algorithms for $PINQ_I$.

Reference	Weights	Indels	The topology of each P_i	Time complexity
Pinter et al. [27]	$\mathbb{R}$	No	Tree	$O^*(6.75^{k+O(\log^2 k)}3^t)$
This paper	$\mathbf{Z}$	**Yes**	**Bounded treewidth**	$\mathbf{O^*(3.7^{k-D+I_A}W)}$
This paper$^\$$	$\mathbf{N}^0$	**Yes**	**Bounded treewidth**	$\mathbf{O^*(3.7^{k-D+I_A}\lfloor \frac{1}{\epsilon} \rfloor)}$

Table 2. Parameterized algorithms for ANQ_I.

Reference	Weights	Indels	The topology of P_1	Time complexity
Blin et al. [7]	$\mathbb{R}$	Yes	Bounded feedback vertex set	$O^*(8.2^{k+I_A})$
Dost et al. [13]	$\mathbb{R}$	Yes	Bounded treewidth	$O^*(8.2^{k+I_A})$
Shlomi et al. [30]	$\mathbb{R}$	Yes	Simple path	$O^*(5.44^{k+I_A})$
Hüffner et al. [20]	$\mathbb{R}$	Yes	Simple path	$O^*(4.32^{k+I_A})$
Pinter et al. [27]	$\mathbb{R}$	No	Tree	$O^*(6.75^k)$
Pinter et al. [27]	$\mathbb{R}$	No	Simple path	$O^*(4^k)$
This paper	$\mathbf{Z}$	**Yes**	**Bounded treewidth**	$\mathbf{O^*(2^{k-D+I_A}W)}$
This paper$^\$$	$\mathbf{N}^0$	**Yes**	**Bounded treewidth**	$\mathbf{O^*(2^{k-D+I_A}\lfloor \frac{1}{\epsilon} \rfloor)}$

Tables 1, 2 and 3 present known FPT algorithms for $PINQ_I$, ANQ_I and GM_I, where tw is the maximum treewidth [8] of a graph in $\mathcal{P}$. The Weights columns refer to the possible values for edge-weights and scores in Δ, excluding $-\infty$, and W denotes the maximum absolute value of any weight. Typically, in our applications W is polynomial in the input size [24]. Entries marked by '$\$$' indicate instances for which we present an *FPT-approximation scheme (FPT-AS)*, that returns a value in $[(1 - \epsilon)OPT, OPT]$, for *any* fixed $\epsilon > 0$.

Our main result is Exact, a randomized $O^*(3.7^{k-D+I_A}W)$ time exact algorithm for $PINQ_I$, which handles a wide class of inputs (see Theorem 1). We then complement Exact by developing an FPT-AS for $PINQ_I$ (see Theorem 2).

Algorithm Exact improves and unifies the previous results as follows.

- We extend the PINQ algorithm presented i n [27], by considering indels and bounded treewidth graphs (see Table 1). Note that a graph with a bounded feedback vertex set has a bounded treewidth [9]. Thus, our results hold also for graphs with bounded feedback vertex sets.
- For inputs with polynomially bounded integral weights, we significantly improve the O^* running times of the best known algorithms for PINQ (due to [27]) and ANQ_I (due to [13] and [20]). For example, using the real data presented in [24], the weights in the table Δ can take integral values in $\{-\infty, 0, \ldots, 4\}$. Applying the best known algorithm (of [13]) for ANQ_I, where P_1 has a bounded treewidth, we get a running time of $O^*(8.2^{k+I_A})$, whereas Exact solves ANQ_I on such inputs in time $O^*(2^{k-D+I_A})$. We note that both algorithms have the same dependency on the treewidth of P_1.

Table 3. Parameterized algorithms for GM_I.

Reference	Weights	Indels	Time complexity
Bruckner et al. [10]	$\mathbb{R}$	Yes	$O^*(k!3^k)$
Dondi et al. [12]	$\{0\}$	Yes	$O^*(2^{O(k-D)})$
Fellows et al. [15]	$\{0\}$	No	$O^*(87^k)$
Pinter et al. [27]	$\mathbb{R}$	No	$O^*(20.25^{k+O(\log^2 k)})$
Betzler et al. [2]	$\{0\}$	Yes	$O^*(29.6^{k-D})$
Betzler et al. [2]	$\{0\}$	No	$O^*(10.88^k)$
Betzler et al. [3]	$\{0\}$	No	$O^*(4.32^k)$
Guillemot et al. [19]	$\mathbb{N}^0$	Yes	$O^*(4^{k-D}W^2)$
Koutis [22]	$\{0\}$	Yes	$O^*(2.54^{k-D})$
Pinter et al. [28]	$\mathbb{N}^0$	Yes	$O^*(2^k W)$
Björklund et al. [6]	$\{0\}$	Yes	$O^*(2^{k-D})$
This paper	**Z**	**Yes**	$\mathbf{O^*(2^{k-D}W)}$
This paper$^\$$	$\mathbf{N^0}$	**Yes**	$\mathbf{O^*(2^{k-D}\lfloor\frac{1}{\epsilon}\rfloor)}$

- We extend the algorithm for GM_I presented in [6] to handle integral weights.
- Exact has the same O^* running time as the best known FPT algorithms for SI, in which the subgraph is a tree [23], or has a bounded treewidth [17]. The same holds for GROUP STEINER TREE [25], and MIN CONNECTED COMPONENTS [28]. Indeed, all of these problems are special cases of $PINQ_I$.

Due to lack of space, some proofs are omitted. The detailed results will be given in the full version of this paper.

Notation: Let $V(\mathcal{P}) = \bigcup_{i=1}^t V_i$ and $E(\mathcal{P}) = \bigcup_{i=1}^t E_i$ be the sets of nodes and edges in $\mathcal{P}$, respectively. Also let $\mathcal{P}^*$ be the set of single-node graphs in $\mathcal{P}$, $V(\mathcal{P}^*) = \bigcup_{P_i \in \mathcal{P}^*} V_i$ and $k^* = |\mathcal{P}^*|$. Let $V(G)$ and $E(G)$ be the node-set and edge-set of a graph G, respectively.

2   Main Technique

In developing algorithm Exact, we combine the algebraic *narrow sieves* technique [5] (see also [21]) with the *divide-and-color* technique [11], which are often used as two separate tools in solving parameterized problems. Our approach contains a novel application of narrow sieves that consists of two monomial-associating procedures, rather than one such procedure, as detailed below. It may be useful in obtaining fast FPT algorithms for other problems that include as special cases "color coding-related" problems (indeed, $PINQ_I$ encompasses GM and SI). We note that the sophisticated algebraic algorithm for the HAMILTONICITY problem by Björklund [4] (see also [5]) also applies (as preprocessing) a combinatorial partitioning phase.

In the narrow sieves technique, we express a parameterized problem by associating monomials with potential solutions. Each monomial either represents a unique correct solution, or an even number of incorrect solutions. Having a polynomial that is the sum of such monomials, we need to determine whether it has a monomial whose coefficient is odd. On the other hand, divide-and-color is a combinatorial technique where, in each step, we have a set S_n of n elements, and we seek a certain subset S_k of k elements in S_n. We randomly partition S_n into two sets: S_n^1 and S_n^2. Thus, we get the problem of finding a subset $S \subseteq S_k$ in S_n^1, and another problem of finding the subset $S_k \setminus S$ in S_n^2.

In solving PINQ_I, we first observe that if the nodes in V can be mapped only to nodes in $V(\mathcal{P}^*)$, PINQ_I can be efficiently solved by a narrow sieves procedure, P_A, that is a straightforward extension of the algorithm for GM given in [6]. On the other hand, if $|\mathcal{P}| = 1$, PINQ_I can be efficiently solved by a *different* procedure, P_B, using a standard application of narrow sieves.

Now, suppose that we have a partition of V into a set A of nodes that can be mapped only to nodes in $V(\mathcal{P}^*)$, and a set B of nodes that can be mapped only to nodes in $V(\mathcal{P}) \setminus V(\mathcal{P}^*)$. For such a scenario, we develop a non-trivial narrow sieves procedure, ManySingles, handling nodes in A in an efficient manner similar to P_A, and nodes in B in a manner similar to P_B. To handle only such scenarios, before each call to ManySingles, we use divide-and-color to partition V into the sets A and B.[4] Indeed, since we build on the results of [6], the correctness of ManySingles crucially relies on the fact that A does not contain nodes that can be mapped to nodes in $V(\mathcal{P}) \setminus V(\mathcal{P}^*)$.

The *combined* application of divide-and-color and ManySingles is efficient only for solutions containing many graphs from $\mathcal{P}^*$.[5] However, solutions containing few graphs from $\mathcal{P}^*$ cannot contain too many graphs from $\mathcal{P}$ (since each solution contains exactly k nodes from $V(\mathcal{P})$). For such solutions, we develop a procedure, FewSingles, handling all the nodes in V in a manner similar to P_B.

Thus, our algorithm proceeds in the following main steps:

1. Examine all choices for the number $n_{\mathcal{P}^*}$ of graphs from $\mathcal{P}^*$ in the solution.
2. If $n_{\mathcal{P}^*}$ is "large":
   (a) Apply divide-and-color to partition V as described above.
   (b) Call ManySingles.
3. Else: Call FewSingles.

3 The Procedures FewSingles and ManySingles

Assume that $\mathcal{P}$ is a set of bounded treewidth graphs, and the weights are nonnegative integers (recall that the weights are the possible values for edge-weights and scores in Δ, excluding $-\infty$). Algorithm Exact (see Sect. 4) only needs the procedures to be correct under these assumptions. For the sake of clarity, we first present a simple version of FewSingles that cannot handle indels.

---

[4] This can be also viewed as applying the color coding technique [1] using only two colors.

[5] For such solutions, the time gained by handling A in a manner similar to P_A prevails the time required for the preceding selection step.

3.1   SimpleFewSingles: A Narrow Sieves Procedure

Assuming that $I_F = I_A = D = 0$, we present a narrow sieves procedure that efficiently finds solutions containing few graphs in $\mathcal{P}^*$. We first define the structure of a potential solution. We then describe the potential solutions, and associate them with monomials. We show how to evaluate some sums of such monomials, and finally, we present the procedure, which heavily relies on such evaluations.

3.1.1   The Structure of a Potential Solution

Recall that any cycle in a solution $\mathcal{S}$ is contained in a subgraph induced by $V_{\mathcal{S}}^i$, for some $1 \leq i \leq t$. Thus, by contracting each of the subgraphs into a single node, and choosing a node as a root, any solution for $\mathrm{PINQ_I}$ can be represented by a rooted tree. We study the mappings of such trees (into graphs in $\mathcal{P}$) below.

A quad $(T, f_{gra}, f_{nod}, f_{con})$ refers to a rooted tree $T = (V_T, E_T)$ on t nodes, $f_{gra} : V_T \rightarrow \mathcal{P}$, $f_{nod} : X \rightarrow V$ and $f_{con} : X \rightarrow 2^X$, where $X = \{(v, p) : v \in V_T, p \in V(f_{gra}(v))\}$. Informally, such a quad describes a structure for a solution as follows. T and f_{gra} specify which graphs to choose from $\mathcal{P}$ and how to connect them; f_{nod} indicates how to map the nodes of graphs chosen from $\mathcal{P}$ to nodes in V; and f_{con} refines our information about how the chosen graphs are connected. Next, we define the quads corresponding to structures of potential solutions for $\mathrm{PINQ_I}$.

Definition 1. *Given $r \in V$, we say that a quad $(T, f_{gra}, f_{nod}, f_{con})$ is r-good if:*

1. $|\{(v, p) \in \mathrm{dom}(f_{nod}) : f_{nod}(v, p) = r\}| =$
$$|\{p \in V(f_{gra}(\mathrm{root}(T))) : f_{nod}(\mathrm{root}(T), p) = r\}| = 1.$$
2. $\forall v \in V_T, \{p, p'\} \in E(f_{gra}(v)) : \{f_{nod}(v, p), f_{nod}(v, p')\} \in E.$
3. $\forall (v, p) \in \mathrm{dom}(f_{nod}) : \Delta(p, f_{nod}(v, p)) \neq -\infty.$
4. $\forall (v, p) \in \mathrm{dom}(f_{con}), (u, p') \in f_{con}(v, p) :$
    (a) *v is the father of u in T, and $\{f_{nod}(v, p), f_{nod}(u, p')\} \in E.$*
    (b) $\forall (u', p'') \in f_{con}(v, p) \setminus \{(u, p')\} : f_{nod}(u, p') \neq f_{nod}(u', p'').$
5. $\forall u \in V_T \setminus \{\mathrm{root}(T)\} : |\{(v, p, p') : (u, p') \in f_{con}(v, p)\}| = 1.$

Condition 1 states that we map only one node in $V(\mathcal{P})$ to r, and this node belongs to the graph mapped to the root of T. Condition 2 requires that the mapping of the graphs in $\mathcal{P}$ to subgraphs of H is correct (i.e., we map edges of graphs in $\mathcal{P}$ to edges in E). By Condition 3, we do not match a node in $V(\mathcal{P})$ with a node in V that cannot be matched according to Δ. Condition 4a states that f_{con} does not contradict the information provided by T on the edges connecting the graphs in $\mathcal{P}$. More precisely, a node v being a father of a node u in T implies that $f_{gra}(v)$ and $f_{gra}(u)$ are connected by an edge. Only then f_{con} may provide information on the connecting edge, where $(u, p') \in f_{con}(v, p)$, for some $p \in V(f_{gra}(v))$ and $p' \in V(f_{gra}(u))$, implies that p and p' are connected by an edge (which, by this condition, is mapped to an edge in E). Condition 4c avoids some quads in which several nodes in $V(\mathcal{P})$ are mapped to the same node in V. Finally, Condition 5 states that for each pair of a node u and its father v in

T, f_{con} provides information on exactly one pair (p, p'), for some $p \in V(f_{gra}(v))$ and $p' \in V(f_{gra}(u))$, indicating that p and p' are connected by an edge.

We now define the score of an r-good quad by the mapping of the edges in $E(\mathcal{P})$, the pairs of matched nodes, and the edges connecting the graphs in $\mathcal{P}$.

Definition 2. *The* score *of an r-good quad $(T, f_{gra}, f_{nod}, f_{con})$ is*

$$\sum_{v \in V_T, \{p,p'\} \in E(f_{gra}(v))} w(\{f_{nod}(v,p), f_{nod}(v,p')\})$$

$$+ \sum_{(v,p) \in \mathrm{dom}(f_{nod})} [\Delta(p, f_{nod}(v,p)) + \sum_{(u,p') \in f_{con}(v,p)} w(\{f_{nod}(v,p), f_{nod}(u,p')\})].$$

3.1.2 Potential Solutions

Let $L = \{1, ..., k + t\}$ be the set of indices used in labeling r-good quads (recall that $t = |\mathcal{P}|$ and $k = \sum_{i=1}^{t} |V_i|$), defining potential solutions of the same score as follows.

Definition 3. *Given an r-good quad $(T, f_{gra}, f_{nod}, f_{con})$ and $\ell : V_T \cup \mathrm{dom}(f_{nod}) \to L$ satisfying $|\mathrm{dom}(\ell)| = k + t$, we say that $(T, f_{gra}, f_{nod}, f_{con}, \ell)$ is an r-solution.*

We now define two sets: $Sol(r, s)$ contains all r-solutions $(T, f_{gra}, f_{nod}, f_{con}, \ell)$ of score s where ℓ is bijective; and $Cor(r, s) = \{(T, f_{gra}, f_{nod}, f_{con}, \ell) \in Sol(r, s) : f_{gra}$ and f_{nod} are injective$\}$. The next lemma implies that each set $Cor(r, s)$ includes enough r-solutions from $Sol(r, s)$, and all these r-solutions are *correct*.

Lemma 1. *The input has a solution of score s iff $\bigcup_{r \in V} Cor(r, s) \neq \emptyset$.*

Note that $(T, f_{gra}, f_{nod}, f_{con}, \ell), (T', f'_{gra}, f'_{nod}, f'_{con}, \ell') \in Sol(r, s)$ are equal iff there is an isomorphism *iso* between the *rooted* trees T and T', such that

1. $\forall v \in V_T : f_{gra}(v) = f'_{gra}(iso(v))$, and $\ell(v) = \ell'(iso(v))$.
2. $\forall (v, p) \in \mathrm{dom}(f_{nod}) : f_{nod}(v, p) = f'_{nod}(iso(v), p)$, $\ell(v, p) = \ell'(iso(v), p)$, and $[\forall (u, p') : (u, p') \in f_{con}(v, p)$ iff $(iso(u), p') \in f'_{con}(iso(v), p)]$.

3.1.3 Associating Monomials with Potential Solutions

Recall that, in the narrow sieves technique, a parameterized problem is solved via associating monomials with potential solutions. Towards defining these monomials, we introduce the variables x, $y_{e,h}$ for all $e \in V(\mathcal{P}) \cup V$ and $h \in V$, and $z_{e,l}$ for all $e \in \mathcal{P} \cup V$ and $l \in L$. Let *ind* denote the number of these variables, i.e., $ind = 1 + (k + |V|)|V| + (t + |V|)|L|$.

We next define the monomials associated with potential solutions. In defining a monomial for an r-solution $sol \in Sol(r, s)$, we store information about sol that allows reconstructing sol iff it is a correct solution (i.e., $sol \in Cor(r, s)$).

Definition 4. $m(T, f_{gra}, f_{nod}, f_{con}, \ell) = x^s \prod_{v \in V_T} z_{f_{gra}(v), \ell(v)}$

$$\prod_{(v,p) \in dom(f_{nod})} [y_{p, f_{nod}(v,p)} z_{f_{nod}(v,p), \ell(v,p)} \prod_{(u,p') \in f_{con}(v,p)} y_{f_{nod}(v,p), f_{nod}(u,p')}].$$

Given an r-solution, x tracks its score (as in [4]); $\prod_{v \in V_T} z_{f_{gra}(v), \ell(v)}$ marks which graphs to choose from $\mathcal{P}$ and how to label them; $\prod_{(v,p) \in dom(f_{nod})} y_{p, f_{nod}(v,p)}$ $z_{f_{nod}(v,p), \ell(v,p)}$ specifies how to map nodes in $V(\mathcal{P})$ to nodes in V and how to label nodes in V; and $\prod_{(v,p) \in dom(f_{nod}), (u,p') \in f_{con}(v,p)} y_{f_{nod}(v,p), f_{nod}(u,p')}$ notes how to connect the graphs chosen from $\mathcal{P}$.

We now claim that correct solutions are associated with unique monomials, and a monomial of an incorrect solution represents an even number of solutions.

Lemma 2. *Pairs* $\{sol, sol'\}$ *of different solutions in* $Cor(r, s)$ *satisfy* $m(sol) \neq m(sol')$, *while* $Sol(r, s) \setminus Cor(r, s)$ *can be partitioned into pairs* $\{sol, sol'\}$ *s.t.* $m(sol) = m(sol')$.

3.1.4 Evaluating the Sum of the Monomials

For each $r \in V$, let $P(r) = \sum_{s \in \{0,...,(|V|+|E|)W\}, sol \in Sol(r,s)} m(sol)$. We will evaluate these polynomials over $\mathbb{F}_q$, the finite field of order q, where $q = 2^{\lceil \log_2(10(2(k+t)+t)) \rceil}$.

By Lemmas 1 and 2, the input has a solution of score s iff there exists a node $r \in V$ such that $P(r)$ has a monomial with an odd coefficient in which the degree of x is s. Since $\mathbb{F}_q$ has characteristic 2, we have the following result.

Lemma 3. *The input has a solution of score s iff there is a node $r \in V$ such that $P(r)$ has a monomial in which the degree of x is s.*

Given $A \subseteq L$, let $P_A(r) = \sum_{sol \text{ is an } r-\text{solution in which } ima(\ell) \subseteq A} m(sol)$. Using inclusion-exclusion, and since $\mathbb{F}_q$ has characteristic 2, we have that $P(r) = \sum_{A \subseteq L} P_A(r)$. Thus, we can evaluate $P(r)$ by using the following lemma.

Lemma 4. *Let $A \subseteq L$ and $a_1, ..., a_{ind-1} \in \mathbb{F}_q$. For all $r \in V$, we can evaluate $P_A(r)(x, a_1, ..., a_{ind-1})$ (assign values to all variables except x) in time $O(W \log W \, |V|^{tw+O(1)} k^{O(1)})$ and space $O(W|V|^{tw+O(1)} k^{O(1)})$ (using dynamic programming).*

3.1.5 Concluding Procedure SimpleFewSingles

SimpleFewSingles first chooses values from $\mathbb{F}_q$ (see below), to be assigned to all the variables, excluding x, of polynomials of the form $P_A(r)$. It evaluates these polynomials, and thus evaluates polynomials of the form $P(r)$. Finally, it determines the maximum score s of a solution by verifying that at least one evaluation of a

---

Procedure. SimpleFewSingles$(\mathcal{P}, H, \Delta)$

1: **forall** $r \in V$ **do** $Sum[r] \Leftarrow 0$. **end for**
2: select $a_1, \ldots, a_{ind-1} \in \mathbb{F}_q$ independently and uniformly at random.
3: **forall** $A \subseteq L, r \in V$ **do** $Sum[r] \Leftarrow Sum[r] + P_A(r)(x, a_1, \ldots, a_{ind-1})$. **end for**
4: return the maximum value s such that (there exists $r \in V$ for which $Sum[r]$ is a nonzero polynomial of degree s), where if no such s exists – reject.

---

polynomial $P(r)$ resulted in a polynomial (whose only variable is x) of degree s. Lemma 4 implies the time and space complexities of SimpleFewSingles, while correctness follows from Lemma 3 and the Schwartz–Zippel lemma [29,32].

Lemma 5. *If there is a solution, then* SimpleFewSingles *returns OPT with probability* $\geq \frac{9}{10}$, *and does not return a higher score otherwise; else, it rejects. It uses* $O(2^{k+t} W \log W |V|^{tw+O(1)} k^{O(1)})$ *time and* $O(W |V|^{tw+O(1)} k^{O(1)})$ *space.*

3.2   Procedures FewSingles and ManySingles

FewSingles extends SimpleFewSingles to handle indels. The input is of the form $(n_E, n_{\mathcal{P}}, \mathcal{P}, H, \Delta, I_F, I_A, D)$, where n_E and $n_{\mathcal{P}}$ indicate that we seek solutions of exactly n_E edges from E, such that $n_{\mathcal{P}}$ graphs in $\mathcal{P}$ are not entirely deleted.

Lemma 6. *If there is a solution s.t.* $|E_S| = n_E$, *where* $\{V_S^1, \ldots, V_S^t\}$ *includes exactly* $n_{\mathcal{P}}$ *nonempty sets, then* FewSingles *returns the maximum score of such a solution with probability* $\geq \frac{9}{10}$, *and not a higher score otherwise; else, it rejects. It uses* $O(2^{k-D+I_A+n_{\mathcal{P}}} W \log W |V|^{tw+O(1)} k^{O(1)})$ *time and* $O(W |V|^{tw+O(1)} k^{O(1)})$ *space.*

ManySingles, extending [6], efficiently finds solutions of many graphs from $\mathcal{P}^*$. Its input is of the form $(n_E, n_{\mathcal{P}^*}, n_{\mathcal{P}}, \mathcal{P}, H, \Delta, I_F, I_A, D)$, where $n_{\mathcal{P}^*}$ indicates that we seek solutions of exactly $n_{\mathcal{P}^*}$ graphs from $\mathcal{P}^*$. ManySingles assumes that there is a set $U \subseteq V$ satisfying $[\forall h \in U : \text{If } p \in V(\mathcal{P}) \setminus V(\mathcal{P}^*) \text{ then } \Delta(p, h) = -\infty]$ and $[\forall h \in V \setminus U : \text{If } p \in V(\mathcal{P}^*) \text{ then } \Delta(p, h) = -\infty]$.

Lemma 7. *If there is a solution without alignment insertions from* U, *satisfying* $|E_S| = n_E$, *in which* $\{V_S^1, \ldots, V_S^t\}$ *includes exactly* $n_{\mathcal{P}}$ *nonempty sets and* $n_{\mathcal{P}^*}$ *one-node sets, then* ManySingles *returns the maximum score of such a solution with probability* $\geq \frac{9}{10}$, *and not a higher score otherwise; else, it rejects. It uses* $O(2^{k-D+I_A+n_{\mathcal{P}}-n_{\mathcal{P}^*}} W \log W |V|^{tw+O(1)} k^{O(1)})$ *time and* $O(W |V|^{tw+O(1)} k^{O(1)})$ *space.*

4   An Exact Algorithm

We now describe our main algorithm (see below). Exact first manipulates the weights to be nonnegative (Step 1). The variable s, initialized to $-\infty$, holds the highest score found so far, corresponding to the original weights. Exact iterates over all choices for $n_E, n_{\mathcal{P}^*}$ and $n_{\mathcal{P}}$, specifying the number of edges, graphs from $\mathcal{P}^*$ and graphs from $\mathcal{P}$, respectively, in the currently searched solution (Step 2).

For each choice, Exact uses a calculation which determines whether $n_{\mathcal{P}^*}$ is "small" or "large" (Step 3), indicating whether it is now preferable (in terms of running time) to call FewSingle or ManySingles. If $n_{\mathcal{P}^*}$ is "small", Exact calls FewSingles to compute the maximum score of a solution complying with $n_E, n_{\mathcal{P}^*}$ and $n_{\mathcal{P}}$ (Step 4). In this step, the term $v(n_E + k - D)$ is used to correctly compare between the score returned by FewSingles and s, since only s corresponds to the original weights. Now, suppose that $n_{\mathcal{P}^*}$ is "large". Before calling ManySingles (Step 11), Exact uses divide-and-color (Steps 6–10) to examine several choices of nodes in V for mapping graphs in $\mathcal{P}^*$, and those used for mapping graphs in $\mathcal{P} \setminus \mathcal{P}^*$. In particular, the number of iterations of Step 6 ensures that, with high probability, Exact examines such a choice that complies with a solution of maximum score. Finally, Exact returns the score s, unless no solution was found, in which case it rejects (Step 15).

Algorithm 2. Exact$(\mathcal{P}, H, \Delta, I_F, I_A, D)$

1: subtract $v = \min(\text{weights})$ from every weight, and initialize $s \Leftarrow -\infty$.
2: **for** $n_E = 0, \ldots, |E|$, $n_{\mathcal{P}^*} = \max\{0, k^* - D\}, \ldots, \min\{k^*, k - D\}$,
       $n_{\mathcal{P}} = n_{\mathcal{P}^*}, \ldots, \min\{t, n_{\mathcal{P}^*} + (k - D + I_A - n_{\mathcal{P}^*})/2\}$ **do**
3:    **if** $2^{n_{\mathcal{P}^*}} \leq \dfrac{(k - D + I_A)^{k - D + I_A}}{n_{\mathcal{P}^*}^{n_{\mathcal{P}^*}} (k - D + I_A - n_{\mathcal{P}^*})^{k - D + I_A - n_{\mathcal{P}^*}}}$ **then**
4:       **if** FewSingles$(n_E, n_{\mathcal{P}}, \mathcal{P}, H, \Delta, I_F, I_A, D)$ returns $s' > s - v(n_E + k - D)$
         then $s \Leftarrow s' + v(n_E + k - D)$. **end if**
5:    **else**
6:       **for** $\dfrac{10(k - D + I_A)^{k - D + I_A}}{n_{\mathcal{P}} \, n_{\mathcal{P}^*} (k - D + I_A - n_{\mathcal{P}^*})^{k - D + I_A - n_{\mathcal{P}^*}}}$ times **do**
7:          initialize $U \Leftarrow \emptyset$ and $\lambda \Leftarrow \Delta$.
8:          **forall** $h \in V$, with probability $\frac{n_{\mathcal{P}^*}}{(k - D + I_A)}$ **do** add h to U. **end for**
9:          **forall** $p \in V(\mathcal{P}) \setminus V(\mathcal{P}^*), h \in U$ **do** $\lambda(p, h) \Leftarrow -\infty$. **end for**
10:        **forall** $p \in V(\mathcal{P}^*), h \in V \setminus U$ **do** $\lambda(p, h) \Leftarrow -\infty$. **end for**
11:        **if** ManySingles$(n_E, n_{\mathcal{P}^*}, n_{\mathcal{P}}, \mathcal{P}, H, \lambda, I_F, I_A, D)$ returns
            $s' > s - v(n_E + k - D)$ **then** $s \Leftarrow s' + v(n_E + k - D)$. **end if**
12:       **end for**
13:    **end if**
14: **end for**
15: **if** $s \neq -\infty$ **then** return s. **else** reject. **end if**

Theorem 1. Exact *solves* PINQ_I *in* $O(3.698^{k - D + I_A} W \log W |V|^{tw + O(1)} k^{O(1)})$ *time and* $O(W |V|^{tw + O(1)} k^{O(1)})$ *space, handling instances with integer weights, where* $\mathcal{P}$ *is a set of bounded treewidth graphs. Its running time for* ANQ_I *is* $O^*(2^{k + I_A - D} W)$, *and for* GM_I, $O^*(2^{k - D} W)$.

Using scaling and rounding, we obtain the next result.

Theorem 2. *There is an FPT-AS for* $\mathrm{PINQ_I}$*, handling instances with nonnegative integer weights, where* $\mathcal{P}$ *is a set of bounded treewidth graphs. It uses* $O(3.698^{k-D+I_A} \lfloor \frac{1}{\epsilon} \rfloor \log(\lfloor \frac{1}{\epsilon} \rfloor)|V|^{tw+O(1)}k^{O(1)})$ *time and* $O(\lfloor \frac{1}{\epsilon} \rfloor|V|^{tw+O(1)}k^{O(1)})$ *space. Its running time for* $\mathrm{ANQ_I}$ *is* $O^*(2^{k+I_A-D}W)$*, and for* $\mathrm{GM_I}$*,* $O^*(2^{k-D}W)$*.*

References

1. Alon, N., Yuster, R., Zwick, U.: Color coding. J. Assoc. Comput. Mach. **42**(4), 844–856 (1995)
2. Betzler, N., Bevern, R., Fellows, M.R., Komusiewicz, C., Niedermeier, R.: Parameterized algorithmics for finding connected motifs in biological networks. IEEE/ACM Trans. Comput. Biol. Bioinf. **8**(5), 1296–1308 (2011)
3. Betzler, N., Fellows, M.R., Komusiewicz, C., Niedermeier, R.: Parameterized algorithms and hardness results for some graph motif problems. In: Ferragina, P., Landau, G.M. (eds.) CPM 2008. LNCS, vol. 5029, pp. 31–43. Springer, Heidelberg (2008)
4. Björklund, A.: Determinant sums for undirected hamiltonicity. In: FOCS, pp. 173–182 (2010)
5. Björklund, A., Husfeldt, T., Kaski, P., Koivisto, M.: Narrow sieves for parameterized paths and packings. CoRR (2010). arxiv:1007.1161
6. Björklund, A., Kaski, P., Kowalik, L.: Probably optimal graph motifs. In: STACS, pp. 20–31 (2013)
7. Blin, G., Sikora, F., Vialette, S.: Querying graphs in protein-protein interactions networks using feedback vertex set. IEEE/ACM Trans. Comput. Biol. Bioinf. **7**(4), 628–635 (2010)
8. Bodlaender, H.L.: A linear-time algorithm for finding tree-decompositions of small treewidth. SIAM J. Comput. **25**(6), 1305–1317 (1996)
9. Bodlaender, H.L., Koster, A.M.C.A.: Combinatorial optimization on graphs of bounded treewidth. Comput. J. **51**(3), 255–269 (2008)
10. Bruckner, S., Hüffner, F., Karp, R.M., Shamir, R., Sharan, R.: Topology-free querying of protein interaction networks. J. Comput. Biol. **17**(3), 237–252 (2010)
11. Chen, J., Kneis, J., Lu, S., Molle, D., Richter, S., Rossmanith, P., Sze, S., Zhang, F.: Randomized divide-and-conquer: Improved path, matching, and packing algorithms. SIAM J. Comput. **38**(6), 2526–2547 (2009)
12. Dondi, R., Fertin, G., Vialette, S.: Maximum motif problem in vertex-colored graphs. In: CPM, pp. 388–401 (2011)
13. Dost, B., Shlomi, T., Gupta, N., Ruppin, E., Bafna, V., Sharan, R.: Qnet: a tool for querying protein interaction networks. J. Comput. Biol. **15**(7), 913–925 (2008)
14. Downey, R.G., Fellows, M.R.: Fixed-parameter tractability and completeness II: on completeness for W[1]. Theor. Comput. Sci. **141**(1–2), 109–131 (1995)
15. Fellows, M.R., Fertin, G., Hermelin, D., Vialette, S.: Upper and lower bounds for finding connected motifs in vertex-colored graphs. J. Com. Sys. Sci. **77**(4), 799–811 (2011)
16. Fionda, V., Palopoli, L.: Biological network querying techniques: Analysis and comparison. J. Comput. Biol. **18**(4), 595–625 (2011)
17. Fomin, F.V., Lokshtanov, D., Raman, V., Saurabh, S., Rao, B.V.R.: Faster algorithms for finding and counting subgraphs. J. Com. Sys. Sci. **78**(3), 698–706 (2012)

18. Garey, M.R., Johnson, D.S.: Computers And Intractability: A Guide To The Theory Of Np-Completeness. W.H. Freeman, New York (1979)
19. Guillemot, S., Sikora, F.: Finding and counting vertex-colored subtrees. Algorithmica 65(4), 828–844 (2013)
20. Hüffner, F., Wernicke, S., Zichner, T.: Algorithm engineering for color-coding with applications to signaling pathway detection. Algorithmica 52(2), 114–132 (2008)
21. Koutis, I.: Faster algebraic algorithms for path and packing problems. In: Aceto, L., Damgård, I., Goldberg, L.A., Halldórsson, M.M., Ingólfsdóttir, A., Walukiewicz, I. (eds.) ICALP 2008, Part I. LNCS, vol. 5125, pp. 575–586. Springer, Heidelberg (2008)
22. Koutis, I.: Constrained multilinear detection for faster functional motif discovery. Inf. Process. Lett. 112(22), 889–892 (2012)
23. Koutis, I., Williams, R.: Limits and applications of group algebras for parameterized problems. In: Albers, S., Marchetti-Spaccamela, A., Matias, Y., Nikoletseas, S., Thomas, W. (eds.) ICALP 2009, Part I. LNCS, vol. 5555, pp. 653–664. Springer, Heidelberg (2009)
24. Lacroix, V., Fernandes, C.G., Sagot, M.F.: Motif search in graphs: Application to metabolic networks. IEEE/ACM Trans. Comput. Biol. Bioinf. 3(4), 360–368 (2006)
25. Misra, N., Philip, G., Raman, V., Saurabh, S., Sikdar, S.: FPT algorithms for connected feedback vertex set. J. Comb. Optim. 24(2), 131–146 (2012)
26. Pinter, R.Y., Rokhlenko, O., Yeger-Lotem, E., Ziv-Ukelson, M.: Alignment of metabolic pathways. Bioinformatics 21(16), 3401–3408 (2005)
27. Pinter, R.Y., Zehavi, M.: Partial information network queries. In: Lecroq, T., Mouchard, L. (eds.) IWOCA 2013. LNCS, vol. 8288, pp. 362–375. Springer, Heidelberg (2013)
28. Pinter, R.Y., Zehavi, M.: Algorithms for topology-free and alignment network queries. J. Discrete Algorithms 27, 29–53 (2014)
29. Schwartz, J.T.: Fast probabilistic algorithms for verification of polynomial identities. J. Assoc. Comput. Mach. 27(4), 701–717 (1980)
30. Shlomi, T., Segal, D., Ruppin, E., Sharan, R.: Qpath: a method for querying pathways in a protein-protein interaction networks. BMC Bioinform. 7, 199 (2006)
31. Sikora, F.: An (almost complete) state of the art around the graph motif problem. Université Paris-Est Technical reports (2012)
32. Zippel, R.: Probabilistic algorithms for sparse polynomials. In: Ng, K.W. (ed.) EUROSAM 1979. LNCS, vol. 72, pp. 216–226. Springer, Heidelberg (1979)

On Kernels for Covering and Packing ILPs with Small Coefficients

Stefan Kratsch and Vuong Anh Quyen[✉]

Technical University, Berlin, Germany
{stefan.kratsch,a.vuong}@tu-berlin.de

Abstract. This work continues the study of preprocessing for integer linear programs (ILPs) via the notion of kernelization from parameterized complexity. Previous work, amongst others, studied covering and packing ILPs under different parameterizations and restrictions on the sparseness of the constraint matrix. Several fairly restricted cases, e.g., if every variable appears only in a bounded number of constraints, were shown not to admit polynomial kernels, and in fact are even W[1]-hard, due to generalizing the SMALL SUBSET SUM(k) problem; this hardness relies on using coefficients of value exponential in the input size. We study these cases more carefully by taking into account also the coefficient size of the ILPs and obtain a finer classification into cases with polynomial kernels and cases that are fixed-parameter tractable (but without polynomial kernel) in addition to the previously established W[1]-hardness for unrestricted coefficients.

1   Introduction

Integer Linear Programs (ILPs) are a powerful language for formulating and solving (NP-)hard combinatorial problems. An important part of successful ILP solvers (like CPLEX) are efficient preprocessing routines that simplify the input ILP before applying branch-and-bound or cutting-plane methods for solving it since the latter are worst-case exponential-time. From a theoretical perspective, this makes it interesting to study the possibility of proving upper and lower bounds for the possible impact of preprocessing. Following previous work [12,13] we study this question using the notion of *polynomial kernelization* to formalize preprocessing (see Sect. 2 for definitions). Concretely, we are interested in the case of covering and packing ILPs as these capture and generalize various parameterized problems known to admit polynomial kernelizations. Formally, by covering and packing ILPs we mean the following normal forms

$$\min c^T x \qquad\qquad \max c^T x$$
$$\text{s.t. } Ax \geq b \qquad\qquad \text{s.t } Ax \leq b$$
$$x \geq 0 \qquad\qquad\qquad x \geq 0$$

Supported by the Emmy Noether-program of the German Research Foundation (DFG), KR 4286/1.

© Springer International Publishing Switzerland 2014
M. Cygan and P. Heggernes (Eds.): IPEC 2014, LNCS 8894, pp. 307–318, 2014.
DOI: 10.1007/978-3-319-13524-3_26

where $A \in \mathbb{N}^{m \times n}$, $b \in \mathbb{N}^m$, and $c \in \mathbb{N}^n$ (here $\mathbb{N}$ denotes the non-negative integers) and all variables are required to be integers; allowing negative coefficients would make the restriction to, e.g., $Ax \leq b$ meaningless. Generalizing well-studied parameterized covering and packing problems, like HITTING SET or INDEPENDENT SET, we consider the decision problems of determining whether there exist feasible solutions with $c^T x \leq k$, respectively $c^T x \geq k$, parameterized by k. We call these problems COVERING ILP(k) and PACKING ILP(k), respectively.

Previous work [12] studied COVERING ILP(k) and PACKING ILP(k) under restrictions on the row- and column-sparseness of the constraint matrix A, i.e., when A has at most r nonzero entries per row and/or at most q nonzero entries per column. These restrictions are equivalent to requiring that there are at most r variables with nonzero coefficients in each constraint and that each variable occurs in at most q constraints with nonzero coefficient. By considering both r and q as either constant, additional parameter, or unbounded this gives rise to covering and packing ILP problems that generalize various covering and packing problems on graphs, hypergraphs, and set families. In this regard, note that for hypergraph problems the edge size corresponds to r and the vertex degree to q (and analogously for problems on set families).

It should not come as a surprise that several of these variants inherit W[1]- or W[2]-hardness, or lower bounds against polynomial kernels, from INDEPENDENT SET and (variants of) HITTING SET. Since these problems can be expressed in an easy way using only small coefficients, this is clearly not an artifact of the generality of allowing large coefficients.[1] If q is bounded, however, then one would hope to reproduce/generalize polynomial kernelizations from bounded-degree graph problems, which in some cases are very easy to obtain. Regrettably, there is a second hardness source for covering and packing ILPs, namely the SUBSET SUM problem or, more accurately, the SMALL SUBSET SUM(k) problem where one seeks a set of at most k numbers to match a specified target number. SMALL SUBSET SUM(k) is known to be W[1]-hard [8,9] and can be expressed using only a constant number of constraints (thereby trivially bounding also the variable occurrences). Thus, if row-sparseness is unbounded then due to the possibility of having huge coefficients even constant column-sparseness q is not helpful to avoid even W[1]-hardness, let alone obtaining polynomial kernels. Note that we study here only the cases with unbounded row-sparseness r since the cases of bounded r and additional parameter r already have either positive results or negative results already for 0/1-coefficients [12].

Our work. In this work, our goal is to analyze more carefully the influence of increasingly large coefficients on the (in-)tractability of COVERING ILP(k) and PACKING ILP(k). We denote by C the largest coefficient in (A, b) and study all cases obtained by the following choices: (1) Letting the column sparseness q be either constant or an additional parameter. (2) Letting C be bounded, or adding parameter C or $\log C$. The former choice corresponds exactly to the two cases for which we only know W[1]-hardness from SMALL SUBSET SUM(k) (when

---

[1] Intuitively, when domains of variables and coefficients are bounded then COVERING ILP and PACKING ILP often behave similar to related hypergraph and set problems.

Table 1. "PK" stands for polynomial kernelization, "No PK" stands for no polynomial kernelization unless $\mathsf{NP} \subseteq \mathsf{coNP/poly}$. All normal-font entries are implied by boldface entries. All cases are FPT except those in the last column (unrestricted C).

Parameterized Complexity of Packing/Covering ILP(k)				
	Constant C	Parameter C	Parameter $\log C$	Unrestricted C
Constant q	PK	**PK**	**No PK**	**W[1]-hard from**
		(Theorems 1 and 2)	**(Theorem 3)**	**Subset Sum(k)** [12]
Parameter q	**No PK**	No PK	No PK	**W[1]-hard from**
	(Theorem 4)			**Subset Sum(k)** [12]

row-sparseness is unbounded). Parameters C or $\log C$ mean that a polynomial kernelization may depend polynomially on the largest coefficient or on the maximum encoding size (in binary) of coefficients. We obtain the following results, which completely settle these questions; see Table 1. Since the further results for PACKING ILP and COVERING ILP in [12] are effectively independent of the coefficient size (e.g., hardness holds already for 0/1-coefficients in several cases), this settles the effect of using bounded C, parameters $\log C$ or C, or unbounded C for all parameterizations of PACKING ILP and COVERING ILP studied in [12].

Theorem 1. q-PACKING ILP $(k + C)$ admits a polynomial kernelization of size $\mathcal{O}(k^{2q} \cdot C^{2q^2} \cdot \log C)$ with $\mathcal{O}(k^q \cdot C^{q^2})$ variables, $\mathcal{O}(k^q \cdot C^{q^2})$ constraints.

Theorem 2. q-COVERING ILP $(k+C)$ admits a polynomial kernelization of size $\mathcal{O}(k^{q+1} \cdot C^q \cdot \log C)$ with $\mathcal{O}(k^q \cdot C^q)$ variables, $\mathcal{O}(k)$ constraints.

Note that the above results also show the fixed-parameter tractability of PACKING ILP and COVERING ILP in the cases that q and C or $\log C$ are considered as constants or parameters since they constitute (not necessarily polynomial) kernelizations for all these cases.

Theorem 3. For all $q \geq 2$, q-PACKING ILP $(k + \log C)$ and q-COVERING ILP $(k + \log C)$ admit no polynomial kernelization unless $\mathsf{NP} \subseteq \mathsf{coNP/poly}$ and the polynomial hierarchy collapses to its third level. If $q = 1$, then q-PACKING ILP $(k + \log C)$ and q-COVERING ILP $(k + \log C)$ admit polynomial kernelizations.

Theorem 4. C-PACKING ILP $(k + q)$ and C-COVERING ILP $(k + q)$ admit no polynomial kernelization unless $\mathsf{NP} \subseteq \mathsf{coNP/poly}$ and the polynomial hierarchy collapses to its third level.

Related work. We have already mentioned the previous work [12,13] on kernelization properties of parameterized ILP problems. Apart from PACKING ILP and COVERING ILP this work addressed also existence of polynomial kernels for the ILP FEASIBILITY problem of, given $A \in \mathbb{Z}^{m \times n}$ and $b \in \mathbb{Z}^m$, determining whether a feasible solution $x \in \mathbb{Z}^n$ exists for $Ax \leq b$. (Note that there the constraints are more general and not necessarily of covering or packing type since A

and b may have negative entries.) We reuse a few of the basic reduction rules for COVERING ILP and PACKING ILP that were observed in [12].

For W[1]/W[2]-completeness of standard problems such as INDEPENDENT SET(k), HITTING SET(k), and SET PACKING(k) we refer to standard textbooks on parameterized complexity [8,11]. Some restricted variants of these problems are known to be fixed-parameter tractable and to admit polynomial kernels: BOUNDED DEGREE INDEPENDENT SET(k) [folklore], d-HITTING SET(k) [2,11], d-SET PACKING(k) [1]. For the latter two, lower bounds ruling out kernels of size $\mathcal{O}(k^{d-\varepsilon})$, for all $\varepsilon > 0$, unless NP $\subseteq$ coNP/poly were proved in [5,6].

W[1]-hardness of SMALL SUBSET SUM(k) was proved by Downey and Fellows [8,9]. Dom et al. [7] proved, amongst others, that SMALL SUBSET SUM($k + d$) has no polynomial kernel in $k + d$, where d bounds the encoding size of the input numbers (i.e., all numbers are less than 2^d). Dom et al. [7] also proved that RED-BLUE DOMINATING SET($|T| + k$) and BOUNDED RANK DISJOINT SET($d + k$) (i.e., SET PACKING($d + k$)) do not have polynomial kernels; we use these as source problems for the lower bounds of Theorems 3 and 4.

Organization. In Sect. 2 we recall definitions and basic concepts of parameterized complexity. Section 3 provides the proofs for Theorems 1 and 2. Section 4 gives the proofs for Theorems 3 and 4. We conclude in Sect. 5. Due to space constraints proofs for statements marked with ★ are deferred to the full version.

2   Preliminaries

Parameterized complexity and kernelization. A *parameterized problem* over some finite alphabet Σ is a language $\mathcal{P} \subseteq \Sigma^* \times \mathbb{N}$. The problem $\mathcal{P}$ is *fixed-parameter tractable* if $(x, k) \in \mathcal{P}$ can be decided in time $f(k) \cdot (|x| + k)^{\mathcal{O}(1)}$, where f is an arbitrary computable function. A polynomial-time algorithm K is a kernelization for $\mathcal{P}$ if, given input (x, k), it computes an equivalent instance (x', k') with $|x'| + k' \leq h(k)$ where h is some computable function; K is a *polynomial kernelization* if h is polynomial bounded (in k). Here equivalence means that $(x, k) \in \mathcal{P}$ if and only if $(x', k') \in \mathcal{P}$. By relaxing the restriction that the created instance (x', k') must be of the same problem and allow the output to be an instance of any language we get the notion of *(polynomial) compression*.

Let P and Q be parameterized problems. We say that P has a *polynomial parameter transformation* to Q ([4]), denoted by $P \leq_{ppt} Q$, if there exists a polynomial time computable function $f : \Sigma^* \times \mathbb{N} \to \Sigma^* \times \mathbb{N}$ and a polynomial p such that for all $(x, k) \in \Sigma^* \times \mathbb{N}$ we have $(x, k) \in P$ if and only if $(x', k') = f(x, k) \in Q$ and $k' \leq p(k)$.

Proposition 1 ([3,4]). *Let P and Q be parameterized problems such that $P \leq_{ppt} Q$. If Q admits a polynomial compression or polynomial kernelization then P admits a polynomial compression. If, possibly under some complexity-theoretic assumption, P admits no polynomial compression then Q admits no polynomial compression or polynomial kernelization under the same assumption.*

3   Kernelizations for ILPs with Fixed Column Sparseness

3.1   Packing ILP Parameterized by the Largest Coefficient

In this subsection, we prove Theorem 1, i.e., that q-PACKING ILP$(k+C)$ admits a polynomial kernelization. We recall that this refers to PACKING ILP with constant column sparseness q and with parameter $k+C$. Before giving the proof, we will define some reduction rules. A reduction rule is safe if after applying it we get a new instance which is equivalent to the original instance. Additionally, we use the following simple lemma that summarizes some elementary reduction steps (a similar statement is used in [12] but we include a proof in the appendix for completeness).

Lemma 1 ($\bigstar$). *Given an instance (A, b, c, k) for* PACKING ILP. *We can in polynomial time reduce to an equivalent instance (A', b', c', k) such that:*

1. *Each variable appears in at least one constraint.*
2. *For every variable x_j and row i of matrix A' we have $A'_{ij} \le b'_i$.*
3. *For every variable x_i we have $1 \le c_i \le k$.*

We call the reductions in the lemma basic rules. Without loss of generality, from now on all instances for PACKING ILP are assumed to satisfy the three conditions of the lemma.

Rule 1. *If two variables x_u, x_v, with $c_u \le c_v$, have the property that they always appear together and their corresponding coefficients in each constraint are always the same, i.e., $a_{iu} = a_{iv}$ for every row i of matrix A, then we can delete x_u from our problem.*

Rule 2. *If there exist k variables which have the property that no pair of them appear together in any constraint then our instance is a YES instance.*

Safeness of these two rules is an easy consequence of Lemma 1. Before continuing with the next reduction we recall the definition of a sunflower and (a simple variation of) the well-known sunflower lemma of Erdős and Rado [10].

Definition 1. *A sunflower of cardinality r with core C is a family of r sets $S_1, \ldots, S_r$ such that $S_i \cap S_j = C$ for all $i \ne j$ and the pairwise disjoint sets $S_1 \setminus C, \ldots, S_r \setminus C$, which are called petals, are required to be non-empty.*

Lemma 2. *Let $r, d, n \in \mathbb{N}$ and $\mathcal{F} = \{S_1, \ldots, S_n\}$ be a family of sets each of size at most d. If $n > d \cdot d! \cdot r^d$ we can find a sunflower of cardinality $r + 1$ in $\mathcal{F}$ in time polynomial in $n + d$. (The extra factor of d allows set size at most d.)*

Definition 2. *Let (A, b, c, k) be an instance of* PACKING ILP. *For each variable x_j we define a set P_j by $P_j = \{i : A_{ij} \ge 1\}$ i.e., P_j is the set of all constraints in which x_j appears.*

Then $x_{i_1}, \ldots, x_{i_r}$ are called sunflower variables if $P_{i_1}, \ldots, P_{i_r}$ form a sunflower of cardinality r, i.e., each constraint involves either all of $x_{i_1}, \ldots, x_{i_r}$ (with nonzero coefficients) or at most one of them. We then call the constraints that involve exactly one sunflower variable petal constraints and the constraints that involve all sunflower variables core constraints.

Rule 3. *Let $s = kq + k$. If $x_{i_1}, \ldots, x_{i_s}$ are sunflower variables with $c_{i_1} \leq \cdots \leq c_{i_s}$ and their corresponding coefficients are always the same when they appear together, then we can delete x_{i_1} from our problem.*

Lemma 3. *Rule 3 is safe.*

Proof. Rule 3 is equivalent to enforcing $x_{i_1} = 0$. Clearly, this additional constraint cannot increase the optimum target function value. Thus, it suffices to show that if our instance has an integer feasible solution x such that $c^T x \geq k$ then there exists another such solution, say x', with $x'_{i_1} = 0$. If $x_{i_1} = 0$ then we have nothing to do. If $x_{i_1} > k$ then we can reduce x_{i_1} to k which preserves feasibility and the cost function is still not smaller than k. Thus, we can restrict to the case that $1 \leq x_{i_1} = h \leq k$.

Let $S = \{\imath : x_i \geq 1\}$, i.e., S is the set of variables with nonzero value, and we have $i_1 \in S$. If $|S| \geq k + 1$ then we simply set $x_{i_1} = 0$. This of course preserves feasibility and since there remain at least k variables with nonzero values, the cost function is still at least k. So, we again restrict ourselves to the case $|S| \leq k$.

Let $P = \bigcup_{j \in S} P_j$, then $|P| \leq \sum_{j \in S} |P_j| \leq |S| \cdot q \leq kq$ since each variable occurs in at most q constraints. On the other hand, each sunflower variable corresponds to one non-empty petal. Consider $s = kq + k$ petals corresponding to $x_{i_1}, \ldots, x_{i_s}$. There are at least k petals disjoint from P, i.e., k sunflower variables corresponding to them do not appear in the same petal constraint with any variable in S. Since $h \leq k$ we can pick h sunflower variables from them, say $x_{j_1}, \ldots, x_{j_h}$. By definitions of S and P we have $x_{j_1} = \cdots = x_{j_h} = 0$.

Now, we construct a new solution x' from x by decreasing x_{i_1} from h to zero and increasing $x_{j_1}, \ldots, x_{j_h}$ from zero to one. Formally, x' is defined as follows:

$$\begin{cases} x'_{i_1} &= 0 \\ x'_{j_1} &= \cdots = x'_{j_h} = 1 \\ x'_j &= x_j \text{ otherwise.} \end{cases}$$

We just need to check the feasibility of x' with constraints that involve sunflower variables, i.e., core constraints and petal constraints. Let us consider a core constraint. Let a be the coefficient value that all of $x_{i_1}, x_{j_1}, \ldots, x_{j_h}$ have in this constraint. Then the left hand size will decrease by $a \cdot h$ when x_{i_1} reduces from h to zero and increase by $a \cdot h$ when h variables $x_{j_1}, \ldots, x_{j_h}$ move up from zero to one. Thus, all core constraints are still satisfied. We have nothing to check with the petal constraints that involve x_{i_1}. Now we consider a petal constraint that involves x_{j_t} for some $t \in \{1, \ldots, h\}$. By the way we choose $x_{j_1}, \ldots, x_{j_h}$, there is no variable in S that appears in this constraint, i.e., all variables in this constraint get value zero. Note that we are now in the case that all conditions in Lemma 1 are satisfied, so increasing one variable in this constraint from zero to one cannot break feasibility. Hence, all petal constraints are also satisfied. We finished checking the feasibility of x' and now only have to check the cost function value:

$$c^T x' - c^T x = c_{j_1} \cdot 1 + \cdots + c_{j_h} \cdot 1 - c_{i_1} \cdot h \geq 0$$

since $c_{j_t} \geq c_{i_1}$ for all t $= 1, \ldots, h$.   □

Now we are ready to prove Theorem 1.

Proof (Theorem 1). We consider an arbitrary instance (A, b, c, k) of q-PACKING ILP$(k+C)$. Let m and n denote the number of constraints and variables respectively and let $r = (kq + k - 1) \cdot C^{q-1}$. We will show that if $n > C^q \cdot q \cdot q! \cdot r^q$ then at least one of Rule 1–3 can be applied.

Assume that $n > C^q \cdot q \cdot q! \cdot r^q$. We define P_j as in Definition 2 and let $\mathcal{F} = \{S \mid S = P_i \text{ for some } i = 1, \cdots, n\}$ the family of sets that appear in the sequence $P_1, \ldots, P_n$. For each $S \in \mathcal{F}$ we define $n(S) = |\{i \mid P_i = S\}|$. By scanning all sets P_j, which can be done in time polynomial in n, we can check if there exists $S \in \mathcal{F}$ such that $n(S) > C^q$. If the answer is yes then there exist $C^q + 1$ variables that always appear together. For each of them, we define its ID by the (nonzero) coefficients in the constraints in which it appears. Since each variable appears in at most q constraints, the length of each ID is at most q. On the other hand, each entry of an ID can get one of C possible values from $\{1, \ldots, C\}$. Hence, there are at most C^q possible IDs. So, by the *pigeon-hole-principle*, we can find two of them which share the same ID and therefore can apply Rule 1.

In the case that $n(S) \leq C^q$ for all $S \in \mathcal{F}$ we have:

$$C^q \cdot q \cdot q! \cdot r^q < n = \sum_{S \in \mathcal{F}} n(S) \leq |\mathcal{F}| \cdot C^q$$

Thus, $|\mathcal{F}| > q \cdot q! \cdot r^q$ and by the Sunflower Lemma, we can find a sunflower of cardinality $r + 1$ in $\mathcal{F}$, say $P_{j_1}, \ldots, P_{j_{r+1}}$, in time polynomial in $|\mathcal{F}| \leq n$, i.e., we can find $r + 1$ sunflower variables. If the set of core constraints is empty, i.e., if the sunflower variables never appear together, then we can apply Rule 2 since $r + 1 \geq k$.

Now we consider the case that the set of core constraints is non-empty. Since each sunflower variable appears in at most q constraints and at least one petal constraint, the number of core constraints is at most $(q-1)$. We define the ID of a sunflower variable by its coefficients in the core constraints. Each entry of an ID can get one of C possible values from $\{1, \ldots, C\}$. The length of each ID is equal to the number of core constraints and therefore at most $(q-1)$. Hence, there are at most C^{q-1} possible IDs. On the other hand, we have $r + 1 = (kq + k - 1) \cdot C^{q-1} + 1$ sunflower variables. Thus, we can find at least $s = kq + k$ sunflower variables that share the same ID, i.e., we can apply Rule 3.

Thus, whenever $n > C^q \cdot q \cdot q! \cdot r^q$ we can determine in polynomial time which one in Rule 1–3 can be applied to reduce the number of variables. Since each iteration reduces the number of variables at least one, in polynomial time we must obtain an equivalent instance with at most $C^q \cdot q \cdot q! \cdot r^q$ variables, i.e., $n \in \mathcal{O}(k^q \cdot C^{q^2})$. Note that each variable appears in at most q constraints so we have $m \leq n \cdot q$ and, therefore, we have at most $\mathcal{O}(k^q \cdot C^{q^2})$ constraints. Since all constraint coefficients are bounded by C and all cost coefficients are bounded by k, the size of this instance is at most $\mathcal{O}(k^{2q} \cdot C^{2q^2} \cdot \log C)$.   □

3.2   Covering ILP Parameterized by the Largest Coefficient

In this subsection, we prove Theorem 2, i.e., that q-COVERING ILP$(k + C)$ admits a polynomial kernelization. For each instance, we also use m, n to denote the number of constraints and variables respectively. We start with a lemma which was mentioned in [12] (Lemma 8 in the full version) and allows us to bound the number of constraints and the cost value of each variable.

Lemma 4 ([12]). *Let (A, b, c, k) be an instance of* COVERING ILP*(k) and let q denote the column-sparseness of A. In polynomial time we can compute an equivalent instance (A', b', c', k) such that:*

1. A' *has at most kq rows (i.e., there are at most kq constraints).*
2. *The cost value c'_i of each variable x_i belongs to $\{1, \ldots, k\}$.*

Now it is easy to prove Theorem 2.

Proof (Theorem 2). By Lemma 4, we can restrict ourselves to the case that $m \le kq$. The only thing we need to do is reducing the number of variables. For each variable we define its ID by its coefficients in all constraints (for the constraints in which it does not appear we let the corresponding coefficients be zero). Thus, each ID has length m and contains at most q nonzero entries. Hence, the number of possible IDs is not greater than $\binom{m}{q} \cdot (C + 1)^q$ (there are $\binom{m}{m-q} = \binom{m}{q}$ ways to choose entries which get value zero and $(C + 1)^q$ ways to assign value in $\{0, \ldots, C\}$ for each of the q remaining entries). Thus, if $n > \binom{m}{q} \cdot (C + 1)^q$, we can find two variables with the same ID, say x_u and x_v with $c_u \le c_v$, then we can delete x_v from our problem. Safeness of this reduction rule can be proved similarly to safeness of Rule 1 (note that we delete x_v instead of x_u because we have a minimization problem). We apply this process until $n \le \binom{m}{q} \cdot (C + 1)^q \le \binom{kq}{q} \cdot (C + 1)^q$ (and we already have $m \le k \cdot q$). So we have in hand an instance with at most $\mathcal{O}(k^q \cdot C^q)$ variables and $\mathcal{O}(k)$ constraints. Since all constraint coefficients are bounded by C and all cost coefficients are bounded by k, the size of this instance is at most $\mathcal{O}(k^{q+1} \cdot C^q \cdot \log C)$.   □

4   Lower Bounds

In this section we discuss the kernelization lower bounds of Theorems 3 and 4. The first case is that we use $\log C$, the maximum encoding size of a coefficient, as a parameter instead of the largest coefficient C. The second case is that we consider C as a constant, i.e., all coefficients are bounded by a constant.

4.1   ILPs Parameterized by the Encoding Size of Coefficients

In this subsection, we prove Theorem 3 by describing a polynomial parameter transformation from COLORED RED-BLUE DOMINATING SET. (The straightforward kernelizations for $q = 1$ are given in the full version.) Recall that in

COLORED RED-BLUE DOMINATING SET we are given an integer k and a bipartite graph $G = (T \cup N, E)$ where the vertices of N are colored by function $col \colon N \to \{1, \dots, k\}$. The problem asks whether there exists a vertex set $N' \subseteq N$ that contains exactly one vertex of each color and such that every vertex in T has at least one neighbor in N'. It was proven in [7] that COLORED RED-BLUE DOMINATING SET parameterized by $|T| + k$ does not admit a polynomial compression.[2] Hence, Theorem 3 is a direct corollary of following lemma.

Lemma 5. *There are polynomial parameter transformations from* COLORED RED-BLUE DOMINATING SET*$(|T| + k)$ to 2-*PACKING ILP*$(k + \log C)$ and to* 2-COVERING ILP*$(k + \log C)$.*

Proof. Let $G = (T \cup N, E)$ be an instance of COLORED RED-BLUE DOMINATING SET and assume that $T = \{t_1, \dots, t_d\}$. Let C_i denote the set of all vertices in N with color i. Set $b = (k + d)^2$ and from now we work in the number system with base b, i.e., a string of numbers is also considered as a number written in the number system with base b. To produce transformations to PACKING ILP and COVERING ILP, we will first consecutively construct coefficients and variable sets of our desired instances.

For each vertex $v \in N$ we define a string $S(v)$ of length $k + 2d$ as follows:

$$S(v) = \underbrace{0\,1\,1\cdots1\,0\,1}_{\text{neighbor-part}}\ \underbrace{0\cdots1\cdots0}_{\text{color-part}}$$

The string contains two parts: color-part with $k + d$ digits and neighbor-part with d digits. As the name suggests, the color-part is for encoding the color of v, i.e., the i-th digit in the color-part gets value one if and only if $v \in C_i$. Similarly, the j-th digit in the neighbor-part gets value one if and only if t_j is a neighbor of v. If v has the color i then we also say that $S(v)$ has color i. Note that up to now we have only k colors while the color-part has $k + d$ digits, it means that the last d digits of $S(v)$ are all zeros.

Now we will construct new strings with new colors. For each $i \in \{1, \dots, d\}$ and $j \in \{0, \dots, k - 1\}$ we define a string a_{ij} as follows:

$$a_{ij} = \underbrace{0\,0\cdots j\cdots0\,0}_{d\text{ digits}}\ \underbrace{0\,0\cdots1\cdots0\,0}_{k+d\text{ digits}}$$

All digits of a_{ij} are zeros except two positions: the i-th digit gets value j and the $(k + d + i)$-th digit gets value one. In this case we say that a_{ij} has color $k + i$ (since the $(k + i)$-th digit in its color-part gets value one). At last, we construct a number t as follows:

$$t = \underbrace{k\cdots k}_{d\text{ digits}}\ \underbrace{1\cdots1}_{k+d\text{ digits}}\ .$$

---

[2] The paper [7] mentions only polynomial kernelizations but applies to polynomial compressions as well (cf. [3]).

For each vertex $v \in N$, we introduce a variable x_v and for each $i \in \{1,\ldots,d\}$ and $j \in \{0,\ldots,k-1\}$ we introduce a variable x_{ij}. To simplify, we use $\sum_{ij}$ instead of $\sum_{i=1}^{d}\sum_{j=0}^{k-1}$. Let us first consider the following system with two linear equations:

$$(*) \begin{cases} \sum_{v \in N} x_v + \sum_{ij} x_{ij} = k + d & (1) \\ \sum_{v \in N} S(v) \cdot x_v + \sum_{ij} a_{ij} \cdot x_{ij} = t & (2) \\ x_v \geq 0 \quad \forall v \in N \text{ and } x_{ij} \geq 0 \quad \forall i \in \{1,\ldots,d\}, j \in \{0,\ldots,k-1\} \end{cases}$$

We will show that $(*)$ has an integer solution if and only if the original instance of COLORED RED-BLUE DOMINATING SET is YES.

Assume that x is a solution of $(*)$. Compare the two sides in constraint (2) at the last digit: We have $\sum_{j=0}^{k-1} x_{dj} = 1$ modulo b. Since $0 \leq \sum_{j=0}^{k-1} x_{dj} \leq \sum_{ij} x_{ij} \leq \sum_{v \in N} x_v + \sum_{ij} x_{ij} \stackrel{(1)}{=} k + d < b + 1$, we must have $\sum_{j=0}^{k-1} x_{dj} = 1$, i.e., there is exactly one index j such that $x_{dj} = 1$. Moreover, there is no carry over into the next digit, i.e., we have $\sum_{j=0}^{k-1} x_{(d-1)j} = 1$ modulo b. Repeating this argument for the remaining digits in the color-part implies that, for each $i = 1,\ldots,d$ there is exactly one index j in $\{0,\ldots,k-1\}$, say $j(i)$, such that $x_{ij(i)} = 1$ and for each color $i = 1,\ldots k$ there is exactly one vertex with color i, say v_i, such that $x_{v_i} = 1$. By (1), all other variables get value zero and we can discard strings corresponding to them. Denote $N' = \{v_1,\ldots,v_k\}$ then (2) becomes

$$\sum_{v \in N'} S(v) + \sum_{i=1}^{d} a_{ij(i)} = t.$$

Consider the left hand side as a sum of $k+d$ numbers in base b, since each digit is at most $k-1$ and $(k+d)\cdot(k-1) < (k+d)^2 = b$ there are no carry overs when computing the sum. Thus identity must hold for each position, i.e.,

$$|N' \cap N(t_i)| + j(i) = k \quad \forall i = 1,\ldots,d.$$

Since $j(i) \leq k-1$ we have $|N' \cap N(t_i)| \geq 1$ for all $i = 1,\ldots,d$ and therefore N' is a solution of the original COLORED RED-BLUE DOMINATING SET instance.

Now assume that $N' \subseteq N$ is a solution of the original COLORED RED-BLUE DOMINATING SET instance. By the problem definition, for each $i \in \{1,\ldots,d\}$ we have $1 \leq |N' \cap N(t_i)| \leq |N'| = k$ and therefore we can find uniquely $j(i) \in \{0,\ldots,k-1\}$ such that $|N' \cap N(t_i)| + j(i) = k$. Then we set $x_{ij(i)} = 1$ for all i and $x_v = 1$ for all $v \in N'$ while letting all other variables get value zero. It is easy to check that with this assignment x is a solution of $(*)$.

We finish the proof by constructing formulations of $(*)$ in packing and covering form. Let us first consider two following instances of 2-PACKING ILP and 2-COVERING ILP with $k' = k + d$:

$$\max \sum_{v \in N} x_v + \sum_{ij} x_{ij}$$

$$\text{s.t.} \sum_{v \in N} S(v) \cdot x_v + \sum_{ij} a_{ij} \cdot x_{ij} \leq t \tag{3}$$

$$\sum_{v \in N} (t - S(v)) \cdot x_v + \sum_{ij} (t - a_{ij}) \cdot x_{ij} \leq t \cdot (k + d - 1) \tag{4}$$

$$x_v \geq 0 \quad \forall v \in N$$

$$x_{ij} \geq 0 \quad \forall i \in \{1, \dots, d\}, j \in \{0, \dots, k - 1\}$$

$$\min \sum_{v \in N} x_v + \sum_{ij} x_{ij}$$

$$\text{s.t.} \sum_{v \in N} S(v) \cdot x_v + \sum_{ij} a_{ij} \cdot x_{ij} \geq t$$

$$\sum_{v \in N} (t - S(v)) \cdot x_v + \sum_{ij} (t - a_{ij}) \cdot x_{ij} \geq t \cdot (k + d - 1)$$

$$x_v \geq 0 \quad \forall v \in N$$

$$x_{ij} \geq 0 \quad \forall i \in \{1, \dots, d\}, j \in \{0, \dots, k - 1\}$$

Equivalence of these ILPs with $(*)$ is easy to prove; a detailed argument is provided in the full version. Therefore they are equivalent to the original instance of COLORED RED-BLUE DOMINATING SET. Note that the number of variables is $|N| + k \cdot d$ which is polynomially bounded in the size of the original instance. This means that the transformations can be done in polynomial time. Finally, every coefficient in the ILP is bounded by $b^{k+2d} = (k + d)^{2(k+2d)}$ thus $\log C = 2(k + 2d) \cdot \log(k + d) \in \mathcal{O}((k + d) \log(k + d))$. This completes the proof.   □

4.2   ILPs with Bounded Coefficients

In this subsection we consider the case that C is considered as a constant and the column-sparseness q is an additional parameter. We prove Theorem 4 by constructing reductions from BOUNDED RANK DISJOINT SET$(d + k)$ and RED-BLUE DOMINATING SET$(|T| + k)$ which were shown by Dom et al. [7] to admit no polynomial compressions unless NP $\subseteq$ coNP/poly.

Lemma 6 (★). *There is a polynomial parameter transformation from* BOUNDED RANK DISJOINT SET*$(d + k)$ to C-PACKING* ILP*$(k + q)$.*

Lemma 7 (★). *There is a polynomial parameter transformation from* RED-BLUE DOMINATING SET*$(|T| + k)$ to C-COVERING* ILP*$(k + q)$.*

5 Conclusion

We have studied the influence of the column-sparseness q and the maximum coefficient size C on the existence of polynomial kernels for PACKING ILP(k) and COVERING ILP(k). We identified several new cases (see Table 1) as admitting a polynomial kernelization or being fixed-parameter tractable but admitting no polynomial kernelization (unless $\mathsf{NP} \subseteq \mathsf{coNP/poly}$). This has shed more light on cases that (with unbounded coefficients) were proved W[1]-hard by reduction from SMALL SUBSET SUM(k) in previous work [12]. Since all further results for PACKING ILP and COVERING ILP in [12] are either lower bounds with 0/1-coefficients or reductions to a polynomial number of variables and constraints, this fully settles cases with C bounded or with parameters $\log C$ or C.

References

1. Abu-Khzam, F.N.: An improved kernelization algorithm for r-set packing. Inf. Process. Lett. **110**(16), 621–624 (2010)
2. Abu-Khzam, F.N.: A kernelization algorithm for d-hitting set. J. Comput. Syst. Sci. **76**(7), 524–531 (2010)
3. Bodlaender, H.L., Jansen, B.M.P., Kratsch, S.: Kernelization lower bounds by cross-composition. SIAM J. Discrete Math. **28**(1), 277–305 (2014)
4. Bodlaender, H.L., Thomassé, S., Yeo, A.: Kernel bounds for disjoint cycles and disjoint paths. Theor. Comput. Sci. **412**(35), 4570–4578 (2011)
5. Dell, H., Marx, D.: Kernelization of packing problems. In: Rabani, Y. (ed.) SODA, pp. 68–81. SIAM (2012)
6. Dell, H., van Melkebeek, D.: Satisfiability allows no non-trivial sparsification unless the polynomial hierarchy collapses. In: STOC (2010)
7. Dom, M., Lokshtanov, D., Saurabh, S.: Incompressibility through Colors and IDs. In: Albers, S., Marchetti-Spaccamela, A., Matias, Y., Nikoletseas, S., Thomas, W. (eds.) ICALP 2009, Part I. LNCS, vol. 5555, pp. 378–389. Springer, Heidelberg (2009)
8. Downey, G.R., Fellows, M.R.: Fixed-parameter tractability and completeness II: on completeness for W[1]. Theor. Comput. Sci. **141**(1&2), 109–131 (1995)
9. Downey, R.G., Fellows, M.R.: Parameterized Complexity. Monographs in Computer Science. Springer, Heidelberg (1998)
10. Erdős, P., Rado, R.: Intersection theorems for systems of sets. J. Lond. Math. Soc. **s1-35**(1), 85–90 (1960)
11. Flum, J., Grohe, M.: Parameterized Complexity Theory. Texts in Theoretical Computer Science. An EATCS Series. Springer, Heidelberg (2006)
12. Kratsch, S.: On polynomial kernels for integer linear programs: covering, packing and feasibility. In: CoRR. arXiv:1302.3496
13. Kratsch, S.: On polynomial kernels for sparse integer linear programs. In: STACS, LIPIcs, vol. 28, pp. 80–91. Schloss Dagstuhl - Leibniz-Zentrum fuer Informatik (2013)

No Small Nondeterministic Read-Once Branching Programs for CNFs of Bounded Treewidth

Igor Razgon[✉]

Department of Computer Science and Information Systems,
Birkbeck, University of London, London, UK
igor@dcs.bbk.ac.uk

Abstract. In this paper, given a parameter k, we demonstrate an infinite class of CNFs of treewidth at most k of their primal graphs such that equivalent nondeterministic read-once branching programs (NROBPs) are of size at least n^{ck} for some universal constant c. Thus we rule out the possibility of fixed-parameter tractable space complexity of NROBPs parameterized by the smallest treewidth of equivalent CNFs.

1 Introduction

Read-once Branching Programs (ROBPs) is a well known representation of Boolean functions. Oblivious ROBPs, better known as Ordered Binary Decision Diagrams (OBDDs), is a subclass of ROBPs, very well known because of its applications in the area of verification [2]. An important procedure in these applications is transformation of a CNF into an equivalent OBDD. The resulting OBDD can be exponentially larger than the initial CNF, however a space efficient transformation is possible for special classes of functions. For example, it has been shown in [3] that a CNF of treewidth k of its primal graph can be transformed into an OBDD of size $O(n^k)$. A natural question is if the upper bound can be made fixed-parameter tractable i.e. of the form $f(k)n^c$ for some constant c. In [8] we showed that it is impossible by demonstrating that for each sufficiently large k there is an infinite class of CNFs of treewidth at most k whose smallest OBDD is of size at least $n^{k/5}$.

In this paper we report a follow up result showing that essentially the same lower bound holds for Non-deterministic ROBPs (NROBPs). In particular we show that there is a constant $0 < c < 1$ such that for each sufficiently large k there is an infinite class of CNFs of treewidth at most k (of their primal graphs) for which the space complexity of equivalent NROBPs is at least n^{ck}. Note that NROBPs are strictly more powerful than ROBPs in the sense that there is an infinite class of functions having a poly-size NROBP representation and exponential ROBP space complexity [4]. In the same sense, ROBPs are strictly more powerful than OBDDs, hence the result proposed in this paper is a significant enhancement of the result of [8].

© Springer International Publishing Switzerland 2014
M. Cygan and P. Heggernes (Eds.): IPEC 2014, LNCS 8894, pp. 319–331, 2014.
DOI: 10.1007/978-3-319-13524-3_27

We believe this result is interesting from the parameterized complexity theory perspective because it contributes to the understanding of parameterized *space* complexity of various representations of Boolean functions. In particular, the proposed result implies that ROBPs are inherently incapable to efficiently represent functions that are representable by CNFs of bounded treewidth. A natural question for further research is the space complexity of read c-times branching programs [1] (for an arbitrary constant c independent on k) w.r.t. the same class of functions.

To prove the proposed result, we use monotone 2-CNFs (their clauses are of form $(x_1 \vee x_2)$ where x_1 and x_2 are 2 distinct variables). These CNFs are in one-to-one correspondence with graphs having no isolated vertices: variables correspond to vertices and two variables occur in the same clause if and only if the corresponding vertices are adjacent. This correspondence allows us to use these CNFs and graphs interchangeably. We introduce the notion of Matching Width (MW) of a graph G and prove two theorems. One of them states that a monotone 2-CNF, whose corresponding graph G has MW at least t, cannot be computed by a NROBP of size smaller than $2^{t/a}$, where a is a constant dependent on the max-degree of G. The second theorem states that for each sufficiently large k there is an infinite family of graphs of treewidth k and max-degree 5 whose MW is at least $b * log n * k$ for some constant b independent of k. The main theorem immediately follows from replacement of t in the former lower bound by the latter one.

The strategy outlined above is similar to that we used in [8]. However, there are two essential differences. First, due to a much more 'elusive' nature of NORBPs compared to that of OBDDs, the counting argument is more sophisticated and more restrictive: it applies only to CNFs whose graphs are of constant degree. Due to this latter aspect, the target set of CNF instances requires a more delicate construction and reasoning.

Due to the space constraints, some proofs are either omitted or replaced by sketches.

2   Preliminaries

In this paper by a *set of literals* we mean one that does not contain both an occurrence of a variable and its negation. For a set S of literals we denote by $Var(S)$ the set of variables whose literals occur in S. If F is a Boolean function or its representation by a specified structure, we denote by $Var(F)$ the set of variables of F. A truth assignment to $Var(F)$ on which F is true is called a *satisfying assignment* of F. A set S of literals represents the truth assignment to $Var(S)$ where variables occurring positively in S (i.e. whose literals in S are positive) are assigned with *true* and the variables occurring negatively are assigned with *false*. For example, the assignment $\{x_1 \leftarrow true, x_2 \leftarrow true, x_3 \leftarrow false\}$ to variables x_1, x_2, x_3 is represented as $\{x_1, x_2, \neg x_3\}$.

We define a Non-deterministic Read Once Branching Program (NROBP) as a connected acyclic read-once *switching-and-rectifier network* [4]. That is, a NROBP

Y implementing (realizing) a function F is a directed acyclic graph (with possible multiple edges) with one leaf, one root, and with some edges labelled by literals of the variables of F in a way that there is no directed path having two edges labelled with literals of the same variable. We denote by $A(P)$ the set of literals labelling edges of a directed path P of Y.

The connection between Y and F is defined as follows. Let P be a path from the root to the leaf of Y. Then any extension of $A(P)$ to the truth assignment of all the variables of F is a satisfying assignment of F. Conversely, let A be a satisfying assignment of F. Then there is a path P from the root to the leaf of Y such that $A(P) \subseteq A$.

Remark. It is not hard to see that the traditional definition of NROBP as a deterministic ROBP with guessing nodes [5] can be thought of as a special case of our definition (for any function that is not constant *false*): remove from the former all the nodes from which the *true* leaf is not reachable and relabel each edge with the appropriate literal of the variable labelling its tail (if the original label on the edge is 1 then the literal is positive, otherwise, if the original label is 0, the literal is negative).

We say that a NROBP Y is *uniform* if the following is true. Let a be a node of Y and let P_1 and P_2 be 2 paths from the root of Y to a. Then $Var(A(P_1)) = Var((A(P_2)))$. That is, these paths are labelled by literals of the same set of variables. Also, if P is a path from the root to the leaf of Y then $Var(A(P)) = Var(F)$. Thus there is a one-to-one correspondence between the sets of literals labelling paths from the root to the leaf of Y and the satisfying assignments of F.

All the NROBPs considered in Sections 3–5 of this paper are uniform. This assumption does not affect our main result because, using the construction described in the proof sketch of Proposition 2.1 of [6], an arbitrary NROBP can be transformed into a uniform one at the price of $O(n)$ times increase of the number of edges. For the technical details, see the appendix of [7].

Given a graph G, its *tree decomposition* is a pair $(T, \mathbf{B})$ where T is a tree and $\mathbf{B}$ is a set of bags $B(t)$ corresponding to the vertices t of T. Each $B(t)$ is a subset of $V(G)$ and the bags obey the rules of *union* (that is, $\bigcup_{t \in V(T)} B(t) = V(G)$), *containment* (that is, for each $\{u, v\} \in E(G)$ there is $t \in V(t)$ such that $\{u, v\} \subseteq B(t)$), and *connectedness* (that is for each $u \in V(G)$, the set of all t such that $u \in B(t)$ induces a subtree of T). The *width* of $(T, \mathbf{B})$ is the size of the largest bag minus one. The treewidth of G is the smallest width of a tree decomposition of G.

Given a CNF ϕ, its *primal graph* has the set of vertices corresponding to the variables of ϕ. Two vertices are adjacent if and only if there is a clause of ϕ where the corresponding variables both occur.

3   The Main Result

A *monotone* 2-CNFs has clauses of the form $(x \vee y)$ where x and y are two distinct variables. Such CNFs can be put in one-to-one correspondence with graphs that

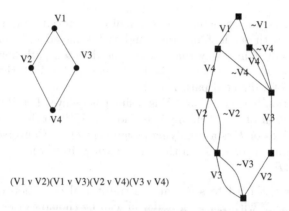

V1

V2   V3

V4

(V1 v V2)(V1 v V3)(V2 v V4)(V3 v V4)

Fig. 1. A graph, the corresponding CNF and a NROBP of the CNF

do not have isolated vertices. In particular, let G be such a graph. Then G corresponds to a 2CNF $\phi(G)$ whose variables are the vertices of G and the set of clauses is $\{(u \vee v)|\{u, v\} \in E(G)\}$. These notions, connected to the corresponding NROBP, are illustrated on Fig. 1.[1] It is not hard to see that G is the primal graph of $\phi(G)$, hence we can refer to the treewidth of G as the primal graph treewidth of $\phi(G)$.

The following theorem is the main result of this paper.

Theorem 1. *There is a constant c such that for each $k \geq 50$ there is an infinite class* **G** *of graphs each of treewidth of at most k such that for each $G \in$ **G**, the smallest NROBP equivalent to $\phi(G)$ is of size at least $n^{k/c}$, where n is the number of variables of $\phi(G)$.*

In order to prove Theorem 1, we introduce the notion of *matching width* (MW) of a graph and state two theorems proved in the subsequent two sections. One claims that if the max-degree of G is bounded then the size of a NROBP realizing $\phi(G)$ is exponential in the MW of G. The other theorem claims that for each sufficiently large k there is an infinite class of graphs of bounded degree and of treewidth at most k whose MW is at least $b * log n * k$ for some universal constant b. Theorem 1 will follow as an immediate corollary of these two theorems.

Definition 1 Matching width. *Let SV be a permutation of $V(G)$ and let S_1 be a prefix of SV (i.e. all vertices of $SV \setminus S_1$ are ordered after S_1). The* matching width *of S_1 is the size of the largest matching consisting of the edges between S_1 and $V(G) \setminus S_1$.[2] The* matching width *of SV is the largest matching width*

---

[1] Notice that on the NROBP in Figure 1, there is a path where v_2 occurs before v_3 and a path where v_3 occurs before v_2. Thus this NROBP, although uniform, is not *oblivious*.

[2] We sometimes treat sequences as sets, the correct use will be always clear from the context.

of a prefix of SV. The matching width *of G, denoted by $mw(G)$, is the smallest matching width of a permutation of $V(G)$.*

Remark. The above definition of matching width is a special case of the notion of *maximum matching width* as defined in [9].

To illustrate the above notions recall that C_n and K_n respectively denote a cycle and a complete graph of n vertices. Then, for a sufficiently large n, $mw(C_n) = 2$. On the other hand $mw(K_n) = \lfloor n/2 \rfloor$.

Theorem 2. *For each integer i there is a constant a_i such that for any graph G the size of NROBP realizing $\phi(G)$ is at least $2^{mw(G)/a_x}$ where x is the max-degree of G.*

Theorem 3. *There is a constant b such that for each $k \geq 50$ there is an infinite class **G** of graphs of degree at most 5 such that the treewidth of all the graphs of G is at most k and for each $G \in **G**$ the matching width is at least $(logn * k)/b$ where $n = |V(G)|$.*

Now we are ready to prove Theorem 1.

Proof of Theorem 1. Let **G** be the class whose existence is claimed by Theorem 3. By Theorem 2, for each $G \in **G**$ the size of a NROBP realizing $\phi(G)$ is of size at least $2^{mw(G)/a_5}$. Further on, by Theorem 3, $wm(G) \geq (logn * k)/b$, for some constant b. Substituting the inequality for $mw(G)$ into the lower bound $2^{mw(G)/a_5}$ supplied by Theorem 2, we get that the size of a NROBP is at least $2^{logn*k/c}$ where $c = a_5 * b$. Replacing 2^{logn} by n gives us the desired lower bound. ∎

From now on, the proof is split into two independent parts: Sect. 4 proves Theorem 2 and Sect. 5 proves Theorem 3.

4   Proof of Theorem 2

Recall that the vertices of graph G serve as variables in $\phi(G)$. That is, in the truth assignments to $Var(\phi(G))$, the vertices are treated as literals and may occur positively or negatively. Similarly for a path P of a NROBP Z implementing $\phi(G)$, we say that a vertex $v \in V(G)$ *occurs* on P if either v and $\neg v$ labels an edge of P. In the former case this is a *positive occurrence*, in the latter case a *negative* one.

Recall that a Vertex Cover (VC) of G is $V' \subseteq V(G)$ incident to all the edges of $E(G)$.

Observation 1. *S is a satisfying assignment of $\phi(G)$ if and only if the vertices of G occurring positively in S form a VC of G. Equivalently, $V' \subseteq V(G)$ is the set of all vertices of G occurring positively on a root-leaf path of Z if and only if V' is a VC of G.*

In light of Observation 1, we denote the set of all vertices occurring positively on a root-leaf path P of Z by $VC(P)$.

The proof of Theorem 2 requires two intermediate statements. For the first statement, let a be a node of a NROBP Z. For an integer $t > 0$, we call a a t-node if there is a set $S(a)$ of size at least t such that for each root-leaf path P, meeting a, $S(a) \subseteq VC(P)$.

Lemma 1. *Suppose that the matching width of G is at least t. Then t-nodes of Z form a root-leaf cut.*

Proof. We need to show that each root-leaf path P passes through a t-node. Due to the uniformity of Z, (the vertices of G corresponding to) the labels of P being explored from the root to the leaf form a permutation SV of $V(G)$. Let SV' be a prefix of the permutation witnessing the matching width at least t. In other words, there is a matching $M = \{\{u_1, v_1\}, \ldots, \{u_t, v_t\}\}$ of G such that all of $u_1, \ldots, u_t$ belong to SV', while all of $v_1, \ldots, v_t$ belog to $SV \setminus SV'$. Let u be the last vertex of SV' and let a be the head of the edge of P whose label is a literal of u. We claim that a is a t-node with a witnessing set $S(a) = \{x_1, \ldots, x_t\}$ such that $x_i \in \{u_i, v_i\}$ for each x_i.

Indeed, observe that for each $\{u_i, v_i\}$ there is $x_i \in \{u_i, v_i\}$ such that $x_i \in VC(P)$ for each root-leaf path P passing through a. Clearly for any root-leaf path Q of Z, either $u_i \in VC(Q)$ or $v_i \in VC(Q)$ for otherwise $VC(Q)$ is not a VC of G in contradiction to Observation 1. Thus if such x_i does not exist then there are two paths Q^1 and Q^2 meeting a such that $VC(Q_1) \cap \{u_i, v_i\} = \{u_i\}$ and $VC(Q_2) \cap \{u_i, v_i\} = \{v_i\}$.

For a root-leaf path Q passing through a denote by Q_a the prefix of Q ending with a and by $\neg Q_a$ the suffix of Q beginning with a. Observe that u_i occurs both in Q_a^1 and Q_a^2. Indeed, assume w.l.o.g. that u_i does not occur in Q_a^1. Then, by uniformity of Z, u_i occurs in $\neg Q_a^1$. Then $P_a + \neg Q_a^1$ (we denote this way the concatenation of two paths) is a root-leaf path with a double occurrence of u_i, a contradiction to Z being read-once. Similarly we establish that v_i occurs in both $\neg Q_a^1$ and $\neg Q_a^2$. It remains to observe that, by definition, u_i occurs negatively in Q_a^2 and v_i occurs negatively in $\neg Q_a^1$. Hence $Q^* = Q_a^2 + \neg Q_a^1$ is a root-leaf path of Z such that $VC(Q^*)$ is disjoint with $\{u_i, v_i\}$, a contradiction to Observation 1, confirming the existence of the desired x_i.

Suppose that there is a root-leaf path P' of Z passing through a such that $S(a) \not\subseteq VC(P')$. This means that there is $x_i \notin VC(P')$ contradicting the previous two paragraphs. Thus being a a t-node has been established and the lemma follows. ∎

For the second statement, let $\mathbf{A}$ and $\mathbf{B}$ be two families of subsets of a universe $\mathbf{U}$. We say that $\mathbf{A}$ *covers* $\mathbf{B}$ if for each $S \in \mathbf{B}$ there is $S' \in \mathbf{A}$ such that $S' \subseteq S$. If each element of $\mathbf{A}$ is of size at least t then we say that $\mathbf{A}$ is a t-*cover* of $\mathbf{B}$. Denote by $\mathbf{VC}(G)$ the set of all VCs of G.

Theorem 4. *There is a function f such that the following is true. Let H be a graph. Let $\mathbf{A}$ be a t-cover of $\mathbf{VC}(H)$. The $|\mathbf{A}| \geq 2^{t/f(x)}$ where x is the max-degree of H.*

The proof of Theorem 4 is provided in Subsection 4.1. Now we are ready to prove Theorem 2.

Proof of Theorem 2. Let N be the set of all t-nodes of Z. For each $a \in N$, specify one $S(a)$ of size at least t such that for all paths P of Z passing through a, $S(a) \subseteq VC(P)$. Let $\mathbf{S} = \{S_1, \ldots, S_q\}$ be the set of all such $S(a)$. Then we can specify *distinct* $a_1, \ldots, a_q$ such that $S_i = S(a_i)$ for all $i \in \{1, \ldots, q\}$.

Observe that $\mathbf{S}$ is covers $\mathbf{VC}(G)$. Indeed, let $V' \in \mathbf{VC}(G)$. By Observation 1, there is a root-leaf path P with $V' = VC(P)$. By Lemma 1, P passes through some $a \in N$ and hence $S(a) \subseteq VC(P)$. By definition, $S(a) = S_i$ for $i \in \{1, \ldots, q\}$ and hence $S_i \subseteq V'$. Thus $\mathbf{S}$ is a t-cover of $\mathbf{VC}(G)$.

It follows from Theorem 4 that $q = |\mathbf{S}| \geq 2^{t/f(x)}$ where x is a max-degree of G and f is a universal function independent on G or t. It follows that Z contains at least $2^{t/f(x)}$ distinct nodes namely $a_1, \ldots, a_q$. ∎

4.1   Proof of Theorem 4

We are going to define a probability distribution of $\mathbf{VC}(G)$ and to show that for a graph G of constant degree the probability of an element of $\mathbf{VC}(G)$ to be a superset of a specific subset of size at least t is exponentially small in t. We then conclude that the number of such subsets covering all the elements of $\mathbf{VC}(G)$ must be exponentially large in t. In the technical details that follow, we do not use the probabilites explicitly but rather present the proof in terms of weighted counting.

Let us define a graph G with *fixed vertices* as (V, E, F) where V and E bear their usual meaning and $F \subseteq V$ is the set of *fixed vertices*. We can also use $V(G), E(G), F(G)$ to denote V, E, F, respectively. A set $S \subseteq V(G)$ is a VC of G if S is a VC of (V, E) and *in addition*, $F \subseteq S$. Then $\mathbf{VC}(G)$ is the set of all VCs of (V, E) that contain F as a subset. We define $G \setminus v$ as (V', E', F') with $(V', E') = (V, E) \setminus v$ (the usual operation of vertex removal from a graph) and $F' = F \setminus \{v\}$. We define G/v as (V', E', F''), where (V', E') are as above and $F'' = F \cup N_G(v)$, where $N_G(v)$ is the set of neighbours of v in (V, E).

Let SV be a permutation of V. Now we are going to define a *decision tree* of $\mathbf{VC}(G)$ w.r.t. SV, denoting it by $T = T_{G,SV}$. It is a rooted binary tree with edges directed from the parent to a child. If a node a of T has two children, we distinguish the *left child* $lch_T(a)$ and the *right child* $rch_T(a)$ (the subscript can be omitted if clear from the context). If a is a unary node, its only child is considered the *left* one and the right child is not defined. We denote by T_a the subtree of T rooted by a. With this notation in mind we define T recursively as follows.

If G is an empty graph then $T_{G,SV}$ consists of a single node. Otherwise, let vf be the first vertex of SV, $SV' = SV \setminus vf$ (the suffix of SV' resulting from the removal of vf), and rt be the root of $T_{G,SV}$. If $vf \in F(G)$ then rt is a unary node, otherwise rt is a binary node. The edge $(rt, lch(rt))$ is labelled with vf and $T_{lch(rt)}$ is $T_{G \setminus vf, SV'}$. If rt is a binary node (the right child of rt is defined) then $(rt, lch(rt))$ is labelled with $\neg vf$ and $T_{rch(rt)} = T_{G/vf, SV'}$.

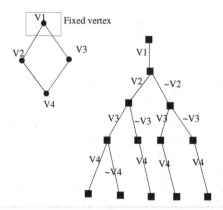

Fig. 2. A tree $T_{C,SVC}$ where C is the graph on the left with $F(C) = \{v_1\}$ and $SVC = (v_1, v_2, v_3, v_4)$. All the edges of $T_{S,SVC}$ are directed to the bottom, hence the arrows on the edges are not shown.

An example of a decision tree as defined above is provided in Fig. 2.

For a root-leaf path P of T, denote by $VC(P)$ the set of vertices occurring positively as labels of the edges of P and let $\mathbf{P}_T$ be the set of all root-leaf paths of T.

Observation 2. *The set* $\{VC(P)|P \in \mathbf{P}\}_T$ *is precisely* $\mathbf{VC}(G)$.

Let $S \subseteq V$. Denote by $\mathbf{P}_{T,S}$ the set if all root-leaf paths P of T such that $S \subseteq VC(P)$. Let (a, b) be an edge of T and let $\mathbf{P}$ be a set of paths of $\mathbf{P}$, all starting from b. Then $(a, b) + \mathbf{P} = \{(a, b) + P|P \in \mathbf{P}\}$ ($(a, b) + P$ denotes the concatenation of a single edge path (a, b) and P).

We say that S is a *distant independent set* (DIS) of G if the distance between any two elements of S in G is at least 3 (the vertices of S are not adjacent and do not have joint neighbours).

Lemma 2. *Suppose that G is not empty and let vf be the first vertex of SV. Assume that S is a* DIS *disjoint with $F(G)$. Then the following statements are true regarding* $\mathbf{P}_{T,S}$.

1. *If* $vf \in S$ *then* $\mathbf{P}_{T,S} = (rt, lch(rt)) + \mathbf{P}_{T_{lch(rt)},S\setminus\{vf\}}$.
2. *If* rt *is a binary node and* vf *is a neighbour of S then* $[\mathbf{P}_{T,S} = (rt, lch(rt)) + \mathbf{P}_{T_{lch(rt)},S}] \cup [(rt, rch(rt)) + \mathbf{P}_{T_{rch(rt)},S\setminus\{vn\}}]$ *where* vn *is the only neighbour of vf in S (due to S being a* DIS).
3. *In all other cases* $\mathbf{P}_{T,S} = (rt, lch(rt)) + \mathbf{P}_{T_{lch(rt)},S}$ *wherever* rt *is a unary node and* $\mathbf{P}_{T,S} = [(rt, lch(rt)) + \mathbf{P}_{T_{lch(rt)},S}] \cup [(rt, rch(rt)) + \mathbf{P}_{T_{rch(rt)},S}]$ *wherever* rt *has two children.*

Proof. Assume that $vf \in S$ and let $P \in \mathbf{P}_{T,S}$. By our assumption about vf, it can occur only as a label on the first edge. Since $vf \in S$, this occurrence must be positive. Consequently, the first edge is $(rt, lch(rt))$. Furthermore, the rest of the labels must be supplied by the suffix of P starting at $lch(rt)$. Hence we conclude that this suffix belongs to $\mathbf{P}_{T_{lch(rt)},S\setminus\{vf\}}$ and hence $P \in (rt, lch(rt)) + \mathbf{P}_{T_{lch(rt)},S\setminus\{vf\}}$. Conversely, let $P \in (rt, lch(rt)) + \mathbf{P}_{T_{lch(rt)},S\setminus\{vf\}}$. Then vf occurs positively on the first edge and the rest of vertices of S occur positively in the subsequent suffix. Thus $S \subseteq VC(P)$ and hence $P \in \mathbf{P}_{T,S}$.

It is straightforward to observe that if $vf \notin S$ then the third statement holds simply owing to the fact the the occurrences of the vertices of S are not contributed by the first edges of paths of $\mathbf{P}_T$. However, if vf is a neighbour of $vn \in S$, it can be noticed that $\mathbf{P}_{T_{rch(rt)},S} = \mathbf{P}_{T_{rch(rt)}S\setminus\{vn\}}$ thus confirming the second statement. Indeed, since $S \subseteq VC(P)$ implies $S \setminus \{vn\} \in VC(P)$ for any $P \in \mathbf{P}_T$, $\mathbf{P}_{T_{rch(rt)},S} \subseteq \mathbf{P}_{T_{rch(rt)},S\setminus\{vn\}}$. For the opposite direction, recall that $T_{rch(rt)} = T_{G/vf,SV'}$ and $vn \in F(G/vf)$. This means that $vn \in VC(P)$ for any path $P \in \mathbf{P}_{T_{rch(rt)}}$. Consequently, $S \setminus \{vn\} \subseteq VC(P)$ implies that $S \subseteq VC(P)$ and hence $\mathbf{P}_{T_{rch(rt)},S\setminus\{vn\}} \subseteq \mathbf{P}_{T_{rch(rt)},S}$. ∎

Let as assign weights to the edges of $T_{G,SV}$ as follows. For a binary node assign weight 0.5 to both its outgoing edges. For a unary node assign weight 1 to its only out-going edge. Denote the weight of an edge e by $w(e)$. For a path P, the weight $w(P)$ of P is a product of weights of its edges, considering the weight of a single vertex path to be 1, and for a set $\mathbf{P}$ of paths, its weight $w(\mathbf{P}) = \sum_{P \in \mathbf{P}} w(P)$.

Observation 3. *Let a be a node of $T_{G,SV}$. Then the following statements hold.*

- $w(\mathbf{P}_{T_a}) = 1.$
- *Let (a, b) be an edge of $T_{G,SV}$ and let $\mathbf{P}$ be a set of paths of $T_{G,SV}$ all starting from b. Then $w((a, b) + \mathbf{P}) = w((a, b)) * w(\mathbf{P})$.*

For $v \in V(G)$, denote $1 - 2^{-(d_G(v)+1)}$ by $p_G(v)$. The following are simple facts regarding $p_G(v)$.

Observation 4. *The following statements hold regarding $p_G(v)$.*

- *Let $u \in V(G) \setminus \{v\}$. Then $p_{G\setminus u}(v) \le p_G(v)$.*
- $0.5 \le p_G(v).$
- *Let c be the max-degree of G. Then $p_G(v) \le 1 - 2^{-(c+1)}$.*

The following is the central statement towards the proof of Theorem 4.

Lemma 3. *Let S be a DIS of G such that $S \cap F(G) = \emptyset$, let SV be an arbitrary permutation of $V(G)$ and let $T = T_{G,SV}$. Then $w(\mathbf{P}_{T,S}) \le \prod_{v \in S} p_G(v)$. (We assume the right-hand part of the inequality to equal 1 if $S = \emptyset$).*

Proof. By induction on $|V(G)|$. If $|S| = 0$ then the theorem clearly holds because $w(\mathbf{P}_{T,S}) \le w(\mathbf{P}_T) = 1$ by Observation 3. So, assume that $|S| > 0$ and hence $|V(G)| > 0$. Let rt be the root of T and let vf be the first vertex of SV.

Suppose rt is a unary node (this means that $vf \in F(G)$ and hence $vf \notin S$). It follows from Lemma 2 and Observation 4 that $w(\mathbf{P}_{T,S}) = w(\mathbf{P}_{T_{lch(rt)},S})$. Recall that $T_{lch(rt)} = T_{G\setminus vf,SV\setminus vf}$ and that S is disjoint with $F(G \setminus vf)$. Hence, the induction assumption stands. Combining it with the first item of Observation 4, we get $w(\mathbf{P}_{T_{lch(rt)},S}) \leq \prod_{v \in S} p_{G \setminus vf}(v) \leq \prod_{v \in S} p_G(v)$ as required.

In the rest of the proof we assume that rt is a binary node. Assume first that $vf \notin S \cup N(S)$. Then S remains non-fixed in both $G \setminus vf$ and G/vf and hence the induction assumption stands for both $w(\mathbf{P}_{T_{lch(rt)},S})$ and $w(\mathbf{P}_{T_{rch(rt)},S})$. Applying the same line of argumentation as in the previous paragraph, we observe that $w(\mathbf{P}_{T_{lch(rt)},S}) \leq \prod_{v \in S} p_G(v)$ and $w(\mathbf{P}_{T_{rch(rt)},S}) \leq \prod_{v \in S} p_G(v)$. By Lemma 2 together with Observation 3, we obtain $w(\mathbf{P}_{T,S}) \leq 0.5 * w(\mathbf{P}_{T_{lch(rt)},S}) + 0.5 * w(\mathbf{P}_{T_{rch(rt)},S})$. Substituting $w(\mathbf{P}_{T_{lch(rt)},S})$ and $w(\mathbf{P}_{T_{rch(rt)},S})$ with $\prod_{v \in S} p_G(v)$, we obtain $w(\mathbf{P}_{T,S}) \leq 0.5 * \prod_{v \in S} p_G(v) + 0.5 * \prod_{v \in S} p_G(v) = \prod_{v \in S} p_G(v)$ as required.

Assume now that $vf \in S$. Observe that $S \setminus \{vf\}$ is not fixed in $G \setminus vf$. Hence, arguing as is the previous two paragraphs, we conclude that $w(\mathbf{P}_{T_{lch(rt)},S \setminus \{vf\}}) \leq \prod_{v \in S \setminus \{vf\}} p_G(v)$. Lemma 2 together with Observation 3 yield $w(\mathbf{P}_{T,S}) \leq 0.5 * w(\mathbf{P}_{T_{lch(rt)},S})$. Substituting $w(\mathbf{P}_{T_{lch(rt)},S \setminus \{vf\}})$, we obtain $w(\mathbf{P}_{T,S}) \leq 0.5 * \prod_{v \in S \setminus \{vf\}} p_G(v)$. By the second item of Observation 4, 0.5 can be replaced by $p_G(vf)$ in the last inequality. That is $w(\mathbf{P}_{T,S}) \leq p_G(vf) * \prod_{v \in S \setminus \{vf\}} p_G(v) = \prod_{v \in S} p_G(v)$ as required.

Finally, suppose that vf is a neighbour of S. That is vf is a neighbour of exactly one vertex $vn \in S$. Observe that S is not fixed in $G \setminus vf$ and $S \setminus \{vn\}$ is not fixed in G/vf. Hence, arguing as above, we conclude that $w(\mathbf{P}_{T_{lch(rt)},S}) \leq p_{G \setminus vf}(vn) * \prod_{v \in S \setminus \{vn\}} p_G(v)$ and that $w(\mathbf{P}_{T_{rch(rt)},S \setminus \{vn\}}) \leq \prod_{v \in S \setminus \{vn\}} p_G(v)$ (notice that we have not replaced $p_{G \setminus vf}(vn)$ by $p_G(vn)$ as retaining the former is essential for the forthcoming reasoning). By Lemma 2 and Observation 3, $w(\mathbf{P}_{T,S}) \leq 0.5 * w(\mathbf{P}_{T_{lch(rt)},S}) + 0.5 w(\mathbf{P}_{T_{rch(rt)},S \setminus \{vn\}})$. Substituting $w(\mathbf{P}_{T_{lch(rt)},S})$ and $w(\mathbf{P}_{T_{rch(rt)},S \setminus \{vn\}})$ and moving $\prod_{v \in S \setminus \{vn\}} p_G(v)$ outside the brackets, we obtain $w(\mathbf{P}_{T,S}) \leq 0.5(p_{G \setminus vf}(vn)+1) * \prod_{v \in S \setminus \{vn\}} p_G(v)$. The last step of our reasoning is the observation that $0.5(p_{G \setminus vf}(vn)+1) = p_G(vn)$. Indeed, $p_G(vn) = (1 - 2^{-(d_G(vn)+1)}) = 0.5(2 - 2^{-d_G(v)}) = 0.5(1 - 2^{-(d_{G \setminus vf}(vn)+1)} + 1) = 0.5(p_{G \setminus vf}(vn) + 1)$. Thus $w(\mathbf{P}_{T,S}) \leq p_G(vn) * \prod_{v \in S \setminus \{vn\}} p_G(v) = \prod_{v \in S} p_G(v)$ as required. ∎

Proof of Theorem 4. To consider H in the theorem statement as a graph with fixed vertices, we represent it as $(V, E, \emptyset)$. Let SV be an arbitrary permutation of $V(H)$ and let $T = T_{H,SV}$.

For the given integer $x > 0$, let a_x be the constant such that $2^{-1/a_x} = (1 - 2^{-(x+1)})$. Let c be the max-degree of H. Then, by the last statement of Observation 4, for any $v \in V(H)$, $p_H(v) \leq 2^{-1/a_c}$.

Let S be a DIS of H. Then, combining the previous paragraph with Lemma 3, we observe that $w(\mathbf{P}_{T,S}) \leq 2^{-|S|/a_c}$.

Let S^* be an arbitrary subset of $V(H)$. Observe that there is a DIS $S \subseteq S^*$ of size at least $|S^*|/(c^2 + 1)$. Indeed, let $S \subseteq S^*$ be a largest DIS which a subset of S. Then each element of $S^* \setminus S$ is at distance at most 2 from an element of S. For each $u \in S$, there are at most $c + c(c - 1) = c^2$ elements of H lying at distance at most 2 from S. Thus $|S^* \setminus S| \leq |S| * c^2$, that is $|S^*| \leq |S| * (c^2 + 1)$ and hence

$|S| \geq |S^*|/(c^2 + 1)$. Since $\mathbf{P}_{T,S^*} \subseteq \mathbf{P}_{T,S}$, $w(\mathbf{P}_{T,S^*}) \leq w(\mathbf{P}_{T,S}) \leq 2^{-|S|/b_c}$, where $b_c = a_c * (c^2 + 1)$.

Let $S_1, \ldots, S_q$ be a t-cover of $\mathbf{VC}(H)$. This means that for each $P \in \mathbf{P}_T$ there is S_i whose vertices occur as positive labels on P. In other words, $\mathbf{P}_T = \bigcup_{i=1}^{q} w(\mathbf{P}_{T,S_i})$ Hence $1 = w(\mathbf{P}_T) \leq \sum_{i=1}^{q} w(\mathbf{P}_{T,S_i}) \leq q * 2^{-t/b_c}$, where the first equality follows from Observation 3. Consequently, $q \geq 2^{t/b_c}$ as claimed. ∎

5 Proof of Theorem 3

Denote by T_r a complete binary tree of height (root-leaf distance) r. Let T be a tree and H be an arbitrary graph. Then $T(H)$ is a graph having disjoint copies of H in one-to-one correspondence with the vertices of T. For each pair t_1, t_2 of adjacent vertices of T, the corresponding copies are connected by making adjacent the pairs of *same* vertices of these copies. Put differently, we can consider H as a labelled graph where all vertices are associated with distinct labels. Then for each edge $\{t_1, t_2\}$ of T, edges are introduced between the vertices of the corresponding copies having the same label. An example of this construction is shown on Fig. 3.

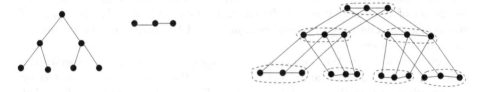

Fig. 3. Graphs from the left to the right: $T_3, P_3, T_3(P_3)$. The dotted ovals surround the copies of P_3 in $T_3(P_3)$.

The following lemma is the critical component of the proof of Theorem 3.

Lemma 4. *Let p be an arbitrary integer and let H be an arbitrary connected graph of $2p$ vertices. Then for any $r \geq \lceil \log p \rceil$, $mw(T_r(H)) \geq (r + 1 - \lceil \log p \rceil)p/2$.*

Before proving Lemma 4, let us show how Theorem 3 follows from it.

Sketch proof of Theorem 3. First of all, let us identify the class $\mathbf{G}$. Recall that P_x a path of x vertices. Let $0 \leq y \leq 3$ be such that $k - y + 1$ is a multiple of 4. The considered class $\mathbf{G}$ consists of all $G = T_r(P_{k-y+1})$ for $r \geq 5\lceil \log k \rceil$. It can be observed that the max-degree of the graphs of $\mathbf{G}$ is 5 and their treewidth is at most k.

Taking into account that starting from a sufficiently large r compared to k, $r = \Omega(\log(n/k))$ can be seen as $r = \Omega(\log n)$, the lower bound of Lemma 4 can be stated as $mw(G) = \Omega(\log n * k)$. ∎

The following lemma is an auxiliary statement for Lemma 4.

Lemma 5. *Let T be a tree consisting of at least p vertices. Let H be a connected graph of at least $2p$ vertices. Let V_1, V_2 be a partition of $V(T(H))$ such that both partition classes contain at least p^2 vertices. Then $T(H)$ has a matching of size p with the ends of each edge belong to distinct partition classes.*

Proof of Lemma 4. The proof is by induction on r. The first considered value of r is $\lceil logp \rceil$. After that r will increment in 2. In particular, for all values of r of the form $\lceil logp \rceil + 2x$, we will prove that $mw(T_r(H)) \geq (x+1)p$ and, moreover, for each permutation SV of $V(T_r(H))$, the required matching can be witnessed by a partition of SV into a suffix and a prefix of size at least p^2 each. Let us verify that the lower bound $mw(T_r(H)) \geq (x+1)p$ implies the lemma. Suppose that $r = \lceil logp \rceil + 2x$ for some non-negative integer x. Then $mw(G) \geq (x+1)p = ((r - \lceil logp \rceil)/2 + 1)p > (r - \lceil logp \rceil + 1)p/2$. Suppose $r = \lceil logp \rceil + 2x + 1$. Then $mw(G) = mw(T_r(H)) \geq mw(T_{r-1}(H)) \geq (x+1)p = ((r - \lceil logp \rceil - 1)/2 + 1)p = (r - \lceil logp \rceil + 1)p/2$.

Assume that $r = \lceil logp \rceil$ and let us show the lower bound of p on the matching width. T_r contains at least $2^{\lceil logp \rceil + 1} - 1 \geq 2^{logp+1} - 1 = 2p - 1 \geq p$ vertices. By construction, H contains at least $2p$ vertices. Consequently, for each ordering of vertices of T_r we can specify a prefix and a suffix of size at least p^2 (just choose a prefix of size p^2). Let V_1 be the set of vertices that got to the prefix and let V_2 be the set of vertices that got to the suffix. By Lemma 5 there is a matching of size at least p consisting of edges between V_1 and V_2 confirming the lemma for the considered case.

Let us now prove the lemma for $r = \lceil logp \rceil + 2x$ for $x \geq 1$. Specify the centre of T_r as the root and let $T^1, \ldots, T^4$ be the subtrees of T_r rooted by the grandchildren of the root. Clearly, all of $T^1, \ldots, T^4$ are copies of T_{r-2}. Let SV be a sequence of vertices of $V(T_r(H))$. Let $SV^1, \ldots, SV^4$ be the respective sequences of $V(T^1(H)), \ldots, V(T^4(H))$ 'induced' by SV (that is their order is as in SV). By the induction assumption, for each of them we can specify a partition SV_1^i, SV_2^i into a prefix and a suffix of size at least p^2 each witnessing the conditions of the lemma for $r-2$. Let $u_1, \ldots, u_4$ be the last respective vertices of $SV_1^1, \ldots, SV_1^4$. Assume w.l.o.g. that these vertices occur in SV in the order they are listed. Let SV', SV'' be a partition of SV into a prefix and a suffix such that the last vertex of SV' is u_2. By the induction assumption we know that the edges between $SV_1^2 \subseteq SV'$ and $SV_2^2 \subseteq SV''$ form a matching M of size at least xp. In the rest of the proof, we are going to show that the edges between SV' and SV'' whose ends do not belong to any of SV_1^2, SV_2^2 can be used to form a matching M' of size p. The edges of M and M' do not have joint ends, hence this will imply existence of a matching of size $xp + p = (x+1)p$, as required.

The sets $SV' \setminus SV_1^2$ and $SV'' \setminus SV_2^2$ partition $V(T_r(H)) \setminus (SV_1^2 \cup SV_2^2) = V(T_r(H)) \setminus V(T^2(H)) = V([T_r \setminus T^2](H))$. Clearly, $T_r \setminus T_2$ is a tree. Furthermore, it contains at least p vertices. Indeed, T^2 (isomorphic to T_{r-2}) has p vertices just because we are at the induction step and T_r contains at least 4 times more vertices than T^2. So, in fact, $T_r \setminus T^2$ contains at least $3p$ vertices. Furthermore, since u_1 precedes u_2, the whole SV_1^1 is in SV'. By definition, SV_1^1 is disjoint with SV_1^2 and hence it is a subset of $SV' \setminus SV_1^2$. Furthermore, by definition,

$|SV_1^1| \geq p^2$ and hence $|SV' \setminus SV_1^2| \geq p^2$ as well. Symmetrically, since $u_3 \in SV''$, we conclude that $SV_2^3 \subseteq SV'' \setminus SV_2^2$ and due to this $|SV'' \setminus SV_2^2| \geq p^2$.

Thus $SV' \setminus SV_1^2$ and $SV'' \setminus SV_2^2$ partition $V([T_r \setminus T^2](H))$ into classes of size at least p^2 each and the size of $T_r \setminus T^2$ is at least $3p$. Thus, according to Lemma 5, there is a matching M' of size at least p created by edges between $SV' \setminus SV_1^2$ and $SV'' \setminus SV_2^2$, confirming the lemma, as specified above. ∎

Acknowledgements. I would like to thank anonymous reviewers for very useful and insightful comments. The research has been partly supported by the EPSRC grant EP/L020408/1.

References

1. Borodin, A., Razborov, A.A., Smolensky, R.: On lower bounds for read-k-times branching programs. Comput. Complex **3**, 1–18 (1993)
2. Bryant, R.E.: Symbolic boolean manipulation with ordered binary-decision diagrams. ACM Comput. Surv. **24**(3), 293–318 (1992)
3. Ferrara, A., Pan, G., Vardi, M.Y.: Treewidth in verification: Local vs. global. In: Sutcliffe, G., Voronkov, A. (eds.) LPAR 2005. LNCS (LNAI), vol. 3835, pp. 489–503. Springer, Heidelberg (2005)
4. Jukna, S.: Boolean Function Complexity: Advances and Frontiers. Springer, Berlin (2012)
5. Razborov, A.: Lower bounds for deterministic and nondeterministic branching programs. In: Budach, L. (ed.) FCT 1991. LNCS, vol. 199, pp. 47–60. Springer, Heidelberg (1991)
6. Razborov, A., Wigderson, A., Chi-Chih Yao, A.: Read-once branching programs, rectangular proofs of the pigeonhole principle and the transversal calculus. In: STOC, pp. 739–748 (1997)
7. Razgon, I.: No small nondeterministic read-once branching programs for CNFs of bounded treewidth. CoRR, abs/1407.0491 (2014)
8. Razgon, I.: On OBDDs for CNFs of bounded treewidth. In: KR, pp. 92–100 (2014)
9. Vatshelle, M.: New width parameters of graphs. Ph.D. Thesis, Department of Informatics, University of Bergen 2012

The Relative Exponential Time Complexity of Approximate Counting Satisfying Assignments

Patrick Traxler[✉]

Software Competence Center Hagenberg, Hagenberg, Austria
patrick.traxler@scch.at

Abstract. We study the exponential time complexity of approximate counting satisfying assignments of CNFs. We reduce the problem to deciding satisfiability of a CNF. Our reduction preserves the number of variables of the input formula and thus also preserves the exponential complexity of approximate counting. Our algorithm is also similar to an algorithm which works particularly well in practice and for which no approximation guarantee is known.

1 Introduction

We analyze the approximation ratio of an algorithm for approximately counting solutions of a CNF. The idea of our algorithm goes back to Stockmeyer. Stockmeyer [19] shows that approximately counting witnesses of any NP-relation is possible in randomized polynomial time given access to a $\Sigma_2 P$-oracle. It is known that we only need an NP-oracle if we apply the Left-Over Hashing Lemma of Impagliazzo et al. [13] which we discuss below. The use of an NP-oracle is necessary, unless P = NP. Stockmeyer's result and its improvement provides us with a first relation between deciding satisfiability and approximately counting solutions, a seemingly harder problem.

The motivation of our results comes from exponential time complexity. Impagliazzio et al. [15] develop a structural approach to classify NP-complete problems according to their exact time complexity. They formulate and prove the Sparsification Lemma for k-CNFs. This lemma allows us to use almost all known polynomial time reductions from the theory of NP-completeness to obtain exponential hardness results. There are however problems for which the sparsification lemma and standard NP-reductions do not yield meaningful results. Relating the exact complexity of approximately counting CNF solutions and the complexity of SAT is such a problem. We show:

Theorem 1. *Let $c > 0$ and assume there is an algorithm for SAT with running time $\tilde{O}(2^{cn})$ where n is the number of variables. For any $\delta > 0$, there is an algorithm which outputs with high probability in time $\tilde{O}(2^{(c+\delta)n})$ the approximation $\tilde{s}$ for the number of solutions s of an input CNF such that*

© Springer International Publishing Switzerland 2014
M. Cygan and P. Heggernes (Eds.): IPEC 2014, LNCS 8894, pp. 332–341, 2014.
DOI: 10.1007/978-3-319-13524-3_28

$$(1 - 2^{-\alpha n}) s \leq \tilde{s} \leq (1 + 2^{-\alpha n}) s$$

with $\alpha = \Omega(\frac{\delta^2}{\log(\frac{1}{\delta})})$.

It is not clear if this approximation problem is in BPP^{NP} because of the super-polynomially small approximation error. An improvement of the approximation error would yield a similar reduction from #SAT to SAT.

Our result also holds for k-SAT. Here, the approximation guarantee depends on the clause width k. We make a case distinction between constant clause width k, i.e. k is independent of the number of variables n, and non-constant k. The algorithm ACOUNT-CONSTANT and ACOUNT are defined in Sect. 4. Both algorithms use an algorithm for deciding satisfiability of k-CNFs. Their running time is $O(n \cdot \log(n) \cdot (n^2 + 2^k \cdot k \cdot n + size(F)))$ times the running time of some algorithm for k-SAT. We note that F is not necessarily a k-CNF.

Theorem 2. *(a) Let $k \geq 5$ be constant and let s be the number of solutions of the input CNF F. The probability that algorithm* ACOUNT-CONSTANT *outputs the approximation $\tilde{s}$ such that*

$$\frac{1}{4} 2^{-n + \frac{\log(n)}{k}} n^{1-4/k} s \leq \tilde{s} \leq 4 s$$

is at least $1/4$.

(b) Let k be such that $4\log(16n) \leq k + 1 \leq n$ and let κ be such that $k + 1 = \kappa \log(512\kappa) 4\log(16n)$. Let s be the number of solutions of the input CNF F. The probability that algorithm ACOUNT *outputs the approximation $\tilde{s}$ such that*

$$\frac{1}{4} 2^{-n/\kappa} s \leq \tilde{s} \leq 4 s$$

is at least $1/4$.

An application of our algorithms is to sample a solution approximately uniformly from the set of all solutions [16]. The reduction in [16] preserves the number of variables. We can get also a result similar to Stockmeyer's result. For any problem in parameterized SNP [15] – an appropriate refinement and subset of NP – we can define its counting version. Every such problem reduces by our result and the sparsification lemma to SAT (or k-SAT) at the expense of an increase of n to $O(n)$ variables. Here, n may be the number of vertices in the graph coloring problem or a similar parameter [15]. We just have to observe that the sparsification lemma preserves the number of solutions.

We also remark that our results allow us to relax the hypothesis that SAT (or k-SAT) is exponential hard to the hypothesis that its approximate counting variant is exponential hard.

1.1   A Practical Algorithm

Stockmeyer's idea was implemented in [12]. Gomes et al. [12] provide an implementation of a reduction which uses a SAT-solver to answer oracle queries. The algorithm of Gomes et al. [12] is almost the same as our algorithm. It preserves the number of variables and the maximum clause width is small. These properties seem to be crucial for a fast implementation, in particular, for the SAT-solver to work fast.

Gomes et al. [12] compare empirically the running time of their algorithm to the running time of exact counting algorithms. Their algorithm performs well on the tested hard instances and actually outperforms exact counting algorithms. The output values seem to be good approximations. A bound on the approximation ratio is not known.

Because there are only small differences between our algorithms and the algorithm of Gomes et al. [12], our bound on the approximation guarantee may be considered as a theoretical justification for the quality of the algorithm of Gomes et al. [12]. One difference is that we need a probability amplification to obtain our result, Theorem 2, and that it only holds for clause width $k \geq 5$ at the moment. Additionally, we consider hash functions with a different probability distribution than in [12].

Algorithms for the k-CNF case with theoretical bounds were proposed by Thurely [20] (approximate counting) and a randomized algorithm by Impagliazzo et al. [14] (exact counting). Thurely achieves an approximation within a factor e^{ϵ} of the number of satisfying assignments in time $\tilde{O}(\epsilon^{-2}c_k^n)$, where e.g. $c_3 = 1.5366$ for 3-SAT.

1.2   Comparison to the Left-Over Hashing Lemma

A possible reduction from approximate counting to satisfiability testing works roughly as follows. We assume to have a procedure which takes as input a CNF F with n variables and a parameter m. It outputs a CNF $F \wedge G_m$ such that the number of solutions of $F \wedge G_m$ times 2^m is approximately the number of solutions of F. We apply this procedure for $m = 1, ..., n$ and stop as soon as $F \wedge G_m$ is unsatisfiable. Using the information when the algorithm stops we can get a good approximation.

The construction of G_m reduces to the following randomness extraction problem. We have given a random point $x \in \{0,1\}^n$ and want a function $h : \{0,1\}^n \to \{0,1\}^m$ such that $h(x)$ is almost uniform. We think of h as m functions $(h_1, ..., h_m)$ and additionally require that each h_i depends only on few variables. We use the latter property to efficiently encode h as a CNF in such a way that the encoding and the input CNF F have the same set of variables. Stockmeyer's result and its improvement can not be adapted in an obvious to get such an efficient encoding. The crucial difference of our approach to the original approach are the bounds on the locality of the hash function. Our analysis is Fourier-analytic whereas the proof of Left-Over Hashing Lemma [13] uses probabilistic techniques.

Impagliazzo et al. [13] show that any pairwise independent[1] family $\mathcal{H}_{\mathrm{ind}}$ of functions of the form $\{0,1\}^n \to \{0,1\}^m$ satisfies the following extraction property:

Lemma 1. *Fix a distribution f over the cube $\{0,1\}^n$ with min-entropy[2] at least $m + \Omega(\log(1/\varepsilon))$ and $y \in \{0,1\}^m$. Then,*

$$\Pr_{h \sim \mathcal{H}_{ind}} (|\Pr_{x \sim f}(h(x) = y) - 2^{-m}| \leq \varepsilon \cdot 10 \cdot 2^{-m}) \geq 0.1.$$

This result, in a slightly more general form [13], is called the *Left-Over Hashing Lemma*. We want for our applications that h, seen as a random function, has besides the extraction property a couple of additional properties. The most important being that h_i is a Boolean function depending on at most k variables. This is what we call a *local hash function*. These hash functions are however not necessarily pairwise independent. This leads to a substantial problem. The proof of the Left-Over Hashing Lemma relies on pairwise independence since it requires an application of Chebyshev's Inequality. In its proof we define the random variable $X = X(h) := \Pr_{x \sim f}(h(x) = y)$. Its expected value is 2^{-m}. This still holds in our situation. Its variance can be however too large for an application of Chebyshev's Inequality. To circumvent the use of Chebyshev's Inequality we formulate the problem in terms of Fourier analysis of Boolean functions. We make use of a close connection between linear hash functions attaining the extraction property and the Fourier spectrum of probability distributions over the cube $\{0,1\}^n$.

1.3 Further Related Work

Calabro et al. [3] give a probabilistic construction of a "local hash function" without the extraction property. They obtain a similar reduction as the Valiant-Vazirani reduction [21]. The extraction property is not necessary for this purpose. Gavinsky et al. [11] obtain a local hash function via the Hypercontractive Inequality. However only for $|A| \geq 2^{n-O(\sqrt{n})}$ where $A \subseteq \{0,1\}^n$ is the set to be hashed. We remark that the motivations and applications in [11] are different from ours. The (Bonami-Beckner) Hypercontractive Inequality, credited to Bonami [2] and Beckner [1], found several diverse applications. See [8,18] for further references.

Other practical algorithms using hashing similar to [12] is discrete integration, a special case of weighted optimization [9,10]. Results which relate different problems to SAT are for example [3–5,22].

---

[1] Pairwise independence means that $\Pr_{h \sim \mathcal{H}_{\mathrm{ind}}}(h(x_1) = y_1, h(x_2) = y_2) = 2^{-2m}$ for any $x_1, x_2 \in \{0,1\}^n$, $x_1 \neq x_2$, and $y_1, y_2 \in \{0,1\}^m$. A Bernoulli matrix with bias $\frac{1}{2}$ induces a pairwise independent family.

[2] See Sect. 2.

2   Preliminaries

We make the following conventions. We use a special $O(\cdot)$ notation for estimating the running time of algorithms. We suppress a polynomial factor depending on the input size by writing $\tilde{O}(\cdot)$. As an example, SAT can be solved in time $\tilde{O}(2^n)$. We assume uniform sampling if we sample from a set without specifying the distribution.

A κ-*junta* is a Boolean function which depends on at most κ out of n variables. We extend this notion to functions $h : \{0,1\}^n \rightarrow \{0,1\}^m$, $h = (h_1, ..., h_m)$, by requiring that $h_i : \{0,1\}^n \rightarrow \{0,1\}$ is a κ-junta for every $i \in [m]$. A Boolean function $f : \{0,1\}^n \rightarrow \mathbb{R}$ is a *distribution* iff all values of f are non-negative and sum up to 1. It has *min-entropy* t iff t is the largest r with $f(x) \leq 2^{-r}$ for all $x \in \{0,1\}^n$. The *relative min-entropy* $\tilde{t}$ is defined as $\tilde{t} := t/n$. A distribution f is t-*flat* iff $f(x) = 2^{-t}$ or $f(x) = 0$ for all $x \in \{0,1\}^n$.

Definition 1. *Let* $0 < p_1, p_2 \leq 1$. *Let* $\mathcal{D}$ *be a distribution over functions of the form* $\{0,1\}^n \rightarrow \{0,1\}^m$. *A random function* h *is called* κ-*local with probability* p_1 *iff*

$$\Pr_{h \sim \mathcal{D}} (h \ is \kappa\text{-}local) \geq p_1.$$

It is called a (t_0, ε)-*hash function (for flat distributions) with probability* p_2 *iff*

$$\Pr_{h \sim \mathcal{D}} (| \Pr_{x \sim f}(h(x) = y) - 2^{-m}| \leq \varepsilon\, 2^{-m}) \geq p_2$$

for every $y \in \{0,1\}^m$ *and every (flat) distribution* f *of min-entropy* t *with* $t_0 \leq t \leq n$.

3   Local Hash Functions: Construction and Analysis

We start with the definition/construction of the two hash functions h and h^c. After this we discuss a basic connection between Fourier coefficients of distributions and the special case of linear hash functions with a one-dimensional range. We generalize this finally to functions with the high-dimensional range $\{0,1\}^m$.

Construction of h with parameter $0 < p \leq \frac{1}{2}$: For $i = 1, ..., m$: Choose a set $S_i \sim \mu_p$. Define $h_i(x) := \bigoplus_{j \in S_i} x_j$. The hash function is $h := (h_1, ..., h_m)$.

Here, $S_i \sim \mu_p$ means that we choose every element in S_i with probability p and do not choose it with probability $1 - p$. This distribution is called the Bernoulli distribution.

Construction of h^c with parameter $k \geq 1$: For $i = 1, ..., m$: Choose a set S_i uniformly at random from $\{S : S \subseteq [n], |S| = k\}$. Define $h_i^c(x) := \bigoplus_{j \in S_i} x_j$. The hash function is $h^c := (h_1^c, ..., h_m^c)$.

3.1 Hashing, Randomness Extraction, and the Discrete Fourier Transform

We start with recalling basics from Fourier analysis of Boolean functions. The *Fourier transform* of Boolean functions is a functional which maps $f : \{0,1\}^n \to \mathbb{R}$ to $\widehat{f} : 2^{[n]} \to \mathbb{R}$ and which we define by $\widehat{f}(S) := \mathrm{E}_{x \sim \{0,1\}^n}(f(x) (-1)^{\bigoplus_{i \in S} x_i})$, $S \subseteq [n]$. We will study the following *normalized Fourier transform* given by $\widetilde{f}(S) := 2^{n-1} \widehat{f}(S)$. We call the values of $\widehat{f}$ *Fourier coefficients* and the collection of Fourier coefficients the *Fourier spectrum* of f.

We can rewrite normalized Fourier coefficients to see the connection to hashing and randomness extraction. We define $\bigoplus_{i \in \{\}} x_i := 0$.

Lemma 2. *Let* $f : \{0,1\}^n \to \mathbb{R}$ *be a distribution. For any* $S \subseteq [n]$,

$$\widetilde{f}(S) = \Pr_{x \sim f}\left(\bigoplus_{i \in S} x_i = 0\right) - \frac{1}{2} = \frac{1}{2} - \Pr_{x \sim f}\left(\bigoplus_{i \in S} x_i = 1\right).$$

We may think of $\bigoplus_{i \in S} x_i$ as a single bit which we extract from f. We are interested in how close to a uniformly distributed bit it is. There is also a combinatorial interpretation of randomness extraction which we are going to use subsequently. We define for non-empty $A \subseteq \{0,1\}^n$ the flat distribution $f_A(x) := \frac{1}{|A|}$ if $x \in A$ and 0 otherwise. We want a random hash function $h : \{0,1\}^n \to \{0,1\}$ such that for every not too small $A \subseteq \{0,1\}^n$ and $b \in \{0,1\}$, $\Pr_h\left(\left|\Pr_{x \sim f_A}(h(x) = b) - \frac{1}{2}\right| \text{ is small}\right)$ is large. This is the same as saying that the probability of the event $|A \cap \{x \in A : h(x) = b\}| \approx \frac{|A|}{2}$ should be large. In words, the hyperplane in $\mathbb{F}_2^n$ induced by h separates A in roughly equal sized parts.

3.2 Analysis of Local Hash Function

In this section we describe our technical tools for analyzing linear local hash functions. We show how to apply them on the example of the two random functions h and h^c. We apply the Hypercontractive Inequality for the analysis of h^c, Lemma 5, and a new inequality, Lemma 3, for the analysis of h, Lemma 4. We start with Lemma 3.

The *support* of a function $g : \{0,1\}^n \to \mathbb{R}$ is the set of all points with a non-zero value and denoted by $\mathrm{Supp}(g)$.

Lemma 3. *Let* $f, g : \{0,1\}^n \to \{-1,0,1\}$, $0 < p \le \frac{1}{2}$, *and* $0 < \alpha \le \frac{1}{9}$. *Let* $\widetilde{A}(\alpha,p)$ *be such that* $\max((1 + 2^{-1/\alpha+8})^{\alpha p}, (1-p)4^{\alpha p}) \le \widetilde{A}(\alpha,p)$. *Then,*

$$\mathrm{E}_{S \sim \mu_p}(\widehat{f}(S)\,\widehat{g}(S)) \le 4^{-n}\,\widetilde{A}(\alpha,p)^n\,(|\mathrm{Supp}(f)| \cdot |\mathrm{Supp}(g)|)^{1-\alpha p}.$$

Lemma 3 is one of the main contributions of our work. It is shown by induction over n. In its proof we work explicitly with the Bernoulli distribution S is chosen from. The purpose is to decompose in the induction step the

n-dimensional functions f and g into $(n-1)$-dimensional functions with the same range $\{-1,0,1\}$.

Lemma 4 is a straight forward application of the previous result, Lemma 3, together with a result of Chor and Goldreich [6].

Lemma 4. *Let* $f : \{0,1\}^n \to \mathbb{R}$ *be a distribution of relative min-entropy* $\tilde{t}$, $2^{\tilde{t}n} \in \{1,...,2^n\}$, *and* $0 < p \le \frac{1}{2}$. *Then,*

$$\mathrm{E}_{S \sim \mu_p}(|\tilde{f}(S)|) \le \frac{1}{2}\sqrt{2}^{-p \cdot n \cdot \tilde{t}/\log(512/\tilde{t})}.$$

Applying the Hypercontractive Inequality in a straight forward way we get the following result.

Lemma 5. *Let* $f : \{0,1\}^n \to \mathbb{R}$ *be a distribution of min-entropy* t *with* $2^t \in \{1,...,2^n\}$, k *be a positive integer, and* $0 < \zeta < 1$. *Then,*

$$\mathrm{E}_{S \sim \binom{[n]}{k}}(|\tilde{f}(S)|) \le \frac{1}{2} n^{-(1-\zeta)k/2} 2^{(n-t)kn^{-\zeta}}.$$

The intuition of Lemmas 4 and 5 is given by Lemma 2. Lemmas 4 and 5 say that the normalized Fourier coefficient is small in case of high min-entropy. Lemma 2 then says that $\bigoplus_{i \in S} x_i$, S as in Lemmas 4 and 5, is good for hashing.

It remains to show how to apply Lemmas 4 and 5 to finally analyze h and h^c.

Lemma 6. (Main Lemma)
Hash Function h. *Let* $0 < \varepsilon < 1$. *Let* $0 < p \le \frac{1}{2}$ *be as in the definition of* h.
Define

$$P(\tilde{t}) := \frac{m}{\varepsilon}\sqrt{2}^{-pn\tilde{t}/\log(512/\tilde{t})}.$$

If there exists $\tilde{t}_0$ *such that* $P = P(\tilde{t}_0) < 1$ *and* $\tilde{t}_0 n + m + 1 \le n$, *then* h *is a* $(\tilde{t}_0 n + m + 1, \varepsilon)$-*hash function for flat distributions with probability at least* $(1-P)^m > 0$.

Hash Function h^c. *Let* $0 < \varepsilon < 1$ *and* $0 < \zeta < 1$. *Let* k *be as in the definition of* h^c. *Define*

$$Q(t) := \frac{m}{\varepsilon} n^{-(1-\zeta)k/2} 2^{(n-t)kn^{-\zeta}}.$$

If there exists t_0 *such that* $Q = Q(t_0) < 1$ *and* $t_0 + m + 1 \le n$, *then* h^c *is a* $(t_0 + m + 1, \varepsilon)$-*hash function for flat distributions with probability at least* $(1-Q)^m > 0$.

We note that h^c is a k-local hash function with probability 1 and that h is for example a $(2pn)$-local hash function with high probability if $p = \Omega(\frac{\log(n)}{n})$.

Our proof works as follows. Assume f has min-entropy t. Conditioning on an event $E \subseteq \{0,1\}^n$ yields a new distribution f' with min-entropy t'. We can not say much about the relation of t and t' in general. If E is however a hyperplane (in the vector space $\mathbb{F}_2^n$) induced by $\bigoplus_{i \in S} x_i$ then our inequalities from above,

Lemmas 4 and 5, tell us that $t' \approx t - 1$ in the expectation. Iterating this step and keeping control of the entropy decay we get our result. This process works as long as we reach some threshold t_0. The proof is an induction over m and the induction step an application of Lemmas 4 and 5.

3.3   Limitations

In this section we discuss that we can only expect small improvements of the Main Lemma.

Rank of Bernoulli Matrices. We will argue that the restriction $p = \Omega(\frac{\log(n)}{n})$ in the construction h is necessary. We recall the combinatorial idea behind hashing. Let M be a Bernoulli matrix with bias p and let $y \in \{0,1\}^m$. The preimage of M, y intersects any large enough subset $A \subseteq \{0,1\}^n$ in approximately $|A| \cdot 2^{-m}$ points. Let us assume $m = n$. If especially $A = \{0,1\}^n$ we expect that the linear system $Mx = y$ has one solution in $\mathbb{F}_2^n$. This is is the case iff M has full rank. The threshold for this property is around $\Theta(\frac{\log(n)}{n})$ [7]. In particular, the probability that M has full rank can get very small and in which case M fails to have the extraction property with high probability. With respect to this consideration it is not surprising that our probabilistic construction of h becomes efficient only if $p = \Omega(\frac{\log(n)}{n})$.

The Isolation Problem. We will argue that the trade-off between the size of A, i.e. the min-entropy of the corresponding flat distribution, and p is close to optimal. We can restrict A to be the solution set of a k-CNF. The following result is due to Calabro et al. [3]: For any distribution $\mathcal{D}$ of k-CNFs over n variables, there is a satisfiable k-CNF F such that $\Pr_{F' \sim \mathcal{D}}(|\mathrm{sol}(F) \cap \mathrm{sol}(F')| = 1) \leq 2^{-\Omega(n/k)}$, where $\mathrm{sol}(F)$ ($\mathrm{sol}(F')$) refers to the set of solutions of F (F'). The corresponding problem of computing F' is the Isolation Problem for k-CNFs [3]. We show how the Main Lemma relates to a solution of this problem. Let G be a k-CNF and let $p = \frac{k}{n}$, $k = \Theta(\kappa \log(\kappa) \log(n))$. The Main Lemma guarantees just that $|\mathrm{sol}(G) \cap \mathrm{sol}(G')|$, G' the CNF-encoding of h, is with high probability within a small interval around $v = 2^{O(n/\kappa)}$. We need to define an appropriate distribution $\mathcal{D}_0$ to apply the mentioned result. Chernoff's Inequality guarantees that h is encodable as a k-CNF G'' with high probability. We extend G'' by constraints (literals) which encode $x_i = 0$ or $x_i = 1$ as follows. Uniformly at random select a set of $\log(v)$ variables. Uniformly at random set the value of these variables. This defines our distribution $\mathcal{D}_0$. With probability at least $2^{-O(n \cdot \log(\kappa)/\kappa)}$ we get a $O(k)$-CNF G' such that $|\mathrm{sol}(G) \cap \mathrm{sol}(G')| = 1$. The reason for this is the following simple to prove fact (Exercise 12.2, p. 152 in [17]): Let $B \subseteq \{0,1\}^n$ be non-empty. There exists a set of variables $I \subseteq [n]$ and $b \in \{0,1\}^I$ such that $|I| \leq \log(|B|)$ and $|\{x \in B : x_i = b_i \, \forall i \in I\}| = 1$. Note that the construction of $\mathcal{D}_0$ depends only on the parameters n, k, and m, but not on the input k-CNF G. We can thus apply the result of Calabro et al. [3].

Comparing the lower and and upper bound we see that we are off by a factor $O(\log(k)^2 \log(n))$ in the exponent.

4 Algorithms and Main Results

The algorithm ACOUNT-CONSTANT is depicted in Fig. 1. It is similar to the algorithm of Gomes et al. [12]. One difference is the output. We output an approximation for the number of solutions. The algorithm of Gomes et al. [12] outputs a lower and an upper bound. Besides the experimental results, Gomes et al. [12] can show that with high probability the output lower bound is indeed smaller than the number of solutions. They give however no estimation for the quality of the output bounds which would be necessary for bounding the approximation ratio.

Input: CNF F over n variables and a parameter k.

1. If F is unsatisfiable then output 0 and stop.
2. For $l = 1, ..., n+1$:
3.   Repeat $8\lceil \log(n) \rceil$ times:
4.   Construct h^c. Select $b \sim \{0,1\}^l$.
5.   Let G be the k-CNF encoding of $h(x) = b$.
6.   Record if $F \wedge G$ is satisfiable.
7.   If unsatisfiability was recorded more than $4\lceil \log(n) \rceil$ times
8.   then output 2^{l-1} and stop.

Fig. 1. Algorithm ACOUNT-CONSTANT with access to a SAT-oracle

We define algorithm ACOUNT similar to ACOUNT but with the only difference that it constructs h. In the construction of h we have to check if h can be encoded by a CNF with small enough clause width. This happens with high probability due to an application of the Chernoff bound.

We stress the fact that our algorithms are easy to implement and that we can amplify the success probability further by repeating the inner loop appropriately.

Finally, Theorem 2 states the analysis of these algorithms. Its proof uses the Main Lemma. Theorem 1 follows from the analysis of ACOUNT together with some simple extra ideas.

References

1. Beckner, W.: Inequalities in Fourier analysis. Ann. Math. **102**, 159–182 (1975)
2. Bonami, A.: Étude des coefficients des Fourier de fonctions de $L^p(G)$. Annales de l'Institut Fourier **20**(2), 335–402 (1970)
3. Calabro, C., Impagliazzo, R., Kabanets, V., Paturi, R.: The complexity of unique k-SAT: an isolation lemma for k-CNFs. J. Comput. Syst. Sci. **74**(3), 386–393 (2008)

4. Calabro, C., Impagliazzo, R., Paturi, R. A duality between clause width and clause density for SAT. In: Proceedings of the 21st Annual IEEE Conference on Computational Complexity, pp. 252–260 (2006)
5. Calabro, C., Impagliazzo, R., Paturi, R.: On the exact complexity of evaluating quantified k-CNF. Algorithmica **65**(4), 817–827 (2013)
6. Chor, B., Goldreich, O.: On the power of two-point based sampling. J. Complex. **5**(1), 96–106 (1989)
7. Cooper, C.: On the rank of random matrices. Random Struct. Algorithms **16**(2), 209–232 (2000)
8. de Wolf, R.: A brief introduction to Fourier analysis on the Boolean cube. Theory Comput. Libr. Grad. Surv. **1**, 6 (2008)
9. Ermon, S., Gomes, C.P., Sabharwal, A., Selman, B.: Taming the curse of dimensionality: discrete integration by hashing and optimization. In: Proceedings of the 30th International Conference on Machine Learning, pp. 334–342 (2013)
10. Ermon, S., Gomes, C.P., Sabharwal, A., Selman, B.: Low-density parity constraints for hashing-based discrete integration. In: Proceedings of the 31th International Conference on Machine Learning, pp. 271–279 (2014)
11. Gavinsky, D., Kempe, J., Kerenidis, I., Raz, R., de Wolf, R.: Exponential separations for one-way quantum communication complexity, with applications to cryptography. SIAM J. Comput. **38**(5), 1695–1708 (2008)
12. Gomes, C.P., Sabharwal, A., Selman, B.: Model counting: a new strategy for obtaining good bounds. In: Proceedings of the 21st National Conference on Artificial Intelligence and the 18th Innovative Applications of Artificial Intelligence Conference (2006)
13. Impagliazzo, R., Levin, L.A., Luby, M.: Pseudo-random generation from one-way functions. In: Proceedings of the 21st Annual ACM Symposium on Theory of Computing, pp. 12–24 (1989)
14. Impagliazzo, R, Matthews, W., Paturi, R.: A satisfiability algorithm for AC0. In: Proceedings of the 23th ACM-SIAM Symposium on Discrete Algorithms, pp. 961–972 (2012)
15. Impagliazzo, R., Paturi, R., Zane, F.: Which problems have strongly exponential complexity? J. Comput. Syst. Sci. **63**(4), 512–530 (2001)
16. Jerrum, M., Valiant, L.G., Vazirani, V.V.: Random generation of combinatorial structures from a uniform distribution. Theor. Comput. Sci. **43**, 169–188 (1986)
17. Jukna, S.: Extremal Combinatorics. Springer, Heidelberg (2001)
18. O'Donnell, R.: Some topics in analysis of boolean functions. In: Proceedings of the 40th Annual ACM Symposium on Theory of Computing, pp. 569–578 (2008)
19. Stockmeyer, L.J.: On approximation algorithms for #P. SIAM J. Comput. **14**(4), 849–861 (1985)
20. Thurley, M.: An approximation algorithm for #k-SAT. In: Proceedings of the 29th International Symposium on Theoretical Aspects of Computer Science, pp. 78–87 (2012)
21. Valiant, L.G., Vazirani, V.V.: NP is as easy as detecting unique solutions. Theor. Comput. Sci. **47**(1), 85–93 (1986)
22. Williams, R.: A new algorithm for optimal 2-constraint satisfaction and its implications. Theor. Comput. Sci. **348**(2–3), 357–365 (2005)

Author Index

Printed in the United States
By Bookmasters

Printed in the United States
By Bookmasters